2016 IEEE Symposium on VLSI Circuits (VLSI-Circuits 2016)

Honolulu, Hawaii, USA
15 – 17 June 2016

IEEE Catalog Number: CFP16VLS-POD
ISBN: 978-1-5090-0636-6

Copyright © 2016 by the Institute of Electrical and Electronics Engineers, Inc
All Rights Reserved

Copyright and Reprint Permissions: Abstracting is permitted with credit to the source. Libraries are permitted to photocopy beyond the limit of U.S. copyright law for private use of patrons those articles in this volume that carry a code at the bottom of the first page, provided the per-copy fee indicated in the code is paid through Copyright Clearance Center, 222 Rosewood Drive, Danvers, MA 01923.

For other copying, reprint or republication permission, write to IEEE Copyrights Manager, IEEE Service Center, 445 Hoes Lane, Piscataway, NJ 08854. All rights reserved.

***This publication is a representation of what appears in the IEEE Digital Libraries. Some format issues inherent in the e-media version may also appear in this print version.**

IEEE Catalog Number:	CFP16VLS-POD
ISBN (Print-On-Demand):	978-1-5090-0636-6
ISBN (Online):	978-1-5090-0635-9
ISSN:	2158-5601

Additional Copies of This Publication Are Available From:

Curran Associates, Inc
57 Morehouse Lane
Red Hook, NY 12571 USA
Phone: (845) 758-0400
Fax: (845) 758-2633
E-mail: curran@proceedings.com
Web: www.proceedings.com

TABLE OF CONTENTS

1.1 ENABLING PROGRESS IN MACHINE LEARNING (INVITED) 1
O. Temam

1.2 ACCELERATING THE SENSING WORLD THROUGH IMAGING EVOLUTION (INVITED) 4
T. Nomoto, Y. Oike , H. Wakabayashi

2.1 A 16NM DUAL-PORT SRAM WITH PARTIAL SUPPRESSED WORD-LINE, DUMMY READ RECOVERY AND NEGATIVE BIT-LINE CIRCUITRIES FOR LOW VMIN APPLICATIONS 8
Y.-H. Chen, K.-C. Lin, C.-W. Wu, W.-M. Chan, J.-J. Liaw, H.-J. Liao , J. Chang

2.2 A 6.05-MB/MM² 16-NM FINFET DOUBLE PUMPING 1W1R 2-PORT SRAM WITH 313 PS READ ACCESS TIME 10
M. Yabuuchi, Y. Sawada, T. Sano, Y. Ishii, S. Tanaka, M. Tanaka , K. Nii

2.3 A 2X LOGIC DENSITY PROGRAMMABLE LOGIC ARRAY USING ATOM SWITCH FULLY IMPLEMENTED WITH LOGIC TRANSISTORS AT 40NM-NODE AND BEYOND 12
Y. Tsuji, X. Bai, A. Morioka, M. Miyamura, R. Nebashi, T. Sakamoto, M. Tada, N. Banno, K. Okamoto, N. Iguchi, H. Hada , T. Sugibayashi

2.4 80KB 10NS READ CYCLE LOGIC EMBEDDED HIGH-K CHARGE TRAP MULTI-TIME-PROGRAMMABLE MEMORY SCALABLE TO 14NM FIN WITH NO ADDED PROCESS COMPLEXITY 14
J. Viraraghavan, D. Leu, B. Jayaraman, A. Cestero, R. Kilker, M. Yin, J. Golz, R. R. Tummuru, R. Raghavan, D. Moy, T. Kempanna, F. Khan, T. Kirihata , S. Iyer

3.1 A 97.99 DB SNDR, 2 KHZ BW, 37.1 µW NOISE-SHAPING SAR ADC WITH DYNAMIC ELEMENT MATCHING AND MODULATION DITHER EFFECT 16
K. Obata, K. Matsukawa, T. Miki, Y. Tsukamoto , K. Sushihara

3.2 A 35µW 96.8DB SNDR 1 KHZ BW MULTI-STEP INCREMENTAL ADC USING MULTI-SLOPE EXTENDED COUNTING WITH A SINGLE INTEGRATOR 18
Y. Zhang, C.-H. Chen, T. He , G. C. Temes

3.3 A 18.5-FJ/STEP VCO-BASED 0-1 MASH DELTASIGMA ADC WITH DIGITAL BACKGROUND CALIBRATION 20
A. Sanyal , N. Sun

3.4 A 13.3 MW 60 MHZ BANDWIDTH, 76 DB DR 6 GS/S CTDELTASIGMAM WITH TIME INTERLEAVED FIR FEEDBACK 22
A. Jain , S. Pavan

4.1 A 128-CHANNEL SPIKE SORTING PROCESSOR FEATURING 0.175 µW AND 0.0033 MM² PER CHANNEL IN 65-NM CMOS 24
S. M. A. Zeinolabedin, A. T. Do, D. Jeon, D. Sylvester , T. T.-H. Kim

4.2 1.74-µW/CH, 95.3%-ACCURATE SPIKE-SORTING HARDWARE BASED ON BAYESIAN DECISION 26
Z. Jiang, J. P. Cerqueira, S. Kim, Q. Wang , M. Seok

4.3 A HIGH-DENSITY CMOS MULTI-MODALITY JOINT SENSOR/STIMULATOR ARRAY WITH 1024 PIXELS FOR HOLISTIC REAL-TIME CELLULAR CHARACTERIZATION 28
J. S. Park, T. Chi, A. Su, C. Zhu, J. H. Sung, H. C. Cho, M. Styczynski , H. Wang

4.4 A FRONT-END ASIC WITH RECEIVE SUB-ARRAY BEAMFORMING INTEGRATED WITH A 32 × 32 PZT MATRIX TRANSDUCER FOR 3-D TRANSESOPHAGEAL ECHOCARDIOGRAPHY 30
C. Chen, Z. Chen, D. Bera, S. B. Raghunathan, M. Shabanimotlagh, E. Noothout, Z. Y. Chang, J. Ponte, C. Prins, H. J. Vos, J. G. Bosch, M. D. Verweij, N. de Jong , M. A. P. Pertijs

5.1 A FULLY-ADAPTIVE WIDEBAND 0.5-32.75GB/S FPGA TRANSCEIVER IN 16NM FINFET CMOS TECHNOLOGY 32
P. Upadhyaya, A. Bekele, D. T. Melek, H. Zhao, J. Im, J. Cho, K. H. Tan, S. McLeod, S. Chen, W. Zhang, Y. Frans , K. Chang

5.2 A 28.3 GB/S 7.3 PJ/BIT 35 DB BACKPLANE TRANSCEIVER WITH EYE SAMPLING PHASE ADAPTATION IN 28 NM CMOS 34
H. Miyaoka, F. Terasawa, M. Kudo, H. Kano, A. Matsuda, N. Shirai, S. Kawai, T. Shibasaki, T. Danjo, Y. Ogata, Y. Sakai, H. Yamaguchi, T. Mori, Y. Koyanagi, H. Tamura, Y. Ide, K. Terashima, H. Higashi, T. Higuchi , N. Naka

5.3 A 32 GB/S RX ONLY EQUALIZATION TRANSCEIVER WITH 1-TAP SPECULATIVE FIR AND 2-TAP DIRECT IIR DFE 36
S. Hwang, S. Moon, J. Song , C. Kim

5.4 A 56GB/S PAM4 WIRELINE TRANSCEIVER USING A 32-WAY TIME-INTERLEAVED SAR ADC IN 16NM FINFET 38
Y. Frans, M. Elzeftawi, H. Hedayati, J. Im, V. Kireev, T. Pham, J. Shin, P. Upadhyaya, L. Zhou, S. Asuncion, C. Borrelli, G. Zhang, H. Zhang , K. Chang

6.1 A 50MHZ 5V 3W 90% EFFICIENCY 3-LEVEL BUCK CONVERTER WITH REAL-TIME CALIBRATION AND WIDE OUTPUT RANGE FOR FAST-DVS IN 65NM CMOS 40
X. Liu, C. Huang , P. K. T. Mok

6.2 95% LIGHT-LOAD EFFICIENCY SINGLE-INDUCTOR DUAL-OUTPUT DC-DC BUCK CONVERTER WITH SYNTHESIZED WAVEFORM CONTROL TECHNIQUE FOR USB TYPE-C 42
W.-H. Yang, C.-H. Lin, K.-H. Chen, C.-L. Wey, Y.-H. Lin, J.-R. Lin, T.-Y. Tsai , J.-L. Chen

6.3 A RECONFIGURABLE SIMO SYSTEM WITH 10-OUTPUT DUAL-BUS DC-DC CONVERTER USING THE LOAD BALANCING FUNCTION IN GROUP ALLOCATOR FOR DIVERSIFIED LOAD CONDITION 44
S.-U. Shin, M.-Y. Jung, K.-D. Kim, S.-H. Park, Y. Huh, C. Shin, S.-H. Park, J.-S. Bang, J.-B. Baek, S.-W. Choi, Y.-M. Ju , G.-H. Cho

6.4 A MICROCONTROLLER WITH 96% POWER-CONVERSION EFFICIENCY USING STACKED VOLTAGE DOMAINS 46
K. Blutman, A. Kapoor, A. Majumdar, J. G. Martinez, J. Echeverri, L. Sevat, A. van der Wel, H. Fatemi, J. Pineda de Gyvez , K. Makinwa

6.5 A FAST, FLEXIBLE, POSITIVE AND NEGATIVE ADAPTIVE BODY-BIAS GENERATOR IN 28NM FDSOI 48
M. Blagojevic, M. Cochet, B. Keller, P. Flatresse, A. Vladimirescu , B. Nikolic

7.1 A BLUETOOTH LOW-ENERGY (BLE) TRANSCEIVER WITH TX/RX SWITCHABLE ON-CHIP MATCHING NETWORK, 2.75MW HIGH-IF DISCRETE-TIME RECEIVER, AND 3.6MW ALL-DIGITAL TRANSMITTER 50
F.-W. Kuo, S. B. Ferreira, M. Babaie, R. Chen, L.-C. Cho, C.-P. Jou, F.-L. Hsueh, G. Huang, I. Madadi, M. Tohidian , R. B. Staszewski

7.2 A 380PW DUAL MODE OPTICAL WAKE-UP RECEIVER WITH AMBIENT NOISE CANCELLATION 52
W. Lim, T. Jang, I. Lee, H.-S. Kim, D. Sylvester , D. Blaauw

7.3 SLEEPTALKER: A 28NM FDSOI ULV 802.15.4A IR-UWB TRANSMITTER SOC ACHIEVING 14PJ/BIT AT 27MB/S WITH ADAPTIVE-FBB-BASED CHANNEL SELECTION AND PROGRAMMABLE PULSE SHAPE 54
G. de Streel, F. Stas, T. Gurné, F. Durant, C. Frenkel , D. Bol

7.4 A 2.4GHZ TERNARY SEQUENCE SPREAD SPECTRUM OOK TRANSCEIVER WITH HARMONIC SPUR SUPPRESSION AND DUAL-MODE DETECTION ARCHITECTURE FOR ULP WEARABLE DEVICES 56
S. J. Kim, C. S. Park, Y. Kim, S.-J. Yun, Y.-J. Hong , S.-G. Lee

7.5 AN 18 µW SPUR CANCELED CLOCK GENERATOR FOR RECOVERING RECEIVER SENSITIVITY IN WIRELESS SOCS 58
Y. Ogasawara, H. Sakurai, R. Fujimoto , K. Sami

8.1 AN ENERGY HARVESTING WIRELESS SENSOR NODE FOR IOT SYSTEMS FEATURING A NEAR-THRESHOLD VOLTAGE IA-32 MICROCONTROLLER IN 14NM TRI-GATE CMOS 60
S. Paul, V. Honkote, R. Kim, T. Majumder, P. Aseron, V. Grossnickle, R. Sankman, D. Mallik, S. Jain, S. Vangal, J. Tschanz , V. De

8.2 LENSLESS SMART SENSORS: OPTICAL AND THERMAL SENSING FOR THE INTERNET OF THINGS (INVITED) 62
P. Gill , T. Vogelsang

8.3 FEATURES OF RETINAL PROSTHESIS USING SUPRACHOROIDAL TRANSRETINAL STIMULATION FROM AN ELECTRICAL CIRCUIT PERSPECTIVE (INVITED) 64
Y. Terasawa, K. Shodo, K. Osawa , J. Ohta

8.4 MULTI-MODAL SMART BIO-SENSING SOC PLATFORM WITH >80DB SNR 35µA PPG RX CHAIN 66
A. Sharma, S. B. Lee, A. Polley, S. Narayanan, W. Li, T. Sculley , S. Ramaswamy

8.5 AN FPGA-ACCELERATED PARTIAL IMAGE MATCHING ENGINE FOR MASSIVE MEDIA DATA SEARCHING SYSTEMS (INVITED) 68
T. Shimizu, Y. Tomita, H. Matsumura, M. Sugimura, H. Yamasaki, D. Thach, T. Miyoshi, T. Baba, Y. Watanabe, A. Ike

9.1 A 66PW DISCONTINUOUS SWITCH-CAPACITOR ENERGY HARVESTER FOR SELF-SUSTAINING SENSOR APPLICATIONS 70
X. Wu, Y. Shi, S. Jeloka, K. Yang, I. Lee, D. Sylvester , D. Blaauw

9.2 WIRELESS POWER TRANSFER SYSTEM WITH ENHANCED RESPONSE AND EFFICIENCY BY FULLY-INTEGRATED FAST-TRACKING WIRELESS CONSTANT-IDLE-TIME CONTROL FOR IMPLANTS 72
C. Huang, T. Kawajiri , H. Ishikuro

9.3 A FULLY INTEGRATED 144 MHZ WIRELESS-POWER-RECEIVER-ON-CHIP WITH AN ADAPTIVE BUCK-BOOST REGULATING RECTIFIER AND LOW-LOSS H-TREE SIGNAL DISTRIBUTION 74
C. Kim, J. Park, A. Akinin, S. Ha, R. Kubendran, H. Wang, P. P. Mercier , G. Cauwenberghs

9.4 A ±36A INTEGRATED CURRENT-SENSING SYSTEM WITH 0.3% GAIN ERROR AND 400µA OFFSET FROM -55 C TO +85 C 76
S. H. Shalmany, D. Draxelmayr , K. Makinwa

9.5 A 114-PW PMOS-ONLY, TRIM-FREE VOLTAGE REFERENCE WITH 0.26% WITHIN-WAFER INACCURACY FOR NW SYSTEMS 78
Q. Dong, K. Yang, D. Blaauw , D. Sylvester

10.1 MOTOR CONTROL USED TO BE BORING (INVITED) 80
A. Tessarolo

10.2 A FULLY INTEGRATED GAN-BASED POWER IC INCLUDING GATE DRIVERS FOR HIGH-EFFICIENCY DC-DC CONVERTERS (INVITED) 82
S. Ujita, Y. Kinoshita, H. Umeda, T. Morita, K. Kaibara, S. Tamura, M. Ishida , T. Ueda

10.3 A TRANSFORMER-BASED DIGITAL ISOLATOR WITH 20KVPK SURGE CAPABILITY AND > 200KV=µS COMMON MODE TRANSIENT IMMUNITY 84
R. Yun, J. Sun, E. Gaalaas , B. Chen

10.4 INNOVATIVE SYSTEM ON CHIP PLATFORM FOR SMART GRIDS AND INTERNET OF ENERGY APPLICATIONS (INVITED) 86
A. Moscatelli

11.1 A 65NM CMOS TRANSCEIVER WITH INTEGRATED ACTIVE CANCELLATION SUPPORTING FDD FROM 1GHZ TO 1.8GHZ AT +12.6DBM TX POWER LEAKAGE 88
S. Ramakrishnan, L. Calderin, A. Puglielli, E. Alon, A. Niknejad , B. Nikolic

11.2 DIGITAL PLL FOR PHASE NOISE CANCELLATION IN RING OSCILLATOR-BASED I/Q RECEIVERS 90
Z.-Z. Chen, Y. Li, Y.-C. Kuan, B. Hu, C.-H. Wong , M.-C. F. Chang

11.3 A CHOPPING SWITCHED-CAPACITOR RF RECEIVER WITH INTEGRATED BLOCKER DETECTION, +31DBM OB-IIP3, AND +15DBM OB-B1DB 92
Y. Xu , P. R. Kinget

11.4 A 180 MW MULTISTANDARD TV TUNER IN 28 NM CMOS 94
J. Xiao, W. Gao, X. Xu, D. Chang, J. Cao, R. Sun, V. Periasamy, N.-Y. Wang, X. Chen, G. Unruh, T. Hayashi, T.-H. Chih, L. Krishnan, K.-K. Huang, S. Dommaraju, G. Wei, B. Shen, A. Venes, D. Koh , J. Y. C. Chang

12.1 FULL CHIP INTEGRATION OF 3-D CROSS-POINT RERAM WITH LEAKAGE-COMPENSATING WRITE DRIVER AND DISTURBANCE-AWARE SENSE AMPLIFIER 96
S. Lee, J. Song, C. Seong, J. Woo, J.-M. Choi, S.-C. Kwon, H.-J. Kim, H.-S. Kang, S. G. Kim, H. G. Jung, K.-W. Kwon , H. Hwang

12.2 EMBEDDED MEMORY AND ARM CORTEX-M0 CORE USING 60-NM C-AXIS ALIGNED CRYSTALLINE INDIUM-GALLIUM-ZINC OXIDE FET INTEGRATED WITH 65-NM SI CMOS 98
T. Onuki, W. Uesugi, H. Tamura, A. Isobe, Y. Ando, S. Okamoto, K. Kato, T. R. Yew, C. B. Lin, J. Y. Wu, C. C. Shuai, S. H. Wu, J. Myers, K. Doppler, M. Fujita , S. Yamazaki

12.3 VERSATILE TLC NAND FLASH MEMORY CONTROL TO REDUCE READ DISTURB ERRORS BY 85% AND EXTEND READ CYCLES BY 6.7-TIMES OF READ-HOT AND COLD DATA FOR CLOUD DATA CENTERS 100
A. Kobayashi, T. Tokutomi, K. Takeuchi

12.4 A 0.9UM² 1T1R BIT CELL IN 14NM SOC PROCESS FOR METAL-FUSE OTP ARRAY WITH HIERARCHICAL BITLINE, BIT LEVEL REDUNDANCY, AND POWER GATING 102
Z. Chen, S. H. Kulkarni, V. E. Dorgan, U. Bhattacharya , K. Zhang

13.1 A 0.6MW 31MHZ 4TH-ORDER LOW-PASS FILTER WITH +29DBM IIP3 USING SELF-COUPLED SOURCE FOLLOWER BASED BIQUADS IN 0.18µM CMOS 104
Y. Xu, S. Leuenberger, P. K. Venkatachala , U.-K. Moon

13.2 3.5MW 1MHZ AM DETECTOR AND DIGITALLY-CONTROLLED TUNER IN A-IGZO TFT FOR WIRELESS COMMUNICATIONS IN A FULLY INTEGRATED FLEXIBLE SYSTEM FOR AUDIO BAG 106
T. Meister, K. Ishida, C. Carta, R. Shabanpour, B. K. Boroujeni, N. Münzenrieder, L. Petti, G. A. Salvatore, G. Schmidt, P. Ghesquiere, S. Kiefl, G. D. Toma, T. Faetti, A. C. Hübler, G. Tröster , F. Ellinger

13.3 A 16-CHANNEL NOISE-SHAPING MACHINE LEARNING ANALOG-DIGITAL INTERFACE 108
F. N. Buhler, A. E. Mendrela, Y. Lim, J. A. Fredenburg , M. P. Flynn

13.4 A FIELD-PROGRAMMABLE MIXED-SIGNAL IC WITH TIME-DOMAIN CONFIGURABLE ANALOG BLOCKS ... 110
Y. Choi, Y. Lee, S.-H. Baek, S.-J. Lee , J. Kim

14.1 A 5.8 PJ/OP 115 BILLION OPS/SEC, TO 1.78 TRILLION OPS/SEC 32NM 1000-PROCESSOR ARRAY ... 112
B. Bohnenstiehl, A. Stillmaker, J. Pimentel, T. Andreas, B. Liu, A. Tran, E. Adeagbo , B. Baas

14.2 28NM FDSOI TECHNOLOGY SUB-0.6V SRAM VMIN ASSESSMENT FOR ULTRA LOW VOLTAGE APPLICATIONS ... 114
R. Ranica, N. Planes, V. Huard, O. Weber, D. Noblet, D. Croain, F. Giner, S. Naudet, P. Mergault, S. Ibars, A. Villaret, M. Parra, S. Haendler, M. Quoirin, F. Cacho, C. Julien, F. Terrier, L. Ciampolini, D. Turgis, C. Lecocq , F. Arnaud

14.3 A 400MV ACTIVE VMIN, 200MV RETENTION VMIN, 2.8 GHZ 64KB SRAM WITH A 0.09 UM² 6T BITCELL IN A 16NM FINFET CMOS PROCESS ... 116
A. Bhavnagarwala, I. Iqbal, A. Nguyen, D. Ondricek, V. Chandra , R. Aitken

14.4 A 350MV-900MV 2.1GHZ 0.011MM² REGULAR EXPRESSION MATCHING ACCELERATOR WITH AGING-TOLERANT LOW-VMIN CIRCUITS IN 14NM TRI-GATE CMOS ... 118
A. Agarwal, S. Hsu, M. Anders, S. Mathew, G. Chen, H. Kaul, S. Satpathy , R. Krishnamurthy

14.5 UNIFIED TECHNOLOGY OPTIMIZATION PLATFORM USING INTEGRATED ANALYSIS (UTOPIA) FOR HOLISTIC TECHNOLOGY, DESIGN AND SYSTEM CO-OPTIMIZATION AT <= 7NM NODES ... 120
S. C. Song, J. Xu, D. Yang, K. Rim, P. Feng, J. Bao, J. Zhu, J. Wang, G. Nallapati, M. Badaroglu, P. Narayanasetti, B. Bucki, J. Fischer , G. Yeap

15.1 A 12-BIT 1.6 GS/S INTERLEAVED SAR ADC WITH DUAL REFERENCE SHIFTING AND INTERPOLATION ACHIEVING 17.8 FJ/CONV-STEP IN 65NM CMOS ... 122
J.-W. Nam, M. Hassanpourghadi, A. Zhang , M. S.-W. Chen

15.2 A 14.6MW 12B 800MS/S 4×TIME-INTERLEAVED PIPELINED SAR ADC ACHIEVING 60.8DB SNDR WITH NYQUIST INPUT AND SAMPLING TIMING SKEW OF 60FSRMS WITHOUT CALIBRATION ... 124
Y.-C. Lien

15.3 AN OSCILLATOR COLLAPSE-BASED COMPARATOR WITH APPLICATION IN A 74.1DB SNDR, 20KS/S 15B SAR ADC ... 126
M. Shim, S. Jeong, P. Myers, S. Bang, C. Kim, D. Sylvester, D. Blaauw , W. Jung

15.4 A 0.44FJ/CONVERSION-STEP 11B 600KS/S SAR ADC WITH SEMI-RESTING DAC ... 128
S.-E. Hsieh , C.-C. Hsieh

16.1 A 35 MW 10 GB/S ADC-DSP LESS DIRECT DIGITAL SEQUENCE DETECTOR AND EQUALIZER IN 65NM CMOS ... 130
A. K. M. D. Hossain, Aurangozeb, M. Mohammad , M. Hossain

16.2 A 125 MW 8.5-11.5 GB/S SERIAL LINK TRANSCEIVER WITH A DUAL PATH 6-BIT ADC/5-TAP DFE RECEIVER AND A 4-TAP FFE TRANSMITTER IN 28 NM CMOS ... 132
B. Raghavan, A. Varzaghani, L. Rao, H. Park, X. Yang, Z. Huang, Y. Chen, R. Kattamuri, C. Wu, B. Zhang, J. Cao, A. Momtaz , N. Kocaman

16.3 A 0.003 MM² 5.2 MW/TAP 20 GBD INDUCTOR-LESS 5-TAP ANALOG RX-FFE ... 134
R. Boesch, K. Zheng , B. Murmann

16.4 A 16GB/S 14.7MW TRI-BAND COGNITIVE SERIAL LINK TRANSMITTER WITH FORWARDED CLOCK TO ENABLE PAM-16 / 256-QAM AND CHANNEL RESPONSE DETECTION IN 28 NM CMOS ... 136
Y. Du, W.-H. Cho, Y. Li, C.-H. Wong, J. Du, P.-T. Huang, Y. Kim, Z.-Z. Chen, S. J. Lee , M.-C. F. Chang

16.5 A LOW-EMI FOUR-BIT FOUR-WIRE SINGLE-ENDED DRAM INTERFACE BY USING A THREE-LEVEL BALANCED CODING SCHEME ... 138
I.-M. Yi, S.-J. Bae, M.-K. Chae, S.-M. Lee, Y.-J. Jang, Y.-C. Cho, Y.-S. Sohn, J.-H. Choi, S.-J. Jang, B. Kim, J.-Y. Sim , H.-J. Park

17.1 A 0.3-2.6 TOPS/W PRECISION-SCALABLE PROCESSOR FOR REAL-TIME LARGE-SCALE CONVNETS ... 140
B. Moons , M. Verhelst

17.2 A 1.40MM² 141MW 898GOPS SPARSE NEUROMORPHIC PROCESSOR IN 40NM CMOS ... 142
P. Knag, C. Liu , Z. Zhang

17.3 A 190GFLOPS/W DSP FOR ENERGY-EFFICIENT SPARSE-BLAS IN EMBEDDED IOT ... 144
R. Dorrance , D. Markovic

17.4 A 58.6MW REAL-TIME PROGRAMMABLE OBJECT DETECTOR WITH MULTI-SCALE MULTI-OBJECT SUPPORT USING DEFORMABLE PARTS MODEL ON 1920X1080 VIDEO AT 30FPS ... 146
A. Suleiman, Z. Zhang , V. Sze

17.5 ADAPTIVE CLOCKING WITH DYNAMIC POWER GATING FOR MITIGATING ENERGY EFFICIENCY & PERFORMANCE IMPACTS OF FAST VOLTAGE DROOP IN A 22NM GRAPHICS EXECUTION CORE.................148
M. Cho, C. Tokunaga, S. Kim, J. Tschanz, M. Khellah , V. De

18.1 A 0.23 µG BIAS INSTABILITY AND 1.6 µG/HZ1/2 RESOLUTION SILICON OSCILLATING ACCELEROMETER WITH BUILD-IN Σ-Δ FREQUENCY-TO-DIGITAL CONVERTER...................150
J. Zhao, X. Wang, Y. Zhao, G. M. Xia, A. P. Qiu, Y. Su , Y. P. Xu

18.2 A BJT-BASED TEMPERATURE-TO-DIGITAL CONVERTER WITH ±60MK (3σ) INACCURACY FROM -70 C TO 125 C IN 160NM CMOS.................152
B. Yousefzadeh, S. H. Shalmany , K. Makinwa

18.3 A 28NM CMOS ULTRA-COMPACT THERMAL SENSOR IN CURRENT-MODE TECHNIQUE.................154
M. Eberlein , I. Yahav

18.4 A 35FJ/STEP DIFFERENTIAL SUCCESSIVE APPROXIMATION CAPACITIVE SENSOR READOUT CIRCUIT WITH QUASI-DYNAMIC OPERATION.................156
H. Omran, A. Alhoshany, H. Alahmadi , K. N. Salama

18.5 A 9.84-73.2 NJ, 0.048 MM² TIME-DOMAIN IMPEDANCE SENSOR THAT PROVIDES VALUES OF RESISTANCE AND CAPACITANCE.................158
Y. Hong, Y. Wang, W. L. Goh, Y. Gao , L. Yao

19.1 A 23MW 24GS/S 6B TIME-INTERLEAVED HYBRID TWO-STEP ADC IN 28NM CMOS.................160
B. Xu, Y. Zhou , Y. Chiu

19.2 A 8.2-MW 10-B 1.6-GS/S 4× TI SAR ADC WITH FAST REFERENCE CHARGE NEUTRALIZATION AND BACKGROUND TIMING-SKEW CALIBRATION IN 16-NM CMOS.................162
Y.-Z. Lin, C.-H. Tsai, S.-C. Tsou , C.-H. Lu

19.3 A 14-BIT 2.5GS/S AND 5GS/S RF SAMPLING ADC WITH BACKGROUND CALIBRATION AND DITHER.................164
A. M. A. Ali, H. Dinc, P. Bhoraskar, S. Puckett, A. Morgan, N. Zhu, Q. Yu, C. Dillon, B. Gray, J. Lanford, M. McShea, U. Mehta, S. Bardsley, P. Derounian, R. Bunch, R. Moore , G. Taylor

19.4 14-BIT 8:9GS/S RF DAC IN 40NM CMOS ACHIEVING >71DBC LTE ACPR AT 2:9GHZ.................166
V. Ravinuthula, W. Bright, M. Weaver, K. Maclean, S. Kaylor, S. Balasubramanian, J. Coulon, R. Keller, B. Nguyen , E. Dwobeng

20.1 A 7-TO-18.3GHZ COMPACT TRANSFORMER BASED VCO IN 16NM FINFET.................168
M. Raj, P. Upadhyaya, Y. Frans , K. Chang

20.2 -197DBC/HZ FOM 4.3-GHZ VCO USING AN ADDRESSABLE ARRAY OF MINIMUM-SIZED NMOS CROSS-COUPLED TRANSISTOR PAIRS IN 65-NM CMOS.................170
A. Jha, A. Ahmadi, S. Kshattry, T. Cao, K. Liao, G. Yeap, Y. Makris , K. K. O

20.3 A 10GB/S, 342FJ/BIT MICRO-RING MODULATOR TRANSMITTER WITH SWITCHED-CAPACITOR PRE-EMPHASIS AND MONOLITHIC TEMPERATURE SENSOR IN 65NM CMOS.................172
S. Saeedi , A. Emami

20.4 A 50.6-GB/S 7.8-MW/GB/S −7.4-DBM SENSITIVITY OPTICAL RECEIVER BASED ON 0.18-µM SIGE BICMOS TECHNOLOGY.................174
T. Takemoto, Y. Matsuoka, H. Yamashita, Y. Lee, K. Akita, H. Arimoto, M. Kokubo , T. Ido

21.1 AN 8.3M-PIXEL 480FPS GLOBAL-SHUTTER CMOS IMAGE SENSOR WITH GAIN-ADAPTIVE COLUMN ADCS AND 2-ON-1 STACKED DEVICE STRUCTURE.................176
Y. Oike, K. Akiyama, L. D. Hung, W. Niitsuma, A. Kato, M. Sato, Y. Kato, W. Nakamura, H. Shiroshita, Y. Sakano, Y. Kitano, T. Nakamura, T. Toyama, H. Iwamoto , T. Ezaki

21.2 A DEAD-TIME FREE GLOBAL SHUTTER CMOS IMAGE SENSOR WITH IN-PIXEL LOFIC AND ADC USING PIXEL-WISE CONNECTIONS.................178
H. Sugo, S. Wakashima, R. Kuroda, Y. Yamashita, H. Sumi, T.-J. Wang, P.-S. Chou, M.-C. Hsu , S. Sugawa

21.3 A 220PJ/PIXEL/FRAME CMOS IMAGE SENSOR WITH PARTIAL SETTLING READOUT ARCHITECTURE.................180
S. Ji, J. Pu, B. C. Lim , M. Horowitz

21.4 A 260µW INFRARED GESTURE RECOGNITION SYSTEM-ON-CHIP FOR SMART DEVICES.................182
S. Oh, N. L. Ba, S. Bang, J. Jeong, D. Blaauw, T. T. Kim , D. Sylvester

22.1 AN INDUCTOR-LESS FRACTIONAL-N INJECTION-LOCKED PLL WITH A SPUR-AND-PHASE-NOISE FILTERING TECHNIQUE.................184
A. Li, Y. Chao, X. Chen, L. Wu , H. Luong

22.2 AN 8.865-GHZ -244DB-FOM HIGH-FREQUENCY PIEZOELECTRIC RESONATOR-BASED CASCADED FRACTIONAL-N PLL WITH SUB-PPB-ORDER CHANNEL ADJUSTING TECHNIQUE.................186
S. Ikeda, H. Ito, A. Kasamatsu, Y. Ishikawa, T. Obara, N. Noguchi, K. Kamisuki, Y. Jiyang, S. Hara, D. Ruibing, S. Dosho, N. Ishihara , K. Masu

22.3 A 2.4-GHZ 6.4-MW FRACTIONAL-N INDUCTORLESS RF SYNTHESIZER 188
L. Kong , B. Razavi

22.4 A PVT-ROBUST -59-DBC REFERENCE SPUR AND 450-FS$_{RMS}$ JITTER INJECTION-LOCKED CLOCK MULTIPLIER USING A VOLTAGE-DOMAIN PERIOD-CALIBRATING LOOP 190
Y. Lee, H. Yoon, M. Kim , J. Choi

22.5 A 0.034MM², 725FS RMS JITTER, 1.8%/V FREQUENCY-PUSHING, 10.8-19.3GHZ TRANSFORMER-BASED FRACTIONAL-N ALL-DIGITAL PLL IN 10NM FINFET CMOS 192
C.-C. Li, T.-H. Tsai, M.-S. Yuan, C.-C. Liao, C.-H. Chang, T.-C. Huang, H.-Y. Liao, C.-T. Lu, H.-Y. Kuo, K. Hsieh, M. Chen, A. Ximenes , R. B. Staszewski

23.1 250MV-950MV 1.1TBPS/W DOUBLE-AFFINE MAPPED SBOX BASED COMPOSITE-FIELD SMS4 ENCRYPT/DECRYPT ACCELERATOR IN 14NM TRI-GATE CMOS 194
S. Satpathy, S. Mathew, V. Suresh, M. Anders, H. Kaul, A. Agarwal, S. Hsu, G. Chen , R. Krishnamurthy

23.2 A COMPACT 446 GBPS/W AES ACCELERATOR FOR MOBILE SOC AND IOT IN 40NM 196
Y. Zhang, K. Yang, M. Saligane, D. Blaauw , D. Sylvester

23.3 A 4FJ/BIT DELAY-HARDENED PHYSICALLY UNCLONABLE FUNCTION CIRCUIT WITH SELECTIVE BIT DESTABILIZATION IN 14NM TRI-GATE CMOS 198
S. Mathew, S. Satpathy, V. Suresh, M. Anders, H. Kaul, A. Agarwal, S. Hsu, G. Chen, R. Krishnamurthy , V. De

23.4 A 0.58MM² 2.76GB/S 79.8PJ/B 256-QAM MASSIVE MIMO MESSAGE-PASSING DETECTOR 200
W. Tang, C.-H. Chen , Z. Zhang

23.5 A MACHINE-LEARNING CLASSIFIER IMPLEMENTED IN A STANDARD 6T SRAM ARRAY 202
J. Zhang, Z. Wang , N. Verma

24.1 A 16-CHANNEL 1.1MM² IMPLANTABLE SEIZURE CONTROL SOC WITH SUB-μW/CHANNEL CONSUMPTION AND CLOSED-LOOP STIMULATION IN 0.18μM CMOS 204
M. Shoaran, M. Shahshahani, M. Farivar, J. Almajano, A. Shahshahani, A. Schmid, A. Bragin, Y. Leblebici , A. Emami

24.2 A MICROELECTRODE ARRAY WITH 8,640 ELECTRODES ENABLING SIMULTANEOUS FULL-FRAME READOUT AT 6.5 KFPS AND 112-CHANNEL SWITCH-MATRIX READOUT AT 20 KS/S 206
X. Yuan, S. Kim, J. Juyon, M. D'Urbino, T. Bullmann, Y. Chen, A. Stettler, A. Hierlemann , U. Frey

24.3 A WEARABLE EAR-EEG RECORDING SYSTEM BASED ON DRY-CONTACT ACTIVE ELECTRODES 208
X. Zhou, Q. Li, S. Kilsgaard, F. Moradi, S. L. Kappel , P. Kidmose

24.4 2.048 MB/S FULL-DUPLEX FREE-SPACE OPTICAL TRANSCEIVER IC FOR A REAL-TIME IN VIVO NEUROFEEDBACK MOUSE EXPERIMENT UNDER SOCIAL INTERACTION 210
G. Hwang, J.-K. Choi, J. Yang, S. Lim, J.-M. Kim, M.-G. Choi, D.-S. Kim, K. Gwak, J. Jeon, H. S. Shin, I.-H. Choi, S. Park , H.-M. Bae

24.5 A 450MV TIMING-MARGIN-FREE WAVEFORM SORTER BASED ON BODY SWAPPING ERROR CORRECTION 212
S. Kim, J. P. Cerqueira , M. Seok

Author Index

FOREWORD

Welcome to the 2016 Symposium on VLSI Circuits

It is our pleasure to welcome you to the 30th Symposium on VLSI Circuits, taking place June 14th-17th, 2016, at the Hilton Hawaiian Village in Honolulu, Hawaii. Founded in 1987, the Symposium on VLSI Circuits is sponsored by the IEEE Solid-State Circuits Society and the Japan Society of Applied Physics, in cooperation with the Institute of Electronics, Information and Communication Engineers of Japan.

The Symposium is held in conjunction with the Symposium on VLSI Technology. The co-location of these two Symposia provides the opportunity to cover a wide variety of technical topics ranging from process technology to systems-on-chip. In an era when close collaboration between transistor, circuits, and system designers is critical to innovation, this conference enables attendees to gain a broad understanding of the latest advances that will drive the industry forward.

As semiconductor scaling continues, the microelectronics industry faces a new inflection point, building upon the heterogeneous integration of leading-edge and mature technologies, and driving "smart" system level applications which will transform the industry. As a premiere international conference on semiconductor technology that defines the pace, progress and evolution of microelectronics, the Symposia on VLSI Technology and Circuits have made this industry transition its focal point, with the conference theme "**Inflections for a Smart Society**" serving as the thread connecting keynote presentations, panel discussions, focus sessions and short courses, reflecting the robust and diverse innovation taking place in the microelectronics industry.

The Symposium on VLSI Circuits begins with two short courses on June 14th. These courses are conducted by industry and academic leaders in their respective fields and provide excellent opportunities for learning and interaction with leading technologists. "**Advanced Wirelines Techniques**," covers 28 – 56 Gb/s design standards, low power CMOS, analog NRZ and silicon photonic transceivers, and integrated electronic-photonic communications circuits. "**Circuit Design in FinFET, FDSOI & Advanced Memory Technologies**," examines the impact of FinFETs in processor design, analog & mixed-signal CMOS and embedded memory designs; as well as FDSOI technology for SRAM and digital logic. For one registration fee, short course registrants will receive materials for both courses and are welcome to move between the courses based on individual preferences.

The plenary session on June 15th features two invited presentations discussing recent advances and new challenges related to VLSI circuits, applications, and technologies. The first presentation "**Enabling Future Progress in Machine-Learning**" will be delivered by Olivier Temam, Google. This presentation will discuss lessons learned from research on architectures for machine-learning, and explain that some of the hurdles ahead largely lie at the circuit level, but can possibly be overcome in the near future. The second presentation, "**Accelerating the Sensing World through Imaging Evolution**" will be delivered by Tetsuo Nomoto, Vice President and Senior General Manager, Sony Semiconductor Solutions Corporation. This presentation will discuss the potential for CMOS image sensors to accelerate the progress of the sensing world by continuously improving image quality, extending detectable wavelengths, and further improving depth resolution and temporal resolution.

The Symposium includes 23 regular sessions covering 97 excellent papers selected from the 375 submitted papers from both industry and universities around the world. Selected papers cover a broad range of subjects including digital processors and circuits, static, dynamic and non-volatile memory, design enablement, power conversion and management, analog circuits, analog/digital conversion, biomedical electronics, sensor interface circuits, and wireline and wireless communication circuits. Augmented with invited industry papers, two newly established focus sessions on June 16th will feature technical papers on "**Innovative Systems for a Smart Society**" and "**Industrial and Power Circuit Directions for a Smart Society**".

Evening panel sessions at the Symposium are well known for their selection of timely topics and enthusiastic discussions with technical leaders on the panel and in the audience. This year the Symposium includes three

Panel Sessions. A joint technology and circuits panel, "**More Moore, More than Moore or Mo(o)re Slowly?**", will address the crucial question of how Moore's Law is being adapted by the IC industry to new business opportunities in the IoT era with a high-profile panel composed of industry executives and experts. Two circuits panels will focus on innovation and co-optimization at the circuit level: "**Top Circuit Techniques: Life With & Without Them,**" will discuss high-impact circuit design techniques and "**It's All A Common Platform – How Do I Build A Differentiated Product?**" will examine how software and hardware co-design, user interface, and other innovations continue to drive competitive products at the circuit level.

In order to provide additional insights into industry and technology trends, the Symposia will include an executive panel on "**Semiconductor Business: Inflections Beyond Scaling**" where a group of executives from prominent semiconductor companies will discuss their views on the industry future.

The excellent technical program is a result of hard work and outstanding efforts of the Technical Program Committees under the leadership of the program chair, Gunther Lehmann, Infineon Technologies, Germany, and co-chair, Makoto Ikeda, The University of Tokyo, Japan. The committee members, all world-wide leaders in the field of VLSI design, have solicited and selected an excellent set of strong papers, and have organized them into a program of interesting technical sessions. We express our sincere thanks to all the members for their highly-skilled efforts. We are sure that you will enjoy the paper presentations, and we invite you to participate in the lively discussions in and outside the sessions.

In 2017, the Symposium will return to Kyoto, Japan, together with the Technology Symposium, and we hope that you will join us again next year.

Jeffrey Gealow   Masato Motomura
Symposium Chair   Symposium Co-Chair

2016 VLSI CIRCUITS SYMPOSIUM COMMITTEES

Symposium Chair:	Jeffrey Gealow	Analog Devices
Symposium Co-Chair:	Masato Motomura	Hokkaido University
Program Chair:	Gunther Lehmann	Infineon Technologies
Program Co-Chair:	Makoto Ikeda	The University of Tokyo
Secretary:	Andreia Cathelin	STMicroelectronics
Secretary Co-Chair:	Makoto Takamiya	The University of Tokyo
Publicity Chair:	Brian P. Ginsburg	Texas Instruments
Publicity Co-Chair:	Kazuhisa Sunaga	NEC
Short Course Chair:	Ken Chang	Xilinx Inc.
Short Course Co-Chair:	Ken Takeuchi	Chuo University
Panel Session Chair:	Pavan K. Hanumolu	University of Illinois, Urbana-Champaign
Panel Co-Chair:	Masao Ito	Renesas Electronics
Electronic Publication Chair:	Geert Van der Plas	Imec
Publication Co-Chair:	Yasumoto Tomita	Fujitsu Lab.
Local Arrangements:	Yoshihisa Kato	Panasonic Corp.
Treasurer:	Stephen Kosonocky	AMD Ft. Collins

EXECUTIVE COMMITTEES

IEEE

Chair:	H.S. Philip Wong	Stanford University
Members:	Ajith Amerasekera	Texas Instruments
	John Chen	nVidia
	Vivek De	Intel Corp.
	Giovanni De Micheli	EPFL
	Charles Dennison	Micron Technology
	Ichiro Fujimori	Broadcom Corporation
	Jeffrey Gealow	Analog Devices, Inc.
	Raj Jammy	Carl Zeiss
	Mukesh Khare	IBM
	Tsu-Jae King Liu	University of California, Berkeley
	Stephen Kosonocky	AMD Ft. Collins
	Gunther Lehman	Infineon Technologies AG
	Ming-Ren Lin	GLOBALFOUNDRIES
	Un-Ku Moon	Oregon State University
	Katsu Nakamura	Analog Devices, Inc.
	Tak Ning	IBM
	Klaus Schruefer	Intel Mobile Communications GmbH
	An Steegan	imec
	Hans Stork	ON Semiconductor
	Jason Woo	University of California, Los Angeles
	Kevin Zhang	Intel

EXECUTIVE COMMITTEES – continued

JSAP

Chair: Tadahiro Kuroda — Keio University

Members:
Eiji Fujii	Panasonic Corporation
Toshiro Hiramoto	The University of Tokyo
Dai Hisamoto	Hitachi, Ltd.
Satoshi Inaba	Toshiba Corporation Storage & Electronic Devices Solutions Company
Hideyuki Kabuo	Socionext, Inc.
Toshihiko Kanayama	AIST
Kinam Kim	Samsung Display Co.
Chih-Yuan Lu	Macronix International Co., Ltd.
Meishoku Masahara	AIST
Yoshio Masubichi	Toshiba Corp.
Akira Matsuzawa	Tokyo Institute of Technology
Masayuki Mizuno	Renesas Electronics Corp.
Yasunori Mochizuki	NEC Corp.
Tohru Mogami	PETRA
Masato Motomura	Hokkaido University
Naoki Nagashima	Sony Corp.

EXECUTIVE COMMITTEES

JSAP (continued)

Masaaki Niwa	Tohoku University
Satoshi Shigematsu	NTT Device Innovation Center
Toshihiro Sugii	Fujitsu Laboratories, Ltd.
Jack Y.-C. Sun	TSMC
Hirotaka Tamaura	Fujitsu Laboratories, Ltd.
Akira Toriumi	The University of Tokyo
Hitoshi Wakabayashi	Tokyo Institute of Technology
Kazuo Yano	Hitachi, Ltd.
Hoi-Jun Yoo	KAIST

TECHNICAL PROGRAM COMMITTEES

NORTH AMERICA/EUROPE

Chair:	Gunther Lehmann	Infineon Technologies
Members:	Vineet Agrawal	Cypress Semiconductor Corp
	Robert Aitken	ARM
	Elad Alon	UC Berkeley
	Edith Beigne	CEA Leti
	Ben Calhoun	University of Virginia
	Andreia Cathelin	STMicroelectronics
	Ken Chang	Xilinx Inc.
	Yun Chiu	University of Texas at Dallas
	John DeBrosse	IBM
	Steven Dillen	Qualcomm Technologies Inc.
	Cyrille Dray	Intel Corp.
	Eric Fogleman	MaxLinear
	Gordon Gammie	MediaTek USA
	Brian P. Ginsburg	Texas Instruments
	Fatih Hamzaoglu	Intel Corp.
	Pavan Kumar Hanumolu	University of Illinois, Urbana-Champaign
	Erwin Janssen	NXP Semiconductors
	Ron Kapusta	Analog Devices
	Kofi Makinwa	Delft University of Technology
	Dejan Markovic	UCLA
	Alyosha Molnar	Cornell University
	Afshin Momtaz	Broadcom Corp.
	Reza Navid	Rambus Inc.
	Jeff L. Nilles	Texas Instruments
	J. Thomas Pawlowski	Micron Technology, Inc.
	Dennis Sylvester	University of Michigan
	Vivienne Sze	Massachusetts Institute of Technology
	Geert Van der Plas	Imec
	Nick Van Helleputte	Imec
	Naveen Verma	Princeton University
	John Wuu	Advanced Micro Devices

JAPAN/FAR EAST

Co-Chair:	Makoto Ikeda	Hokkaido University
Members:	Kenichi Agawa	Toshiba Corp.
	Jonathan Chang	TSMC
	Koji Fujii	NTT Device Innovation Center
	Masanori Hashimoto	Osaka University
	Makoto Ikeda	The University of Tokyo
	Yutaka Hirose	Panasonic Corporation
	Hiroki Ishikuro	Keio University
	Masao Ito	Renesas System Design Co., Ltd.
	Kouichi Kanda	Fujitsu Laboratories, Ltd.
	Ming-Dou Ker	National Chiao Tung University
	Jri Lee	National Taiwan University
	Nicky Lu	Etron Technology, Inc.
	Noriyuki Miura	Kobe University
	Masanori Natsui	Tohoku University
	Srikanth Nimmagadda	Intel Technology India Pvt. Ltd.
	Hiromasa Noda	Micron Memory Japan
	Jun Ohta	Nara Institute of Science and Technology
	Yusuke Oike	Sony Corp.
	Kenichi Okada	Tokyo Institute of Technology
	Takeshi Okumoto	Socionext
	Hyunchol Shin	Kwangwoon University
	Yun-Shiang Shu	MediaTek Inc.
	Jae-Yoon Sim	Pohang University of Science and Technology
	Kyomin Sohn	Samsung Electronics Co., Ltd.
	Kazuhisa Sunaga	NEC Corp.
	Makoto Takamiya	The University of Tokyo
	Ken Takeuchi	Chuo University
	Yasumoto Tomita	Fujitsu Laboratories, Ltd.
	Hayato Wakabayashi	Sony Corp.
	Masanao Yamaoka	Hitachi, Ltd.
	Hiroyuki Yamauchi	Fukuoka Institute of Technology
	C. Patrick Yue	Hong Kong University of Science and Technology

Enabling Future Progress in Machine-Learning

Olivier Temam
Google (Inria)[1]

Abstract

Amazing progress in machine-learning, largely based on deep neural networks, has started to make applications once considered impossible, such as real-time translation or self-driving cars, a reality. However, even if, on some restricted problems, machine-learning is getting close to human-level performance, we are still far from the capabilities of the human brain. Machine-learning researchers themselves acknowledge that the progress observed in the past 10 years has been largely due to rapid increase in computing performance, allowing to tackle larger neural networks and larger training sets. So the computer systems and circuits communities can play a very significant role in enabling future progress. While GPUs have been a major driver of this recent progress, both the slowing rate of improvement of standard CMOS technology and the need for even faster progress suggest to at least explore alternative approaches. In this talk, we will discuss lessons learned from research on architectures for machine-learning, and that some of the hurdles ahead largely lie at the circuit level, but can possibly be overcome in the near future.

Overview on Machine-Learning

Neural networks research has been controversial, going through a typical hype curve cycle. One of the first neural network algorithms, the Perceptron [13], was presented as brain-inspired, but its capabilities were very limited, only capable of doing linear classification [11], though the multi-layer perceptron was later shown to be capable of non-linear classification. Towards the end of the 1980s and the beginning of the 1990s, such neural networks became fairly efficient, even leading to hardware neural network accelerators, such as the ETANN from Intel [9]. However, at the time, hardware neural networks had at least three limitations; (1) the application scope of neural networks was fairly restricted; (2) the clock frequency of processors was increasing fast enough that an accelerator could be outperformed by a software neural network run on a processor after a few technology generations; (3) competitive machine-learning algorithms emerged, especially Support Vector Machines (SVM) [5]. In addition, Cybenko's theorem [6], which stipulated that a neural network with a single hidden layer could approximate any continuous function with infinite precision, also suggested that deeper and larger neural networks would bring diminishing returns. This combination of factors created the condition for the temporary demise of neural networks.

Still, researchers such as Yoshua Bengio [2], Geoffrey Hinton [14] or Yann LeCun [10] kept pushing the notion of neural networks in the community, and around 2006, neural network models with large and wide layers (and at the time also combined with auto-encoders), i.e., so-called Deep Neural Networks (or DNNs), were shown to achieve competitive results on some applications with respect to state-of-the-art machine-learning techniques [1]. As GPUs started to allow the training of larger neural networks, on larger training sets, the performance of these deep neural networks kept increasing, and they have now consistently been shown to achieve state-of-the-art performance on a broad range of applications.

Different teams at Alphabet, especially the Brain team at Google, or DeepMind have been exploring various applications of deep neural networks, ranging from object recognition [16] to speech recognition [15], real-time scene understanding, and gaming [12].

---

[1] The author's current affiliation is Google, but the work described in this article has been done while the author was at Inria, France.

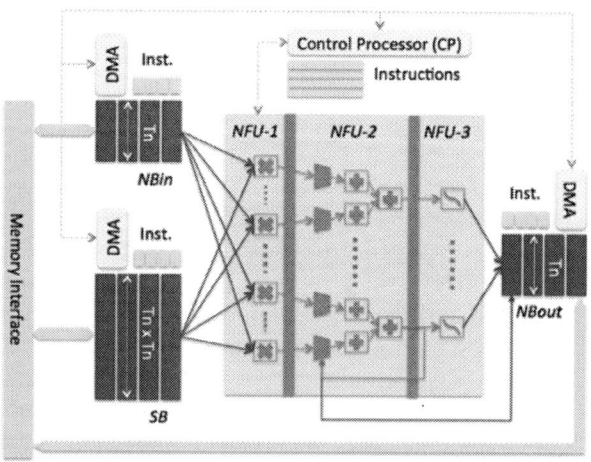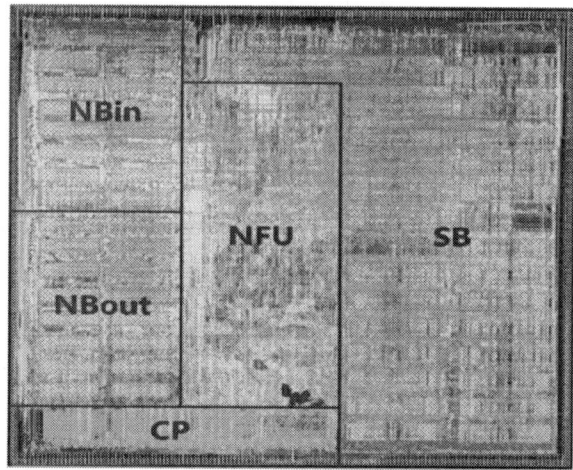

Hardware neural network accelerator (DianNao).

Research Opportunities in Hardware for Machine-Learning

There is a remarkable convergence of trends in applications, machine-learning and hardware, which creates opportunities for hardware machine-learning acceleration. We now know that machine-learning has become ubiquitous in a broad range of Cloud services and embedded devices. Simultaneously, as we mentioned in the previous section, deep neural networks have become the state-of-the-art machine-learning algorithms. And roughly at the same time, technology constraints have started to progressively initiate a shift towards heterogeneous architectures and the notion of hardware accelerators.

Now, the key challenge for hardware accelerators is the tradeoff between flexibility and efficiency. Too narrow a scope, and an accelerator has little appeal; too broad a scope and its efficiency decreases. This is where the trend in machine-learning is so remarkable. It creates a fairly unique situation where it is possible to reconcile accelerator efficiency and flexibility. It is possible to achieve efficiency because only a single algorithm (or few variations thereof), i.e.,deep neural networks, needs to be implemented in hardware. And yet, this algorithm has a broad application scope.

These converging trends and the notion that they lead to the concept of hardware neural network accelerators was first presented during a keynote at ISCA in 2010 [17]. A hardware implementation of one of the popular deep neural network algorithms, at the fringe between machine-learning and neuroscience, was subsequently explored [8]. Later on, a first accelerator for hardware neural networks, squarely targeting machine-learning, was introduced and shown to achieve high performance with a small area and power footprint, together with defect tolerance properties [18]. The main limitation of that accelerator was the memory bandwidth, so this work was later extended by combining the same accelerator with local memories in order to capture the locality properties of deep neural networks and overcome memory bandwidth limitations (DianNao accelerator, see Figure) [3]. The partitioning properties of deep neural network layers were subsequently leveraged to show the scalability of the concept, and that a multi-chip version of that accelerator (multiple connected instances of the accelerator) can be used to overcome on-die memory capacity limitations, and to further improve the performance of both neural network inference and training [4]. Finally, a version of this accelerator directly tied to a vision sensor, and tuned for convolutional neural networks (image recognition), was shown to also avoid memory bandwidth limitations and achieve low energy [7].

In summary, this research showed that hardware neural networks carry the promise of significant gains in performance, energy and area over conventional CPUs and GPUs. But it also highlights the importance of overcoming on-die memory capacity and memory bandwidth limitations. Even though architectural techniques could successfully be used to partially overcome these limitations, circuit-level innovations can bring a new wave of significant performance improvements on that front.

References

[1] Y. Bengio, P. Lamblin, D Popovici, H. Larochelle. (2007) Greedy layer-wise training of deep networks - Advances in neural information processing systems.

[2] Y. Bengio. (2009). Learning deep architectures for AI - Foundations and trends in Machine Learning.

[3] T. Chen, Z. Du, N. Sun, J. Wang, C. Wu, Y. Chen, O. Temam. (2014) DianNao: A Small-Footprint High-Throughput Accelerator for Ubiquitous Machine-Learning, International Conference on Architectural Support for Programming Languages and Operating Systems (ASPLOS).

[4] Y. Chen, T. Luo, S. Liu, S. Zhang, L. He, J. Wang, L. Li, T. Chen, Z. Xu, N. Sun, O. Temam. (2014) DaDianNao: A Machine-Learning Supercomputer, ACM/IEEE International Symposium on Microarchitecture (MICRO).

[5] C. Cortes and V. Vapnik. Support-Vector Networks. In Machine Learning, pages 273–297, 1995.

[6] G. Cybenko. (1989) Approximations by superpositions of sigmoidal functions, Mathematics of Control, Signals, and Systems, 2 (4), 303-314.

[7] Z. Du, R. Fasthuber, T. Chen, P. Ienne, L. Li, T. Luo, X. Feng, Y. Chen, and O. Temam. (2015) ShiDianNao: shifting vision processing closer to the sensor. SIGARCH Comput. Archit. News 43, 3.

[8] A. Hashmi, H. Berry, O. Temam, M. Lipasti. (2011) Automatic Abstraction and Fault Tolerance in Cortical Microarchitectures, ACM/IEEE International Symposium on Computer Architecture (ISCA).

[9] M. Holler, S. Tam, H. Castro, and R. Benson. An electrically trainable artificial neural network (ETANN) with 10240 float- ing gate synapses. In Artificial neural networks, pages 50–55, Piscataway, NJ, USA, 1990. IEEE Press.

[10] Y. LeCun, L. Bottou, Y. Bengio, P Haffner. (1988) Gradient-based learning applied to document recognition - Proceedings of the IEEE.

[11] M. Minsky, S. Papert. (1969). Perceptrons. MIT Press.

[12] V. Mnih, K. Kavukcuoglu, D. Silver, A. Graves, I. Antonoglou, D. Wierstra, M. Riedmiller. (2013) Playing Atari with Deep Reinforcement Learning view publication. NIPS 2013 Deep Learning Workshop.

[13] F. Rosenblatt. (1957), The Perceptron--a perceiving and recognizing automaton. Report 85-460-1, Cornell Aeronautical Laboratory.

[14] D. E. Rumelhart, G. E. Hinton, and R. J. Williams. (1986) Learning internal representations by error propagation. In Parallel distributed processing: explorations in the microstructure of cognition, vol. 1.

[15] H. Sak, A. Senior, and F. Beaufays. (2014)Long short-term memory based recurrent neural network architectures for large vocabulary speech recognition. arXiv preprint arXiv:1402.1128.

[16] C. Szegedy, W. Liu, Y. Jia, P. Sermanet, S. Reed, D. Anguelov, D. Erhan, V. Vanhoucke, A. Rabinovich. (2015) Going deeper with convolutions. Proceedings of the IEEE Conference on Computer Vision and Pattern Recognition.

[17] O. Temam. (2010) The Rebirth of Neural Networks, International Symposium on Computer Architecture (ISCA).

[18] O. Temam. (2012) A Defect-Tolerant Accelerator for Emerging High-Performance Applications, ACM/IEEE International Symposium on Computer Architecture (ISCA).

Accelerating the Sensing World through Imaging Evolution

Tetsuo Nomoto, Yusuke Oike, Hayato Wakabayashi

Sony Semiconductor Solutions Corporation 4-14-1 Asahi-cho, Atsugi-shi, Kanagawa, 243-0014 Japan
E-mail: Tetsuo.Nomoto@jp.sony.com

Abstract

The evolution of CMOS image sensors (CIS) and the future prospect of a "Sensing" world utilizing advanced imaging technologies promise to improve our quality of life by sensing everything, everywhere, every time. Charge Coupled Device image sensors replaced video camera tubes, allowing the introduction of compact video cameras as consumer products. CIS now dominates the market for digital still cameras created by its predecessor and, with the advent of column-parallel ADCs and back-illuminated technologies, outperforms them. CIS's achieve better signal to noise ratio, lower power consumption, and higher frame rate. Stacked CIS's continue to enhance functionality and user experience in mobile devices, a market that currently comprises over several billion units per year. CIS imaging technologies promise to accelerate the progress of a sensing world by continuously improving sensitivity, extending detectable wave-lengths, and further improving depth resolution and temporal resolution.

1. Introduction

With the rise of smartphones, image sensors have become a familiar part of our lives as imaging devices, selling over four billion units per year. With image sensors beginning to be installed in automobiles in recent years, the use of cameras is shifting from their intrinsic role of "recording particular moments" to object recognition and decision-making. In the coming years, this trend will lead to new growth industries such as IoT, drone, robot, medical, self-driving cars and infrastructure, with increasing demand for image sensors.

This presentation looks back on the transition from CCD to CMOS image sensors (CIS) [1], discusses the details of technology development, including the superiority of CIS and the stacked device structure that enables miniaturization and multi-functionality, introduces efforts for further enhancing image sensors in the future, and describes future developments that will stimulate the sensing market.

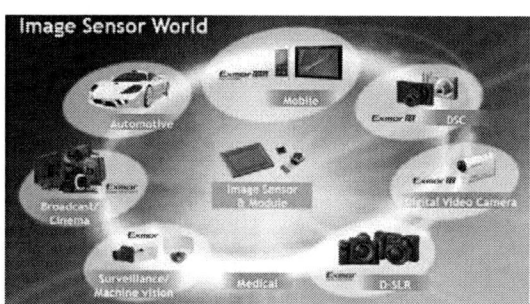

Fig. 1 Image sensor world

2. Transition from CCD to CMOS Image Sensors

Previous cameras that employed video camera tubes did not lend themselves to miniaturization and high resolution. CCD image sensors depicted in Fig. 2(a) were put into use to make small video cameras for consumer use a reality. A CCD image sensor transfers the signal electrons from photo diodes (PD) to a vertical charge-coupled device (vertical CCD) at the same time, with the vertical CCD sequentially transferring the

signal electrons. A horizontal CCD scans one row at a time to output the voltage signal of each pixel via an output amplifier. The simple pixel structure of a CCD allows for pixel shrinking and high pixel resolution in response to advances in semiconductor microfabrication technologies, contributing to the evolution of Digital Still Cameras (DSC).

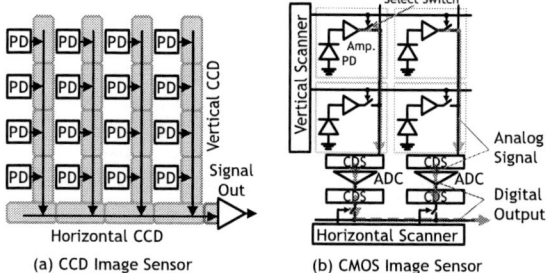

(a) CCD Image Sensor   (b) CMOS Image Sensor

Fig. 2 Image sensor architecture: (a) CCD image sensor, (b) CMOS image sensor

CCDs drive all pixels simultaneously to read out signals, making it difficult to achieve high pixel resolution and a high frame rate together with low power consumption. On the other hand, CISs can read out signals with simultaneous access to one row and one column as illustrated in Fig. 2(b), meaning that high pixel resolution and a high frame rate can be achieved together with low power consumption by carrying out parallel AD conversion of the signals of each row. This feature CISs is advantageously utilized as HDTV video cameras become higher in definition. It is the Column-parallel ADC architecture that has paved the way for this high frame rate operation of CISs. As illustrated in Fig. 2(b), correlated double sampling (CDS) and AD conversion are carried out on readout analog signals in a column-parallel manner, allowing for high-speed, low-power digital output by parallel processing while minimizing the noise mixing paths of the analog signals. [2] In the case of DSCs, stronger demand for high resolution has driven requirements for reduced pixel size. In addition to the low-noise, high-speed output of CISs, it was also necessary to achieve sufficient sensitivity using small pixels. What played a major role in improving the sensitivity of CISs was the introduction of Back-illuminated CIS (BI CIS) illustrated in Fig. 3. With its simple pixel structure and fewer metal layers, CCDs allow for superior light gathering by the on-chip lens even in small pixels; on the other hand, with multiple transistors in pixels and signal lines needed to control them, CISs must include more metal layers, making the aperture aspect ratio higher in small pixels, which in turn makes light gathering difficult. With the orientation of the metal layers and photo diodes reversed, BI CIS is structured to receive light on the substrate side. This has dramatically improved sensitivity in small pixels[3-4] by eliminating structures that may obstruct light gathering, expanding adoption in DSCs.

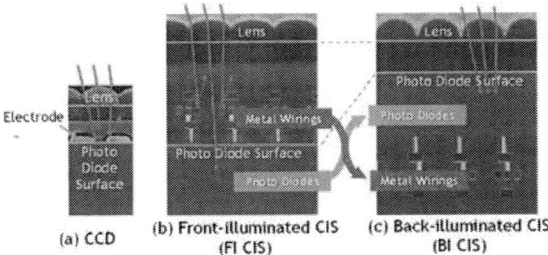

Fig. 3 Device structures: (1) CCD, (b) Front-illuminated CIS, (c) Back-illuminated CIS

As CISs became capable of being operated at high frame rates and high definition, they also became capable of capturing fast-moving subjects by taking advantage of their high-speed. This arise requirement for the global shutter operation. All pixels in a CCD have the same exposure timing since the signal charge of all pixels are simultaneously transferred to the vertical CCD, providing global shutter operation; on the other hand, in a CIS, as illustrated in Fig.4(a), exposure timing varies by row because signal charges stored in photo diodes are transferred to the floating diffusion amplifier on a line-by-line basis before being read out. This creates distortion when shooting fast-moving subjects. Although many cameras incorporate a mechanical shutter for still pictures, distortion (focal-plane distortion) is inevitable in movies. As illustrated in Fig. 4(b), we have incorporated an analog memory within the pixel for storing the signal charge and implemented a structure for causing all pixels to simultaneously transfer the signal charge, thereby successfully obtaining images free of focal-plane distortion even with a CMOS image sensor.[5]

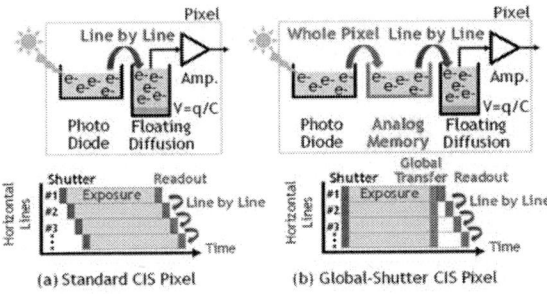

Fig. 4 Pixel structure for rolling/global shutter operations: (a) Standard CIS pixel, (b) Global-shutter CIS pixel

3. Evolution of Image Sensors

Once imaging of high image quality became possible with small pixels, there arose a growing demand for Smartphones that could provide image quality comparable to conventional digital cameras with ease of use without sacrificing small, thing form factors. In the case of the previous versions of BI CISs as illustrated in Fig. 5(a), enhanced functionality would necessitate an increase in chip size, hindering the miniaturization of mobile devices. As a means to solve this issue, we have developed Stacked CIS technology, illustrated in Fig. 5(b), in which the pixels are stacked over the signal processing circuit section. Conventional BI CISs have a structure in which the pixels and peripheral circuit are integrated on the same chip and the chip is bonded to a supporting substrate. In contrast Stacked BI CIS devices are structured such that the peripheral circuitry is integrated onto the supporting substrate.[6]

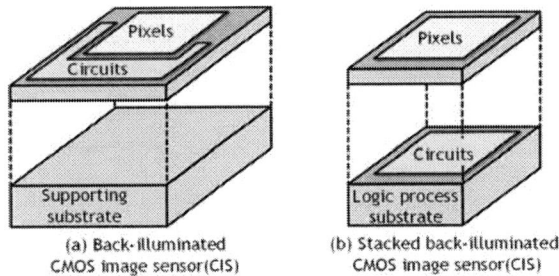

Fig. 5 Device structure: (a) BI CIS, (b) Stacked BI CIS

With the Stacked CIS structure, it has become possible to integrate more advanced signal processing functions into a small CIS. For example, a technique known as high dynamic range (HDR) exists for synthesizing images with long and short exposure times to prevent underflow and overflow. As illustrated in Fig. 6, controlling exposure to long and short exposure times in the sensors would cause them to output a bright and dark pair of images, which requires image synthesis via a dedicated image signal processing before being input into a general Application Processor. Our Stacked CIS allows part of these image processing functions to be incorporated into the CIS, making it possible to build a camera system with a more versatile ISP or AP, contributing to the miniaturization of camera modules.

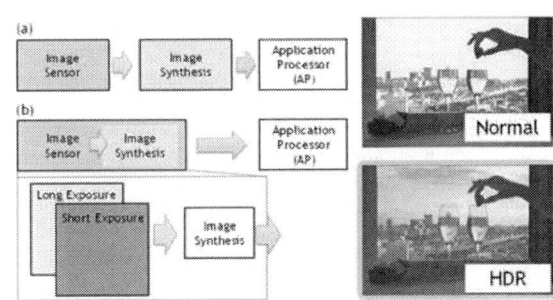

Fig. 6 Example of camera system configuration: (a)

Conventional CIS and Image Synthesis, (b) Stacked CIS and standard AP.

Such integration of hetero-process technologies is the key to solving these trade-offs, as illustrated in Fig. 7, between the process requirements of sensors, which must optimize image quality, and those of the peripheral logic circuits, which are required to perform at high speed with low power consumption. In this way, Stacked CISs continue to enhance functionality and user experience in mobile devices, a market that currently comprises over several billion units per year.

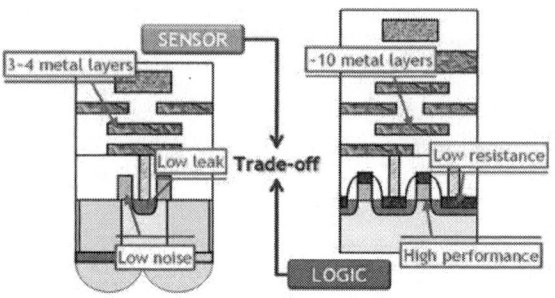

Fig. 7 Trade-off between process technologies optimized for sensor pixels and logic circuits

Furthermore, the stacked structure promotes parallelization of signal processing. For example, capturing an image at a higher resolution and frame rate than those of human vision requires wide band data transmission and memory. One implementation example of heterogeneous technologies to meet such requirements is the System in package (SiP) illustrated in Fig. 8. With DRAM chips integrated within the same package and the expanding data bandwidth between the Stacked CIS and the DRAM, our device can store high-resolution, high-frame rate image data.

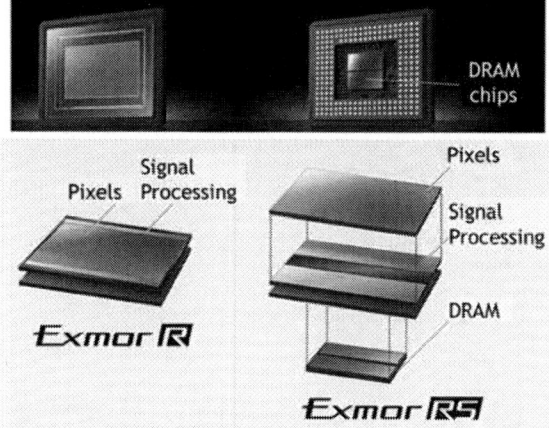

Fig. 8 Example: integration of heterogeneous process technologies for CMOS image sensor.

In addition, to achieving high-resolution, slow-motion images free of distortion even with quick-moving subjects using the pixels equipped with the global shutter function described above, we have also enabled the types of hetero-process technology integration illustrated in Fig. 9. Applying Chip-on-chip integration with micro bump interconnection as

illustrated in Fig. 9(a) to the image sensor[7], we have successfully integrated multiple diced logic chips utilizing the properties of global shutter pixels.[8] The technologies illustrated in Fig. 9(b) will allow us to integrate more chips optimized for multiple processes with greater flexibility in the future.

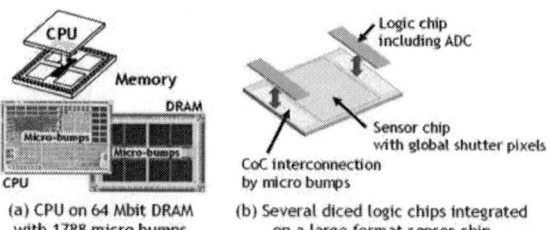

(a) CPU on 64 Mbit DRAM with 1788 micro bumps

(b) Several diced logic chips integrated on a large format sensor chip

Fig. 9 Trade-off between process technologies optimized for sensor pixels and logic circuits

These technologies will enable image sensors to evolve further while simultaneously satisfying the needs for high functionality, high performance, and those for miniaturization of camera modules.

4. Sensing World

CCD and CIS have driven the performance improvement of various cameras, with the former developed with the goal of "Exceeding Film Quality" and the latter that of "Exceeding Human Vision". [9] Image sensors can convert many targets into electric signals, with such targets by no means limited to subjects perceptible to human vision. We are living in a world in which an increasing amount of data and information can be used to make decisions without human intervention to assist the behavior and operation of humans and machines, as areas such as IoT, Big Data, Deep Learning, and Artificial Intelligence suggest. We refer to such use of future image sensors as "Sensing," and our next goal is "Exceeding Human Senses."

5. Sensing Technology

The traditional role played by image sensors has been to convert subjects perceptible to human vision into two-dimensional information. "Sensing" means to simultaneously provide information of an extra dimension in addition to this two-dimensional information, i.e., provision of 2D+α information.

For example, adding distance information (α=Z) to two-dimensional information allows for gesture input control. Once hands and fingers are recognized as two-dimensional information, front/rear and left/right movement can be determined by entering the distance information of each point. There are multiple techniques to simultaneously obtain two-dimensional information and distance information. In the Time-of-Flight (TOF) method, for example, high-speed blinking infrared light is used to irradiate the subject, with the observed flight time of the light reflected back from the subject used to measure the distance between the light source and the subject.[10]

Systems can be configured to provide nutritional and freshness information of foods by adding wave-length information (α=λ) to the two-dimensional information.[11] Conventional imaging senses visible light such as red, green, and blue at the fixed resolution of the three-color mosaic. However, there are technologies that allow easy implementation of many types of color filters, such as Plasmonic color filters, that can be combined with signal processing to make the resolution selectable in 50 nm increments.[12]

The time resolution ($\alpha=t$) of the two-dimensional information can also be dramatically enhanced. For example, if a sensor can take 1,000 two-dimensional pictures per second, application to automobiles and robots will quickly expand more.[13] As another example, mounting a high-speed vision sensor on a robot arm for computer vision will allow for highly accurate feedback positioning , promising improvements in throughput and productivity.

Going forward, we are going to increase the types of information detectable together with the two-dimensional information, thereby aiming to expand the "Sensing" applications of image sensors.

6. Summary

Image sensors have continued to evolve as applications expand—from video cameras to digital still cameras and smartphones—including the transition from CCD to CIS and technical evolution of BI CIS and Stacked BI CIS. In terms of image quality, they have attained high sensitivity and high-speed operation that exceeds human vision, with some recent cameras being able to capture scenes not perceptible to human vision. With mass production of Stacked BI CISs, people will learn to utilize the freedom in process selection it offers, facilitating proposals of new, innovative functionality. To meet increasingly broad requirement, we will continue to propose a wide variety of functions from the image sensor side. In the growing "Sensing world", image sensors will also perform functions of sensing and recognizing information in addition to the conventional visual functions.

Reference

1) T. Hirayama, "The Evolution of CMOS Image Sensors," in Proc. of IEEE A-SSCC, pp. 5-8, Nov. 2013.

2) Satoshi Yoshihara, et al., "A 1/1.8-inch 6.4MPixel 60frames/s CMOS Image Sensor With Seamless Mode Change", IEEE Journal of Solid-state circuits, vol. 41 No12, Dec 2006.

3) Shin Iwabuchi, et al., "A Back-Illuminated High-Sensitivity Small-Pixel Color CMOS Image Sensor with Flexible Layout of Metal Wiring", ISSCC Dig. Tech. Papers, pp.302-303, Feb 2006

4) H. Wakabayashi, et al., "A 1/2.3-inch 10.3Mpixel 50frams/s Back Illuminated CMOS Image Sensor", ISSCC Dig. Tech. Papers, pp.410-411, Feb 2010.

5) Masaki Sakakibara, et al., "An 83dB-Dynamic-Range Single-Exposure Global-Shutter CMOS Image Sensor with In-Pixel Dual Storage", ISSCC Dig. Tech. Papers, pp.380-381, Feb 2012.

6) Shunichi Sukegawa, et al., "A 1/4-inch 8Mpicel back-Illuminated Stacked CMOS Image Sensor", ISSCC Dig. Tech. Papers, pp.12-14, Feb 2013.

7) T. Ezaki et al., "A 160Gb/s Interface Design Configuration for Multichip LSI," IEEE ISSCC Dig. Tech. Papers, pp.140-141, Feb 2004.

8) Yusuke Oike, et al., "An 8.3M-pixel 480fps Global-Shutter CMOS Image Sensor with Gain-Adaptive Column ADCs and 2-on-1 Stacked Device Structure", Symp. VLSI Circuits Dig. Tech. Papers, June 2016.

9) Tomoyuki Suzuki, "Challenge of Image-Sensor Development",ISSCC Dig. Tech. Papers, pp.27-30, Feb 2010.

10) Daniel Van Nieuwenhove, et al., "A 15um CAPD Time-of-Flight pixel with 80% modulation contrast at 100MHz", International Image Sensor Workshop, 7-18, June 2015.

11) Masaru Kashiwazaki, et al., "Study on non-destructive measurement of strawberry fruit firmness", Journal of the Japanese society of agricultural machinery, pp.90-97, 2009.

12) Sozo Yokogawa, et al., "Investigation of the Optical Properties of Plasmonic Color Filter for CMOS Image Sensor Application", SPIE 8457-66, Aug 2012.

13) Masatoshi Ishikawa, "High-Speed Image Sensor Technologies", ISSCC Forum, pp.516-517, Feb 2010.

A 16nm Dual-Port SRAM with Partial Suppressed Word-line, Dummy Read Recovery and Negative Bit-line Circuitries for Low V_{MIN} Applications

Yen-Huei Chen, Kao-Cheng Lin, Ching-Wei Wu, Wei-Min Chan, Jhon-Jhy Liaw, Hung-Jen Liao and Jonathan Chang
Memory Solution Division (MSD), Taiwan Semiconductor Manufacturing Company, Taiwan
Email: yhchenu@tsmc.com, TEL: 886-3-5636688-7038259

Abstract

A total solution for 8T dual-port (DP) SRAM to improve its operating voltage range (V_{MIN}/V_{MAX}) is proposed. Partial suppressed word-line (PSWL) technique improves the static noise margin (SNM) when both ports (A, B ports) access at the same time. Dummy read recovery (DRR) and negative bit-line (NBL) techniques are introduced to eliminate the dummy read induced write recovery failure and write contention failure, respectively. The silicon results show that the VDD operation window can be improved from 220mV to 570mV in 16nm FinFET technology.

Introduction

In the System-on-Chip (SoC) design, more than several megabit of SRAM cells are embedded in an application processer[1][2]. Continuous scaling down of transistor results in increasing random threshold voltage variations in the deep-submicron technology. It significantly degrades SRAM cell operating margin and limits the minimum operating voltage (V_{MIN}). DP-SRAM cell provides two independent ports (A, B port) to realize a parallel operation. Fig.1 shows the default function of DP SRAM with two ports (A, B ports) access at the same row for Cell-1 A-port write and Cell-2 B-port read operation [3]. The read disturb write failure happened at Cell-1 that is performing one port (A-port) at write operation and the other port (B-port) is also accessed due to dummy read. Negative bit-line (NBL) technique is reported to improve the read disturb write issue in the previous works [4]. However, NBL can only increase the PG strength of write "0" side but not helping on the opposite recovery side. Thus DP SRAM cell may still suffer severe write recovery failure if the write word-line pulse is too short to recover the storage node. On the other hand, Cell-2 suffered serious read disturbance because both A and B-port word-lines are activated at the same time that leads to significant static noise margin (SNM) degradation. In this paper, we proposed Dummy Read Recovery (DRR) scheme to improve the read disturb write and Partial Suppressed Word-Line (PSWL) scheme to mitigate the read disturbance issues.

Partial Suppressed Word-line scheme

Fig. 2 shows the comparison of DC SNM versus VDD between one port versus two ports turning on with 6 sigma DP-SRAM weak bit from statistical model at FSG 125C. The blue curve represents the SNM with one-port activation and the red curve represents the SNM with two-port activation. There is a significant SNM degradation from one-port "ON" to two-port "ON" condition. The V_{MAX} of the DP-SRAM is limited by this two-port "ON" condition. The green curve shows the SNM improvement by the proposed partial suppressed word-line technique. The simulation result shows that the V_{MAX} can be significantly improved. Fig. 3 shows the block diagram of the proposed design solutions which include PSWL scheme, A/B ports address decoder, pulse generators and the DRR blocks. If the same

row address of A/B ports are activated, the overlapping of A, B ports word-line signals turn on the pull down PMOS and enable PSWL function. The PSWL waveforms are shown in fig. 4 with different A, B word-line pulse skew conditions.

Dummy Read Recovery

Fig. 5 shows a simultaneously access the same row but different column DP-SRAM which performs A-port write "0" and B-port dummy read operation. At the beginning, the data "1" is stored on node (S) and the node (SB) is "0". Because B_WL is asserted earlier than A_WL by a period of Tdr, the B_BLB is discharged by cell current during the dummy read period of Tdr. Then, write "0" operation is asserted on the A-port, the A_WL is activated and A_BL is discharged to GND ("0") by the write driver. There is a contention between PU1 and APG1 on the storage node (S). In order to achieve successful write, the node SB has to be recovered by pull-up PMOS (PU2) to the power rail as shown in the timing waveforms in fig. 6(a) and (b). However, for long bit-line configuration such as 128 bits per column with longer Tdr results in lower B_BLB. The PU2 will suffer a heavier bit-line load of B_BLB and will not be able to recover SB node to power rail before A_WL turns off. Fig. 6(c) shows the longer dummy read period (Tdr). The cell current discharged B_BLB to ground induced SB node is unable to fully recover to power rail and leads to A-port write recovery failure. Fig. 5 shows the proposed DRR scheme composes of a column signal controlled enable switch (PB0) and A/B port cross talk PMOSs (PB1 and PB2). The A port DRR is disabled by the selected column signal (A_Y) and the B port DRR is enabled by the selected column signal (B_Y) for the A-port write and B-port dummy read condition. A_DB drives data "0" to A_BL that triggers PB2 to turn on to provide the charging current to assist the recovery of the discharged B_BLB. Therefore, the dummy read induced write failure can be mitigated as shown in fig. 6(d).

Si Results

A DP-SRAM test chip was manufactured by 16nm FinFET technology. The test chip is built from sixteen 128kb (8192x16) DP-SRAM macros that include four 256x128 sub-arrays. The area overhead of all proposed design schemes (PSWL, DRR, NBL) is 3.33%. Design features of the test chip are shown in Table I. Fig. 7 shows the silicon results of proposed design solutions versus conventional design without design solution. The operation voltage window of the conventional design is 220mV (V_{MAX}-V_{MIN}) at 95 percentile yield spec. The write V_{MIN} can be enhanced by DRR and NBL techniques for write contention and recovery improvement. In addition, PSWL technique can mitigate the read disturbance issue at higher voltage region thus the operation voltage window can be enlarged to more than 570mV (V_{MAX}-V_{MIN}) and the overall V_{MIN} can be improved by 80mV with the proposed design solutions.

References

[1] M. Naruse et al., ISSCC Digest, pp.260-261, 2008
[2] S.-Y. Wu et. al., IEDM Tech. Dig., pp. 224-227, 2013
[3] Y. Ishii, et al, ASSCC, Digest of Tech., 2010
[4] D. P. Wang et al., IEEE SOC Conf., pp.211-214, 2007

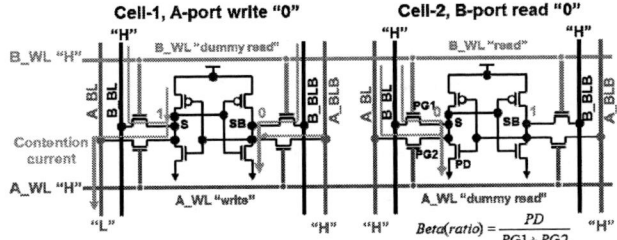

Fig. 1 Dual-port SRAM simultaneous read/write operation: (a) A-port write and B-port half-select, (b) A-port half-select and B-port read or half-select.

$$Beta(ratio) = \frac{PD}{PG1+PG2}$$

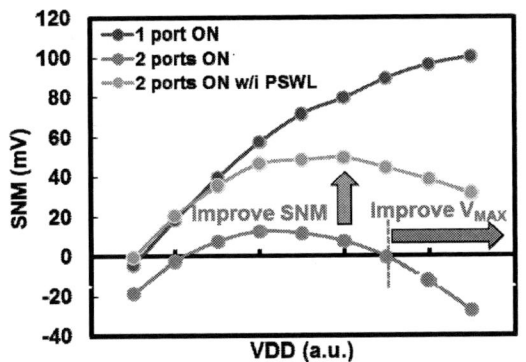

Fig. 2 DC SNM comparison with SWL for 6 sigma worse bit

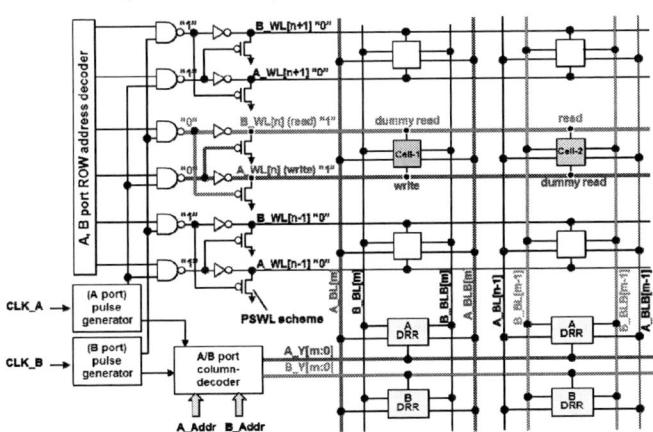

Fig. 3 Block diagram of proposed design techniques

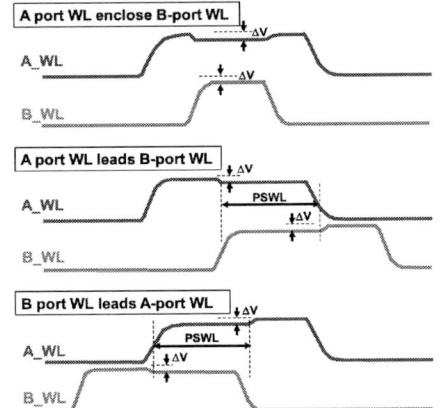

Fig. 4 PSWL waveforms with different clock skew of A, B ports

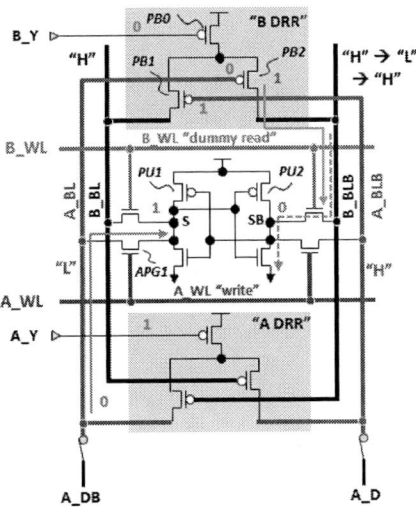

Fig. 5 Circuitry and operation of DRR block

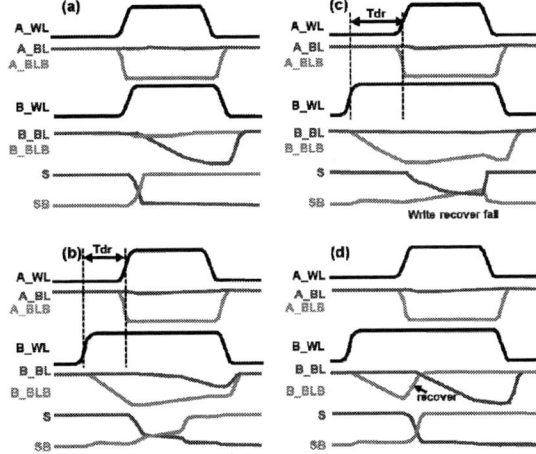

Fig. 6 Dummy read period disturb issue: (a) zero dummy read skew, (b) negative skew, (c) longer 'Tdr' dummy read induced write recovery failure, (d) with enabling DRR scheme

Technology	16nm HK-MG FinFET
Metal scheme	1P7M
Supply voltage	Core: 0.85V, IO: 1.8V
SRAM macro configuration	8192x16 MUX16, 256 bits/BL, 128 bits/WL
SRAM capacity	2Mb (128kb x 16)
Area overhead	3.33%

Table I: Design features of DP-SRAM macro

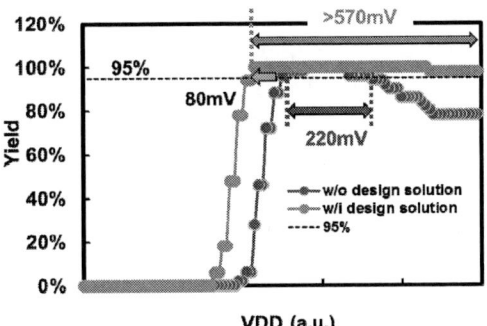

Fig. 7 Si results of V_{MIN}/V_{MAX} between without design solution and with proposed design solutions

978-1-5090-0636-6/16 $31.00 © 2016 IEEE   9   2016 Symposium on VLSI Circuits Digest of Technical Papers

A 6.05-Mb/mm² 16-nm FinFET Double Pumping 1W1R 2-port SRAM with 313 ps Read Access Time

Makoto Yabuuchi, Yohei Sawada, Toshiaki Sano[†], Yuichiro Ishii, Shinji Tanaka, Miki Tanaka[†] and Koji Nii

Renesas Electronics Corporation, Tokyo, 187-8588, Japan

[†]Renesas System Design Corporation, Tokyo, 187-8588, Japan

E-mail: makoto.yabuuchi.ub@renesas.com

Abstract

High-density and low-leakage 1W1R 2-port (2P) SRAM is realized by 6T 1-port SRAM bitcell with double pumping internal clock in 16 nm FinFET technology. Proposed clock generator with address latch circuit enables robust timing design without sever setup/hold margin. We designed a 256 kb 1W1R 2P SRAM macro which achieves the highest density of 6.05 Mb/mm². Measured data shows that a 313 ps of read-access-time is observed at 0.8 V. Standby leakage power in resume standby (RS) mode is reduced by 79% compared to the conventional dual-port SRAM without RS.

Keywords: 1W1R, 2-port, SRAM, 16-nm, Double Pumping

Introduction

The demand for higher performance with limited power budget is increasing especially in advanced SoCs such as application processors and automotive infotainment systems [1, 2]. Image processing with 4K/8K high resolution is realized by high-performance embedded graphic engines. Autonomous driving assistant system (ADAS) also requires higher performance in order for the car to recognize its surroundings at a time. Parallel processing is the key for these applications, in which the embedded memories with multi-port types are needed. 1-write/1-read (1W1R) 2-port (2P) SRAM is typically used and thus smaller macros with lower active/standby powers have become more crucial than ever.

Double Pumping Internal Clock Generation

1W1R 2P SRAM operation can be achieved by utilizing three types of memory bitcell shown in Fig. 1. An 8T SRAM bitcell including isolated read-port is widely used [3]. An 8T dual port (DP) SRAM bitcell realizes higher speed since complementary bitline (BL) pairs with differential sense amplifier enables higher access time but consumes excessive area. The other technique is to use 6T 1-port (1P) SRAM with double pumping clock scheme to reduce area [4].

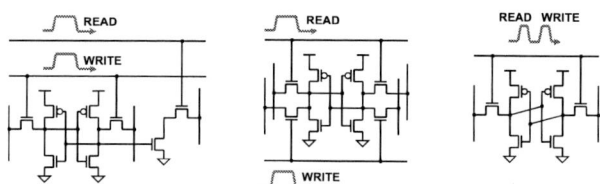

Fig. 1: Realization of 1W1R 2-port SRAM operation.

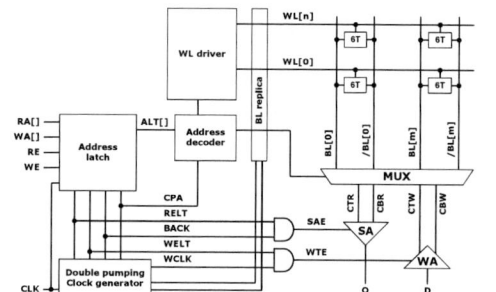

Fig. 2: Block diagram of proposed double pumping 1-write/1-read (1W1R) 2-port (2P) SRAM macro.

In this work, we introduce double pumping clock scheme for internal clock generator to achieve robust timing design without strictly severe setup/hold margin. Fig. 2 depicts the block diagram of proposed double pumping 1W1R 2P SRAM macro using 6T 1P SRAM bitcell. To generate optimum internal read and write clock in serial, the double pumping clock generator and the address latch circuit to relax setup/hold timing are introduced. Fig. 3 shows the schematic of double pumping internal clock generator and its timing chart. Replica BL is used as an adoptive delay element so that the delay can be controlled flexibly for SRAM compiler. The feedback loop with bypassed clock buffer, which minimizes clock delay, generates internal read/write clock pulses (CPA) and write enable pulse (WCLK) automatically within a single system clock cycle (CLK.) Fig. 4 shows the address latch circuits for double pumping. Type 1 is a simple circuit with multiplexer after the latch outputs. In this type, both setup/hold of SELA signal on the CPA falling/rising edge need to be strictly controlled to keep the timing margin. On the other hand, type 2 helps to relax the setup/hold timing without any circuit overhead because only hold margin of SELA signal on the CPA rising edge needs to be optimized. Thus, the type 2 address latch circuit is applied to our double pumping scheme. Simulated waveforms at the worst PVT condition are shown in Fig. 5. The system clock frequency is 1.2 GHz.

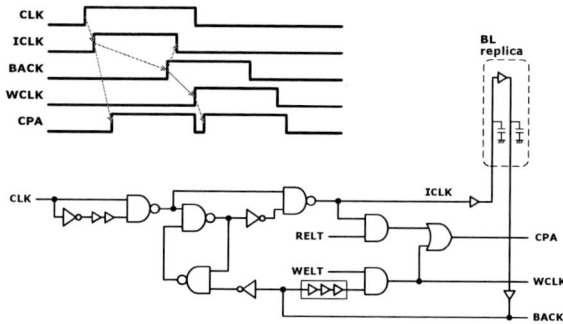

Fig. 3: Schematic of double pumping clock generator and its timing chart.

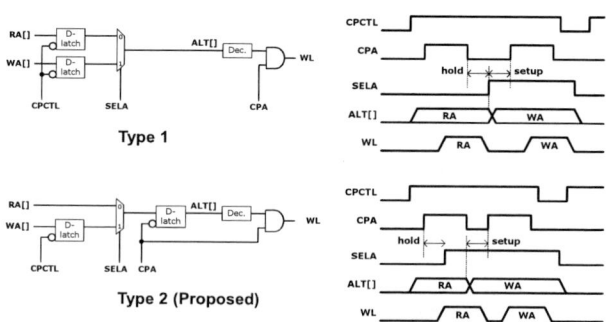

Fig. 4: Proposed address latch circuit.

978-1-5090-0636-6/16 $31.00 © 2016 IEEE

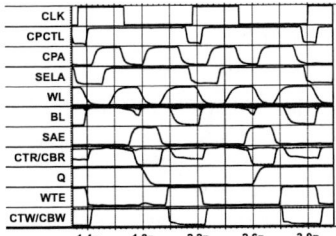

Fig. 5: Simulated waveforms at the worst PVT condition.

Advanced Resume Standby in 16 nm FinFET

In advanced planar bulk CMOS process, VDD lowering in the standby mode is effective to reduce the leakage power. However, FinFET devices have good drain-induced barrier lowering (DIBL) so that the leakage power is less dependent on the supply voltage. Therefore, we introduce the VSS source bias control for SRAM array in the resume standby (RS) mode [5]. Fig. 6 illustrates proposed RS circuits. The footer switches for power gating are inserted for peripheral and cell array, respectively. The LCVSS source of peripheral is simply cut off by core NMOS footer switch, whereas the ARVSS source of cell arrays keeps the intermediate voltage by core NMOS diode and PMOS source follower circuits to retain the stored data with sufficient cell bias. Thus, proposed circuit contributes to robust cell bias against process corner variations at the worst conditions of VDD and temperature as shown in Fig. 6. The PMOS source follower prevents unexpected degradation of cell bias caused by unbalanced leakage of PMOSs in cell array and NMOS diode in the slow-N/fast-P (SF) corner condition.

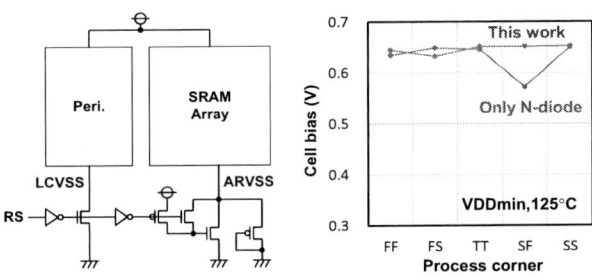

Fig. 6: Resume standby circuit for 16nm FinFET double pumping 1W1R 2P SRAM and its simulation results.

Test chip design and evaluation in 16 nm FinFET

Fig. 7 shows a photograph of test chip, layout plot image, and features of proposed 256 kb double pumping 1W1R 2P SRAM macro using 16 nm FinFET technology [6]. To compare the effectiveness of utilizing 6T 1P SRAM rather than 8T DP SRAM, we also implemented 256 kb DP SRAM macro for reference. The bit density of 1W1R 2P SRAM reaches 6.05 Mb/mm², which is the highest record compared to the past literatures. Fig. 8 indicates histogram of measured read-access-times of 66 samples at 0.8 V and 25°C. VDD dependence (4-chip median.) of read access time is also shown in Fig. 8, achieving 313 ps at 0.8 V typical supply voltage, and 203 ps at 1.1 V overdriven voltage. Fig. 9 indicates distributions of measured leakage power and active power dependence on supply voltage, VDD. By introducing proposed RS circuit, the standby power is reduced to 52% at 0.88 V overdrive and 125°C high temperature. It was reduced by 79% compared to the conventional 8T DP SRAM. Fig. 10 summarizes the macro density *vs.* read access time in the published literature.

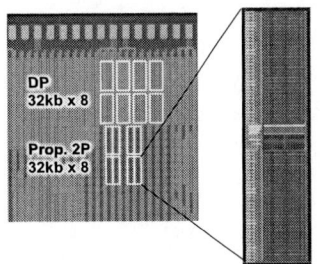

Fig. 7: Test-chip photo, layout plot, and features.

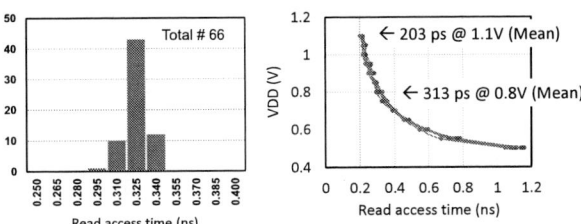

Fig. 8: Histogram of measured read-access-time at 0.8 V and 25°C and VDD dependence (4-chip median.)

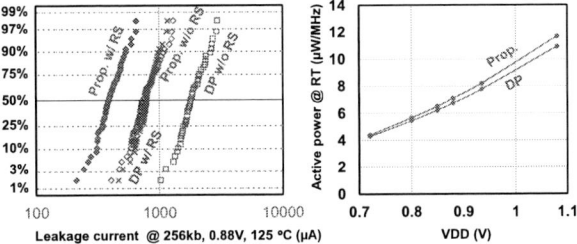

Fig. 9: Measured leakage current distribution w/ and w/o RS modes, and active power *vs.* supply voltage.

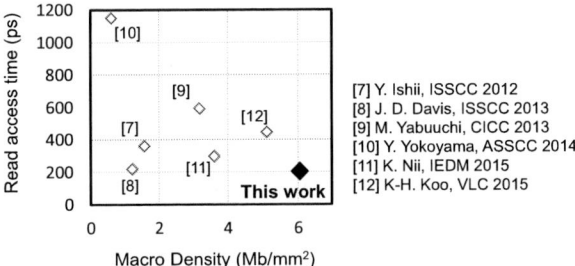

Fig. 10: Comparison of density *vs.* read access time.

Conclusion

A double pumping 1W1R 2P SRAM was demonstrated in a 16 nm FinFET process. Proposed double pumping clock generator and address latch circuit achieved robust timing design. Developed a 256 kb 1W1R 2P SRAM macro achieved the highest density of 6.05 Mb/mm², where we observed 313 ps of read-access-time at 0.8 V. Standby power in resume standby mode was 338 uW at 25°C, resulting in 79% reduction compared to the conventional 8T DP SRAM.

References

[1] T. Yamauchi et al., VLC, 2015. [2] Mochizuki et al., ISSCC, 2016 to be presented. [3] L. Chang, et al., VLT, 2005. [4] C-W. Wu et al., ASSCC, 2104. [5] N. Maeda, et al., JSSC Vol. 48, No. 4, Apr. 2013. [6] Y-H. Chen et al., ISSCC, 2014 [7] Y. Ishii et al., ISSCC, 2012. [8] J. D. Davis et al., ISSCC, 2003. [9] M. Yabuuchi et al., CICC, 2013. [10] Y. Yokoyama et al., ASSCC, 2014. [11] K. Nii et al., IEDM, 2015. [12] K-H. Koo et al., VLC, 2015.

A 2x Logic Density Programmable Logic Array using Atom Switch Fully Implemented with Logic Transistors at 40nm-node and beyond

Y. Tsuji, X. Bai, A. Morioka, M. Miyamura, R. Nebashi, T. Sakamoto, M. Tada, N. Banno,
K. Okamoto, N. Iguchi, H. Hada, and T. Sugibayashi
NEC Corp., Tsukuba, Ibaraki, Japan. E-mail: y-tsuji@az.jp.nec.com

ABSTRACT

Programmable Logic (PL) with a high logic density is demonstrated by cross-bar (xbar) of atom switches, which are programmed through logic transistors. The PL has 4 4-input LUTs to minimize area-delay product owing to small area & capacitance of atom switch. Xbar with 50% and 100% populations mixed and programming lines shared architecture achieves a 2x higher logic density comparing to a commercial PL chip on same technology node of 40 nm. 3x higher operation frequency and 40% lower power consumption are also assessed.

Keywords: Atom switch, Non-volatile, Programmable logic, Cross bar, Logic density, Scalability

INTRODUCTION

PL has been increasingly a preferred digital implementation platform of Internet of Things (IoT) from end-device to data-server [ref.1]. Technology node of shipping chips of PL with SRAM switches reaches to 14 nm. In contrast of the difficulties to push the frontier node further, ReRAM devices including atom switch [ref.2] are expected to be a post scaling technology. In this paper, we show the scalability of atom switch based PL comparing to a widely-used commercial SRAM-based PL chip for the first time.

High Logic Density PL with Atom Switch

Atom switch, integrated in back-end-of-line (BEOL), is one of conductive bridge ReRAMs and with high on/off ratio (Figs.1&2). The formation/deformation of metal (Cu) bridge is controllable with programming voltages. Serially connected atom switches with a shared select transistor (1T2R or CAS: Complementary atom switch) contributes to low programming voltage and high off-state reliability [ref.3]. It also brings small switch cost since the transistor is separated from data path. Reduction in programming voltage of each atom switch has realized a full logic transistor implementation and scalability of PL (PL-a, b in Fig.1).

1T2R-switches-based xbars constitute memories of look up table (LUT), interconnection multiplexers (IMUXs) and switch multiplexers (SMUXs) in a PL cell. The cells are aligned and connected through SMUXs in Fig.3. 4 4input-LUTs in the cell (cluster size CS=4), 4 channels (NC=4) and 4 segment length (SL=4) of connecting resource is found to minimize area-delay product for MCNC benchmark application mapping on the PL (Fig.4(a)).

The derived CS is smaller than that on SRAM-based PL (CS=12) [ref.4]. It is caused by small switch cost and small overhead for connecting LUTs through internal wire among cells compared to inner one. The small CS has advantage on less restriction on application mapping. The xbar for IMUX and SMUX are partially depopulated by 50% to reduce the cell area, while its mappability is confirmed by MCNC benchmark (Fig.4(b)). In 50%-depopulated xbar, the 1T2R switches are placed in a lattice-like pattern but the programming line is connected to the select transistor alternately (Fig.5).

Figure 6 shows a PL cell using atom-switch based xbar. Xbar switches for IMUX and SMUX are adjacently put to share input signal lines from outside of PL cell. Xbar is also used for memory of LUT connecting data-input to VDD or GND. The aspect size of the xbar for the memory is designed for mesh connection of global write control lines, and common write driver circuits are placed on the peripheral of the PL arrays for programming atom switches as shown in Fig.7.

Figure 8 shows layouts of various atom-switch based PLs. PL-a has 2 LUT clustering, 3 channels and 4 segments (CS=2, NC=3, SL=4) homogenous architecture on 65-nm technology node. PL-b has 4 LUT clustering, 4 channels and 4 segments depopulated routing architecture on 40-nm node, and shows 30% compactness of logic density relative to PL-a (Fig.1). The density is found to be 2x higher than SRAM-based PL of a commercial chip at the same technology node of 40 nm.

Chain pattern with data transfer functionality is mapped on the PL-a chip for performance characterization (Fig.9). Data transfer speed and power consumption are measured on various patterns. Evaluated parameters such as capacitance and resistance are used for SPICE and static timing analysis (STA) on placement and routing tools for target applications. Figure 10 shows Shmoo plot on PL-a where an application (16bit-ALU) was configured. Maximum operation frequency and power consumption are shown in Fig. 11, and the experimental data are exactly reproduced with the in-house STA tool.

As shown in Fig.12, atom switch has scalability on 40-nm node and beyond due to small Cu bridge size (~10 nm [ref.2]). The performances of PL-b, which is now fabricated, are estimated on the base of experimental and simulation results of PL-a. Due to scaling, both internal data transfer speed and inner one increase by 7% and 1% respectively in spite of 2x larger cluster size. The power consumptions of data transfer is expected to decrease by 23~26% (see Fig.13).

Performances of PLs (a commercial one, PL-a, PL-b) for an application (16bit-ALU) are summarized in Fig.14. PL-a with 65-nm node marks the better performance than the 40-nm node SRAM based commercial one. It is caused by post-scaling technology using atom switch, in Fig. 2, with small ON resistance (~1 kΩ) and small capacitance (0.14 fF). And excellent performances (3x higher operation frequency and 40% lower power consumption compared to a commercial chip) are assessed by evaluating PLA-b.

SUMMARY

Nonvolatile programmable logic using logic transistor based 1T2R switch shows 2x higher logic density at same technology node, and extends performance (1/5 smaller power-delay product) comparing to a commercial chip.

ACKNOWLEDGEMENT

A part of this work was performed of the METI R&D Program ("Leading New Technology for Energy and Environment") supported by NEDO.

REFERENCES

[1] J. Rodriguez-Andina et al., IEEE Trans. on Industrial Informatics 11, pp.853 (2015). [2] M. Aono et al., Proc. of the IEEE 98, pp.2228 (2010). [3] M.Tada et al., IEDM, pp.689 (2011). [4] E. Ahmed et al., FPGA, pp.3-12 (2000). [5] M. Miyamura et al., FPGA, pp.236-239 (2015).

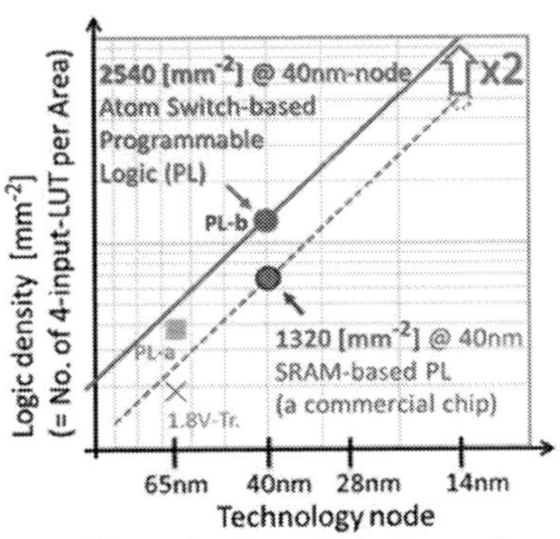

Fig. 1: Logic density vs. Technology node.

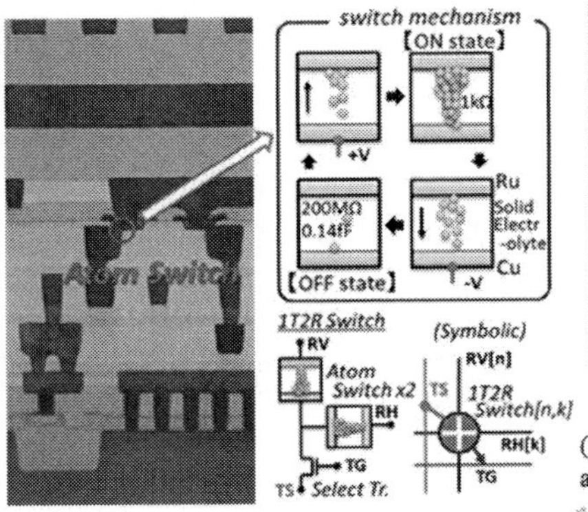

(←) Fig. 2: TEM of atom switch in BEOL, switch mechanism, and 1T2R (complimentary type) switch structure.

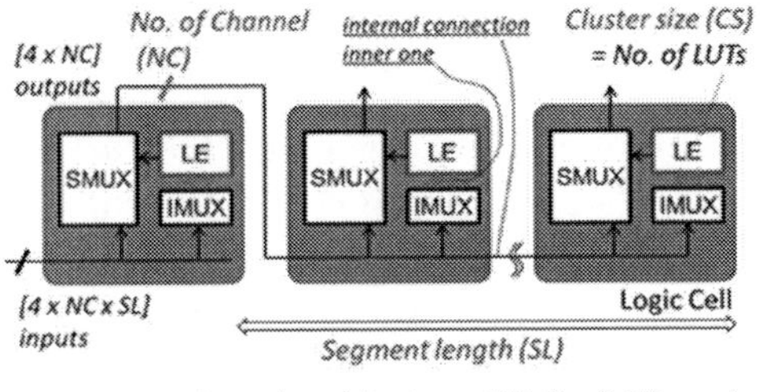

(↑) Fig. 3: Basic architecture of PL. Logic Element (LE) in a cell includes several LUTs and Flip-Flops.

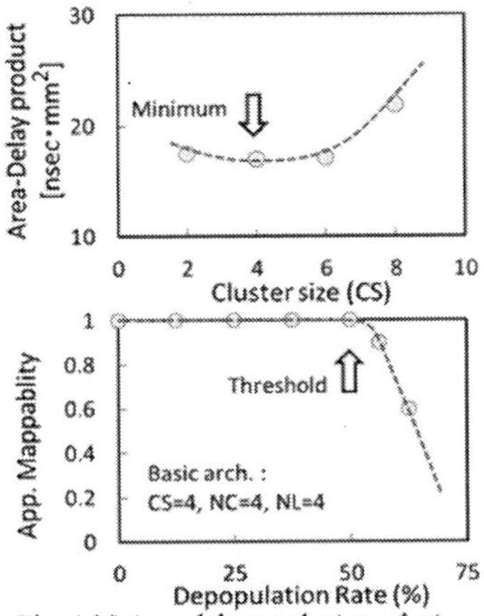

Fig. 4 (a) Area-delay product vs. cluster size. (b) Application mappability vs. depopulation rate of switches in xbar.

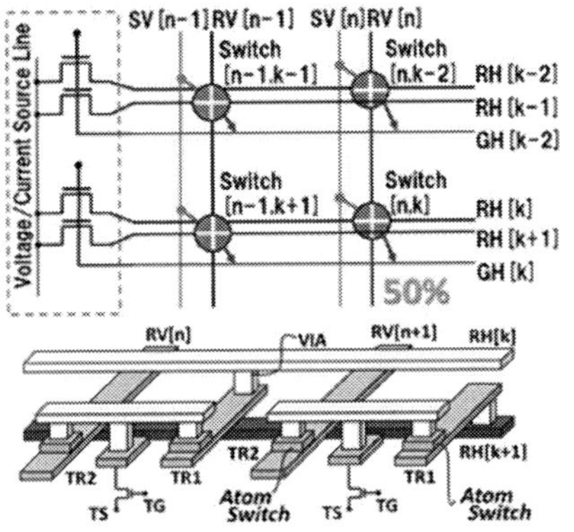

Fig. 5: Circuit and layout of switch-depopulated xbar in case [No. of switches]/[No. of cross points] is 50%.

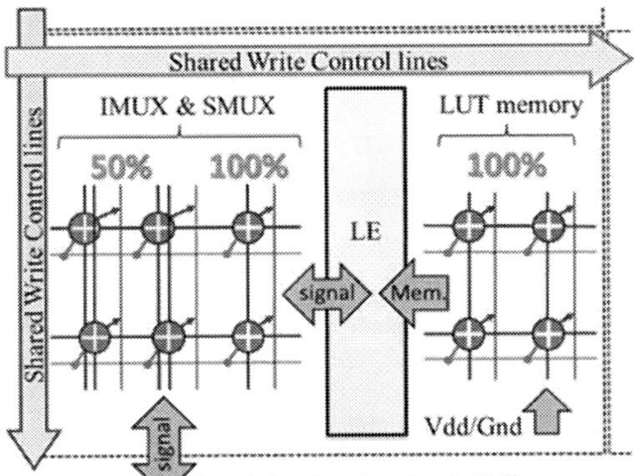

Fig. 6: Cross bar (xbar) switch for a programmable logic cell.

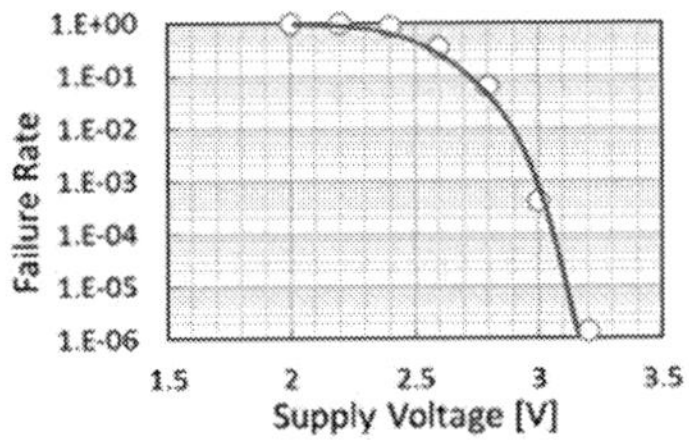

Fig. 7: Failure rate dependence on supply voltage for programming 1T2R structured switch.

(→) Fig. 8: Layouts of programmable logic cell on 65nm node with 1.8V/logic transistors, and on 40nm node with logic transistors.

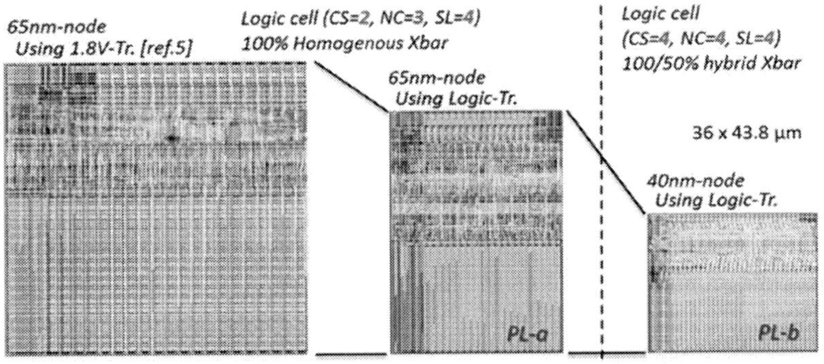

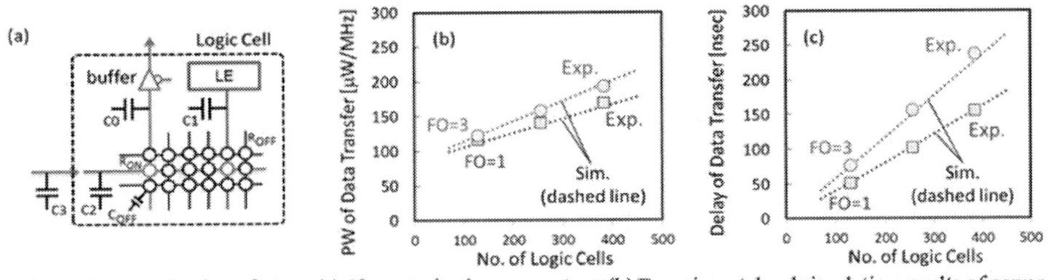

Fig. 9: Characterization of PL-a: (a) Characterization parameters. (b) Experimental and simulation results of power consumption and (c) delay of data transfer among cells. Switch pattern changes load capacitance (= fan-out FO).

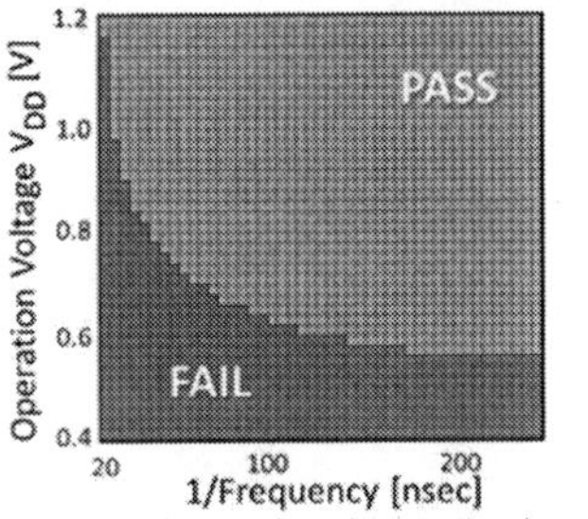

Fig. 10: Shmoo plot of an application (ALU) configured on PL-a chip.

	Max Op. Freq.(MOF)	Dynamic Power (DPW)
Sim.	37 MHz	43.1 uW/MHz
Exp.	40 MHz	41.5 uW/MHz

Fig. 11: STA tool and experimental results of ALU on PL-a at VDD=1.0V.

	MOF	DPW	DPW/MOF
Commercial	0.5	1.07	2.14
PL-a	1	1	1
PL-b*	1.56	0.63	0.40

Fig. 14: Performance of ALU on a commercial chip, PL-a, and PL-b (*simulated value) at VDD=1.0V. Arbitrary unit relative to PL-a's values.

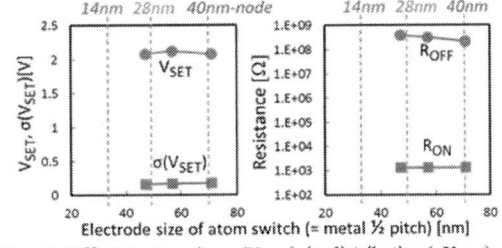

Fig. 12: Off to On Set voltage (V_{SET}), its distribution (σV_{SET}), and On/Off resistances vs. Electrode size of atom switch.

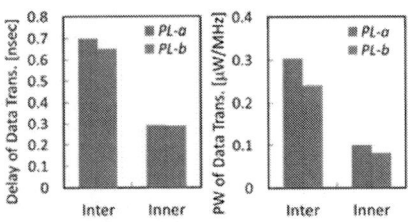

Fig. 13: Expected performance (delay & power) of internal/inner data transfer on PL-a/b with SPICE.

80Kb 10ns Read Cycle Logic Embedded High-K Charge Trap Multi-Time-Programmable Memory Scalable to 14nm FIN with no Added Process Complexity

Janakiraman Viraraghavan[1], Derek Leu[2], Balaji Jayaraman[1], Alberto Cestero[2], Robert Kilker[3], Ming Yin[2], John Golz[2], Rajesh R. Tummuru[1], Ramesh Raghavan[1], Dan Moy[2], Thejas Kempanna[1], Faraz Khan[2,4], Toshiaki Kirihata[2], Subramanian Iyer[4]

GLOBALFOUNDRIES ([1]BANGALORE INDIA, [2]NY USA). IBM ([3]MN USA). [4]UCLA CA USA

Abstract

An 80Kb logic Embedded Multi-Time Programmable Memory (MTPM) employs charge trapping and de-trapping behavior in 32nm/22nm High-K transistor, resulting in no added process complexity. Multi-step verification with overwrite protection employs block-write and signal margin degradation (~30%) to satisfy 10 year retention at 105° C.

Introduction

Conventional embedded nonvolatile memories (eNVMs) use floating gate [1], charge trap MONOS [2], magnetic [3], or resistive element [4], which may not be preferred for high performance logic or low cost foundry technologies due to additional mask or process complexity and voltage incompatibility. Logic process compatible electrical fuse [5] cannot be reprogrammed, and scaling saturates beyond 22nm.

Technology

Programming is achieved by electron injection into the HiK gate dielectric in standard logic NMOS [6] by applying a word-line (WL) voltage of VPP (~2V), a source-line (SL) of VSL1 (~1.5V), and grounding (0V) the bit-line (BL). This enables efficient trapping of electrons in the HfO_2 / interfacial layer, resulting in increase of the threshold voltage (V_{TH}). The typical V_{TH} shift (ΔV_{TH}) in 32nm and 22nm, Fig.1, is ~200mV with a ~10ms programming pulse and can be increased to ~300mV with a ~100ms pulse. The trapped electrons may be de-trapped by driving the WL to VWL (~-1V) and the SL to VSL2 (~2V). The multi-time programming feature, high density, relatively low voltage and zero mask adder or process steps are significant advantages for embedded applications.

Macro Architecture

The memory array, Fig. 2, is organized as 256 rows and 320 columns, resulting in 80Kb density. In order to improve the ΔV_{TH} detectability, a single bit is stored in a twin cell, which consists of two logic NMOS transistors (NMOSt and NMOSc). The cell is controlled by word-line (WL), source-line (SL), bit-line-true (BLt) and bit-line-complement (BLc). A logic 1 (or 0) is stored by shifting the V_{TH} of NMOSt (or NMOSc). WL drivers include a voltage switch to select between VPP and a main voltage (VDD ~1V) in programming and read mode respectively. The column decoder selects 1 of 4 BL pairs thus providing an 80 bit data-line (DL) read out. A SL switch coupling to VSL1, VDD, and ground (GND) is provided per DL segment (4 columns) such that the SLs, only in the selected segment, are raised to VSL1 and VDD during programming and read respectively, while grounding the SLs in unselected DL segments. The macro supports a default bit function using a preprogrammed indicator bit.

Multi-step-programming with Block Write Algorithm

Shifting the V_{TH} of all cells to the same extent may either cause an insufficient V_{TH} shift with an initially lower V_{TH} (cell 0,b for example), or can potentially damage the dielectric of cell with an initially higher V_{TH} (cell 0,a), through excessive charge injection. Multi-step-programming approach with Over-Write-Protection (OWP), Fig. 3, is performed in incremental time steps while verifying the contents of the cell in the read cycle and re-programming only those cells that fail verification. In order to avoid a false OWP due to shallow trap electrons (unusable for steady-state V_{TH} shift), the approach also employs a block write algorithm: Step-1: program multiple cells in a same selected DL segment sequentially as a block, step-2: sequentially verify the written bits, and rewrite only those bits that fail to read the expected value, step-3: repeat step-2 process "m" times, and step-4: perform a final margin write with ~30% signal margin for retention by activating either EN_T, EN_C, or both, Fig. 2.

Slew Sense Amplifier

A high gain slew sense amp (SSA), Fig. 4, is used to sense the V_{TH} difference (ΔV_{DIF}) between the twin cells. SA nodes are pre-discharged to GND and left floating while simultaneously ramping the WL to VDD to enable charging of SAt and SAc through the twin cells. As the WL ramps up, the NMOS with lower V_{TH} turns ON in saturation while the other is still in sub-threshold. This results in an abrupt increase in voltage on one side (SAt or SAc), turning ON the pull up PMOS P0/P1 through inverters I0/ I1 respectively while the cross-coupled NMOS stack drives the other side low. Monte-Carlo simulation shows that the SSA can sense ~10% less ΔV_{DIF} compared to the Cross-Coupled Sense Amp (CCSA).

Hardware Results

Fig. 5 shows the bitmap change (32nm) for the multi-step verification checker-board (CKB) pattern write (a) without OWP, and (b) with OWP. The bitmap (a) shows overwrite fails (OW fails) for multi-step writes > 130ms due to excessive charge trap. The bitmap (b) with OWP circuit shows a perfect bitmap in 20ms, demonstrating successful protection. The indicator bit, successfully emulates the default bit state of 0 (Fig. 5c). High temperature (25, 125, 180, 240 ^{O}C) bake tests, Fig. 5d, for 1000 hours using 32Kb module (32nm) show a projected 10 year V_{TH} degradation of ~30% at 125^OC. Fig. 6a shows the 4X write functionality using CKB pattern. Fig. 6b shows the endurance studies using 32nm device macros, demonstrating 10 cycles with some hysteresis. 22nm study, Fig. 6c, shows the benefit of trap forming process (multiple writes) improving the endurance up to 100X. The MTPM read cycle shmoo (32nm), Fig. 7a, with 32nm SSA shows <10ns read cycle at 1V, and is functional down to 0.65V at 20ns , which is significantly better than MTPM using CCSA (22nm), Fig. 7b.

Scalability to FIN Technology

14nm FIN device, Fig. 8a, shows a similar ΔV_{TH} programmability. The ΔV_{DIF} increases as the programming time increases, separating out the 0 and 1 data histograms while not altering the distribution ($\sigma \Delta V_{TH}$), Fig. 8b. Read Shmoo (c) and 4X write functionality (d) showing "M', "T", "P", and "M" have been confirmed. Table I summarizes the key features of the MTPM prototypes.

References:
[1] H. Kojima et. al., IEDM, 2007, pp. 677–680.
[2] Y. Taito et., al., ISSCC, pp. 132-133, Feb. 2015.
[3] M. Jefremow et. al., ISSCC, pp. 216-217, Feb. 2013.
[4] M. Ueki et. al, VLSI Tech., pp. 108-109, 2015.
[5] G. Uhlmann et. al., ISSCC, pp. 406-407, Feb. 2008.
[6] C. Kothandaraman et. al., IEEE IRPS, pp. MY2.1-2.4, 2015

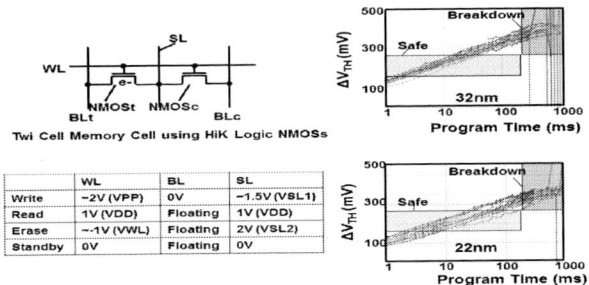

Fig. 1: Twin Cell Charge Trap Memory.

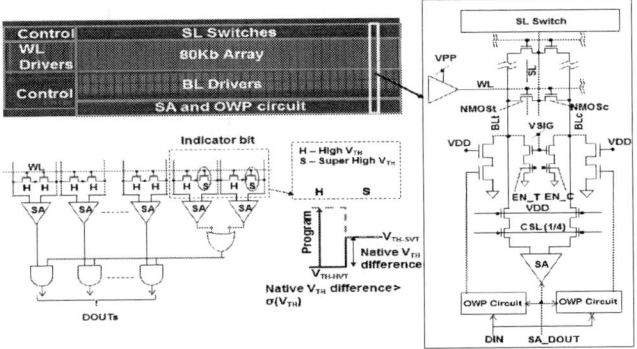

Fig. 2: 80Kb MTPM Macro Architecture.

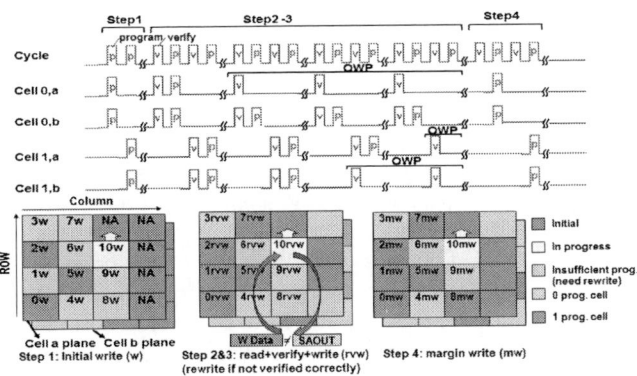

Fig. 3: Multi-Step Write and Block Write Algorithm.

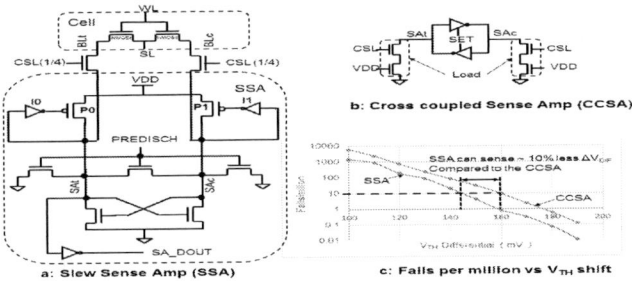

Fig. 4: Slew Sense Amplifier (SSA).

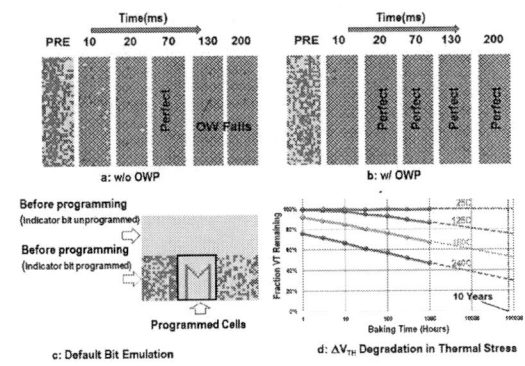

Fig. 5: 32nm Hardware Results.

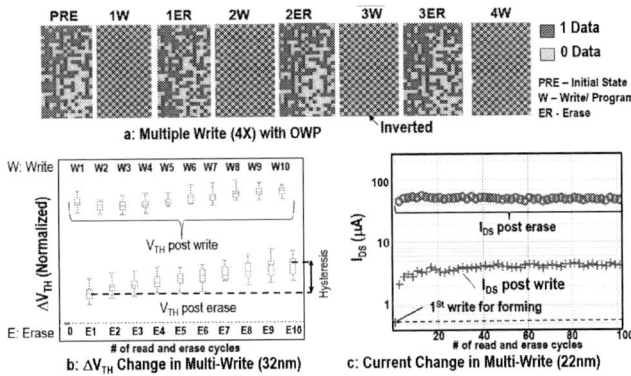

Fig. 6: Multi-Time-Programming Results

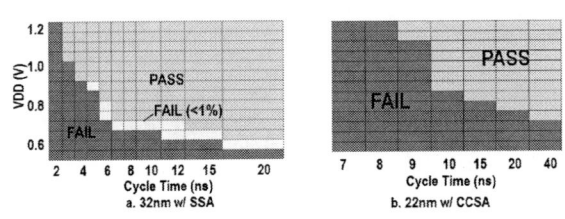

Fig. 7: Read Cycle Time Shmoo.

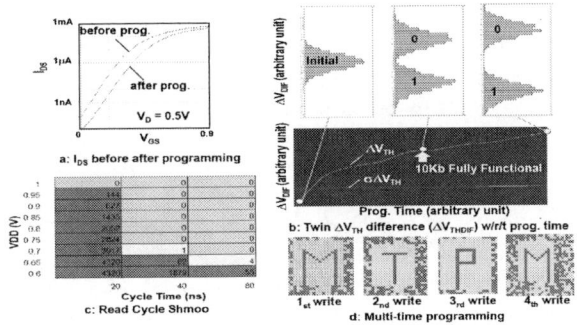

Fig. 8: 14nm Bulk FIN MTPM Hardware Results.

Table I: Summary of the MTPM Prototypes.

Technology	32nm SOI	22nm SOI	14nm Bulk
Cell	0.109µm² with 1.4nm Gox NMOS	0.144µm² with 1.2nm Gox NMOS	0.1411µm² with FIN NMOS
Macro Density	80Kb	64Kb	40Kb
Density/mm²	~2Mb/mm²	~2.5Mb/mm²	~1.3Mb/mm²
Activation Energy	~1.4eV	~2.4eV	NA

A 97.99 dB SNDR, 2 kHz BW, 37.1 μW Noise-Shaping SAR ADC with Dynamic Element Matching and Modulation Dither Effect

Koji Obata, Kazuo Matsukawa, Takuji Miki, Yusuke Tsukamoto, Koji Sushihara

Panasonic Corporation, Osaka, Japan
obata.koji@jp.panasonic.com

Abstract

A 97.99 dB SNDR, 2 kHz bandwidth noise-shaping SAR ADC was fabricated in 28 nm CMOS process. By integrating residue of 12 bit SAR AD conversion with 3rd order integrator, Σ modulation is achieved and noise floor of AD conversion is shaped. Distortion due to mismatch of capacitive DAC is eliminated by introducing dynamic element matching (DEM) technique and by utilizing modulation dither effect. The ADC consumes 37.1 μW with 100 kHz sampling speed and achieves Schreier's figure of merit (FoMs) of 175.3 dB.

Keywords: SAR ADC, Σ modulation, Noise-shaping and Dynamic element matching (DEM)

Introduction

In recent years, analog front ends (AFEs) to obtain very small signals such as electroencephalogram (EEG) and electromyography (EMG) are developed. In such AFEs, very small signal is amplified more than 1000 times before analog to digital converter (ADC) and the amplified signal is digitized using around 12 bit ADC[1]. Required signal bandwidth for such AFEs is several kHz. In addition to high gain and low noise, low power consumption is required for amplifiers when such AFEs are applied to mobile usage. Therefore complicated architecture is adapted for such AFE and silicon area is large. By developing very high resolution, that is high dynamic range, with ultra-low power consumption ADC, amplifiers before the ADC can be simplified. This approach leads AFEs to low power and small area.

Successive approximation register (SAR) ADC is known as the most power efficient ADC. However effective number of bit (ENOB) of SAR ADCs is determined by mismatch of capacitive DAC, comparator noise and so on. To enhance ENOB of SAR ADCs some techniques of other ADC architectures such as noise-shaping technique is combined[2,3]. In this paper, Σ modulation using 3rd order integrator and dynamic element matching (DEM) technique are introduced into SAR ADC and a new noise-shaping SAR ADC is proposed. The proposed SAR ADC has higher resolution than conventional SAR ADCs and overcomes the limitation of SAR ADCs maintaining power efficiency. The proposed ADC was fabricated in 28nm CMOS process and 97.99 dB SNDR (16.0 ENOB) with 2 kHz bandwidth is measured using 37.1 μW under 1.55 V and 0.75 V power supplies.

Proposed ADC architecture

The block diagram of the proposed noise-shaping SAR ADC is shown in Fig. 1. SAR ADC part has 12 bits, and 12 bit DAC consists of upper and lower DACs which are connected by a capacitor. Residue of the SAR ADC is integrated by the modulator and the integrated residue is used as the reference voltage of the next SAR conversion. Σ modulation is achieved by this process and noise-shaping characteristic is realized. In the proposed ADC, 3rd order integrator is used as the modulator. In addition to the Σ modulation, DEM technique is adapted to upper DAC to eliminate effects of capacitor mismatch.

In the modulator, complete 3rd order integration characteristic is achieved using 3 integrators. Residue of SAR AD conversion is integrated by these integrators and so input voltage to them is small. These integrators are stable because of this feature and further high order integration is also possible. Re-sampling is not necessary and remaining charge in the capacitive DAC is integrated. Only 3 additional clocks to integrate residue are added to conventional SAR AD conversion. Dedicated reference voltage is not required to the integration and charge/discharge of capacitors is done with power supply and ground like SAR ADC. No power is consumed for reference voltage generation.

As the operational amplifier (opamp) in the integrators, resistive load type is adapted. There is no need of common mode feedback (CMFB) and therefore high speed start up is possible. Power of opamps is only on at the integration time and power consumption of modulator is lowered. Furthermore, since this type of opamps is very simple, area of them also becomes small. Figure 2 shows the operational sequence of the proposed noise-shaping SAR ADC. By making operational frequency of SAR conversion higher than it of integration, further speed up is possible.

To eliminate the effect of mismatch of upper capacitive DAC, DEM is adapted. A new DEM algorithm of dual pointer bi-directional data weighted averaging is developed based on the conventional data weighted averaging and application of DEM to the noise-shaping SAR ADC becomes possible using the new algorithm. An example of the pointer operation of the new DEM algorithm is shown in Fig. 3. Two pointers are prepared for plus and minus sides of DAC and these pointers are updated every comparison in SAR conversion. Either plus side or minus side pointer is updated. Reference voltage for the comparator changes every sampling because of Σ modulation. This change becomes dither for lower capacitive DAC and DEM is not required for lower DAC.

Measurement results and Summary

The proposed ADC was fabricated in 28 nm CMOS process. Chip photograph is shown in Fig. 4. The ADC occupies 160 μm × 610 μm. Figure 5 and 6 show FFT spectrums with 100 kHz sampling and 1 MHz sampling, respectively. For 100 kHz sampling, 97.99 dB SNDR (16.0 ENOB) and 111.8 dB SFDR for 2 kHz bandwidth is measured with only 37.1 μW power consumption. For 1 MHz sampling, on the other hand, 93.95 dB SNDR (15.3 ENOB) and 108.0 SFDR for 20 kHz bandwidth with 493.1 μW is measured. Figure 7 shows a FFT spectrum without DEM. Many large distortions appear and these distortions disappear with DEM. Figure 8 and 9 show

978-1-5090-0636-6/16 $31.00 © 2016 IEEE

input signal power versus SNR and SNDR for 100 kHz sampling and 1 MHz sampling, respectively. Table 1 shows performance comparison with state-of-the-arts high SNDR ADCs. The highest Schreier's figure of merit (FoMs) of 175.3 dB is achieved. The proposed ADC can reduce power consumption with required bandwidth and is very suitable for sensor front end usage.

References

[1] J. Xu, et. al., "A Wearable 8-Channel Active-Electrode EEG/ETI Acquisition System for Body Area Networks," IEEE JSSC, vol. 49, no, 9, pp.2005-2016, Sept. 2014.

[2] J. A. Fredenburg, et. al., "A 90-MS/s 11-MHz-Bandwidth 62-dB SNDR Noise-Shaping SAR ADC," IEEE JSSC, vol. 47, no. 12, pp.2898-2904, Dec. 2012.

[3] Z. Chen, et. al., "A 9.35-ENOB, 14.8 fJ/conv.-step Fully-Passive Noise-Shaping SAR ADC," Digest of VLSI Circuit Symposium 2015, pp.C64-C65, June 2015.

[4] H. Park, et. al., "A 0.7-V 100-dB 870-uW Digital Audio SD Modulator," Digest of VLSI Circuit Symposium 2008, pp.178-179, June 2008.

[5] Y. Yang, et. al., "A 114dB 68mW Chopper-Stabilized Stereo Multi-Bit Audio A/D Converter," ISSCC 2003, pp.56-57, Feb. 2003.

[6] K. Nguyen, et. al., "A 106dB SNR Hybrid Oversampling ADC for Digital Audio," ISSCC 2005, pp.176-177, Feb. 2005.

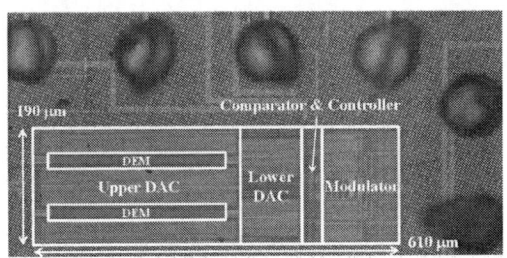

Fig. 4. Chip photograph.

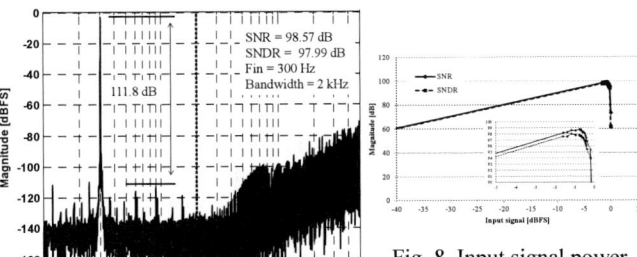

Fig. 5. FFT spectrum with 100 kHz sampling.

Fig. 8. Input signal power versus SNR/SNDR with 100 kHz sampling.

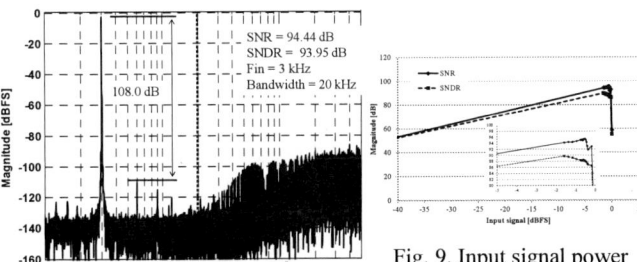

Fig. 6. FFT spectrum with 1 MHz sampling.

Fig. 9. Input signal power versus SNR/SNDR with 1 MHz sampling.

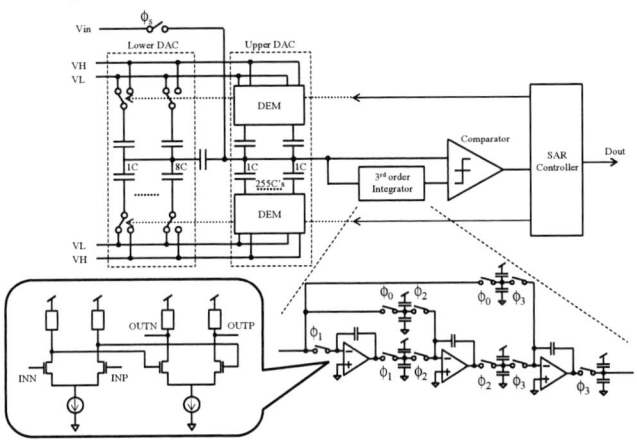

Fig. 1. Block diagram of the proposed noise-shaping SAR ADC.

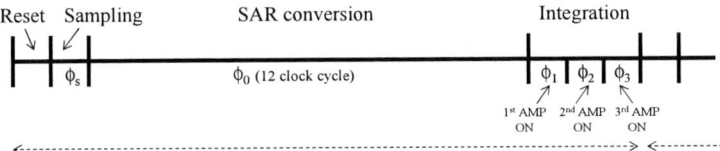

Fig. 2. Operational sequence of the proposed ADC.

Fig. 7. FFT spectrum when DEM is not used with 100 kHz sampling.

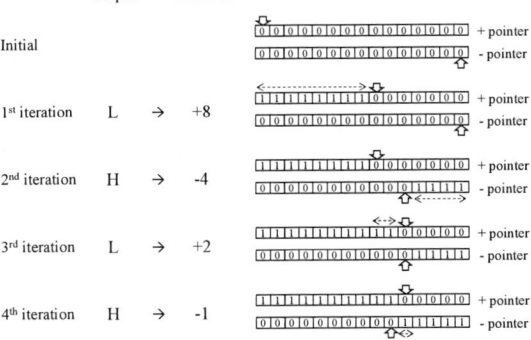

Fig. 3. An example of pointer operation of dual pointer bi-directional data weighted averaging for 4 bit DAC.

Table 1. Performance comparison.

	This work		[4]	[5]	[6]
Technology	28 nm		180 nm	350 nm	350 nm
Architecture	NS SAR		ΔΣ	ΔΣ	ΔΣ
Supply voltage [V]	1.55/0.75	1.8/1.1	0.7	5/1.8	
Sampling rate [MHz]	0.1	1	5	6.144	5.12
Bandwidth [kHz]	2	20	25	20	20
SNR [dB]	98.57	94.44	100		106
SNDR [dB]	97.99	93.95	95	105	97
SFDR [dB]	111.8	108.0			
Area [mm²]	0.116		2.16	5.62	0.82
Power [µW]	37.1	493.1	870	68000	18000
FoMs [dB]	175.3	170.0	169.6	159.7	157.5

A 35μW 96.8dB SNDR 1 kHz BW Multi-Step Incremental ADC Using Multi-Slope Extended Counting with a Single Integrator

Yi Zhang, Chia-Hung Chen, Tao He, and Gabor C. Temes

School of EECS, Oregon State University, Corvallis, Oregon, USA

E-mail: zhangy7@eecs.oregonstate.edu

Abstract

A multi-step incremental ADC (IADC) with multi-slope extended counting is presented. In the proposed IADC, the accuracy is enhanced by reconfiguring it as a multi-slope ADC in two additional steps. For the same accuracy, the conversion cycle is shortened by a factor of about 2^9 as compared to the single-step IADC. Fabricated in 0.18-μm CMOS process, the prototype ADC operates at 642 kHz and achieves a peak SNDR = 96.8 dB and DR = 99.7 dB over a 1 kHz bandwidth. The power consumption is 35 μW, which results in an excellent Schreier FoM of 174.6 dB.

Introduction

Incremental ADCs (IADCs) have found wide applications in sensor interface circuitry since they provide low-latency high-accuracy conversion and easy multiplexing among multiple channels [1]. High power efficiency was reported in [2], although the ADC converted only DC signals. For high accuracy with larger bandwidth, hybrid schemes with extended counting techniques have been reported in [3-4]. In [5], an extra order of noise shaping was achieved by multi-step operation and hardware recycling, thus enhancing the power efficiency. However, all these reported schemes achieved the extended accuracy by recycling the residue voltage from the coarse quantization. This makes the fine quantization vulnerable to non-ideal effects, such as charge injection, clock feedthrough and parasitic coupling. Therefore the operation requires careful shielding and protection of the quantization residue.

In this work, instead of manipulating the residue voltage, we proposed to cancel the quantization residue of the IADC by using a multi-slope extended counting technique with a single integrator, which enhances the power efficiency and makes the conversion much less susceptible to non-ideal effects.

ADC Architecture

The block diagram of the proposed architecture is shown in Fig. 1. The conversion cycle between adjacent resets is divided into three steps. They require $M_1=256$, $M_2=32$ and $M_3=32$ clock periods, respectively. During the first step, the ADC uses a first-order low-distortion $\Delta\Sigma$ loop with input feedforward paths [1]. At the end of the first conversion step, the quantization residue is stored at the integrator output V_{INT}. In the second step, the input path is disconnected, and the ADC is configured as a single-slope ADC with slope $G_2 = 1/M_2$. This scales the DAC step size by $1/M_2$. For the final step, the circuit remains configured as a slope ADC, but with a finer step size achieved using a slope $G_3 = 1/(M_2 \cdot M_3)$. The quantization residue at V_{INT} is cancelled in the second and third steps by being reduced in fixed steps, until $V_{INT} = 0$ is reached, as shown in the top right part of Fig. 1. The V_{INT} is not connected to any switch for finer quantization. Once a

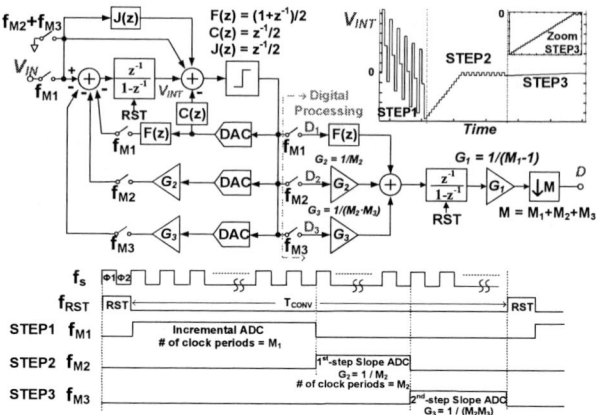

Fig. 1. Block diagram of the proposed multi-step incremental ADC with multi-slope extended counting and its timing diagram.

conversion cycle is completed, the residual error stored as V_{INT} is bounded by $|V_{FS}/(M_2 \cdot M_3)|$, equivalent to SQNR $\approx 20\log_{10}(M_1 \cdot M_2 \cdot M_3)$. For a 16-bit accuracy, this reduces the length of the conversion cycle by a factor of more than 2^9 compared to that of the single-step IADC.

A single-bit internal quantizer is used for good linearity, and an FIR feedback path $F(z)=0.5+0.5z^{-1}$ with a compensation path $C(z) = 0.5z^{-1}$ is introduced to reduce the voltage step size at the integrator output V_{INT}. A feedforward path $J(z)=0.5z^{-1}$ is added to cancel the input signal introduced by $C(z)$, and therefore to reduce the swing of V_{INT}. The digital decimation process reconstructs the signal using only a single digital counter shared among the three steps.

Circuit Implementation

The switched-capacitor circuit implementations for the three steps are shown in Fig. 2(a)-(c). For the first step, the equal-weighted two-tap FIR feedback DAC with its compensation path $C(z)$ is realized as illustrated in Fig. 2(a). A delayed input feedforward path $J(z)$ is implemented with C_4-C_5 to cancel the signal content introduced by the FIR feedback compensation path $C(z)$ in V_{INT}. A capacitive passive adder using C_1-C_5 is employed for the summation at the input of the quantizer. Chopper stabilization is activated for the first and second step. In the second and third step, as shown in Fig. 2(b)-(c), the slope factor $G_2 = 1/M_2$ is realized by the ratio C_U/C_F, whereas the smaller coefficient $G_3 = 1/(M_2 \cdot M_3)$ is obtained by a capacitive T-network. It gives an equivalent capacitance C_{DAC3} equal to $C_U/(k+2)$, where k is chosen as (M_3 - 2). A unit capacitor $C_U = 100$ fF was used in the design. The opamp in the shared integrator was implemented with a two-stage circuit using cascode Miller compensation. Bootstrapped switches were used in the input signal path.

978-1-5090-0636-6/16 $31.00 © 2016 IEEE   18   2016 Symposium on VLSI Circuits Digest of Technical Papers

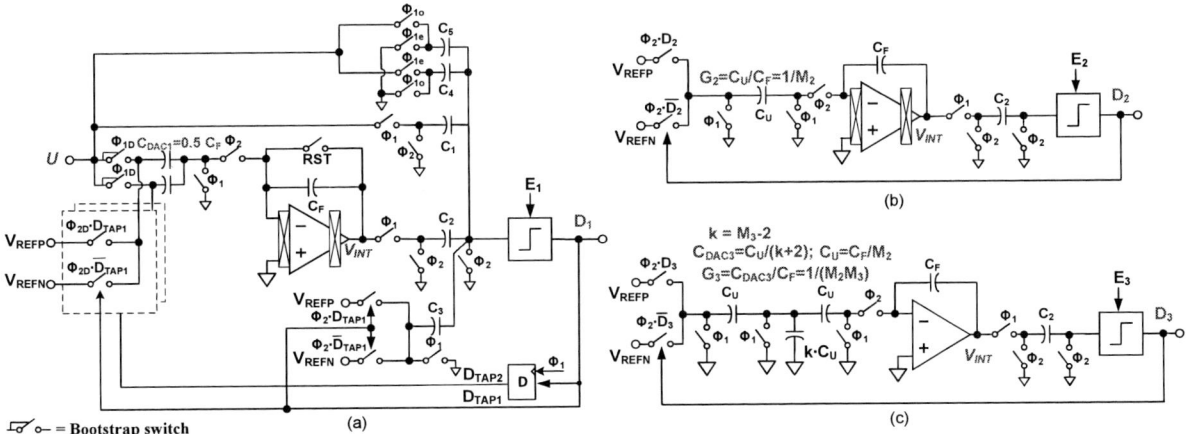

Fig. 2. Switched capacitor circuit implementation of the proposed ADC: (a) first step as a first-order IADC with FIR feedback path; (b) second step as single slope ADC for coarse quantization; (c) third step as single slope ADC for finer quantization.

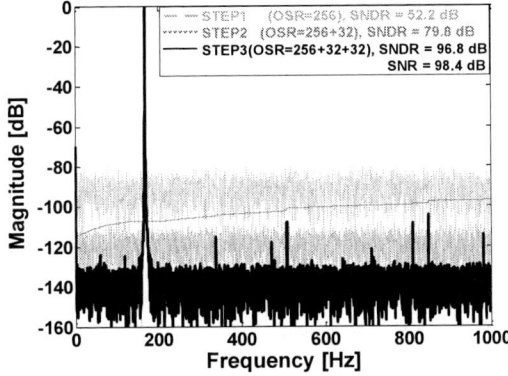

Fig. 3. Measured PSD of ADC output for the three steps with -0.44dBFS, 170Hz input (In-band: 2^{14}-point FFT).

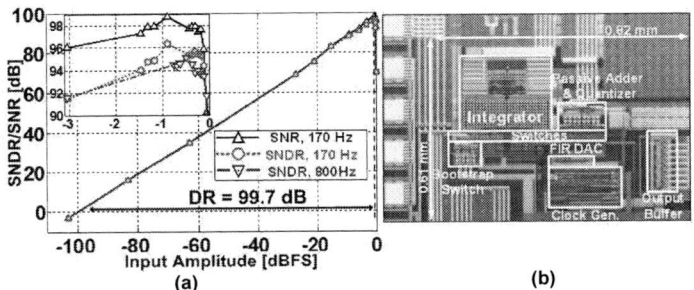

Fig. 4. (a) Measured SNR/SNDR versus the input amplitude for 170 Hz and 800 Hz input signals; (b) Die photograph

Measured Results

The prototype ADC was fabricated in a 0.18-μm CMOS process. It occupies an active area of 0.5 mm². Clocked at 642 kHz, the ADC consumes 35 μW from a 1.5 V supply. The analog part consumes 26 μW and the digital part 9 μW. Figure 3 shows the measured output power spectral densities (PSDs) for each step with a 170 Hz, -0.44 dBFS input signal. The multi-slope extended counting in the second and third steps enhances the peak SNDR to 96.8 dB, and the SNR to 98.4 dB, over a 1 kHz signal band. The SNR and SNDR are plotted as functions of the input sine wave amplitude in Fig. 4(a). The dynamic range is 99.7 dB. The microphotograph of the chip is shown in Fig. 4(b).

Table I summarizes the measured performance, and compares it with state-of-art IADCs of different architectures for various bandwidths. The prototype ADC achieves a Walden FoM$_W$ of 0.32 pJ/conv-step and a Schreier FoM$_S$ of 174.6 dB, both excellent values.

Acknowledgements

This work was sponsored by the Semiconductor Research Corporation (SRC) and the NSF Center for Design of Analog-Digital Integrated Circuits (CDADIC). The chip fabrication was provided by Asahi Kasei Microdevices (AKM).

References

[1] J. Markus, J. Silva, and G. C. Temes, "Theory and applications of incremental delta sigma converters," *IEEE Trans. Circuits Syst. 1*, vol.51, no. 4, pp. 678–690, Apr. 2004.

Table I. IADC performance comparison for wideband application

Parameter	This Work	[5]	[3]	[7]	[6]	[4]
Architecture	IADC1 + Multi-Slope	IADC2 +IADC1	IADC2 + SAR	Single IADC	CT IADC	IADC1 + Cyclic
Technology	0.18 μm	65 nm	0.18 μm	0.16μm	0.18 μm	0.8 μm
Area (mm²)	0.5	0.2	3.5	0.45	0.337	1.3
V$_{DD}$ (V)	1.5	1.2	1	1.2	1.2/1.8	1.2
Diff. input range	2 V$_{PP}$	2.2 V$_{PP}$	2 V$_{PP}$	0.7 V$_{PP}$	-	2.4 V$_{PP}$
Sampling Freq.	642 kHz	192 kHz	45 MHz	750kHz	320 kHz	256 kHz
Bandwidth	1 kHz	250 Hz	500 kHz	667 Hz	4 kHz	8 kHz
Power	34.6 μW	10.7 μW	38.1mW	20 μW	34.8 μW	150 μW
PSRR	>102dB@ 50Hz	100dB @DC	-	-	90dB@ 918.9Hz	-
SNDR (dB)	96.8	90.8	86.3	81.9	75.9	80
DR (dB)	99.7	99.8	90.1	81.9	85.5	82
FoM^{1}w (pJ/conv)	0.32	0.76	1.46	1.48	0.85	1.15
FoM^{2}s (dB)	174.6	173.5	161.3	157.1	166.1	159.3

1. FoMw = Power/($2^{(SNDR-1.76)/6.02}$ x 2 x BW).   2. FoMs = DR + 10 x log$_{10}$ (BW / P)

[2] Y. Chae, K. Souri, and K. A. Makinwa, "A 6.3 μW 20 bit incremental zoom-ADC with 6 ppm INL and 1 μV offset," *IEEE J. Solid-State Circuits*, vol. 48, no. 12, pp. 3019–3027, Dec. 2013.

[3] A. Agah, et al., "A high-resolution low-power oversampling ADC with extended-range for bio-sensor arrays," *IEEE J. Solid-State Circuits*, vol.45, no. 6, pp. 1099–1110, Jun. 2010.

[4] P. Rombouts, W. De Wilde, and L. Weyten, "A 13.5-b 1.2-V micropower extended counting A/D converter," *IEEE J. Solid-State Circuits*, vol. 36, no. 2, pp. 176–183, Feb. 2001.

[5] C.-H. Chen, Y. Zhang, T. He, P. Chiang and G. C. Temes, "A micro-power two-step incremental analog-to-digital converter," *IEEE J. of Solid-State Circuits*, vol. 50, no.8, pp. 1796-1808, 2015.

[6] S. Tao, A. Rusu, "A power-efficient continuous-time incremental sigma-delta ADC for neural recording systems," *IEEE Tran. Circuits Syst.I*, vol. 62 no. 6, pp. 1489-1498, June, 2015.

[7] C. Chen, Z. Tan and M.A.P. Pertijs, "A 1V 14b self-timed zero-crossing-based incremental ΔΣ ADC," in *IEEE ISSCC Dig. Tech. Papers*, Feb. 2013, pp. 274–275.

A 18.5-fJ/step VCO-Based 0-1 MASH ΔΣ ADC with Digital Background Calibration

Arindam Sanyal and Nan Sun

The University of Texas at Austin, TX, USA; arindam3110@utexas.edu, nansun@mail.utexas.edu

Abstract

A scaling-friendly and energy-efficient 0-1 MASH ΔΣ ADC is proposed in this work. An 8b SAR is used as the 1st stage for coarse quantization. A ring VCO is used as the 2nd stage for fine quantization. The proposed ADC uses digital background calibration to track VCO gain variation across PVT. A 40nm CMOS prototype achieves a Walden FoM of 18.5 fJ/conv-step while operating from 1.1V supply.

Introduction

With technology scaling, traditional voltage-domain ADCs face severe challenges due to reduction in transistor intrinsic gain and supply voltage. This has brought to the fore time-domain (TD) quantizers which can leverage technology scaling to build high performance ADCs. Ring VCO is widely used as a TD ADC due to its highly digital nature and simplicity of design [1]-[6]. It also provides an intrinsic 1st-order noise shaping. However, the VCO's frequency tuning gain is nonlinear and highly sensitive to PVT variations, which significantly undermines the ADC accuracy and robustness. In addition, FoMs of existing VCO-based ADCs are around 100-fJ/ step [1]-[6], which presents a scope for improvement.

This work presents a scaling-friendly and energy-efficient 0-1 MASH ΔΣ ADC. It combines a coarse SAR with a fine VCO. The VCO is effective at quantizing small voltages in TD. Since the VCO only sees a small SAR residue, the VCO nonlinearity is greatly suppressed and does not need any correction. The PVT variation of the VCO tuning gain can still cause SAR quantization noise leakage, degrading SNDR. To address this issue, a simple, low-power, and fast-convergence digital background calibration technique is developed. It enables the precise tracking of the VCO gain and high ADC linearity. A prototype in 40nm CMOS achieves 74.3 dB SNDR with a BW of 2MHz while consuming only 350μW.

Proposed SAR-VCO 0-1 MASH ΔΣ ADC

Fig. 1 shows the architecture of the proposed ADC. During $\Phi 1$, the input is sampled across the bottom-plates of the DAC array. During $\Phi 2$, the sampled input is quantized by the SAR. The residue is fed to a pseudo-differential dual-VCO at $\Phi 3$. The VCO performs phase domain integration, and the output is differentiated digitally before being combined with the SAR output to generate the ADC output. To reduce power consumption of the SAR, the bi-directional single-sided switching scheme of [7] has been adopted. The SAR DAC has no redundancy as the VCO can absorb SAR decision errors as long as they are not too large to result in VCO phase overflow. This relaxes the precision requirement of the SAR comparator and saves power. The SAR in turn reduces the VCO swing and obviates the need for any VCO nonlinearity calibration.

Fig. 2 shows the circuit diagram of the 2nd stage VCO. Each VCO consists of a 7-stage pseudo-differential ring inverter chain. Although the VCOs quantize the SAR residue V_{res} only during $\Phi 3$, they are not stopped when $\Phi 3$ goes low. This is to prevent charge leakage which can corrupt the phase information held by the VCOs. Instead, the VCOs are biased with I_{cm} and run at a fixed frequency during $\overline{\Phi 3}$, which is kept low to save power and reduce phase noise. The digital logic for

the 2nd stage runs at the ADC sampling frequency. There is no phase-overflow counter running at high VCO frequency, which significantly lowers the power consumption of the 2nd stage compared to [6].

Fig. 3 shows the ADC model, where G represents the SAR residue voltage attenuation due to parasitic capacitors, K_{VCO} is the VCO tuning gain, G_d is the digital gain, and Rn is a dither. The ADC output can be derived as:

$$d_{out} = V_{in} + (q_1 - Rn)\left(1 - \frac{GK_{vco}}{G_d}\right) + \frac{q_2(1 - z^{-1})}{G_d}$$

Both Rn and SAR conversion error q_1 can be cancelled if G_d matches the analog inter-stage gain GK_{VCO}. The cancellation of SAR error allows the use of a low power comparator. The final quantization noise at the ADC output d_{out} comes solely from the VCO (q_2) and is first-order shaped.

The VCO gain K_{VCO} is sensitive to PVT variations and can cause mismatch between G_d and GK_{VCO}, leading to noise leakage. To solve this issue, an efficient and fast background calibration technique is developed to precisely track the inter-stage gain GK_{VCO}. An *on-chip* pseudo-random number generator (PRNG) injects a dither Rn into the SAR DAC. GK_{VCO} is extracted from the difference between the d_2 averages for $Rn = 1$ and $Rn = 0$ [8]. The hardware cost of this background calibration scheme is low. It requires only an extra LSB capacitor in the DAC, two digital averagers, a MUX, and a subtractor (see Fig. 1). Its convergence speed is very fast. This is because the unknown ADC input V_{in}, which is the primary source of perturbation in the background calibration loop, is substantially attenuated by the 1st-stage 8-b SAR [9]. The part in d_2 that is uncorrelated with Rn is very small.

Measurement Results

The proposed ADC is implemented in 40nm CMOS. Fig. 4 shows the measured spectrum with a 2.2V differential input at 500kHz and the sampling frequency of 36MHz. Calibration improves the SNDR from 64.5 dB to 74.3 dB at the OSR of 9. The ADC consumes 350μW. Fig. 5 shows the SNDR and SNR sweep versus input amplitude. The ADC has a dynamic range of 75.7 dB. The measured histogram of d_2 is shown in Fig. 6. The shift in d_2 distributions from $Rn=1$ to $Rn=0$ can be clearly seen, and the difference of the two averages gives the inter-stage gain GK_{VCO} of 1.3. Fig. 7 shows that the proposed background calibration has a very fast convergence speed and requires only 10^3 samples (or 25μs) to converge. Fig. 8 shows the die photo. Table I compares the proposed hybrid SAR-VCO ADC with state-of-the-art VCO-based ADCs. It can be seen that this work has a FoM of 18.5 fJ/conv-step which represents a significant improvement over the prior art.

References

[1] M. Park and M. H. Perrott., *ISSCC*, pp. 170-172, 2009.
[2] G. Taylor and I. Galton, *VLSI*, pp. 166-167, 2012.
[3] K. Reddy, et. al., *ISSCC*, pp. 152-154, 2012.
[4] S. Rao, et. al., *VLSI*, pp. 68-69, 2013.
[5] K. Reddy, et. al., *VLSI*, pp. 256-257, 2015.
[6] A. Sanyal, et. al., *CICC*, pp. 1-4, 2014.
[7] L. Chen, et. al., *ESSCIRC*, pp. 219-222, 2014.
[8] E. Siragusa and I. Galton, JSSC, pp. 2126-2138, 2004.
[9] N. Sun, et. al, *ESSCIRC*, pp. 269-272, 2012.

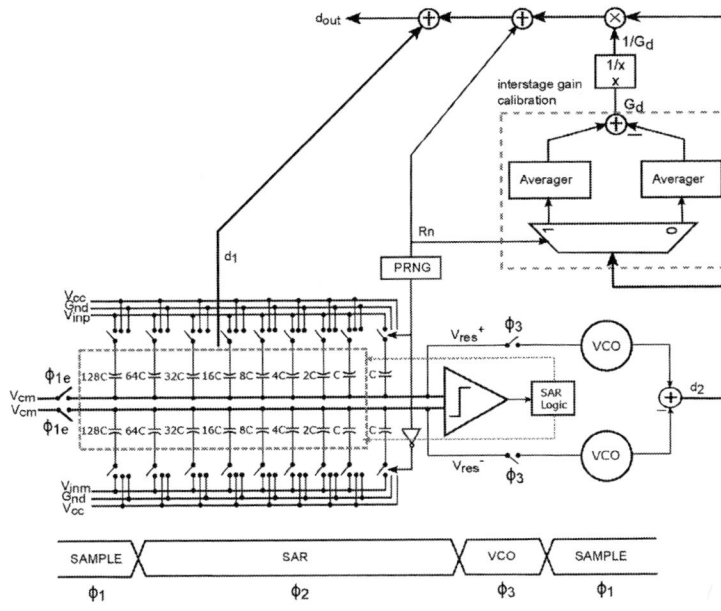

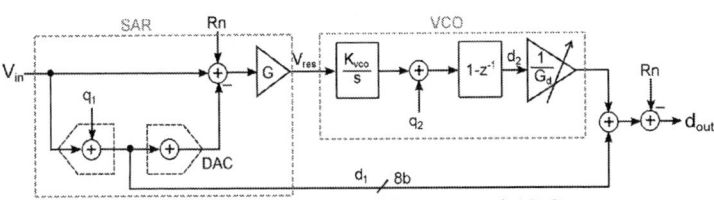

Fig.1. Circuit diagram of the proposed ADC.

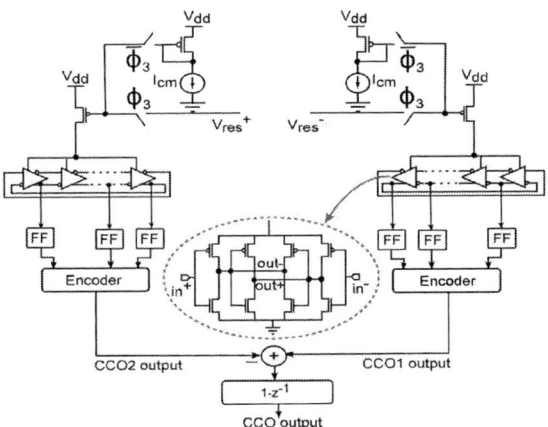

Fig.2. Circuit diagram showing the 2nd stage VCO.

Fig. 3. Block diagram of the proposed ADC.

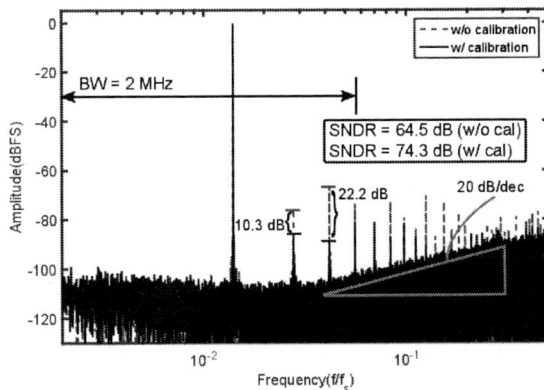

Fig. 4. Measured ADC spectrum with and without calibration.

Fig. 5 Measured SNDR and SNR vs input amplitude.

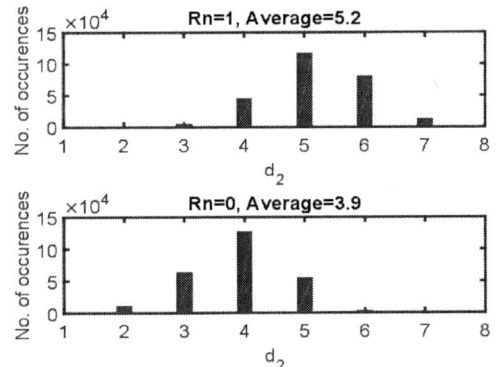

Fig. 6. Measured d_2 histogram for Rn=1 and Rn=0.

Fig. 7. Measured SNDR convergence curve.

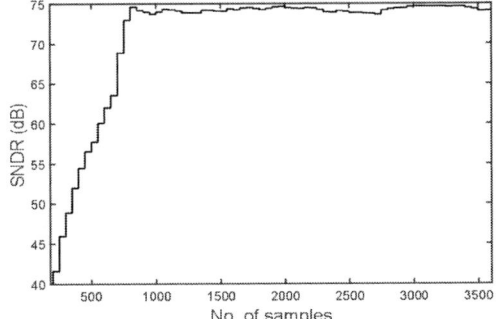

Fig. 8. Die photo

Table I. Performance summary and comparison.

	[2]	[3]	[4]	This work	
Process (nm)	65	90	90	**40**	
Power (mW)	11.5	16	4.1	**0.35**	
Area (mm²)	0.07	0.36	0.16	**0.03**	
f_s (MHz)	1300	600	640	**36**	
BW (MHz)	5.1	10	5	**3**	**2**
SNDR (dB)	75	78.3	74.7	**71.4**	**74.3**
FoM$_S$ (dB)[1]	161	163	165	**171**	**172**
FoM$_W$ (fJ/step)[2]	246	120	92	**18.5**	**21.3**

[1]FoM$_S$ = SNDR + 10log$_{10}$(BW/Power)

[2]FoM$_W$ = Power/(2*BW)/2^ENOB

A 13.3 mW 60 MHz Bandwidth, 76 dB DR 6 GS/s CT$\Delta\Sigma$M with Time Interleaved FIR Feedback

Ankesh Jain and Shanthi Pavan

Indian Institute of Technology, Madras, India

Abstract---We present a wideband single-bit CT$\Delta\Sigma$M that uses a 2x time-interleaved quantizer and FIR DAC. Time interleaving reduces power dissipation and regeneration errors of the FIR DAC when compared to a full rate implementation. Fabricated in a low leakage 65nm CMOS, the prototype modulator operates at 6 GS/s and achieves 67.6/76 dB SNDR/DR in a 60 MHz bandwidth while consuming 13.3 mW. The FoM is 56.5 fJ/conv-step.

Introduction

A single-bit CT$\Delta\Sigma$M is attractive due to its simplified ADC, which results in reduced power dissipation. Since the feedback DAC is inherently linear, DEM is not needed. However, the full scale feedback waveform renders the CT$\Delta\Sigma$M unduly sensitive to clock jitter. Another issue with 1-bit operation is the higher sampling frequency needed to achieve the desired SQNR, which reduces the regeneration time available for the latches in the ADC. The resulting data-dependent delay of the quantizer results in an increased in-band noise floor, much like the effect of clock-jitter. These problems can be addressed by the use of an FIR feedback DAC. Thanks to FIR filtering, the step-size of the feedback waveform is reduced. This benefits the CT$\Delta\Sigma$M in two ways: apart from lowering jitter sensitivity, the modulator's linearity is improved since the loop filter now processes a much smaller signal. Further, the chain of flip-flops in the FIR DAC reduce the effect of latch metastability. Thanks to the semi-digital implementation of the FIR DAC, element mismatch does not cause nonlinearity; it only modifies the frequency response of the FIR filter, which is benign. At GHz clock rates, even the FIR DAC approach begins to run into difficulties due to the limited time available for regeneration. To address this issue, the design reported in this work uses a time-interleaved 1-bit ADC and FIR DACs to address the speed limitations of a conventional ADC+FIR-DAC design. Implemented in a 65nm-LP process, the CT$\Delta\Sigma$M achieves SNR/SNDR/DR of 68.8/67.6/76 dB in a 60 MHz BW while sampling at 6GS/s. Thanks to time interleaving, it consumes only 13.3 mW and occupies 0.07 mm^2, resulting in a FoM of 56.5 fJ/lvl.

Architecture and Circuit Design

Fig. 1(a) shows the block diagram of a 1-bit CT$\Delta\Sigma$M with FIR feedback. Part (b) shows the conventional implementation of the ADC and FIR DAC (for a 4-tap example, where all taps are equal). At high clock rates, data dependent jitter at the tap outputs degrades in-band SNDR. Fig. 1(c) shows the proposed time-interleaved ADC and FIR DAC. clk_e and clk_o, which clock the interleaved halves, are derived from clk by a /2 divider. Time-interleaving has two benefits - (a) reduction of the data-dependent jitter at the tap outputs due to increased regeneration time (see Fig. 1(d)) and (b) reduction of the FIR DAC power by 50%. The latter is because the same number of flip-flops now operate at half the frequency.

Fig. 2 shows the simplified single-ended schematic of the fourth order 1-bit CT$\Delta\Sigma$M with f_s=6 GHz, resulting in OSR=50. A CIFF-B loop filter architecture is chosen, so that the fast $1/s$ and $1/s^2$ paths around the quantizer are

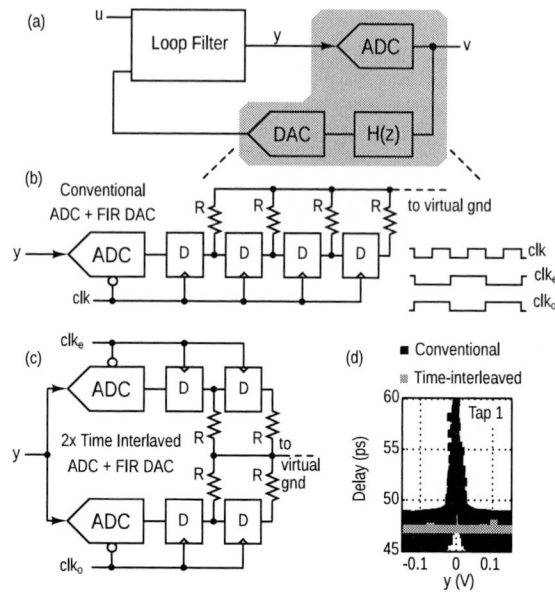

Fig. 1: (a) CT$\Delta\Sigma$M with an FIR DAC (b) Conventional (c) Time interleaved (d) Data dependent delay at the output of tap-1.

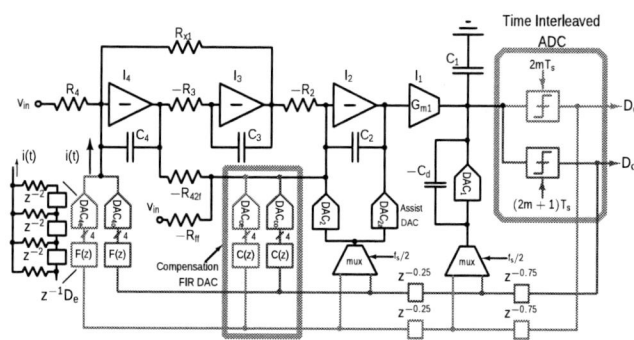

Fig. 2: Simplified single-ended CT$\Delta\Sigma$M architecture.

decoupled from the precise $1/s^4$ path. Thanks to feedforward, the gain of the input integrator I_4 is large in the signal band: this reduces the noise and distortion of the rest of the loop filter when referred to the input. The integrator I_1, which drives the 2x time-interleaved ADC is realized using Gm-C techniques, consisting of G_{m1} and C_1 [1]. This way, the parasitic capacitances of G_{m1} and the quantizer are no longer a burden, as they form part of C_1. Since the 1b-quantizer output only depends on the sign of its input, variations of C_1 do not compromise robustness (as they would if a multibit quantizer was used). The direct path around the quantizer is implemented using capacitive division through C_d and C_1. The other integrators (I_2, I_3, I_4) are realized using active-RC techniques. I_2, which is in the $1/s^2$ path of the loop filter, is sped up through the use of opamp assistance via DAC_{2a}. R_{42f} feeds the output of I_4 to I_2's input to realize the $1/s^3$ path. Input feedforward through R_{ff} reduces the output swing of I_3. The ADC uses 2x time interleaving. The resulting 3 Gb/s bit streams D_e and D_o [1] drive two 4-tap FIR DACs DAC_{4e} and DAC_{4o} respectively (also operating at half rate), which form the main (outermost) DAC. The FIR DACs are implemented using semi-digital techniques. The two 4-tap half-rate FIR DACs are equivalent to a single 8-tap FIR DAC operating at the full rate. The delay introduced by DAC_4 is compensated by DAC_c, whose transfer function is $C(z)$. DAC_c reuses the flip-flops of the DAC_4. DAC_4 and DAC_c are resistive to achieve low noise. DAC_1 and DAC_2 are current steering DACs which complete the (fast) $1/s$ and $1/s^2$ paths around the quantizer. They are driven at full rate by multiplexing between D_e and D_o. At high clock rates such as this, the decimation filter will anyway have to be implemented in poly-phase form, so the outputs D_e and D_o need not be recombined into a full rate bitstream.

Gain and timing mismatch in the time-interleaved FIR sub-DACs degrade CT$\Delta\Sigma$M performance. Analysis shows that the error due to timing skew among the two sub-DACs has a first order high pass spectrum. Gain mismatch, on the other hand, can alias shaped noise from $f_s/2$ into the signal band. This is addressed by calibration (implemented off-chip). The CT$\Delta\Sigma$M full rate output sequence is expressed in terms of its half rate outputs D_e and D_o as $D_e(z^2) + (1 + \alpha)D_o(z^2)z^{-1}$, where $\alpha(\ll 1)$ models gain mismatch between the sub-DACs. Measurements across temperature (see Fig. 4) show that α determined at 30ºC can be used across temperature without loss of performance.

Measurement Results

The prototype ADC, fabricated in a low leakage 65nm standard CMOS process, occupies only 0.07 mm^2. The

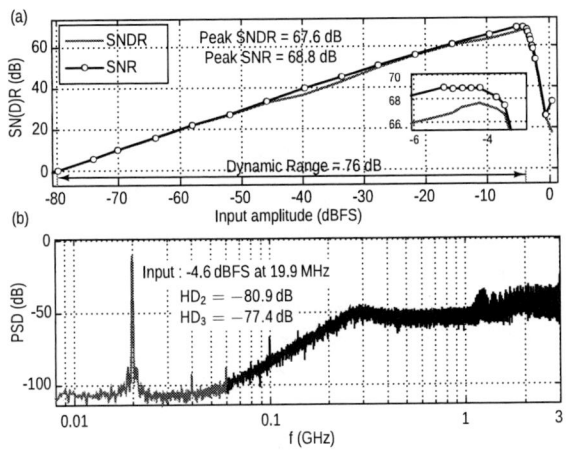

Fig. 3: (a) SN(D)R as a function of input amplitude and (b) PSD for a -4.6 dBFS tone at 19.9 MHz.

SN(D)R, plotted as a function of input amplitude, is shown in Fig.3(a). The peak SNR/SNDR in a 60 MHz bandwidth are 68.8/67.6 dB respectively. The dynamic range is 76 dB. The PSD for a -4.6 dBFS tone at 19.9 MHz is shown in Fig. 3(b). The CT$\Delta\Sigma$M, clocked at 6 GHz, consumes 13.3 mW

[1]the subscripts e and o denote even and odd respectively.

(including power drawn from the references) from a 1.4 V supply. The resulting Walden and Schreier Figures of Merit are 56.5 fJ/lvl and 172.5 dB respectively.

Gain error and rise-fall asymmetry of the interleaved sub-DACs is corrected in the digital domain, when the CT$\Delta\Sigma$M operates at 30ºC. To test the robustness of calibration, measurements are made by sweeping temperature over the 0-70ºC range, but by using coefficients derived at the nominal temperature of 30ºC. As seen in Fig. 4, the peak SNDR (after calibration) remains largely unchanged, indicating that a one point correction is sufficient to account for gain error mismatch of the interleaved sub-DACs. The layout snapshot

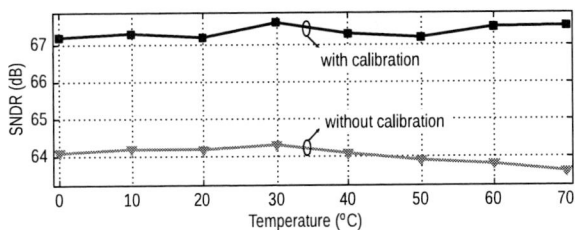

Fig. 4: Peak SNDR variation with temperature - the ADC is calibrated at 30ºC, and the *same* coefficients are used across temperature.

and die micrograph (inset) is shown in Fig.5(a). The performance of our CT$\Delta\Sigma$M is compared against state-of-the-art modulators in Fig. 5(b).

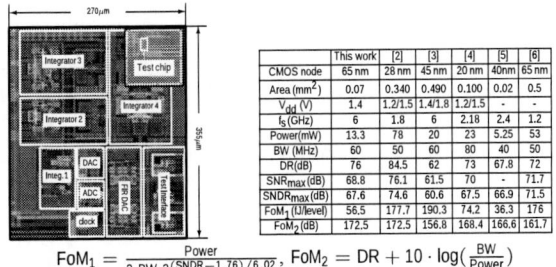

	This work	[2]	[3]	[4]	[5]	[6]
CMOS node	65 nm	28 nm	45 nm	20 nm	40nm	65 nm
Area (mm^2)	0.07	0.340	0.490	0.100	0.02	0.5
V_{dd} (V)	1.4	1.2/1.5	1.4/1.8	1.2/1.5	-	
f_s(GHz)	6	1.8	6	2.18	2.4	1.2
Power(mW)	13.3	78	20	23	5.25	53
BW (MHz)	60	50	60	80	40	50
DR(dB)	76	84.5	62	73	67.8	72
SNR$_{max}$(dB)	68.8	76.1	61.5	70		71.7
SNDR$_{max}$(dB)	67.6	74.6	60.6	67.5	66.9	71.5
FoM$_1$ (fJ/level)	56.5	177.7	190.3	74.2	36.3	176
FoM$_2$(dB)	172.5	172.5	156.8	168.4	166.6	161.7

$$FoM_1 = \frac{Power}{2 \cdot BW \cdot 2^{(SNDR-1.76)/6.02}}, \quad FoM_2 = DR + 10 \cdot \log\left(\frac{BW}{Power}\right)$$

Fig. 5: (a) Chip layout snapshot and microphotograph (inset) and (b) Comparison table.

References

[1] M. Bolatkale, L. Breems, R. Rutten, and K. Makinwa, ``A 4 GHz CT$\Delta\Sigma$M with 70 dB DR and 74 dBFS THD in 125 MHz BW,'' *IEEE Journal of Solid-State Circuits.*, vol. 46, no. 12, 2011.

[2] D. Yoon, S. Ho, and H. Lee, ``An 85 dB-DR 74.6 dB SNDR 50 MHz BW CT MASH $\Delta\Sigma$ modulator in 28 nm CMOS,'' *Digest of Technical Papers, ISSCC*, 2015.

[3] V. Srinivasan, V. Wang, P. Satarzadeh, B. Haroun, and M. Corsi, ``A 20 mW 61 dB SNDR (60 MHz BW) 1-b third order CT$\Delta\Sigma$M clocked at 6 GHz in 45 nm CMOS,'' *Digest of Technical Papers, ISSCC*, 2012.

[4] S. Ho, C.-L. Lo, J. Ru, and J. Zhao, ``A 23 mW, 73 dB DR, 80 MHz BW CT$\Delta\Sigma$M in 20 nm CMOS,'' *IEEE Journal of Solid-State Circuits.*, vol. 50, no. 4, 2015.

[5] S. Loeda, J. Harrison, F. Pourchet, and A. Adams, ``A 10/20/30/40 MHz feed-forward FIR DAC CT$\Delta\Sigma$M with robust blocker performance for radio receivers,'' *Symposium on VLSI Circuits, 2015.*, 2015.

[6] K. Reddy, S. Dey, S. Rao, B. Young, P. Prabha, and P. K. Hanumolu, ``A 54 mW 1.2 GS/s 71.5 dB SNDR 50 MHz BW VCO-based CT$\Delta\Sigma$M using dual phase/frequency feedback in 65 nm CMOS,'' *Symposium on VLSI Circuits,*, 2015.

A 128-Channel Spike Sorting Processor Featuring 0.175 μW and 0.0033 mm² per Channel in 65-nm CMOS

Seyed Mohammad Ali Zeinolabedin[1], Anh Tuan Do[2], Dongsuk Jeon[3], Dennis Sylvester[4], Tony Tae-Hyoung Kim[1]

[1]VIRTUS, Nanyang Technological University, Singapore, [2]Institute of Microelectronics, Singapore, [3]Graduate School of Convergence Science and Technology, Seoul National University, Korea, [4]University of Michigan, Ann Arbor, USA

Abstract

This paper presents a power and area efficient processor for real-time neural spike-sorting. We propose a robust spike detector (SD), a feature extractor (FE), and an improved k-means algorithm for better clustering accuracy. Furthermore, time-multiplexing architecture is used in SD for dynamic power reduction. A customized 39kb 8T SRAM is also implemented to minimize leakage and storage area. The proposed processor consumes 0.175 μW/ch with leakage of 0.03 μW/ch at 0.54 V and area of 0.0033 mm²/ch.

Introduction

Multi-electrode intracranial recording technology is required for many applications such as neural prosthetics and neuroscience research [1]. The first critical step in decoding the brain signals is detecting spikes and assigning them to an individual neuron source. This process is called spike sorting (Fig. 1). Traditionally, neural signals from a recording chip are transmitted to a nearby computer for sorting. However, this approach faces practical limitations due to the requisite high data rates and power consumption [1]. On-chip spike sorters (SS) exhibit much better power efficiency with shorter lag time, which is essential for real-time multi-channel neural signal processing [1-5]. SS typically consists of (1) a spike detector (SD) to detect and align spikes; (2) a feature extractor and dimensionality reduction (FE & DR) to extract information-rich features from noisy data; (3) a classifier to assign the detected spike to a neuron ID; and (4) a training engine (TE) to train the chip and store cluster means to a memory to be used by the classifier [1] (Fig. 1). This work improves clustering accuracy by enhancing detection, feature extraction, and clustering algorithms. At the same time, various advanced circuit techniques are employed to significantly reduce the dynamic and leakage power consumption.

Spike Sorting Operation and Algorithm

Prior to normal operation, on-chip SS must be trained to identify mean values of clusters, each representing one neuron source. Once trained, the TE writes cluster means to a memory for subsequent clustering. During normal operation, TE is turned off and features of the detected spikes are compared with cluster means and assigned to a cluster with the minimum distance.

SD, FE and DR: We propose an integer coefficient detector (ICD), $y_D[n]$, that offers better detection accuracy than the widely used absolute thresholding (AT) and nonlinear energy operator (NEO) approaches (Fig. 2). In fact, ICD not only filters out the noise which reduces the probability of false alarm (PFA) but also improves the detection by strengthening the signal. All detected spikes are aligned to the maximum slope for better clustering accuracy [1]. After spike detection, an integer coefficient FE, $y_{FE}[n]$, is also proposed and executed utilizing the aligned spikes. Then FE is followed by DR to reduce the number of features, which is critical in reducing SRAM size (and hence overall system power/area) needed in clustering. Extracted features provide better isolation between different clusters compared to the original data (Fig. 3(a)). Extensive simulation on 16 widely-used datasets in [2] reveals that reducing 48 features to 4 features (indexed 8, 11, 18, and 25) provides the best clustering performance using standard k-means compared to existing discrete derivative (DD) technique [1] with either 24 or 4 features (Fig. 3b). Furthermore, it reduces the required memory capacity. For instance, for a 3-neuron input signal, only 156 bit storage is required for four 13-bit features compared to 936 bits for the DD counterpart with 24 features. In addition, the SRAM size is reduced by 6× (from 234kb) when only using 4 features versus 24.

Proposed Clustering Algorithm: Iterative k-means algorithm is a powerful software-based clustering algorithm, but it requires several iterations and a large memory to store the full data set. In a real-time implementation, iteration is not feasible as data continuously streams in. Furthermore, the number of clusters (k) should be user-specified; however, determining the number of clusters (neurons) for spike sorting is challenging. Therefore, the iterative k-means algorithm is not suitable for real-time hardware implementations. We propose an improved k-means algorithm (Fig. 4) in which the number of clusters (k) is not necessarily required to be provided. Instead, the approach forms a new cluster for new data if its weighted distance to the existing clusters is larger than that between the existing ones. This allows all clusters means to be adjusted and converge even if initial points are purposely assigned from the same cluster (Fig. 5). In the case when k is specified clustering accuracy further improves. Analysis results over various datasets [2] demonstrate that the proposed k-means performance is comparable to iterative k-means with 100 iterations running in MATLAB (Fig. 6) and it doesn't require storage of the full dataset.

Hardware Implementation

Since the FE, DR, classifier and memory are only active when a spike is detected, they are all clock-gated to save power. SD is the main contributor to dynamic power (80%) because it is always active and employs a large number of D flip-flops (5625) to store the incoming data stream of 128 channels. In the interleaving architecture [3], all SD DFFs are clocked concurrently and thus its dynamic power increases quadratically with channel count (both frequency and load grow linearly). A time-multiplexing architecture (Fig. 7) is designed to avoid data transition of all registers by clock gating and multiplexing. Thus, its dynamic power is reduced by 74% for 128 channels (Fig. 8). After a spike is detected, SD is clock-gated and FE, DR, the classifier, and SRAM are activated to process the next 48 data points of that particular channel. After that, cluster means of the corresponding channel are fetched from the SRAM to the local registers of the classifier. The classifier then searches for a cluster with the shortest l_1-distance to the new spike and finally the index of that cluster is sent out as the neuron ID. A subthreshold 8T SRAM is designed to minimize area and leakage power of the system (Fig. 9). An auto-biased bit-line keeper provides reliable sensing margin at near- and sub-threshold operation so that a single power supply can be used for both the digital core and SRAM.

Measurement Results

The chip was fabricated in 65-nm CMOS process and occupies 0.414 mm² (Fig. 10). The functionality of the chip was verified using datasets in [2] and clustering accuracy is similar to Fig. 6. The minimum operating voltage is 0.54 V while operating at 3.2 MHz (Fig. 11). The design consumes 0.175 μW/channel and 0.003 mm²/ch which are 2.6× and 10× smaller than previous designs respectively. Power improvements are mainly due to the time-multiplexing SD, the reduction of the number of features in FE & DR block, and the sub-threshold SRAM. The area improvement is due to the proposed algorithm and use of SRAM rather than a register file. Table I compares this work with other state-of-the-art spike sorting designs. To our knowledge, this is the first real-time multi-channel spike-sorting chip that includes SD, AL, FE, DR and clustering.

References

[1] S. Gibson etal., *IEEE Signal Process. Mag.*, Jan. 2012.

[2] R. Q. Quiroga, etal *Neural Comp.*, Aug. 2004.

[3] R. Ollson and K. Wise, *IEEE JSSC*, Dec 2005.

[4] M. Chae et al., *ISSCC Dig. Tech. Papers*, Feb. 2008.

[5] T.T. Liu, and J. M. Rabaey, IEEE JSSC, Apr. 2013.　　　　[6] V. Karkare, S. Gibson, and D. Marković, IEEE JSSC, Jun. 2013.

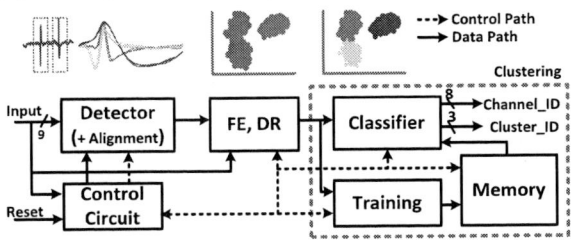

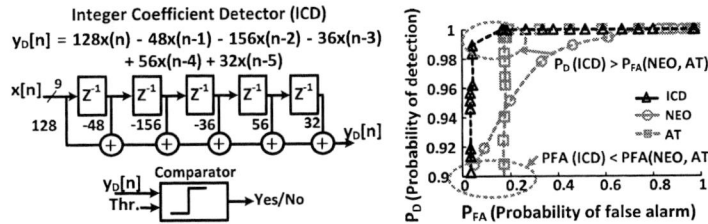

Fig. 1. Architecture of on-chip SS. Clustering consists of a training engine and a classifier which often require accessing to memory.

Fig. 2. Proposed integer coefficient detector (ICD) outperforms AT and NEO methods by reducing the number of false alarms at the same improving detection probability.

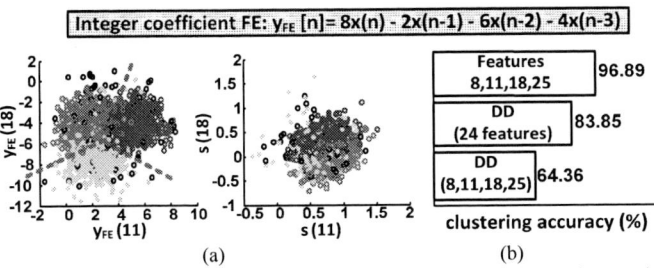

(a)　　　　　　　(b)

Fig. 3. (a) Features space for a very noisy input signal. The outputs of y_{FE} and original spike samples whose indexes are 11 and 18 are shown. (b) Averaged clustering accuracy for different feature extraction methods. (K-means algorithm used to do clustering).

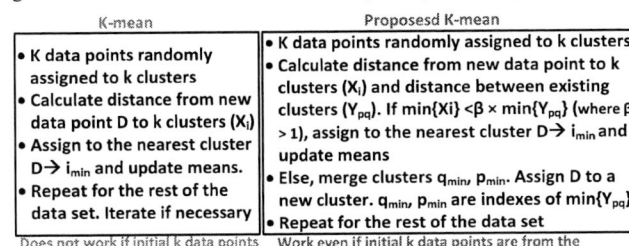

Fig. 4. Proposed hardware friendly k-means algorithm that is tolerant to wrongly assigned initial clusters and does not need iterations and storage of the full dataset.

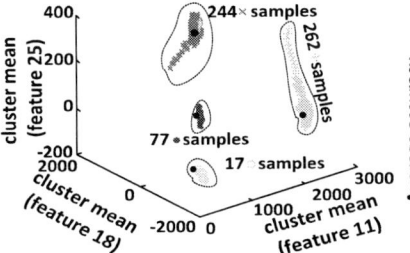

Fig. 5. Cluster means converged to 4 clusters during the training phase even when initial data points are purposely chosen from the same cluster.

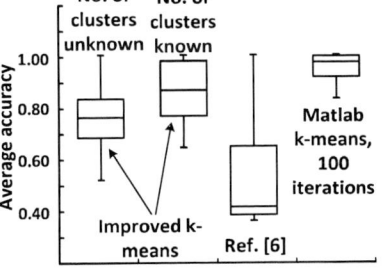

Fig. 6. Clustering accuracy performance of improved k-means compared to Ref. [6] and Matlab k-means.

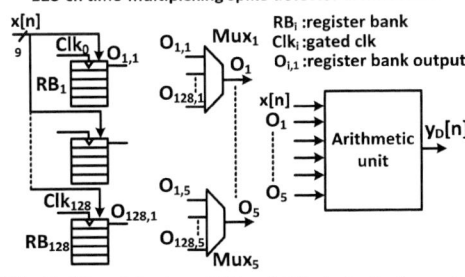

Fig. 7. SD architecture. RB_i and Clk_i indicate the register banks and gated-clocks for each channel. Arithmetic unit performs the multiplierless calculation.

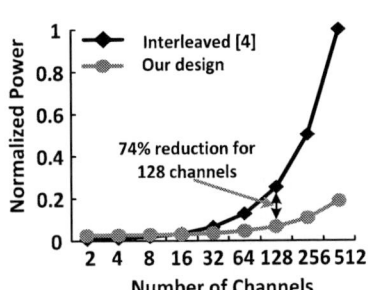

Fig. 8. Power comparison between interleaved [4] and time-multiplexing architecture.

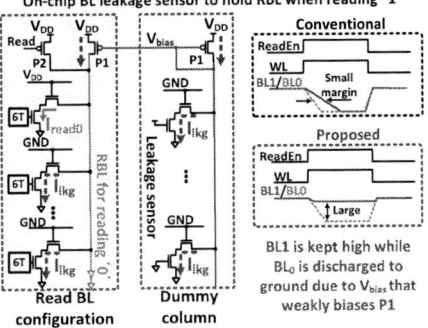

Fig. 9. Sub-threshold 8T SRAM with BL leakage compensation to improve sensing reliability at ultra-low voltage condition.

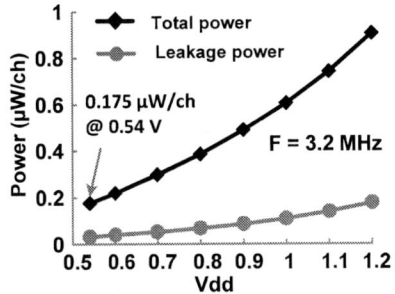

Fig. 11. Measured total and leakage power results vs Vdd.

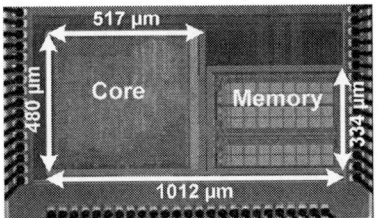

Fig. 10. 128-channel spike sorting chip micrograph.

Technology : 65 nm CMOS
Supply voltage : 0.54 V
Total power : 22.4 µW
Leakage power : 3.8 µW
Operating freq. : 3.2 MHz
Average accuracy: 72% ~ 87%
Number of Ch. : 128
Data reduction : 257
Total area : 0.414 mm²

TABLE I. COMPARISON

Reference	[3]	[4]	[5]	[6]	This work
No. of Chs.	32	128	64	16	128
Detection	Y	Y	Y	Y	Y
FE	N	Y	Y	N	Y
Clustering	N	N	N	Y	Y
VDD (V)	3	3.3	0.25	0.27	0.54
Power (µW/Ch)	75	100	0.46	4.68	0.175
Area (mm²/Ch)	0.11	1.58	0.03	0.07	0.003
Power Density (µW/mm²)	682	63.3	15.3	66.8	58.33
Process (nm)	500	350	65	65	65

1.74-μW/ch, 95.3%-Accurate Spike-Sorting Hardware based on Bayesian Decision

Zhewei Jiang, Joao Pedro Cerqueira, Seongjong Kim, Qi Wang, Mingoo Seok, Columbia University, New York, NY, USA

Abstract: This paper presents algorithm/hardware co-design for real-time unsupervised spike sorting hardware for reducing power and improving sorting accuracy. We devise an algorithm based on Bayesian decision, which enables high accuracy while using noisy and simple time-domain features. Those simple features significantly reduce computation complexity, memory requirement, and thus the required number of cycles per sorting. The latter, coupled with the sparsity of spikes in time, makes the hardware idle for most of time, and thus we employ aggressive power gating and balloon latches to sleep most of the circuits and wake them up only when a spike is detected for maximal power savings. The hardware prototyped in a 65nm achieves higher accuracy at lower power than the existing arts.

I. Introduction

Online neural spike detection and sorting (clustering spikes by waveform features) is an important step between neural signal sensing and motor-intention decoding for closed-loop neural prosthetic systems. Implanted electrodes sense activities of multiple neurons while decoding often needs single-unit spiking rates. Therefore, implantable hardware for real-time spike detection and sorting [1,2] is critical to close a loop and to reduce communication cost between an implant and an external receiver [3,4].

One of the state-of-the-art hardware for spike sorting implements the sequential leader algorithm which computes sums of differences between samples of incoming and centroid waveforms [1]. While the simple time-domain feature allows it consume low power (4.68μW/ch) and compact area (0.07mm²/ch), the hardware suffers from low clustering accuracy (78.2% while sorting 96k spikes) mainly due to the use of less robust time-domain features.

In this work, we pursue algorithm/hardware co-design to increase clustering accuracy and minimize power dissipation. We propose a novel sorting algorithm that can self-learn a decision metric based on Bayesian boundary [5] from incoming spike waveforms. While the algorithm also uses very simple time-domain feature (i.e., min and max values of a spike), the self-learning capability enables the algorithm to robustly sort spikes. When mapped onto hardware architecture, indeed, the simple time-domain features can largely reduce computation and memory requirement, allowing the hardware to finish sorting a spike in 2 clock cycles, while the conventional sequential leader algorithm can take hundreds of cycles. The reduced computing complexity coupled with spike sparsity in time allows us to have the hardware in a sleep mode for most of the time via power gating switches and balloon latches (BL).

We prototype test chips for the proposed sorter having one channel, which can sort 420 spikes/s at the sorting accuracy of 95.3% while consuming 1.74μW at VDD of 0.6V, both significantly improved from the prior arts [1].

II. Algorithm and Implementation

As shown in Fig. 1, the hardware starts with a threshold based spike detector at the input. The detector is preconfigured with a threshold derived from channel noise, and operates on the continuous sample stream from the ADC. Then, a feature extractor extracts the peak and trough of action potentials of each spike waveform for the following training or sorting process.

The first half of the training process is to identify Bayesian decision boundary. For each detected spike, the hardware updates the two histograms of the peak and trough values (Fig. 1) in the Feature Space Distribution Memory (FSDM). After the specified number of spikes are used for updating histograms, the Bayesian Prior Controller traverses the FSDM for finding the local minima of each feature distribution and store them as Bayesian decision boundaries in the Boundary Memory (BM). The boundaries are used to orthogonally partition the 2-dimension (peak, trough) feature space, with each partition identifiable by a pair of indexes (Fig. 1).

The second half of the training is to update the confidence level of the cluster status of each partition of the 2-D feature space. The specific steps are as follow. First of all, the Index Pair Lookup (Fig. 2) compares the features of an incoming spike with the Bayesian boundaries in the BM. This locates the specific partition that the spike belongs in. The pair of indexes is then fed to a CAM (Fig. 3). If the CAM finds the pair of indexes in it, it increases the confidence level of the entry by setting the associated 2b indicator (00-vacant, 01-outlier, 10-weak cluster, 11-strong cluster). If no match, it places the index pair in the first vacant entry, with indicator set to 01. The controller periodically decreases all entries' indicators once per N spikes to remove outliers from strongly-recognized clusters (Fig. 3).

The spike sorting process can start after a specified amount of training. It performs the same computation for finding an index pair, but the hardware no longer updates the CAM indicators. The result of the CAM which represents distances to partitions enters the Adjacency Checker, which then finds the closest partition that is a valid cluster. The checker performs min function with a vector mask that selects clusters only with 10 and 11 indicators. The partition index is found via the comparison paths (Fig. 4).

The proposed architecture can greatly reduce memory requirement and computational complexity. While the conventional sequential leader algorithm needs to store a full waveform per each cluster centroid, the proposed architecture stores only a pair of indexes and a few boundaries per each cluster. Assuming 48 samples/spike, 8b ADC resolution, 4 clusters/ch, and 2b/index, and 6b/boundary, the former requires 1,536(=48·8·4) bits while the latter can require only 36(=6·2+4·6) bits. Similarly, the proposed architecture requires eight 6b comparisons and single 6b equality operation per sorting while the sequential leader algorithm needs 48 8b additions and 45 14b additions per sorting.

Thanks to the low computational complexity, the proposed architecture takes only 2 clock cycles for sorting a spike. As incoming spikes are sparse in time, the hardware can be idle for most of the time. In order to minimize the power dissipation during idle time, as shown in Fig. 6, the hardware aggressively employ power gating switches (PGS) and BLs. Therefore it can have most of the modules in a sleep mode and wake them up only when detecting a new spike (Fig. 5). We design a BL in thick oxide devices for leakage savings (Fig. 7). The control signals for PGS are bootstrapped to 1V via level converters for reducing wakeup time (T_{2WKU}). Fig. 8 shows the operating waveforms of a BL.

III. Measurements and Comparisons

We prototype test chips for the proposed sorter in a 65nm high-Vt. The proposed architecture achieves the average accuracy of 95.3% in sorting 96k spikes in four datasets. The waveforms (2 to 4 clusters) are measured from mice sensory thalamus. For the same waveforms, the sequential leader algorithm can achieve the accuracy of only 78.2% (Fig. 9). At VDD of 0.6V and the throughput of 420 spikes/s/ch, the proposed hardware consumes 1.74 μW, 2.7X smaller than the prior art's power dissipation of 4.68 μW/ch at the same throughput. Note that the power consumption scales with throughput (Fig. 11). The proposed architecture takes 0.116mm² (Fig. 10).

Acknowledgement: The work is supported by WFPF and CAPES.

[1] V. Karkare, et al., "A 75-μW, 16-Channel...," JSSC, 2013
[2] T. C. Chen, et al, "A Biomedical ...," ISSCC, 2009
[3] M. Chae, et al., "A 128-Channel 6mW...," ISSCC, 2008
[4] S. Mitra, et al., "24-Channel Dual-Band ...," ISSCC, 2013
[5] P. Domingos, et al., "On the Optimality ...," ML, 1997

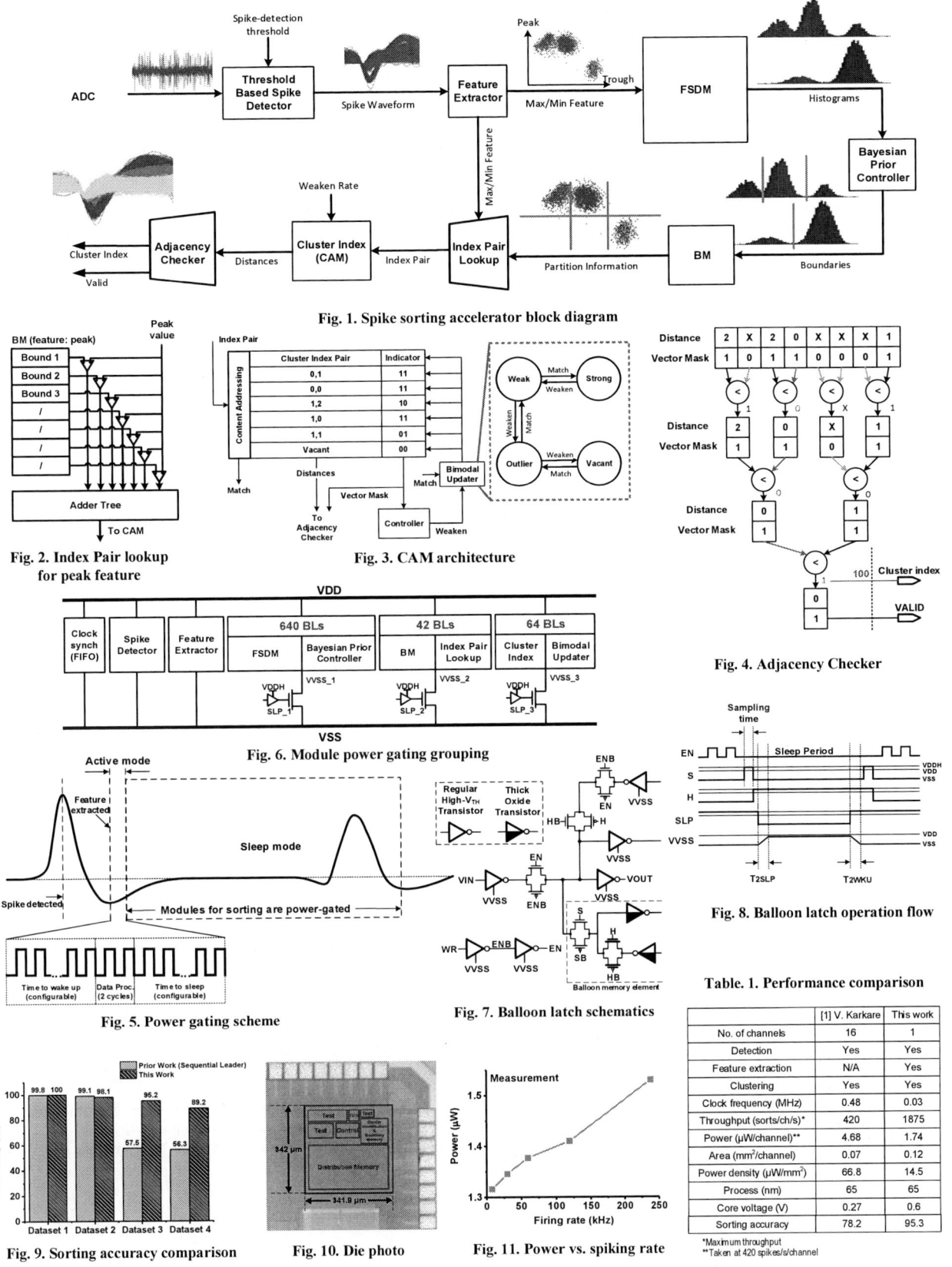

Fig. 1. Spike sorting accelerator block diagram

Fig. 2. Index Pair lookup for peak feature

Fig. 3. CAM architecture

Fig. 4. Adjacency Checker

Fig. 6. Module power gating grouping

Fig. 5. Power gating scheme

Fig. 7. Balloon latch schematics

Fig. 8. Balloon latch operation flow

Fig. 9. Sorting accuracy comparison

Fig. 10. Die photo

Fig. 11. Power vs. spiking rate

Table. 1. Performance comparison

	[1] V. Karkare	This work
No. of channels	16	1
Detection	Yes	Yes
Feature extraction	N/A	Yes
Clustering	Yes	Yes
Clock frequency (MHz)	0.48	0.03
Throughput (sorts/ch/s)*	420	1875
Power (µW/channel)**	4.68	1.74
Area (mm²/channel)	0.07	0.12
Power density (µW/mm²)	66.8	14.5
Process (nm)	65	65
Core voltage (V)	0.27	0.6
Sorting accuracy	78.2	95.3

*Maximum throughput
**Taken at 420 spikes/s/channel

978-1-5090-0636-6/16 $31.00 © 2016 IEEE

A High-Density CMOS Multi-Modality Joint Sensor/Stimulator Array with 1024 Pixels for Holistic Real-Time Cellular Characterization

Jong Seok Park, Taiyun Chi, Amy Su, Chengjie Zhu, Jung Hoon Sung*, Hee Cheol Cho*, Mark Styczynski and Hua Wang

Georgia Institute of Technology, Atlanta, GA, USA, *Emory University, Atlanta, GA, USA

Abstract

This paper presents a fully integrated 1024-pixel world-first joint multi-modality sensor/stimulator array in CMOS for holistic real-time cell characterization. Each pixel supports extracellular voltage recording, optical detection, and cellular impedance measurement, as well as current-mode cellular stimulation. Four independent on-chip temperature sensors monitor the ambient temperature variation. The chip is implemented in a 130nm CMOS process with a pixel size of 58μm×58μm, achieving the largest array-size and smallest pixel for a multi-modality joint sensor/stimulator array in CMOS. The electrical and biological measurements demonstrate the utility of this high-density multi-modality array in cell-based assays for drug and chemical screening.

Introduction

Fully characterizing the cell/tissue response is essential for cell-based assays in high-content drug development and chemical screening. However, cells are highly complex systems often with concurrent multi-physics responses when subjected to external stimuli, which cannot be captured by conventional single-modality sensors, e.g., with electrical [1] or optical [2] only detections. Thus, there is an unmet need for new multi-modality sensor arrays comprised of pixels each capable of detecting multi-physics cellular responses [3] [4]. Moreover, high spatiotemporal resolution is critical to precisely capture detailed cell/tissue structures and transient cell responses for pharmacodynamics and pharmacokinetics studies. In addition, high-precision cellular electrical stimulation is required to characterize electrogenic cells, such as cardiac tissues and neurons/neuron-networks.

To address these challenges, we present a world-first joint sensor/stimulator array in CMOS supporting multi-modality sensing and stimulation on on-chip cultured cells/tissues with the largest multi-modality array size and smallest pixel size.

CMOS Multi-Modality Joint Sensor/Stimulator Array

The CMOS array system is comprised of 4 pixel groups each with 256 pixels and one temperature sensor shown in Fig.1. All the 1024 pixels can be randomly accessed and independently configured for the three in-pixel sensing modalities (voltage/optical/impedance) or for cell stimulation. Four signal conditioning blocks contain variable-gain-amplifiers and low-pass-filters and process the detected multi-modality cellular responses.

Each pixel contains 4 photodiodes (12μm×12μm each) for optical detection and one gold-plated electrode (28μm×28μm) shared by voltage recording, impedance measurement with voltage (V)-excitation/current (I)-sensing, and stimulation. For optical detection, the photocurrent is integrated on the parasitic capacitance and buffered by an in-pixel source follower. For voltage recording, two pixels (one for sensing and the other for reference) are selected, and the detected voltage signals are capacitively coupled to the in-pixel common-source amplifier and further amplified differentially at the pixel-group (Fig.2a). For impedance measurement, one pixel is selected for V-excitation (100kHz to 2MHz) and another for I-sensing. The

AC excitation voltage is fed to the pixel and results in an AC current conducting through the cells/medium. This AC current is detected by the I-sensing pixel and amplified by a trans-impedance amplifier in pixel group and down-converted at the signal conditioning block. The V-excitation and I/Q down-conversion local-oscillator (LO) signals are generated by the on-chip clock generation block. When sensing the resulting AC cellular current, the I/Q LOs are sequentially applied to perform quadrature detection of the impedance.

In the cellular stimulation mode, the biphasic stimulation current is generated by a current digital-to-analog converter (DAC) in each pixel group. The stimulation current is fed to the selected pixel with another pixel as the reference.

Measurement Results

The multi-modality cellular joint sensor/stimulator array chip is implemented in a 130nm CMOS process. A fully packaged module and its zoom-in views are shown in Fig.6. The chip is first characterized with electrical measurements by directly wire-bonding the selected pixel electrode to the PCB. For the voltage recording mode, the measured voltage gain, bandwidth, and input referred voltage noise are shown in Fig.2b and 2c. The low cut-off frequency is 0.1Hz, while the high cut-off is adjustable from 600Hz to 26kHz to accommodate both cardiac and neural extracellular potentials (sampling rate > 15kS/s). Figure 2d shows the measured reconfigurable biphasic current stimulation pulses. We further perform a real-time loop-back test of the cellular current stimulation and voltage recording in the PBS buffer solution. One pixel in pixel group 1 is used for stimulation, and 3 other pixels each in pixel groups 1, 3, and 4 are enabled for concurrent voltage recording. The measured local medium voltages are shown in Fig.3.

Next, human ovarian cancer cells (HeyA8-F8) are cultured on-chip. We first perform 2D optical shadow imaging of HeyA8-F8 cells to illustrate the cell localization with a high spatial resolution. The recorded optical image matches well with the reference image (Fig.4). Then, we add luciferin to trigger cellular bioluminescence responses. Real-time recordings at 3 pixels show desired bioluminescence decays.

To demonstrate cell-based drug/chemical detections, a cell adhesion/detachment assay with HeyA8-F8 cells is performed. Surface adhesion is essential for mammalian cell growth and tissue formation, and cell migration assays for cancer study. HeyA8-F8 cells are first cultured on-CMOS with reliable surface attachment. Cell detachment is then triggered by adding 1mL accutase, which is an enzyme mixture with proteolytic/collagenolytic activities. 2D impedance mapping is measured before and 1-hour after the accutase administration. The real-time decrease in cellular impedance indicates the cells detachment process (Fig.5), aligning well with the accutase mechanistic effect.

Reference

[1] B. Eversmann, et al., IEEE JSSC, Dec. 2003.
[2] R. Field, et al., IEEE JSSC, Apr. 2014.
[3] J. Park, et al., IEEE ISSCC, Feb. 2015.
[4] T. Chi, et al., IEEE TBCAS, Dec. 2015.

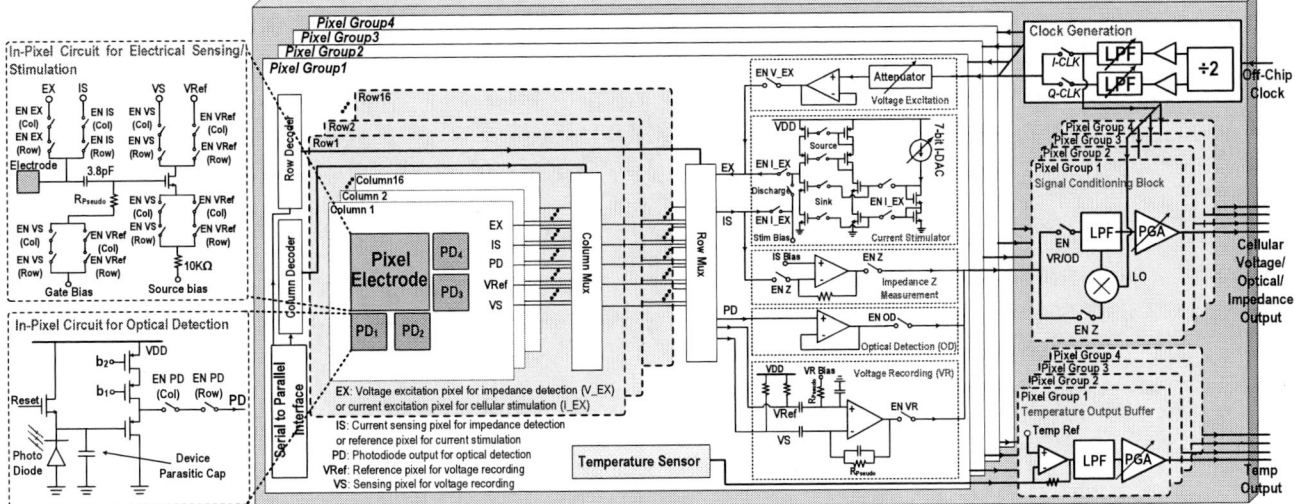

Fig. 1. Schematic of the high-density CMOS multi-modality sensor/stimulator array for holistic cell characterization. The array provides 1024 pixels, supporting extracellular voltage recording, optical detection, impedance measurement, thermal monitoring, and current stimulation.

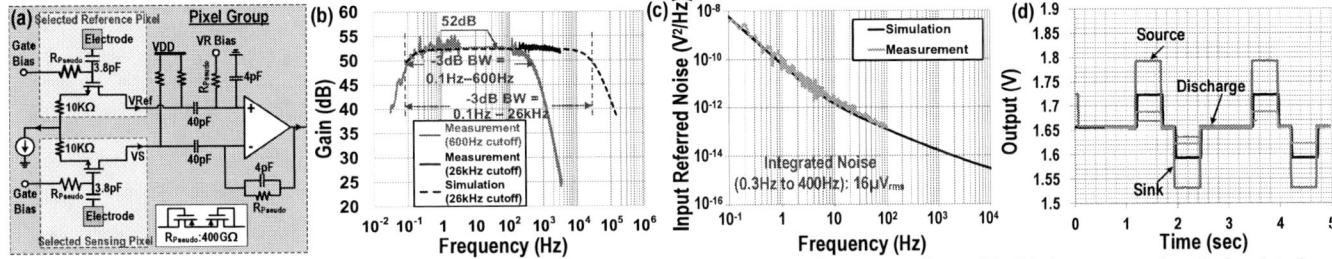

Fig. 2. (a) The fully differential low-noise amplifier configuration for extracellular voltage recording with (b) its measured and simulated voltage gain/3dB cut-off and (c) measured input referred noise. (d) Measured biphasic stimulation pulses with different strengths.

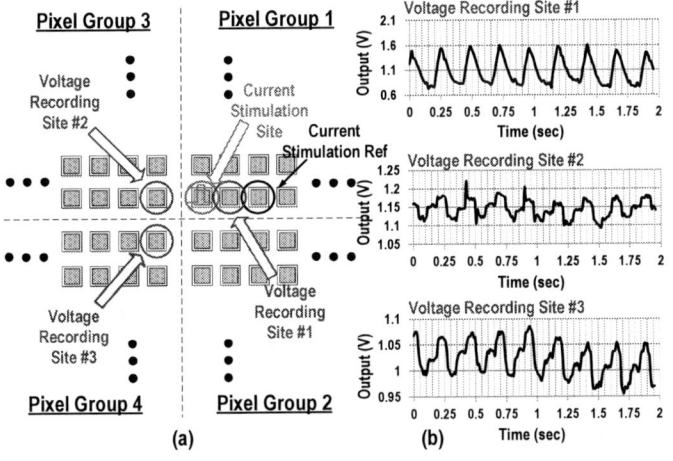

Fig. 3. (a) The loop-back test configuration of the current stimulation and voltage recording in the PBS buffer solution and (b) real-time recorded local culture medium voltages at three different sites.

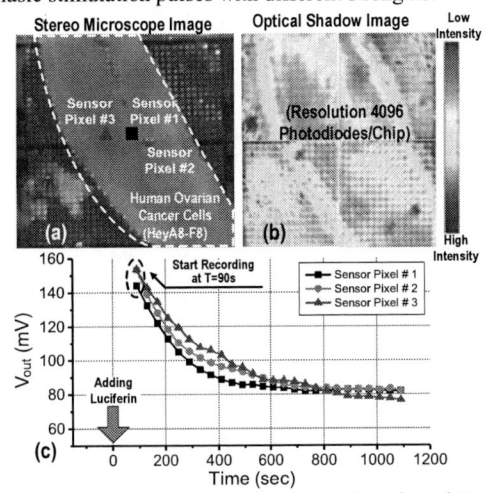

Fig. 4. (a) The reference image of the on-chip cultured HeyA8-F8 cells. (b) Measured optical shadow image by the CMOS chip. (c) The real-time bioluminescence measurement results for 3 pixels.

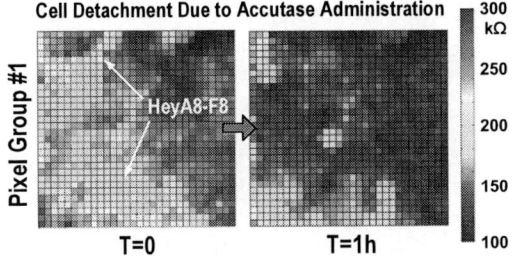

Fig. 5. Measured real-time 2D impedance mapping for cell adhesion/detachment assay with accutase administration.

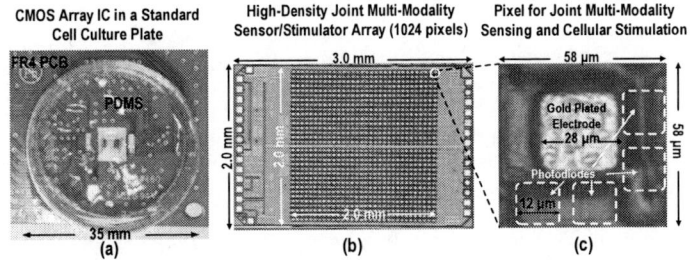

Fig. 6. (a) Photograph of a packaged module. (b) Chip microphotograph. (c) Zoom-in view of one pixel with one gold plated electrode and 4 photodiodes.

978-1-5090-0636-6/16 $31.00 © 2016 IEEE   29   2016 Symposium on VLSI Circuits Digest of Technical Papers

A Front-end ASIC with Receive Sub-Array Beamforming Integrated with a 32 × 32 PZT Matrix Transducer for 3-D Transesophageal Echocardiography

C. Chen[1], Z. Chen[1], D. Bera[3], S. B. Raghunathan[2], M. Shabanimotlagh[2], E. Noothout[2], Z.Y. Chang[1], J. Ponte[4], C. Prins[4], H.J. Vos[2,3], J.G. Bosch[3], M.D. Verweij[2,3], N. de Jong[2,3], M.A.P. Pertijs[1]

[1]Electronic Instrumentation Lab., Delft University of Technology, Delft, The Netherlands
[2]Lab. of Acoustical Wavefield Imaging, Delft University of Technology, The Netherlands
[3]Dept. of Biomedical Engineering, Thoraxcenter, Erasmus MC, Rotterdam, The Netherlands
[4]Oldelft Ultrasound, Delft, The Netherlands

Abstract

This paper presents a power- and area-efficient front-end ASIC that is directly integrated with an array of 32 × 32 piezoelectric transducer elements to enable the next-generation miniature ultrasound probes for real-time 3-D transesophageal echocardiography. The 6.1 × 6.1 mm² ASIC, implemented in a low-voltage 0.18 μm CMOS process, effectively reduces the number of cables required in the probe's narrow shaft by means of 96 sub-array beamformers, which have a compact element-matched layout and employ mismatch-scrambling to enhance the dynamic range. The ASIC consumes less than 230 mW while receiving and its functionality has been successfully demonstrated in a 3-D imaging experiment.

Challenges & System Approach

Transesophageal echocardiography (TEE) utilizes an ultrasound transducer mounted on the tip of a gastroscopic tube to make ultrasonic images of the heart from the esophagus. For real-time 3-D imaging, a 2-D array of 1000+ independent transducer elements is needed, presenting an interconnection challenge due to the limited number of cables that fit in the tube (Fig.1a). Integrating the transducer array with an ASIC that locally processes the signals is an efficient way to reduce the channel count. Its power consumption is limited by self-heating to about 0.5 mW/channel [1], while the circuits per element should fit within the λ/2 element pitch (150 μm for our 5-MHz probe) required to minimize grating lobes [2]. Both requirements are beyond the state-of-the-art [3][4][5]. Here, we present an ASIC that is optimized in both architecture and circuit implementation to fulfill these stringent constraints.

We use an array of 32 × 32 PZT elements, with separate transmit and receive elements (Fig. 1b), which directly connect to a matrix of bondpads on the ASIC using an interconnect layer [6] (Fig. 1d). An 8 × 8 central sub-array is wired out to transmit channels in the external imaging system using metal traces in the ASIC that run underneath 96 un-connected elements to bondpads on the chip's periphery. All other 864 elements connect directly to 96 sub-array receiver circuits, whose outputs are fed to the system's receive channels. Despite the missing elements in the receiver aperture, the point spread function (PSF) is comparable with a fully-populated receiver, as shown by simulations in [7]. This configuration allows the use of a dense low-voltage IC technology, thus saving power and area. Compared to [4], which uses the majority of elements to transmit and a sparse array to receive, it achieves better receiving sensitivity and lower side-lobes. Moreover, it also helps to reduce the overall in-probe heat dissipation, as transmit circuits normally consumes more power [5].

Each of the 96 sub-array receivers interfaces with a 3 × 3 transducer sub-array of 450 μm × 450 μm, and delays and sums the associated received signals, thus realizing a 9-fold channel reduction (Fig.1c), similar to the approach used in [6] for a much smaller array with a larger pitch. The delays are programmable in steps of 30 ns up to 210 ns, allowing the sub-array's directivity to be steered across a range of ± 37°.

Circuit Implementation

Fig. 2 shows the schematic of a 3 × 3 sub-array receiver, which includes 9 LNAs, 9 analog delay lines, a time-gain-compensator (TGC) and a cable driver. The LNA is a revised version of the design described in [8], which utilizes a compact inverter-based OTA to achieve a high power efficiency. The OTA's bias point is set during the transmit phase when the LNA is not needed. A switchable capacitive feedback network to provide 3 gain levels for dynamic range enhancement. Albeit single-ended, its capability of supply and ground noise rejection is enhanced by sharing a positive and a negative regulator with other 8 LNAs in the same sub-array.

The LNA output is AC-coupled to a flipped source follower that drives the analog delay line. This consists of pipeline-operated S/H memory cells running at a sampling rate of 33 MHz. The outputs of all 9 delay lines are joint together to form charge summation. A delay stage index rotator determines the sequence in which the memory cells are used. It consists of an 8-stage shift register (D_1-D_8) in which the 4-bit binary indices of memory cells (1-8) are rotated. Upon startup, register D_n is preset to n. D_1 stores the index of the memory cell used for sampling the input signals, while D_2-D_8 store the indices of candidate memory cells for readout. A 3-bit selection code, provided by a built-in SPI interface, decides which of these candidates is used, allowing the delay depth of the individual delay line to be programmed. One-hot codes expanded from the selected 4-bit indices are re-timed by non-overlapped clocks to control the S/H switches in the memory cells. The SPI interfaces in all sub-arrays are normally loaded in parallel, but can also be configured as a daisy-chain to load different delay-patterns to individual sub-arrays, which enables near-field focusing.

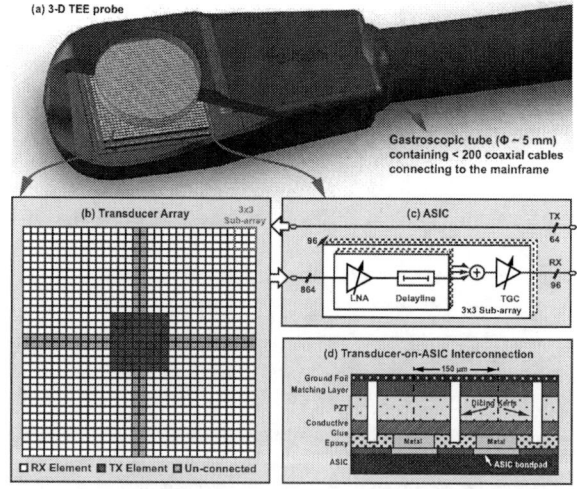

Fig. 1: (a) A miniature 3-D TEE probe; (b) the proposed transducer array configuration; (c) the ASIC architecture; (d) the interconnection between the transducer and the ASIC.

The S/H memory cells suffer from charge injection and clock feed-through errors, the mismatch of which introduces a ripple pattern with a period of 8 delay steps (30 ns) at the output of the delay lines. This limits the dynamic range of the signal chain and manifests itself as tones in the output spectrum (Fig. 3b). To mitigate this interference, we propose a mismatch-scrambling technique (Fig. 3a) by adding an extra memory cell and a redundant index register D_9. A pseudo-random number generator (PRNG) generates a bit sequence (PRBS) that decides whether the index of D_8 or D_9 shifts into D_1, while the other index shifts into D_9. Thus, delay cells are randomly taken out and inserted back into the sequence. This randomizes the ripple pattern and converts the interfering tones into broadband noise (Fig. 3c).

The TGC amplifies the summed signal at the joint delay-line outputs with programmable gain steps that interpolate between the gains steps of the LNA. Finally, a class-AB super source follower drives a cable capacitance up to 300 pF.

Measurement Results

Fig. 4 shows the photographs of the ASIC and the fabricated prototype with integrated PZT matrix transducer.

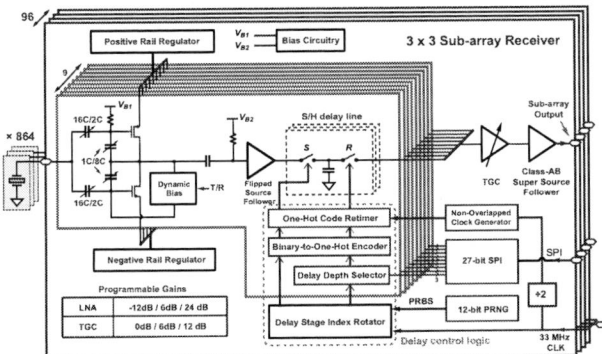

Fig. 2: Schematic of 3 x 3 sub-array receive circuits.

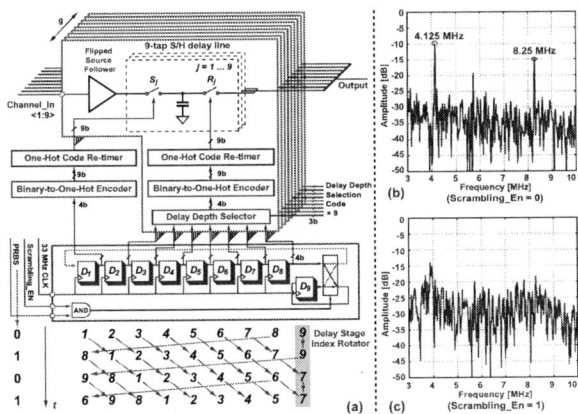

Fig. 3: Sub-array beamformer with mismatch-scrambling

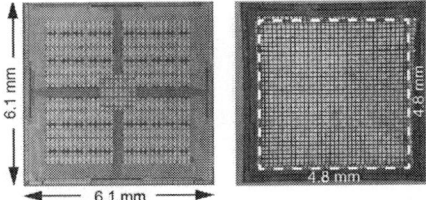

Fig. 4: Photographs of the ASIC (left) and the fabricated prototype with integrated transducer (right).

The measured electrical performance of the ASIC is summarized in Table I. Table II gives a system-level comparison with prior works on ASICs for 3-D ultrasound imaging. This work achieves the best power-efficiency in receiving and the highest integration density, with an element-match layout with a $<\lambda/2$ element pitch. To demonstrate the 3-D imaging capability of the prototype, a pattern of seven needles was placed at a distance of approximately 16 mm in front of the transducer array (Fig. 5a, b; the dotted circle depicts a needle which was slightly behind the other needles). A spherical wave was transmitted and a 3-D volume image was re-constructed. This volume dataset was rendered to a frontal view (Fig. 5c), clearly showing the needle points.

References

[1] Z. Yu, *et al*, IEEE T-UFFC, Jul. 2012
[2] R. Cobbold, "Foundations of biomedical ultrasound", 2007
[3] A. Bhuyan, *et al*, ISSCC 2013, Feb. 2013
[4] A. Bhuyan, *et al*, IEEE TBCAS, Dec. 2013
[5] K. Chen, *et al*, VLSI 2014, Jun. 2014
[6] C. Chen, *et al*, IEEE T-UFFC, Jan. 2016
[7] S. B. Raghunathan, *et al*, IUS 2014, Sept. 2014
[8] C. Chen, *et al*, ESSCIRC 2015, Sept. 2015
[9] K. Chen, *et al*, IEEE JSSC, Nov. 2013

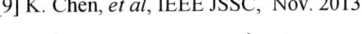

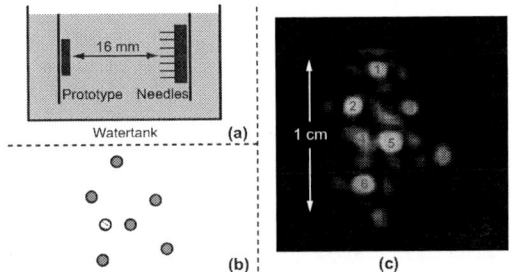

Fig. 5: (a) Imaging experiment setup; (b) Layout of the 7-needle phantom; (c) Volume-rendered 3-D image.

TABLE I. ASIC PERFORMANCE SUMMARY

	Supply voltage	Analog: 1.8 V	Digital: 1.4 V
R X	Total power	228.9 mW	
		Analog: 190 mW	Digital: 38.9 mW
	-3 dB Bandwidth	6 MHz	
	Input-referred noise density @ 5 MHz	w/o mismatch-scrambling: 3 mPa/√Hz	
		w/ mismatch-scrambling: 6 mPa/√Hz (worst case*)	
	RX sensitivity	~ 3 µV/Pa @ LNA input	
	Gain steps	-12/-6/0/6/12/18/24/30/36 dB	
	HD2	43 dBc @ 300 mV$_{p-p}$ output, 5 MHz	
T X	Max. TX pulse voltage	30 V$_{p-p}$	
	Pressure @max. TX voltage	160 kPa @ 5 cm	

*The measured input-referred noise with the mismatch-scrambling function enabled varies with different delay patterns because of a systematic mismatch in the layout of S/H delay lines, which could be optimized by a better layout.

TABLE II. SYSTEM-LEVEL COMPARISON WITH PRIOR WORKS

	[3]	[4]	[5]	[6]	This work
Process	1.5 µm HV	0.25 µm HV	0.18 µm HV	0.18 µm LV	0.18 µm LV
Transducer	CMUT	CMUT	CMUT	PZT	PZT
Array size	16 × 16	32 × 32	16 × 16	9 × 12	32 × 32
Center freq.	5 MHz	5 MHz	5 MHz	5 MHz	5 MHz
Element Pitch	250 µm	250 µm	250 µm	200 µm	150 µm
Pitch ≤ λ/2?	No	No	No	No	Yes
Beamform Function	TX	TX	Off-chip	RX Sub-array	RX Sub-array
# of TX el.	256	960	256	N/A	64
# of RX el.	32	64	256	81	864
Integration method	Flip-chip bonding Via Interposer	Flip-chip bonding	Flip-chip bonding Via Interposer	Direct Integration	Direct Integration
ASIC size	10 × 6 mm²	9.2 × 9.2 mm²	6 × 5.5 mm²	3.2 × 3.8 mm²	6.1 × 6.1 mm²
RX power/el.	9 mW	4.5 mW	1.4 mW	0.44 mW	0.27 mW

A Fully-Adaptive Wideband 0.5-32.75Gb/s FPGA Transceiver in 16nm FinFET CMOS Technology

Parag Upadhyaya, Ade Bekele, Didem Turkur Melek, Haibing Zhao, Jay Im, Junho Cho, Kee Hian Tan, Scott McLeod, Stanley Chen, Wenfeng Zhang, Yohan Frans, Ken Chang

Xilinx, Inc., San Jose, CA

paragu@xilinx.com

Abstract— **This paper describes the design of a low power fully-adaptive wideband, flexible reach transceiver in 16nm FinFET CMOS embedded within FPGA. The receiver utilizes a 3-stage CTLE with a segmented AGC to minimize parasitic peaking and 15-tap DFE to operate over both short and long channels. The transmitter uses a swing boosted CML driver architecture. Low noise wideband fractional N LC PLLs combined with linear active inductor based phase interpolators and high speed clocking are utilized for low jitter clock generation. The transceiver achieves >1200mV$_{dpp}$ TX swing with <190 fs RJ and 5.39 ps TJ to achieve BER < 10^{-15} over a 30 dB loss backplane at 32.75 Gb/s, while consuming 577 mW.**

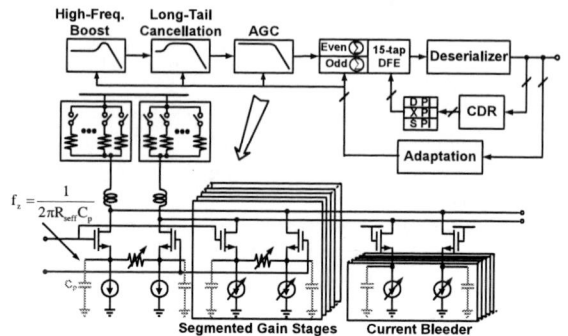

Figure 1: Auto adaptive receiver front-end with 3 stage CTLE with parasitic peaking mitigated AGC

High-speed backplane transceivers embedded within FPGAs are used to address ever increasing demand for bandwidth in communication and storage systems supporting wide range of line rates over variety of channels ranging from low loss to the maximum possible insertion loss [1-2]. Transceivers within FPGA are the early adopter of new digital process such as 16nm FinFET and they must be highly flexible and programmable to support many key protocols like 10G-KR, PCIe Gen1/2/3/4, OIF-CEI-25G, CEI-28G, 802.3bj, and 32G-FC per channel basis. Like most of the CMOS processes, the 16nm FinFET process is tailored for high-speed digital but not for high-precision analog design. Several technology challenges including higher gate capacitance and back-end of line layer (BOEL) parasitics, and higher flicker noise over planer process like 20nm SOC, must be overcome. This requires new circuit techniques for receiver (RX), transmitter (TX), and clocking circuits.

Figure 1 shows the fully adaptive receiver front-end with 3-stage inductively peaked CTLE, phase interpolator (PI) based CDR, and a 15-tap 1-bit unrolled DFE [1]. To support both high and low loss channels, automatic gain control circuit (AGC) is adapted. A source degeneration has been used for gain control [2] to support large dynamic range. Unfortunately, large parasitic (C_p) at tail nodes of the gain stages introduces a zero (f_z) in the transfer function. At lower gain codes, this results in unwanted parasitic peaking (up to 7 dB). This peaking produces a larger than desired amplitude at the input of the DFE and erroneous output for the error slicer driving the adaptation loop and lead to lack of convergence in the adaptation loop as well as Clock and Data Recovery (CDR) offset and result in bit errors. A programmable cap at the output could be used to suppress the parasitic peaking [2]. However, over process, voltage and temperature (PVT) achieving a flat gain transfer curve is difficult. To mitigate this the AGC is segmented into several source degenerative differential pairs consisting of coarse adjustment based on the enabled pairs and fine adjustment based on the source degeneration resistors. To reduce the step size and achieve

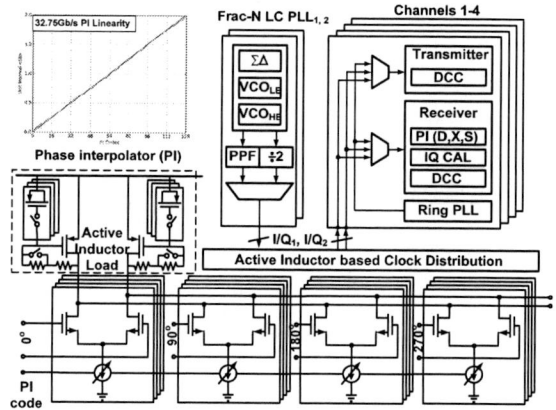

Figure 2: Active inductor based low power linear PI

good voltage swings, the AGC codes is interleaved. By doing so, at lower codes or lower gain settings only smaller parasitic capacitance associated with used gain pair is seen, significantly reducing parasitic peaking (to below <1.5 dB) while maintaining monotonicity and good linearity.

The clocking scheme for the transceiver for phase interpolator (PI) based digital CDR is shown in Figure 2. Significant power is traditionally consumed to provide low jitter high speed clock distribution [1]. Active inductor load based CML are both area and power efficient for low jitter clock distribution [3-4]. For even lower power operation, an active inductor based linear phase interpolator is implemented. Given that three PIs (D: data, X: crossing, S: eyescan) are used in the receiver, as shown in Figure 1, the power reduction is significant. To achieve linearity requirement over a very wide frequency range of operation, the PI uses segmented differential pairs and programmable tuned active inductor loads. The input clock to the PI is duty cycle and I/Q phase error corrected to insure good linearity.

Figure 3 illustrates the temperature compensated fractional-N LC PLL architecture that overcomes high flicker noise of the FinFETs process, where core PMOS is ~40x worse than 20nm planer process. The LC VCO uses NMOS

978-1-5090-0636-6/16 $31.00 © 2016 IEEE

only topology with regulation on both supplies via NMOS regulator and NMOS current sources. Low pass filtering is done to reduce flicker noise contribution from bandgap bias used for reference generation (Vref) and that of the regulator OTA to achieve very low phase noise over wide frequency range supporting 8-16.25GHz operation and overcoming intrinsic higher gain of varactor in FinFET technology, as shown in Figure 3.

The transmitter (TX) of the transceiver must overcome supply scaling, higher parasitic and support wide voltage swing levels. To meet these requirements, a merged mux/pre-driver is used to drive a 3-tap swing boosted CML TX driver shown in Figure 4. A current source from a higher supply-1.8V, is used to adjust the output common mode of the TX driver higher, allowing a higher swing desired for long reach applications. This technique allows TX swing to increase by greater than 30% compared to [1].

Figure 5 shows the quad transceiver micrograph, fabricated in a 16nm FinFET process, and the power breakdown of the transceiver at 32.75Gb/s. The clocking power is significantly lower than that of TX and RX, showing the effectiveness of the low power clocking circuit.

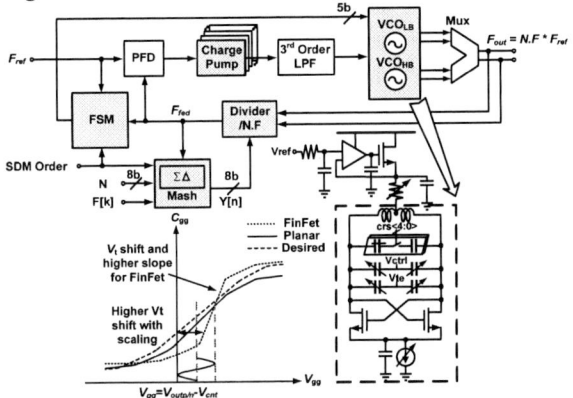

Figure 3: Temperature compensated wide range low noise fractional N PLL mitigating higher flicker noise and varactor intrinsic gain of FinFET process.

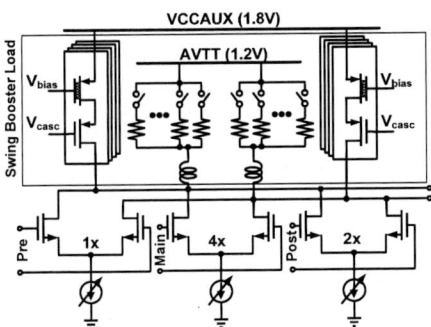

Figure 4: Swing boosted TX driver architecture

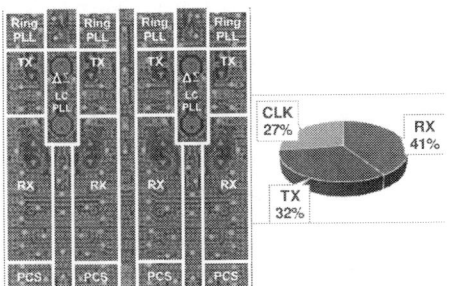

Figure 5: Quad micrograph and power breakdown

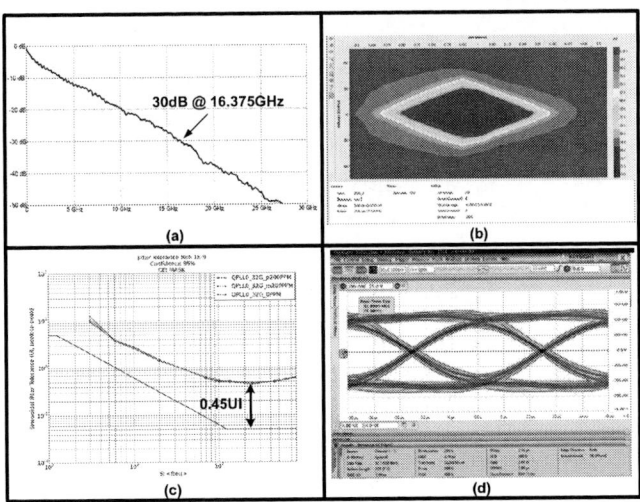

Figure 6: (a) Channel loss (excluding package) and (b) corresponding stressed RX eye scan at 32.75Gb/s with all lane running to account for crosstalk (c) RX jitter tolerance with +/-200 PPM at 32.75Gb/s and (d) TX eye at 32.75Gbps with swing boost.

Technology	CMOS 16nm FinFET
Power Supply (V_{avcc}, V_{avtt}, V_{aux})	0.9 V, 1.2V, 1.8 V
Frequency range	500 Mb/s – 32.75 Gb/s
Transceiver Quad area	2.625 mm × 2.218 mm
LC PLL range	8-16.375 GHz
Ring PLL range	2-6.25 GHz
TX PRBS7 jitter at 32.75Gb/s	TJ: 5.39 ps, RJ: 190 fs
32.75Gb/s RX JTOL @ 30MHz	0.45 UI
@ 100MHz	0.6 UI
Channel loss at 32.75Gb/s	30 dB
Measured BER at 32.75Gb/s	$< 10^{-15}$
Power at 32.75Gb/s with DFE	577mW/ch (17.6pJ/b)

Table 1. Performance Summary

Figure 6 (a) shows the channel loss and (b) shows the corresponding internal RX eyescan with +/-200ppm for the 32.75Gb/s after 30dB loss. These measurement exclude package losses and account for crosstalk of other transceiver lanes. Figure 6(c) show the RX jitter tolerance at 32.75Gb/s with 0.45UI margin at 30MHz and 0.6UI at 100MHz. The output TX eye at 32.75Gb/s utilizing TX swing boost is shown in Fig. 6(d). Table 1 summarizes the transceiver performance.

Acknowledgments
The authors thank Sai Lalith Chaitanya Ambatipudi and Jayesh Patil for data collection and the entire transceiver team.

[1] P. Upadhyaya, et.al, "A 0.5-to-32.75Gb/s Flexible-Reach Wireline Transceiver in 20nm CMOS," *ISSCC*, pp 56-57, 2015.
[2] Y. Frans, et.al, "A 0.5–16.3 Gb/s Fully Adaptive Flexible-Reach Transceiver for FPGA in 20 nm CMOS," *IEEE Journal of Solid-State Circuits*, Vol. 50, No.8, pp. 1932-1944, 2015.
[3] J. Jaussi, et.al, "A 205mW 32Gb/s 3-Tap FFE/6-Tap DFE Bidirectional Serial Link in 22nm CMOS," *ISSCC*, pp 440-441, 2014.
[4] H. Kimura, et.al., "A 28 Gb/s 560 mW Multi-Standard SerDes With Single-Stage Analog Front-End and 14-Tap Decision Feedback Equalizer in 28 nm CMOS," *IEEE Journal of Solid-State Circuits*, Vol. 49, No.12, pp. 3091-3103, 2014.

A 28.3 Gb/s 7.3 pJ/bit 35 dB Backplane Transceiver with Eye Sampling Phase Adaptation in 28 nm CMOS

Hiroki Miyaoka[1], Futoshi Terasawa[1], Masahiro Kudo[1], Hideki Kano[1], Atsushi Matsuda[1], Noriaki Shirai[1], Shigeaki Kawai[1], Takayuki Shibasaki[2], Takumi Danjo[2], Yuuki Ogata[2], Yasufumi Sakai[2], Hisakatsu Yamaguchi[2], Toshihiko Mori[2], Yoichi Koyanagi[2], Hirotaka Tamura[2], Yutaka Ide[1], Kazuhiro Terashima[1], Hirohito Higashi[1], Tomokazu Higuchi[1], and Naoaki Naka[1]

[1]Socionext, Yokohama, Japan, [2]Fujitsu Laboratories, Kawasaki, Japan

E-mail : miyaoka.hiroki@socionext.com

Abstract

28.3 Gb/s transceiver with 35 dB channel loss equalization is presented. The transmitter deploys 3-tap feed forward equalizer (FFE). The driver employs the hybrid architecture of low voltage differential signaling (LVDS) and source-series-terminated (SST) driver which enables the low power consumption and output signal amplitude fine tune. The receiver comprised with continuous time linear equalizer (CTLE) and 2-tap loop unrolled decision feedback equalizer (DFE). It saves the power consumption by not applying DFE at the eye edge, and increases the eye margin with adaptive sampling clock phase adjustment capability. The transceiver is composed of one PLL and four lanes, occupies 1.67 mm^2 and consumes 829 mW (7.3 pJ/bit).

Introduction

The rapid growth of network traffic demands the increase of data center network bandwidth. The data rate of 100G-BASE KR4 and CEI-25G-LR exceeds 25 Gb/s, and transceivers are required to compensate the loss of 30 dB at the Nyquist frequency. At the same time, the power consumption of the network system needs to be reduced. Therefore, transceivers needs to satisfy the requirements of high-speed data rate, large channel loss compensation and low power consumption. This paper describes the 28.3 Gb/s, 35 dB loss compensation transceiver with 7.3 pJ/bit and 1.67 mm^2.

Transceiver Architecture

Figure 1 shows the block diagram of the transceiver. The transceiver is comprised of 4-lane of transmitters (TX) and receivers (RX), and LC-VCO PLL. In the TX circuit, incoming 32-bit parallel data is multiplexed to 28.3 Gb/s differential serial data. In contrast, in the RX circuit, incoming 28.3 Gb/s differential serial data is demultiplexed to 32-bit parallel data.

A LC-VCO oscillating at 28.3 GHz supplies the clock signals to each lane. The TX uses only two of the four phases, whereas the RX uses all the quadrature phases. Since the clocks for each lane are fed from a single PLL, the power dissipation and total area required are lower than transceivers having a PLL in each lane.

The TX consists of a FIFO, a 32:16 multiplexer (MUX), a 16:2 MUX, and source-series-terminated (SST) drivers with 3-tap feed forward equalizer (FFE). To mitigate the electro static discharge (ESD) protection diode parasitic capacitance, T-coils are employed at the analog front-end to improve return loss.

The RX consists of a CTLE ([1]), a 2-tap loop unrolled DFE, a 2:16 demultiplexer (DEMUX), a 16:32 DEMUX, a

clock-recovery unit (CRU) and an adaptation logic [2]. The CRU employs a half rate phase interpolator (PI) based digital CDR. Equalization is realized with the CTLE and the 2-tap loop unrolled DFE. The adaptation logic controls equalization performance of the CTLE and DFE, adjusts variable gain amplifier (VGA) gain, cancels offset at CTLE and DFE, and generates eye sampling phase shift code.

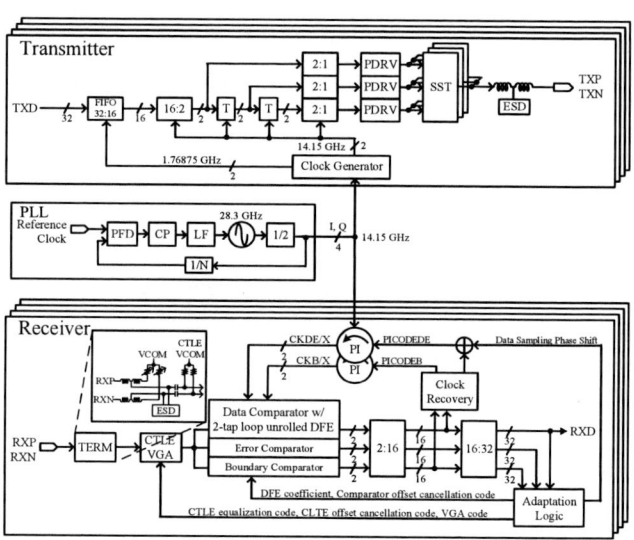

Fig. 1 Block diagram of the transceiver.

LC-VCO PLLs are required to have following characteristics. (1)low power (2)wide frequency tuning range (3)small area (4) low phase noise. By employing multiple coils enables wide frequency tuning range at the cost of area [3]. To satisfy all those four-requirement, the LC-VCO employs the circuits shown in Figure 2(a). It has one coil and capacitor array for coarse frequency tuning. The digital calibration circuits optimize capacitor array and realize 6.5 GHz of frequency tuning range. The drain and source node of the switch nmos for controlling capacitor array need to be biased at half of the supply voltage to protect from break down. As shown in the example of large noise bias circuit, since the source of the transmission gate is biased at the half of the supply voltage, the device channel noise increases. Therefore, the switch transistors for adjusting bias settings are fully turned on to realize low phase noise by suppressing device noise.

Though the SST driver is suitable for low power consumption, the output signal amplitude is limited by the supply voltage. To mitigate this drawback, SST-LVDS Hybrid Driver is proposed (Fig. 2(b)). The slices for current control

978-1-5090-0636-6/16 $31.00 © 2016 IEEE

are added in parallel with the slices for the resistors. The output signal amplitude is expressed as

$$V_{diff} = \frac{VDD}{2} + 50 * I_0, \tag{1}$$

where the termination resistor for the driver is 50 Ohm, the receiver side is 100 Ohm, the current of the current slices is I_0, and the supply voltage is VDD. The SST-LVDS Hybrid Driver enables to control over the output signal amplitude by adjusting current. It allows the fine tuning of output signal amplitude and FFE coefficient.

Figure 2(c) shows the eye diagram at the CTLE output. The ISI lower than fb/4 is equalized by the CTLE. Figure 2(d) shows the eye diagram of the equivalent DFE output. By changing the DFE equalization coefficient to control signal amplitude, eye opening is improved. However, since the data-sampling phase is offset from the eye center as shown with gray square mark in Figure 2(d), eye margin decreases. The scheme to adjust sampling phase to the eye-center is proposed to make the eye margin larger. The Vamp (Fig. 2(c)) is determined based on the CTLE output eye. The Vamp is the amplitude of the 0011 output of the CTLE. It is determined with cross point of the solid lines 011 and 110, and that of the dashed-lines 100 and 001. Since the data-sampling phase at the CTLE output locates at 0.5 UI from boundary sampling phase, the slew rate is expressed as

$$slew = \frac{\left(V_{amp}/2\right)}{0.5}. \tag{2}$$

As shown in Figure 2(d), the amount of phase shift (PHSFT) to the optimum data sampling phase is given as

$$PHSFT = \frac{DFE_{coef}}{slew} = \frac{DFE_{coef}}{V_{amp}}, \tag{3}$$

where DFE_{coef} is the DFE equalization coefficient.

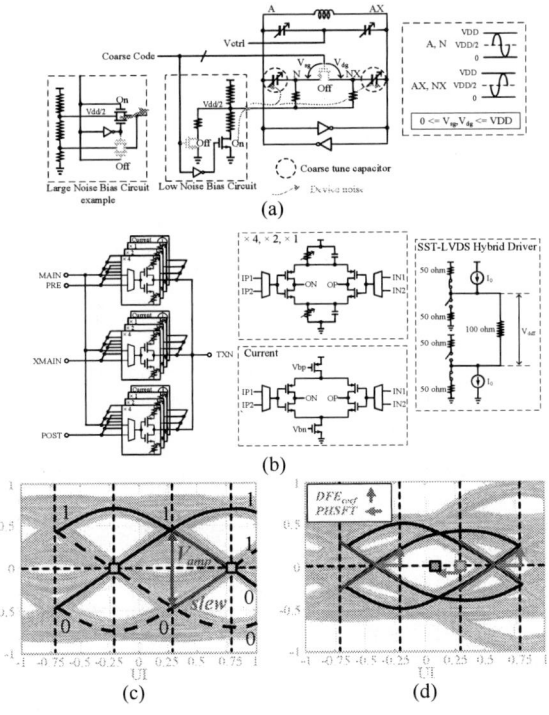

Fig. 2 (a) VCO circuit, (b) SST-LVDS hybrid driver, (c) CTLE output, (d) Equivalent DFE output.

Measurement Results

The Figure 3(a) shows that the TX eye diagram at the condition of 28.3 Gb/s PRBS9 data pattern. The total jitter is 7.12 ps at BER 10^{-15}, RMS jitter is 237 fs. Figure 3(b) the jitter tolerance test results with PRBS31. The insertion loss at the Nyquist frequency is 35.3 dB. The jitter from the pulse pattern generator (PPG) is 0.3 UI, and the tolerance is 0.17 UI at the jitter tolerance test. Compared to the target mask, there is a 0.12 UI margin. The RX internal eye diagram is shown in Figure 3(c). By applying the data sampling phase adaptation scheme, eye can be opened and the sampling phase is located on the eye center.

Test chips were fabricated in a 28 nm CMOS process (Fig. 3(d)). The transceiver occupies 1.67 mm^2.

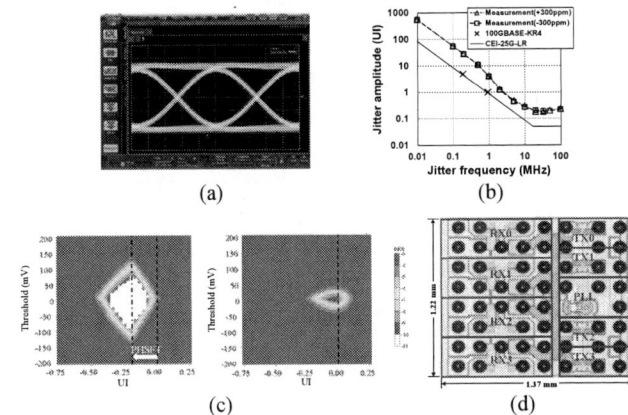

Fig. 3 (a) TX eye, (b) Jitter Tolerance, (c) RX eye w/ PHSFT(left), w/o PHSFT(right), (d) Chip photograph.

The power consumption of the LC-VCO PLL and 4-lane transceiver macro is 829 mW (7.3 pJ/bit) at 28.3 Gb/s. The performance of the transceivers are compared in Table I. It obviously shows that this work is the best performance in terms of the power efficiency and the area.

TABLE I Performance summary and comparison

	[3]	[4]	This work
Technology	28 nm	28 nm	28 nm
MAX Data Rate	28.05 Gb/s	28.125 Gb/s	28.3 Gb/s
Supply voltage	1.8 V, 0.9 V	1.25 V, 1 V	1.5 V, 0.96 V
TX EQ topology	4-tap FFE	5-tap FFE	3-tap FFE
TX jitter RMS	185 fs	230 fs	237 fs
RX EQ topology	CTLE + 36-tap DFE	CTLE + 14-tap DFE	CTLE + 2-tap DFE
Channel loss at Nyquist	40 dB	40 dB	35 dB
Power efficiency	25.1 pJ/bit	10.5 pJ/bit	7.3 pJ/bit
Area	24.8 mm^2 (8lanes)	2.48 mm^2 (4 lanes)	1.67 mm^2 (4 lanes)

References

[1] S. Parikh, et al., *IEEE ISSCC*, pp. 28-29, Feb. 2013.
[2] Y. Hidaka, et al., *IEEE JSSC*, vol. 44, pp. 3547-3559, Dec. 2009.
[3] T. Kawamoto, et al., *IEEE ISSCC*, pp. 54-55, Feb. 2015.
[4] B. Zhang, et al., *IEEE ISSCC*, pp. 52-53, Feb. 2015.

A 32 Gb/s Rx Only Equalization Transceiver with 1-tap Speculative FIR and 2-tap Direct IIR DFE

Sewook Hwang, Sungjun Moon, Junyoung Song, and Chulwoo Kim

Korea University, Seoul, Korea

e-mail: ckim@korea.ac.kr

Abstract

This paper presents an Rx only equalization (ROE) technique that eliminates all Tx equalizations to reduce power dissipation, circuit complexity, and cost. The proposed Rx consists of a CTLE, a 1-tap speculative FIR and 2-tap direct IIR DFE. A simpler Tx architecture owing to the ROE facilitates a wide bandwidth and energy efficient dual-mode (differential and single-ended) Tx operation. The proposed Tx consists of dual-mode 2:1 serializer/pre-drivers and main drivers. The transceiver was fabricated in a 65 nm CMOS technology. The Rx achieves BER < 10^{-12} over a -22 dB loss PCB channel at 32 Gb/s with 0.62 pJ/b energy efficiency, and occupies 0.024 mm^2. The Tx has only 0.77 pJ/b and 0.40 pJ/b energy efficiency at 32 Gb/s in differential mode, and 32 Gb/s/pin in single-ended mode, respectively, and occupies only 0.002 mm^2.

Introduction

Recently, high-speed serial I/Os for data centers or future processor/memory interfaces have been introduced in the literature [1-6]. Main goals for these I/Os are maximizing overall bandwidth and better energy efficiency. However, the continuous bandwidth increase in I/Os causes severe intersymbol interference (ISI) and necessitates energy-efficient equalization techniques. On the Tx side, an equalization with de-emphasis is usually utilized with impedance modulation [1] or segmentation technique [2]. First technique requires an additional impedance modulation loop to maintain the output impedance during de-emphasis. [1]. Second technique slices the Tx with many segments and increases complexity and power dissipation [2-3]. On the Rx side, a continuous time linear equalizer (CTLE) and a decision feedback equalizer (DFE) are commonly used. An infinite impulse response (IIR) DFE [4] comes into the spotlight due to its powerful long-tail ISI cancellation capability, low power, and simpler architecture than a finite impulse response (FIR) DFE [5-6]. However, it is difficult to increase the bandwidth due to the slow feedback path of the IIR DFE. A hybrid DFE was introduced to compensate the feedback delay of the IIR DFE, but the maximum bandwidth was still 10 Gb/s [4] and therefore more circuit and system level optimizations are required.

Rx Only Equalization (ROE)

Fig. 1 shows the overall architecture of the proposed transceiver. The proposed Tx can operates as pseudo-DIFF-mode or independent SE-mode exclusively. Tx clocking (CLK) receives an external half-rate clock, and Rx CLK also receives an external half-rate clock, but the clock timing is controlled manually for the ROE.

The comparison between conventional equalization and the proposed ROE is shown in Fig. 2(a). The conventional case sets the main-cursor at the maximum point of the impulse response. Tx de-emphasis cancels the pre-cursor. Consequently, the amplitude of the main-cursor is reduced due to the de-emphasis. The remaining post-cursor is cancelled by the Rx DFE. On the other hand, the proposed case aggressively shifts the clock sampling point which depends on the pre-cursor amplitude. The pre-cursor amplitude becomes zero and the amplitude of the main-cursor is reduced as much as the pre-cursor. The amplitude reductions in both cases are the same with a simple linear impulse response model. However, in reality, the proposed equalization has a better amplitude (44% improvement) due to the nonlinear exponential curve ($e^{-t/RC}$) of the impulse response. An analysis with ideal DFE model for both cases are shown in Fig. 2(b). Impulse response of the proposed case has no pre-cursor ripple and better amplitude, as a result, the eye height is improved by 243%.

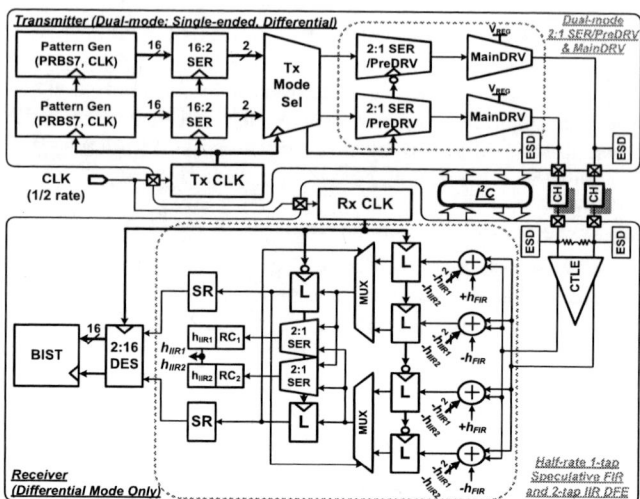

Fig. 1 Overall architecture of the proposed transceiver.

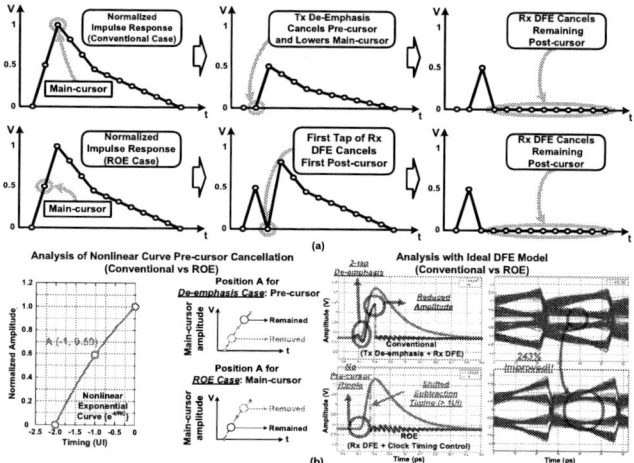

Fig. 2 (a) Concept of the proposed ROE, and (b) analysis results.

Fig. 3 shows the circuit implementation of the proposed DFE. The 1-tap speculative FIR DFE is a natural choice because the timing of the 1st post-cursor is critical and the subtraction strength of the 1st post-cursor must be strong due to the ROE. To reduce the feedback delay of the proposed DFE, a summer and an offset generator are merged to the latch. Coefficient of the 1st FIR tap is directly controlled by the voltage input (h_{FIR}). Previously, binary-weighted taps were used to control the coefficients of the 1st and 2nd IIR taps [4]. However, these taps increase the parasitic capacitance of the summer and degrade the maximum speed. The proposed 2:1 serializer (SER) shown in Fig. 3(b) has a merged IIR filter with output CM level control in order to realize the efficient IIR DFE coefficient control. Output CM level of the IIR filter is controlled by P_CNT[7:0] and N_CNT[7:0]. High CM level decreases the gain of the summer and vice versa. In addition, to reduce the critical path of IIR DFE, output of the two MUXes are directly applied to two 2:1 SERs as shown in Fig. 1. The SR latch or second latch is not presented in the IIR DFE feedback path to reduce the critical path further. However, due to the direct feedback of the IIR DFE, the timing variation or lower data rate

978-1-5090-0636-6/16 $31.00 © 2016 IEEE

may create an overlap between the 1st and 2nd post-cursors (FIR DFE and IIR DFE) and increase the 1st post-cursor coefficient. Balancing the coefficients between the 1st and 2nd cursors can simply eliminate this issue.

Dual-mode Tx

Fig. 4(a) shows the relationship between the pre-driver (PreDRV) output and the main-driver (MainDRV) output. Usually, the swing level of PreDRV output starts from ground which is lower than a threshold voltage of the MainDRV's input NMOS. A signal in this region cannot turn-on the MainDRV and distorts the MainDRV output. To prevent this problem, the proposed 2:1 SER/PreDRV increases the minimum output level as shown in Fig. 4(a). Always turned on PMOSs and PMOS active inductors maintain the minimum voltage level of the 2:1 SER/PreDRV outputs. This method also can relax the slew-rate requirement of the PreDRV. In addition, the node V_P shown in Fig. 4(b) creates a differential configuration and rejects a common mode noise while enhances the output slew-rate of the 2:1 SER/PreDRV. Impedance of MainDRV is digitally controlled by the signals $Z_{CON,TOP}[4:0]$ and $Z_{CON,BOT}[4:0]$ and the voltage regulator is shared by two MainDRVs as shown in Fig. 4(c).

Measurement

Fig. 4(d) shows the measured Tx eye diagrams of DIFF and SE-modes at 32 Gb/s with PRBS7 pattern. In DIFF-mode, total jitter and random jitter are 10.90 ps_{pp} and 295.83 fs_{rms}, respectively. Fig. 5 shows the measured bathtub curves of the Rx and three test channel conditions. A 32 Gb/s Rx with PRBS7 pattern was tested with -22 dB loss CH1. The proposed ROE widens the eye width up to 0.38 UI with 10^{-9} BER. Measured bathtub curves at 28 Gb/s with CH2 and CH3 are

also shown in Fig. 5. Eye widths for CH2 and CH3 are 0.56 UI and 0.29 UI with 10^{-9} BER, respectively. Table I is a performance summary and comparison with state-of-the-arts. This work demonstrates 32 Gb/s Tx and Rx fabricated in a 65 nm CMOS process. The proposed Tx has substantial energy efficiency improvement and occupies the smallest area. The proposed Rx realizes the NRZ type IIR DFE at 32 Gb/s thanks to the direct feedback of the IIR DFE. Fig. 6 shows the die micrograph and power breakdowns of Tx and Rx.

References

[1] T.-C. Hsueh *et al.*, ISSCC, pp. 444-445, Feb. 2014.
[2] Y.-H. Song *et al.*, ISSCC, pp. 446-447, Feb. 2014.
[3] J. Kim *et al.*, ISSCC, pp. 1-3, Feb. 2015.
[4] S. Shahramian *et al.*, JSSC, pp. 1722-1735, Jul. 2015.
[5] J. W. Jung *et al.*, JSSC, pp. 515-526, Feb. 2015.
[6] J. Bulzacchelli *et al.*, JSSC, pp. 3232-3248, Dec. 2012.

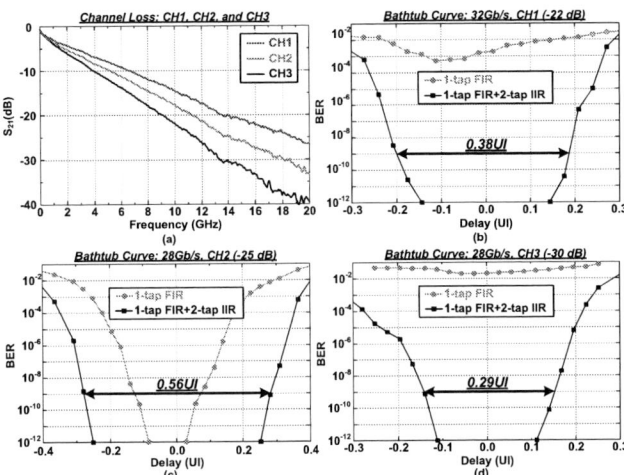

Fig. 5 (a) Measured S_{21} of three channels, (b) measured bathtub curves at 32 Gb/s, (c), and (d) measured bathtub curves at 28 Gb/s.

TABLE I PERFORMANCE COMPARISON

	This Work	[1]	[2]	[3]		This Work	[4]	[5]	[6]
	Tx	Tx	Tx	Tx***		Rx	Rx	Rx	Rx
Architecture	1/2 rate	1/2 rate	1/4 rate	1/4 rate	Architecture	1/2 rate	1/2 rate	1/2 rate	1/2 rate
Signaling	NRZ DIFF / NRZ SE	NRZ DIFF	NRZ SE	NRZ DIFF	NRZ DIFF	EQ Composition	CTLE, 1-tap FIR, 2-tap IIR	CTLE, 2-tap IIR	CTLE, 2-tap FIR	CTLE, 15-tap FIR
# of taps	No EQ	1-tap	2-tap	4-tap	# of taps	1-tap			
Supply (V)	1.2 / 1.2	1.05	0.9/1.5	1/0.5	Signaling	NRZ DIFF	NRZ DIFF	NRZ DIFF	NRZ DIFF
Data Rate (Gb/s)	32	25.6	6.4	16	28	40	Data Rate (Gb/s)	32	28	10	25	28
RJ (ps$_{rms}$)	0.296	0.391 / 0.377	0.82	1.9	-	0.33	0.51	Supply (V)	1.2	1.2	1	1	1.05
TJ (ps$_{pp}$)	10.90	14.67 / 17.50	16.06	24.3	29.1	10.72	12.89	Channel Loss (dB)	22	25	24	24	35
Power (mW)	24.7*	25.3*	19.8	14.08	16.8	195	518	Eye Width (UI)	0.29 @ 10^{-12}	0.50 @ 10^{-12}	0.33 @ 10^{-12}	0.44 @ 10^{-12}	0.35 @ 10^{-9}
Efficiency (pJ/b)	0.77 / 0.40	1.1	2.2	1.05	6.95	12.89	Power (mW)	21.4*	19.8*	4.1*	5.8*	80**
Technology (nm)	65	22	65	14	Efficiency (pJ/b)	0.67	0.62	0.41	0.23	2.85
Area (mm²)	0.002	0.005	0.006**	0.028	Technology (nm)	65	28	45	32
						Area (mm²)	0.024***	0.009	0.01	0.44

* Power of (Tx Mode Sel with CLK Distribution + 2:1 SER/PreDRV + MainDRV + Voltage Regulator)
** One Tx channel without voltage regulator
*** [4] also supports 40Gb/s PAM4 without equalization

* Power of (CTLE + DFE without CLK Distribution)
** DFE power only
*** CTLE (0.022mm²) + DFE (0.002mm²)

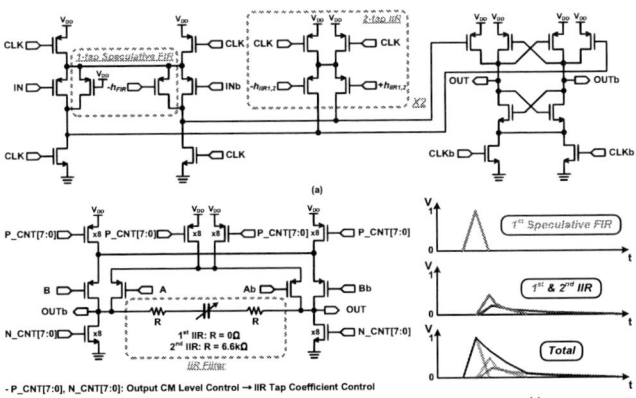

Fig. 3 (a) Latch with 1-tap speculative FIR offset generation and 2-tap IIR subtractions, (b) 2:1 SER and merged IIR filter with output CM level control, and (c) impulse responses of DFE.

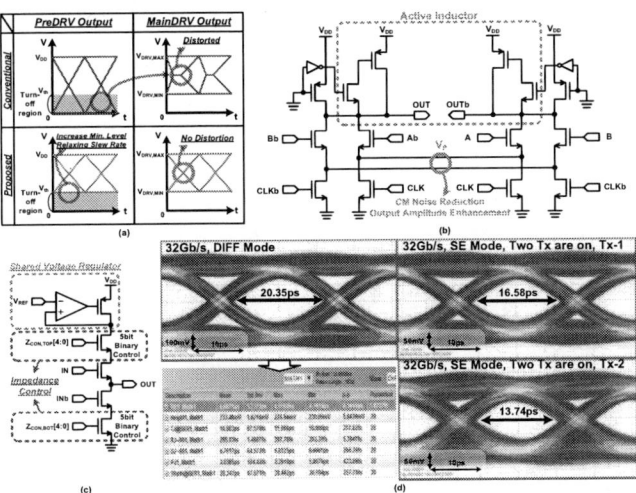

Fig. 4 (a) Pre-driver comparison, (b) 2:1 SER/PreDRV, (c) MainDRV, and (d) measured 32 Gb/s Tx eye diagrams (DIFF and SE modes).

Fig. 6 Die micrograph and power breakdown at 32 Gb/s.

A 56Gb/s PAM4 Wireline Transceiver using a 32-way Time-Interleaved SAR ADC in 16nm FinFET

Yohan Frans, Mohamed Elzeftawi, Hiva Hedayati, Jay Im, Vassili Kireev, Toan Pham, Jaewook Shin, Parag Upadhyaya, Lei Zhou, Santiago Asuncion, Chris Borrelli, Geoff Zhang, Hongtao Zhang, Ken Chang

Xilinx, Inc., San Jose, CA

yohanf@xilinx.com

Abstract— **A 56Gb/s PAM4 wireline transceiver testchip is implemented in 16nm FinFET. The CML transmitter incorporates an auxiliary current injection at the output nodes to maintain PAM4 amplitude linearity. The receiver consists of continuous-time linear equalizers with constant DC-gain and a 28GSa/s 32-way time-interleaved SAR ADC. The transceiver achieves 1e-8 BER over a backplane channel with 25dB loss at 14GHz while consuming 550mW power, excluding DSP.**

In order to meet increasing bandwidth demand, a new 56Gb/s electrical interface standard was recently proposed [1]. The interface must support legacy channels (i.e., backplane) that were initially designed for NRZ signaling only up to 28Gb/s. These legacy channels often have a very large insertion loss beyond 14GHz with significant reflections. PAM-4 signaling is chosen in order to limit the frequency content of the signal below 14GHz. ADC [3][4] is used in the receiver so that Digital Signal Processing (DSP) can be applied to correct for inter-symbol-interference from these legacy channels. Since a Forward Error Correction (FEC) mechanism is employed in the system, the target BER for this interface is not very stringent (e.g., BER < 1e-4 is currently proposed as standard [1]).

The PAM4 transmitter front-end (Figure 1a) is realized by the current summing of 2X driver (MSB) and 1X driver (LSB). The transmitter must maintain linearity between the four output levels while delivering >1V diff-pp swing. This is achieved by injecting current from an auxiliary supply (1.8V) to the outputs to raise its common mode, and by placing a small cascode device above the tail current source to increase the output impedance, which further improve DC linearity and reduce AC distortion. Open-loop compensation using a replica of driver input differential pair helps maintain optimum output common-mode over PVT. A combined 4-to-2-MUX/pre-driver circuit (Figure 1b) incorporates a pseudo H-bridge scheme with positive feedback. This consumes less power than a conventional CML circuit. The circuit has a high gain at zero crossing, which help suppress clock switching noise at the driver output.

Figure 2 shows the receiver block diagram. The analog front-end (4 stages of CTLE and AGC) provides signal equalization and conditioning which reduces the resolution and full-scale-range requirement of the ADC [2]. The 28GSa/s ADC converts the differential analog input into 8-bit digital values. The ADC outputs are sampled periodically and stored in a 64Kb (8K symbols) storage. An off-chip FPGA is used to take these 8K symbols, performs DSP, and generates equalized symbols. The DSP inside the FPGA consists of 24-tap FFE and 1-tap DFE. The FPGA also performs equalization, adaptation, clock recovery (CDR), and ADC offset/gain/skew calibrations based on sampled ADC outputs.

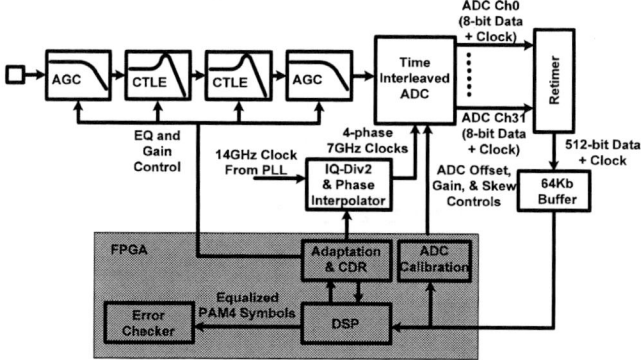

Figure 2: Receiver Block Diagram

The CTLE (Figure 3) is designed to have a constant-DC gain (~0dB) and programmable high-frequency peaking while the AGC has a 10dB programmable DC gain range. Compared to a constant high-frequency gain CTLE (with programmable DC gain), the constant-DC gain CTLE can either reduce the required AGC's gain at high-loss channels and/or improve the linearity of subsequent stage at low-loss channels. Furthermore, this approach minimizes interaction between AGC adaptation loop and CTLE adaptation loop.

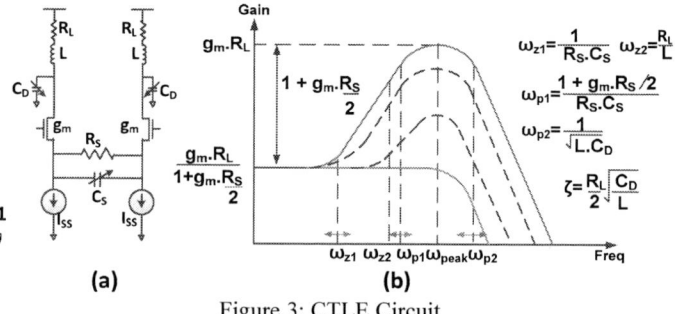

Figure 3: CTLE Circuit

Figure 4a shows the 8-bit 28GSa/s, 600mV diff-pp full-scale range, 32-way time-interleaved SAR ADC used in the receiver. There are two stages of time-interleaving. The first stage is a 4-

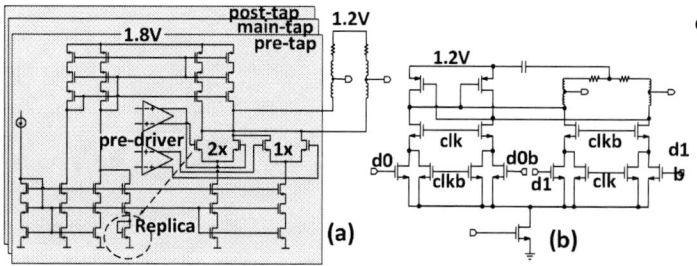

Figure 1: TX Front-End (a) and 4-to-2-MUX/Pre-driver (b)

978-1-5090-0636-6/16 $31.00 © 2016 IEEE

way time-interleaver, where the input is sampled and held using four-phase, non-overlapping 7GHz clocks. The second stage is an 8-way time-interleaver, where each of the signals sampled by the 7GHz clocks are further sampled and held using 8-phase 875MHz clocks and converted to digital values using 8 instances of 875MHz SAR ADC. The output of the 32 instances of SAR ADCs are then re-timed to a single 875MHz clock domain and sent to a 64Kb storage. Figure 4b shows the ADC clocking timing diagram. The timing skew calibration of the 7GHz clocks and the gain/offset calibration of the 875MHz ADC instances are performed using pseudo-random data input, in contrast to the use of sinusoidal tones for calibration in [3]. Our approach allows live data calibration where scrambled pseudo-random data is common in most high-speed electrical interfaces.

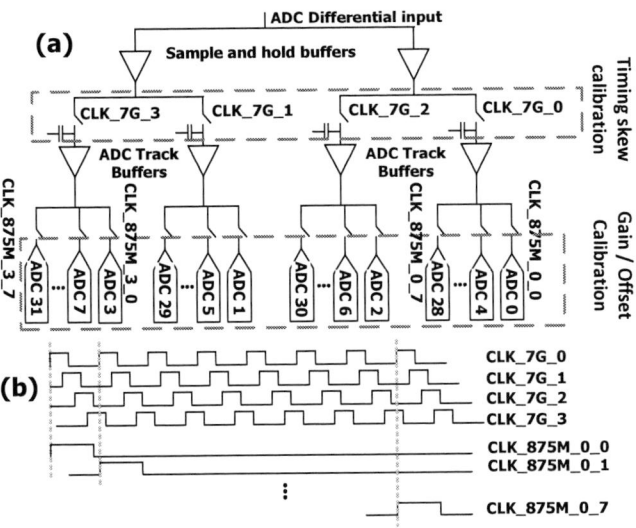

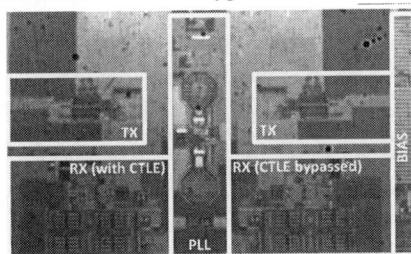

Figure 4: 28GSa/s Time-Interleaved SAR ADC (a) and Clock Timing Diagram (b)

Two TX/RX lanes with common PLL, clock distribution, and bias block is fabricated (Figure 5) in 16nm FinFET. In order to perform direct measurements of the ADC performance, the CTLE in the second RX lane is bypassed.

Figure 5: Die Photo

Figure 6 shows the transmitter output eye diagram over ~5dB channel. The transmitter is configured to transmit PRBS7 pattern with ~4.5dB post-tap equalization. The average of the four output levels show good linearity. The ADC performance is measured by feeding sinusoidal inputs at various frequencies, capturing the ADC output using 64Kb storage, and performing FFT. The ADC achieves ENOB of 6.3 at 180HMz and 4.9 at 14GHz, as illustrated in Figure 7.

The link performance is tested by connecting the transmitter output to the receiver over backplane channels. Figure 8 shows eye diagrams at the CTLE/AGC outputs (captured using the ADC as real-time digital oscilloscope) and the post-DSP histograms of the four PAM4 levels (sampled at the CDR lock

point) over 200K symbols. The CTLE/AGC outputs show open eye with ~6dB channel (Figure 8a) and closed eye with 25dB channel (8b). In both cases, the DSP open the eye in the post-DSP PAM4-level histograms. The estimated BER is around 10^{-8} based on extrapolation of the histograms.

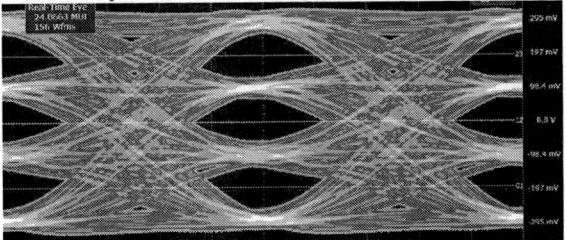

Figure 6: Transmitter PRBS7 Output Eye Diagram

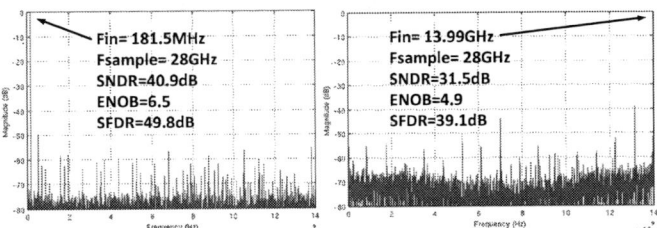

Figure 7: ADC Performance

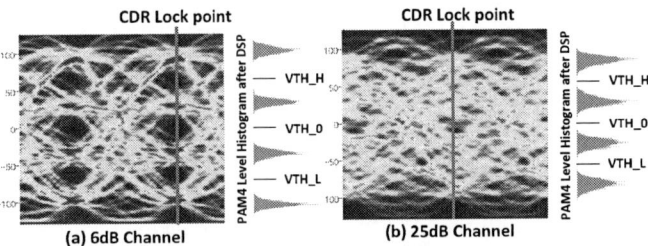

Figure 8: Eye diagram at ADC output and post-DSP PAM4 level for ~6dB channel (a) and 25dB channel (b)

Technology	CMOS 16nm FinFET
Power Supply (V_{avcc}, V_{avtt}, V_{aux})	0.9V, 1.2V, 1.8V
Dual Transceiver Active Area	2.8mm^2
Max TX Swing	1.2V diff-pp
TX RJ (PRBS7, Major Transition)	200fs
ADC ENOB	6.5@0.18GHz, 4.9@14GHz
ADC Power (including ADC clocks)	280mW
BER at 56Gb/s	~1x10^{-8}
Power per lane at 56Gb/s (Does not include DSP)	550mW (140mW TX, 370mW RX, 40mW PLL/Clock Distribution)

Table 1. Performance Summary

Acknowledgments

The authors thank Xilinx serdes design and validation teams for contributing to their circuit design and silicon measurements.

References

[1] Optical Internetworking Forum (OIF), "CEI-56G-LR-PAM4 Long Reach Implementation Agreement Draft Text", 2015.
[2] E-Hung Chen, et.al., "Power Optimized ADC-Based Serial Link Receiver," in Solid-State Circuits, IEEE Journal of , April 2012.
[3] Lukas Kull, et.al., "A 90GS/s 8b 667mW 64× Interleaved SAR ADC in 32nm Digital SOI CMOS", ISSCC 2014
[4] Delong Cui, et.al., "A 320 mW 32 Gbps 8-bit ADC-Based PAM4 Receiver with Programmable Gain Control and Analog Peaking in 28nm CMOS", ISSCC 2016

A 50MHz 5V 3W 90% Efficiency 3-Level Buck Converter with Real-Time Calibration and Wide Output Range for Fast-DVS in 65nm CMOS

Xun Liu[1], Cheng Huang[1,2] and Philip K. T. Mok[1]

[1]Hong Kong University of Science and Technology, Hong Kong SAR, China; [2]Keio University, Yokohama, Japan

xliuam@ust.hk, eemok@ust.hk

Abstract

In this paper, a 50-MHz 5-V input 3-W output 3-level buck converter is presented. A real-time flying capacitor (C_F) calibration is proposed to ensure a constant $V_g/2$ voltage across the C_F, which is essential to ensure the reliability and maintain the advantages of 3-level converters. A 0.6-4.2V wide output range, a 90% peak efficiency and a 23-29ns/V reference tracking response are observed in measurements with 65nm process. A significantly reduced V_O ripple is achieved after enabling the proposed calibration.

Introduction

3-level buck converters are highly attractive in high-voltage applications in both academia and industry. As shown in Fig. 1, it stacks 2 power transistors, so the input and output voltage can be as high as twice of the power transistors' nominal voltage without breakdown issues. The higher voltage rating makes it suitable for dynamic-voltage-scaling (DVS) or RF envelope-tracking applications. At the same time, it has extra benefits brought by a flying capacitor C_F, which introduces an additional voltage level. Compared to the traditional 2-level converters, 3-level converters provide a much smaller inductor current ripple, and thus the power inductor and output capacitor can be significantly reduced to save costs, reduce the form factor, and boost the reference tracking response.

However, all these advantages of the 3-level structure rely on the 3-level operation that the voltage of C_F is essential to maintain as $V_g/2$. In real silicon implementation, V_{CF} can easily deviate from $V_g/2$ and even go to 0 or V_g due to imperfect conditions [1], such as parasitic capacitance of C_F, time mismatches between the power transistors' driving signals and current-drawing circuits, as shown in Fig. 1. This may result in breakdown and reliability issues for the power transistors. The inductor current ripple and V_O ripple will also increase significantly. The system stability, such as sub-harmonic oscillation [2], and the system bandwidth may also be affected due to the much larger noise. Thus a calibration keeping V_{CF} always equal to $V_g/2$ is essential to ensure proper operation and maintain the advantages of 3-level converters. However, this issue is not considered in most prior works [3], [4]. Although a C_F balancing technique is proposed in [5], the V_{CF} regulation is only executed every 32 clock cycles, which is slow and may still bring voltage stress and reliability issues. The regulation of V_{CF} also depends on loading conditions because of the limited resolution.

Proposed V_{CF} Real-Time Calibration

Fig. 1 shows the block diagram of the proposed 3-level converter. The power transistors (P_{1-2} and N_{1-2}) are implemented by standard I/O devices to achieve a 0.6-V~4.2-V output V_O from 5-V input V_g for high-voltage applications, such as RF power amplifier supply or battery/USB powered devices. Different from a conventional 3-level converter, C_F is not free-switching but real-time calibrated by the proposed V_{CF} calibration scheme to maintain $V_g/2$.

The detailed circuitry of the proposed V_{CF} real-time calibration is shown in Fig. 2. The calibration signals V_{NN}, V_{NP}, V_{PN} and V_{PP} are first generated from V_A, V_B, V_g and $V_g/2$ to fit into the inputs of the amplifier with switching noise filtered. A fully differential difference amplifier (DDA) then amplifies the difference between V_{CF} and $V_g/2$ as the difference between V_{EA_CALI1} and V_{EA_CALI2}, whose common mode voltage are regulated by the common mode feedback (CMFB) circuits based on V_{EA}. Through these mechanisms, V_{EA_CALI1} and V_{EA_CALI2}, deviating from V_{EA} differentially, generate modulated D and D_S. The charge and discharge time for C_F is thus changed to ensure $V_{CF} = V_g/2$. In this way, the real-time calibration always maintaining V_{CF} at $V_g/2$ under any conditions is realized.

The timing diagram of the calibration process is sketched in Fig. 3. Under ideal conditions, D and D_S have the same duty cycle but with a 180° phase shift. Considering V_{CF} is smaller than $V_g/2$ due to the non-idealities mentioned before, as an example. For a conventional 3-level converter, the voltage on V_X becomes higher/lower than $V_g/2$, causing the power transistors to sustain over-voltage stress. I_L, and thus V_O ripple, will increase. With the proposed calibration, V_{EA_CALI1} and V_{EA_CALI2} are split from V_{EA} to increase D_1 and decrease D_3, and thus C_F has a longer charging than discharging phase. V_{CF} will return to $V_g/2$ after several cycles, with a small $|V_{EA\_CAL} - V_{EA}|$ as the compensation for the non-idealities in steady-state. As a result, the over-voltage stress and reliability issues are relieved, and I_L and V_O ripple will be significantly reduced, with the possibility of higher system bandwidth. Since D and D_S are modified differentially, $V_O = (D_1+d)V_g - 2dV_{CF} \approx DV_g$, meaning that the calibration loop will not affect the main control loop's stability since V_O is independent of d.

Measurement Results

The proposed 3-level buck converter occupies 1875μm x1250μm in a 65nm process. A 5-nF integrated C_F is implemented, stacking MIM, MOM and MOS capacitors. As shown in the measured waveforms in Fig. 4(a), before calibration enables, V_{CF} is about 1V which is far from $V_g/2 = 2.5V$. After enabling the calibration, V_{CF} returns to $V_g/2$ immediately. As a result, up to 69% V_O ripple reduction by the V_{CF} calibration is observed, as shown in Fig. 4(b). It also helps reduce the switching-frequency noise in the control loop, and thus the system bandwidth can be extended without worrying about potential sub-harmonic oscillations. Fig. 5 shows the dynamic performance of the proposed 3-level converter. With the reduced noise and the well-optimized type-III compensator, the up/down reference tracking-times are only 56ns and 44ns, respectively, between 1.5V and 3.4V. This is much faster than the buck converter with a similar switching frequency designed for DVS [6]. Thanks to the real-time calibration, V_{CF} is always constant during the transients. Fig. 6 shows the measured efficiency versus I_O. Over 90% peak efficiency is achieved. Fig. 7 shows the chip micrograph. A comparison with prior arts is shown in Table I. The proposed converter achieves a higher efficiency, a wider output range and much faster dynamic performance than prior arts with a higher voltage rating.

978-1-5090-0636-6/16 $31.00 © 2016 IEEE

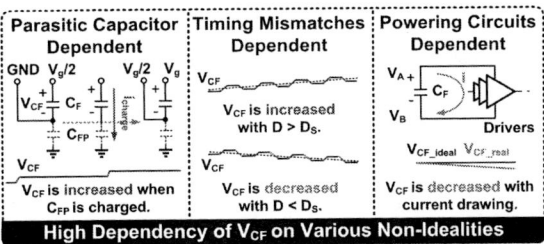

Acknowledgment

This work was supported in part by a grant from the Research Grant Council of Hong Kong SAR, under Project 16207014.

References

[1] T. A. Meynard, et al., *IEEE TIE*, pp. 356-364, 1997.
[2] C. K. Tse, et al., *Proc. IEEE*, pp. 768-781, 2002.
[3] G. Villar, et al., *IEEE PESC*, pp. 4229-4235, 2008.
[4] W. Kim, et al., *IEEE JSSC*, pp. 206-219, 2012.
[5] P. Kumar, et al., *IEEE CICC*, 2015.
[6] L. Cheng, et al., *IEEE JSSC*, pp. 2778-2799. 2014.

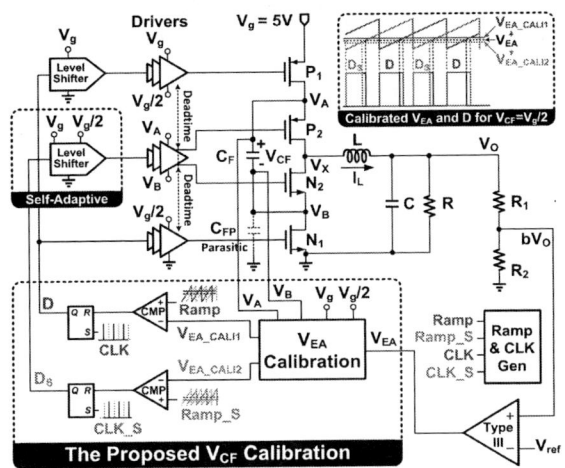

Fig. 1 V_{CF} variations and system block diagram.

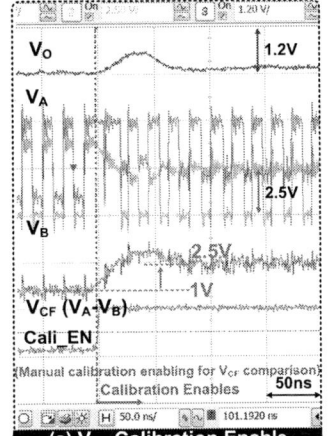

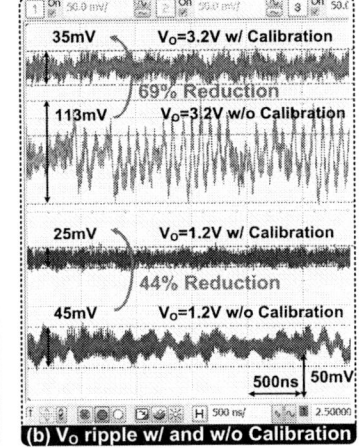

Fig. 4 Measured waveforms of (a) V_{CF} (b) ripple voltage.

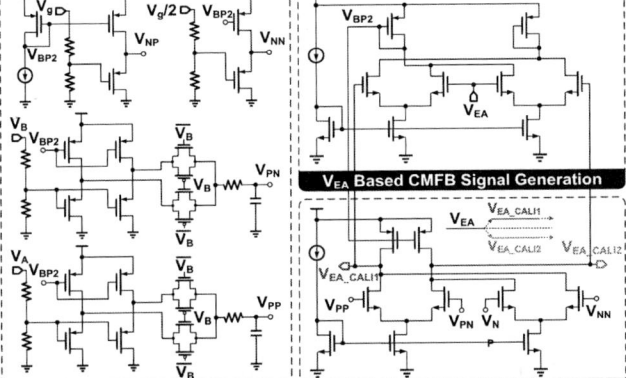

Fig.2. Circuitry of the proposed V_{CF} real-time calibration.

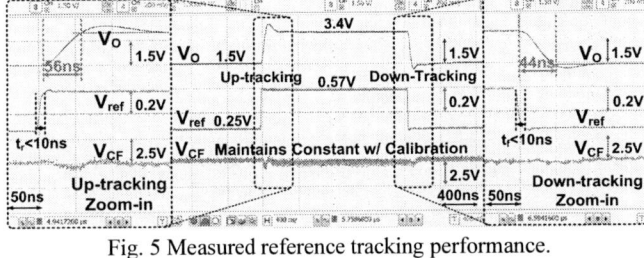

Fig. 5 Measured reference tracking performance.

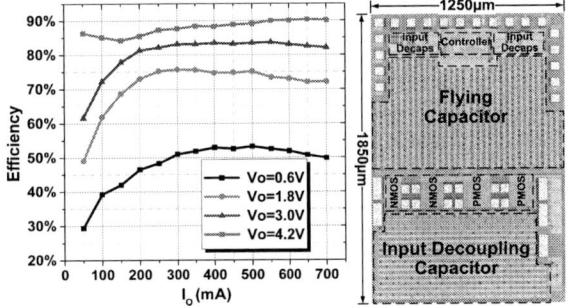

Fig. 6 Measured efficiency.   Fig.7. Chip Photo.

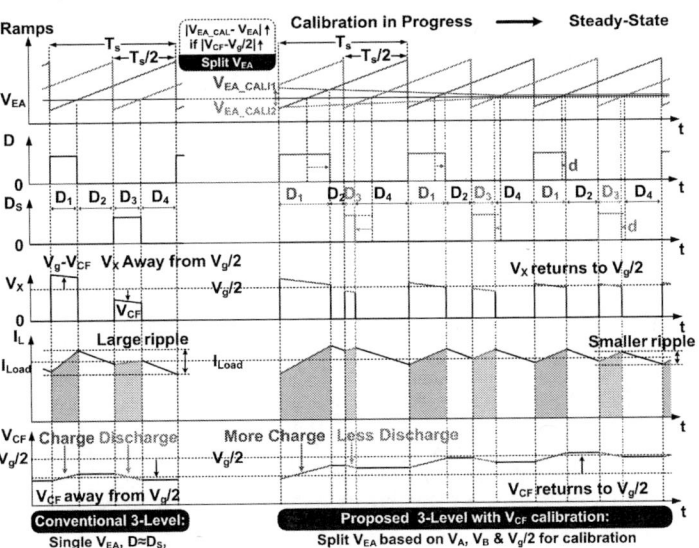

Fig.3. Timing diagram of the proposed V_{CF} real-time calibration.

TABLE I. COMPARISON WITH PRIOR ART

Publication	[3] PESC'08	[4] JSSC'12	[5] CICC'15	[6] JSSC'14	This Work
CMOS Tech.	250 nm	130 nm	22 nm	130 nm	65 nm
Topology	3-Level	3-Level	3-Level	Buck	3-Level
Max. P_{OUT}	0.1W	1W	0.25W	3.6W	3W
Freq.	37.5MHz	50-200MHz	250MHz	30MHz	50MHz
C_{FLY}	5.07nF	18nF	5nF (MIM Only)	N/A	5nF
C_{OUT}	25.09nF	10nF	5nF	1µF	10nF
L	26.7nH	1×4nH	1.5nH	330nH	100nH
V_g	3.6V	2.4V	3.3V	3.3V	5V
V_O	1V	0.4~1.4V	0.4~1.2V	0.45~2.4V	0.6~4.2V
V_O-Range/V_{IN}	28%	42%	53%	59%	72%
Max. Efficiency	69.7%	77%	72%	86.6%	90%
Up-tracking	N/A	Open-loop	286ns/V	670ns/V	29ns/V
Down-tracking			N/A	1560ns/V	23ns/V
V_{CF} Calibration	No	No	Once 32 cycles	N/A	Real-time calibration

95% Light-load Efficiency Single-Inductor Dual-Output DC-DC Buck Converter with Synthesized Waveform Control Technique for USB Type-C

Wen-Hau Yang[1], Chiun-He Lin[1], Ke-Horng Chen[1], Chin-Long Wey[1], Ying-Hsi Lin[2], Jian-Ru Lin[2], Tsung-Yen Tsai[2], and Jui-Lung Chen[3]

[1]National Chiao Tung University, [2]Realtek Semiconductor Corp., [3]Vanguard Semiconductor Corp, Hsinchu, Taiwan

Abstract

The proposed single-inductor dual-output (SIDO) converter can provide wide range in duty ratio control to convert input voltage 5-20V to dual output voltages 3.3V and 1.2V when its switching frequency is raised to 10MHz for compact size solution. The proposed synthesized waveform control (SWC) technique can emulate the inductor current without being affected by switching noise. Thus, the minimum allowable duty ratio can be lowered to 6% to meet the requirement of USB-C in one-stage low duty ratio conversion. Moreover, the switching frequency is dynamically decreased by the derived DC loading information from the SWC technique. Not only the output power MOSFET but also the main power MOSFET switch in a load-dependent switching frequency for power saving. 67% more power reduction can be obtained. 95% and 83% efficiency are achieved at light and heavy loads, respectively, when the silicon is limited within 1400μm*1350μm.

I. Introduction

A USB link delivering 4.5-W (900 mA at 5 V) to 100-W (5 A at 20 V) induces a potential for smartphones and tablets to be quickly charged by directly using USB with high power delivery (PD) specification revision 2.0. Design challenge is voltage conversion ratio from 6% (V_{IN}=20V and V_{OUT}=1.2V) to 66% (V_{IN}=5V and V_{OUT}=3.3V). Fig. 1 shows a conventional two-stage structure in which first and second stages respectively convert the USB input voltage (5-20V) to 4.5V and then to 3.3V and 1.2V. Obvious drawbacks are low efficiency and high cost. Thus, single-inductor dual-output (SIDO) converter can be employed to convert wide range input voltage of 5-20V to 3.3V and 1.2V directly [1] - [4]. As we know, compact size SIDO converter can be achieved by using one small 1μH inductor if it operates at the switching frequency f_S higher than 10MHz. However, less than 100ns switching period T_S is allowable for controlling two outputs. Thus, it is hard to fully turn on the power MOSFET in the worst case of conversion when converting from 20V to 3.3V and 1.2V where the duty ratios D_1 and D_2 of the two outputs are 16.5% and 6%, respectively.

High switching frequency causes large switching noise as shown in Fig. 2. Large switching noise will cause large perturbation at D_1 and D_2 values. Correspondingly, the output voltage ripples increase. How to accurately derive D_1 and D_2 becomes more difficult in high switching operation and low duty ratio conditions. In this work, accurate D_2 for guaranteeing regulated 1.2V from 20V is a challenge because noise disturbing region t_{d2} causes large perturbation at D_2. Furthermore, high switching operation induces low efficiency at light loads. In Fig. 3, although the prior arts with the function of varying the switching frequency of each output switch [3] or with skipping function [2] can reduce the switching frequency of each output switch, the maximum power reduction is only 33% because the main switches M_H and M_L occupy 67% of the entire switching power loss in one switching period. It is necessary to further reduce the switching frequency of M_H and M_L according to output loading condition.

II. Proposed SIDO Converter with the SWC Technique

The developed SIDO converter in Fig. 4 has a synthesized waveform controlled (SWC) technique, which can adjust the system frequency by the valley DC extraction (VDCE) circuit and synthesize the AC inductor current by fast AC emulator (FACE) circuit. The VDCE circuit generates V_{DC} from the DC value of inductor current to adaptively adjust the power switching frequency of M_H and M_L. At light loads, the variable switching frequency is adopted to reduce the value of CLK and to enhance light-load efficiency. As a result, 67% power loss can be decreased. Besides, the skip energy control (SEC) technique in [2] manipulates error signals V_{E1}-V_{E2} and generates

energy control signals V_{C1}-V_{C2} to gain the advantage of power loss reduction of each output switch. All power switches can have a suitable switching frequency corresponding to its loading. The FACE circuit rebuilds the inductor AC current ripple by detecting the two terminals of the inductor V_{X1} and V_{X2} to get rid of switching noise.

III. Circuit Implementations

A. VDCE circuit

In Fig. 5(a), the DC inductor current information V_{DC} is generated by the replica of M_{senL} from the low-side MOSFET M_L where an operational amplifier forces the source voltages of M_L and M_{senL} equally. The sample-and-hold (S/H) switch acquires the correct inductor dc information and eliminates the noise coupling from the switch node V_{X1} before M_L turns off. Because the DC inductor-current information is introduced, the compensation network is simplified to proportional-integral (PI) compensator, which contains only a resistor and a capacitor. In Fig. 5(b), when the V_{DC} is smaller the V_{trig}, the voltage-controlled oscillator (VCO) adaptively generates an appropriate clock signal CLK determined by the V_{DC} to decrease switching power loss at light loads. On the other hand, at normal and heavy loads, the VCO generates constant CLK according to one fixed value of V_{fix} for low voltage ripple.

B. FACE circuit

In Fig. 6, the proposed FACE circuit releases the noise coupled from the switch node, V_{X1}, and thus obtains an almost noise-free current ripple by integration. When high-side switch M_H turns on, V_{X1} and V_{X2} are V_{IN} and V_{Oi} (i=1, 2), respectively. Two dynamic-basing folded flipped voltage follower (DFVF) circuits can convert the voltage difference between V_{IN} and V_{Oi} (i=1, 2) to the emulated inductor current V_{ac} with positive slope. On the other hand, V_{X1} and V_{X2} are 0 and V_{Oi} (i=1, 2), respectively, when low-side switch M_L turns on. Similarly, the emulated inductor current V_{ac} with negative slope is derived. Moreover, the capacitor C_{ac} and the switch M_{sw} adopted in FACE can effectively filter out the switching noise. The switch, M_{sw}, which turns on shortly before the end of period in every cycle, resets the voltage on the capacitor to prevent the noise from being accumulated. The DFVF circuit is dynamically biased by the current source I_B which is function of V_{DC} to ensure low duty ratio of 6% at heavy loads. Wide input common mode range (ICMR) and low output impedance ensures high linearity and accuracy.

IV. Experimental Results

Fig. 7(a)-(d) show waveforms of high duty ratio conversion and low duty ratio conversion. All outputs can be properly regulated with V_{IN} equal to 5V and 20V. Low duty ratio of 6% at the output of 1.2V is realized by the SWC technique in Fig. 7(a). In Fig. 7(c), improved light-load efficiency by reducing system frequency when I_{O2}=400mA and I_{O1} changes from 500 to 150mA where the system frequency rapidly decreases from 10MHz to 8MHz. Maximum output ripple is kept below 20mV. Fig. 8 shows that the developed SWC SIDO achieves the 95% of light-load efficiency when the system frequency is adaptively adjusted from 10MHz to 0.2MHz. In contrast, the light-load efficiency of conventional SIDO is less than 70%.

In Fig. 9, the CCAH control technique in [5] varies its switching frequency f_{sw}. Unfortunately, the change of switching frequency is opposite to the change of output loading. At heavy loads, decreased f_{sw} causes large output voltage ripple. Green mode (GM) in [2] uses the burst mode function to keep efficiency but its f_{sw} keeps constant. It still faces large output voltage ripple. In contrast, the SWC technique decreases the f_{sw} for high light-load efficiency and keeps small output voltage ripple simultaneously. Table I summarizes the performance of the developed SWC SIDO and the prior arts with the light load control mechanisms. Results show that the developed SWC SIDO is the only solution which achieves the lowest conversion ratio of 6% with high efficiency for requirement of the USB PD specification revision 2.0. Fig. 10 shows the SWC chip micrograph.

References

[1] Danzhu Lu, et al., *ISSCC Dig. Tech. Papers*, pp. 82-83, Feb. 2014.
[2] Yi-Ping Su, et al, *ASSCC Dig. Tech. Papers*, pp. 65-68, Nov. 2014.
[3] Min-Yong Jung, et al., *ISSCC Dig. Tech. Papers*, pp. 309-311, Feb. 2015.
[4] Yi-Ping Su, et al., *ISSCC Dig. Tech. Papers*, pp. 312-314, Feb. 2015.
[5] Xiaocheng Jing, et al, *IEEE JSSC*, pp. 2350 – 2362, Oct. 2011.

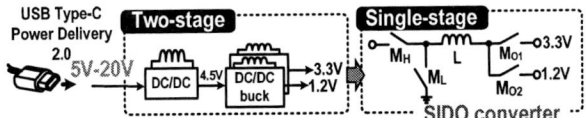

Fig. 1. Comparison of power conversion from USB Type-C to internal circuits by two-stage and single-stage SIDO convertor.

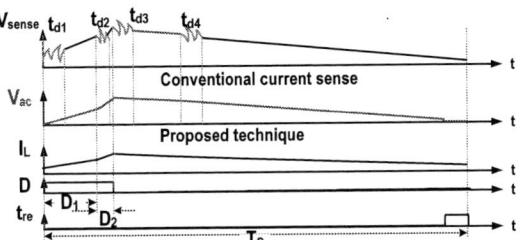

Fig. 2. Switching noise causes the minimum allowable duty cycle.

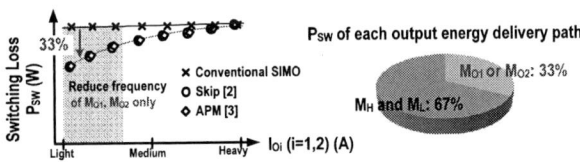

Fig. 3. Reduced switching frequency at each output switch can have only 33% power reduction because the main power switches M_H and M_L still operate at high switching frequency.

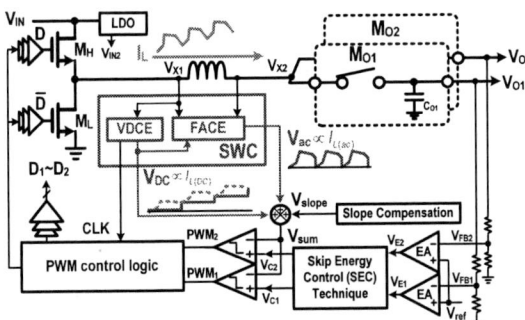

Fig. 4. Architecture of the proposed SIDO DC-DC buck converter with the SWC technique.

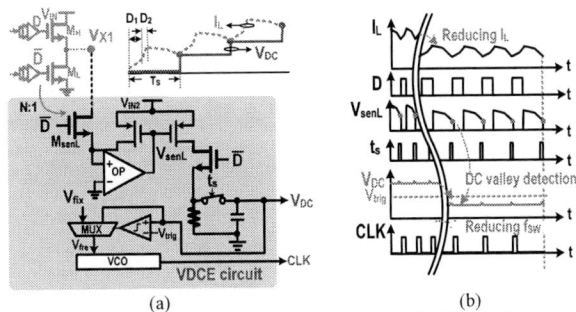

Fig. 5. (a) Circuit implementation of VDCE circuit. (b) Timing diagrams of VDCE circuit.

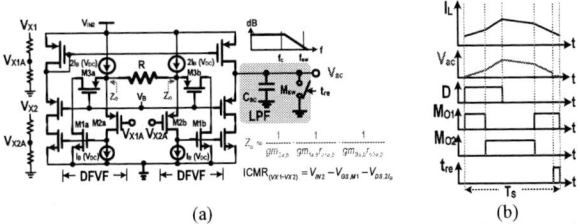

Fig. 6. (a) Proposed FACE circuit can speed up the ac emulation current by dynamic biasing current controlled the V_{DC} from the VDCE circuit. (b) Timing diagrams

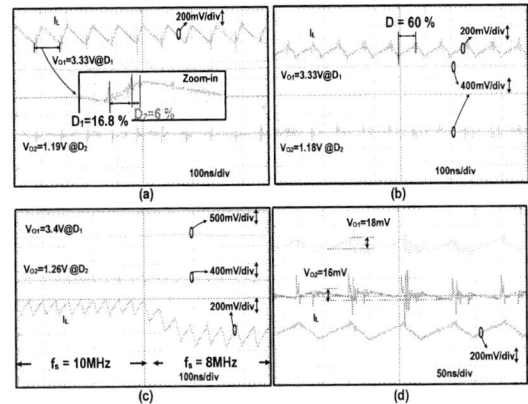

Fig. 7. Measurement results. (a) V_{IN}=20V, I_{O1}=255mA, and I_{O2}=650mA. (b) V_{IN}=5V, I_{O1}=650mA, and I_{O2}=140mA. (c) V_{IN}=5V, I_{O2}=400 mA, and I_{O1}=500-150mA. (d) V_{IN}=5V, I_{O1}=450 mA, and I_{O2}=250 mA.

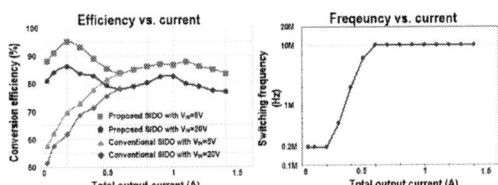

Fig. 8. Performance statistics.

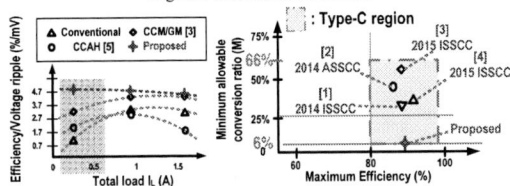

Fig. 9. Comparison with state-of-the-art literatures.

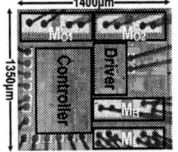

Fig. 10. Chip micrograph.

Table I: Specifications and comparison with the prior arts.

	Proposed	[2]	[3]	[5]
Adopt light load control	SWC	GM	APM	CCAH
Topology	2 Buck outputs	4 Buck outputs	10 Buck outputs	2 Boost outputs
Input voltage (V)	5V~20V	2.7V~5.5V	5V	1.8V~2.4V
Output voltage (V)	1.2V and 3.3V	0.6V~5V	N.A.	3V~3.6V
Max. efficiency	95%	85.2%	88.7%	91.6%
Ripple (mV)	<20	<40	<40	<200
Cross regulation (mV/mA)	0.087	0.0432	0.1	0.0714
Load regulation (mV/mA)	0.11	0.02	0.17	N.A.
Switching frequency	10 MHz @1.5A, 0.2MHz @0.2A	1.1 MHz	0.7 MHz	1 MHz

A Reconfigurable SIMO System with 10-Output Dual-Bus DC-DC Converter using the Load Balancing Function in Group Allocator for Diversified Load Condition

Se-Un Shin, Min-Yong Jung, Ki-Duk Kim, Sang-Hui Park, Yeunhee Huh, Changsik Shin, Se-hong Park, Jun-Suk Bang, Jong-Beom Baek, Sung-Won Choi, Yong-Min Ju, and Gyu-Hyeong Cho

KAIST, Daejeon, Korea Email: sswsin@kaist.ac.kr

Abstract

This paper presents a reconfigurable SIMO system with 10-output dual-bus DC-DC converter having two buses for heavy and light load outputs, respectively. The converter controls the load condition for each bus to be well balanced. Under diversified load condition, a group allocator assigns each output to the corresponding bus properly depending on its load current. Due to such load balancing function, severe regulation issues which could occur under diversified load condition are resolved with output voltage ripples below 25mV, and over 81% efficiency is achieved under wide range of load (0-300mA).

Introduction

In the field of power management IC, the reduction of external component cost and PCB complexity is a trend. For such purpose, a single-inductor multiple-output converter (SIMO) was successfully developed with comparator-based control recently [1-2]. However, the SIMOs still have technical hurdles for realistic mixture of heavy and light loads, i.e. diversified load condition as depicted in Fig. 1, which frequently occurs in mobile applications where some blocks are in active mode and others are in idle and where loads for the blocks vary in real time. In such a situation, for a SIMO using comparator-based control with fixed frequency, some outputs could fail in regulation for very light load because of too-short on-duty period ($D_{O1,DESIRED}$) for output switching [3], and the efficiency significantly degrades due to the switching loss that is dominant under light load condition as shown in Fig. 2. To handle these issues for the light load output, adaptive pulse modulation (APM) [4] could be utilized, which skips adaptively the switching of light load outputs. However, the skipping of APM introduces non-periodic switching of heavy load outputs which causes poor cross regulation and generates large ripple when load currents are diversified as shown in Fig 2 [3]. Moreover, the resultant EMI for heavy load powering becomes hard to filter out and this could cause serious problems for mobile applications. Besides, the huge amount of discontinuous current flow of the SIMO under diversified load condition generates the switching noise due to the parasitic inductance especially at the light load output as shown in Fig. 3. It causes additional regulation problem and adds supply voltage stress to the loading blocks because switching noise voltage (V_{SWN}) becomes much larger than that of the output ripple. In order to prevent such switching noise problem and ensure stable powering, light loads must be separately regulated.

In this paper, to overcome the aforementioned issues, a SIMO system with dual-bus multiple-output (DBMO) topology is proposed as

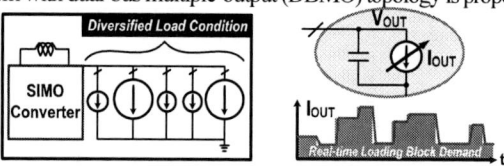

Fig. 1 SIMO converter under diversified load condition.

● Issues of Comparator-based Control w/ Light Loads

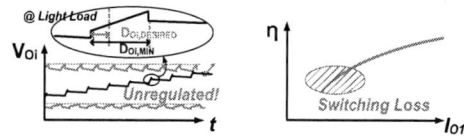

● Issues of APM Control or Pulse Skip w/ Heavy Loads

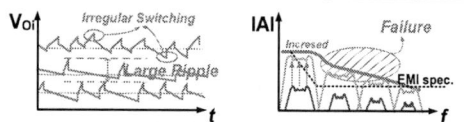

Fig. 2 Issues of SIMO converter under diversified load condition.

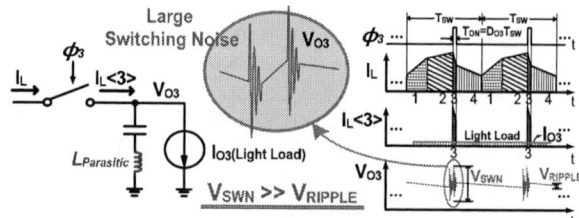

Fig. 3. The switching noise problem of SIMO converter.

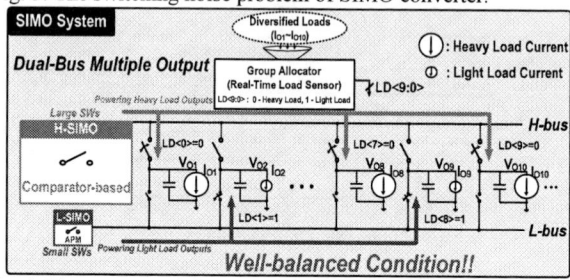

Fig. 4. The conceptual diagram of DBMO converter.

in Fig. 4. The DBMO has two buses (H-bus/L-bus) for heavy/light load outputs connected to H-SIMO/L-SIMO, respectively. The group allocator (GA) properly assigns each output to either the H-bus or the L-bus while balancing the load condition for each bus properly. Therefore, the on-duty period of the light load output is guaranteed well by the L-bus, eliminating the regulation issues. As heavy load switching becomes periodic in the H-bus, the EMI problem is also eased because a single tone is easy to filter out. Moreover, owing to the optimal sizing of switches, the size of L-SIMO becomes much smaller than that of H-SIMO, resulting in only small additional chip area. Above all, because of the reconfigurable structure, DBMO is capable of always operating near the peak efficiency region.

Proposed Dual-Bus Multiple-Output Topology (DBMO)

Fig. 5 shows the overall block diagram of the DBMO with 10 outputs. The H-bus and the L-bus are the power lines coming from H-SIMO and L-SIMO, respectively. GA periodically estimates all ten load currents by processing the two quantities, the duty cycles of the output switches, and the inductor currents. GA judges whether the load is heavy or light, generating a 10-bit table ('LD<9:0>') – '0' means heavy and '1' means light. According to the table, heavy and light loads are distributed to H-SIMO and L-SIMO, respectively, in real time. Each output error current, reflecting each output error ($V_{REFi}-V_{Oi}$), is generated by the shared gm-cells in a global controller and is allocated to H-SIMO or L-SIMO depending on the value of LD<i>. The sum of heavy load error currents with LD<i>=0 ($I_{ES,H}$) is utilized for controlling the H-SIMO. Main duty for the H-SIMO is generated so that $I_{ES,H}$ becomes zero by using the sawtooth voltage V_{SAW} and the control voltage V_P which is used to fix the output switching frequency $\Phi_{O,SW}$. Single-boundary comparator is used for the main duty of H-SIMO, and it easily implements fixed frequency control with a well-defined spectrum having fast transient response and low cross regulation as well [1-2]. Meanwhile, each output of H-SIMO operates under comparator-based control by simply detecting the polarity of I_{ERRi} with LD<i>=0. On the other hand, for the L-SIMO, APM [4] is used, and the sum of light load error currents with LD<i>=1 ($I_{ES,L}$) is utilized for the control of L-SIMO. The main duty is generated so that $I_{ES,L}$ becomes zero in the peak-current mode. The APM is power-efficient for light load outputs because this method monitors each output error current I_{ERRi} with LD<i>=1 and supplies inductor current to the most urgent output with maximum error selector, resulting in variable output switching frequency. Besides, the switches of L-SIMO are smaller than

978-1-5090-0636-6/16 $31.00 © 2016 IEEE

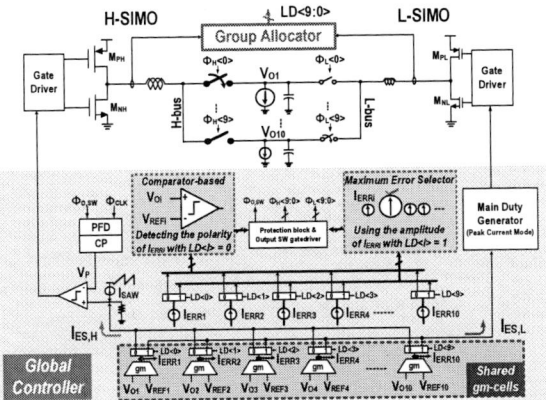

Fig. 5. The overall structure of the DBMO converter

those of H-SIMO in size in our DBMO architecture because the switches for heavy loads and light loads are separately designed. Since the outputs change their buses according to their load condition thanks to the reconfigurable structure, outputs assigned to their suitable buses enable to achieve high efficiency under diversified load condition.

Group Allocator (GA)

Fig. 6 shows the structure of the proposed GA, which senses the inductor currents of H-SIMO and L-SIMO indirectly using OTA_H and OTA_L with series connected resistor $R_{F,H/L}$ and capacitor $C_{F,H/L}$ in parallel to each inductor. The set of $R_{F,H/L}$ and $C_{F,H/L}$ has the same time constant with the inductor itself in order to extract the emulated inductor current. On the other hand, to sense the target load current, another indirect method is also used. It is done by placing small switches (SW_{Hi}', SW_{Li}') in the controller at the outputs of OTA_H and OTA_L so that the small switches correspond to the main switches of H-SIMO (SW_{Hi}) and L-SIMO (SW_{Li}), respectively, with the same number. For example, if V_{O1} is selected as the target output in Fig. 6 while H-SIMO operates, SW_{H1}' in the controller is on and off in synchronization with SW_{H1}. It is repeated continuously for multiple cycles with a solely operating switch SW_{H1}' while the other switches in the controller are all in off-state. Note that the main switches normally operate to regulate the multiple outputs in this case, and the switching states of the main switches are not the same as those of the small switches where only one switch is operating. As a result, the pieced current I_{PIECE} is supplied to C_{FILT} repetitively for multiple cycles through one switch only when the corresponding main output switch is on-state. Meanwhile, the buffer configured OTA_F discharges C_{FILT}, and this buffered current indirectly indicates the target load current. The buffer current is smoothed with low ripple using C_{FILT} and the limited bandwidth of the buffer. Then, this current is amplified ten times; the resultant current I_{SEN} is compared to a threshold current I_{BD} using the hysteresis current comparator and judged to update the heavy/light result table (in this example, LD<0>). The hysteresis current comparator is adopted because it is used when two currents, I_{SEN} and I_{BD}, are in similar level, eliminating unstable operation. After the judgment of LD<0>, GA repeats this operation to the other target output one by one. Since the GA has only one load current sensor for both SIMOs, the proposed method is simple and power-efficient.

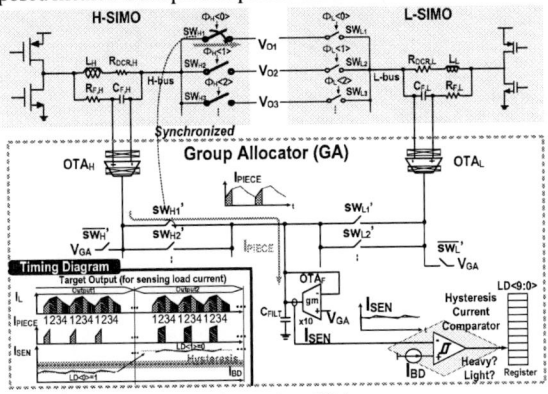

Fig. 6. Circuit implementation and operation of GA.

Measurement Results

Waveforms in Fig. 7(a) verify the motivation of this work by showing regulation problems of a conventional SIMO converter under diversified load condition. The left of Fig. 7(a) shows the case of comparator-based control with very light load output, and it results in regulation failure. The right of Fig. 7(a) shows the case of APM control under diversified load condition, and the large ripple of the outputs is due to the steady state cross regulation caused by irregular switching.

The proposed 10-output DBMO converter is fabricated in a 1P4M 0.18µm CMOS process. Fig. 7(b) shows the regulation performance with the measured voltage ripple waveforms for 6 outputs; the left 3 outputs in H-SIMO and the right 3 outputs in L-SIMO. Due to the load balancing function, all the ripple values are below 25mV for both SIMOs. In Fig. 8(a), the bus change operation for the output V_{O6} is shown: the left is for the case from the L-bus to the H-bus and the right is for the case reverse versa. The difference level of V_{O6} between in the H-bus and the L-bus is due to the DC error-voltage [1] which the comparator control originally has and the load regulation. This meets the target specs. Undershoots/overshoots of the output waveforms are caused by cross regulation due to the status change of each bus, resulting in lower than 0.1 (mV/mA) for both SIMOs. Fig. 8(b) shows I_{SEN} sensed by the GA from the load current I_{O1}. Fig. 9(a) shows the efficiency plot versus load current I_{O7}. The proposed SIMO system implemented with the DBMO converter supports a load balancing function for the first time under diversified load condition from 0mA to 300mA. Moreover, over 81% efficiency was achieved throughout the whole range of load thanks to the reconfigurable structure which allocates the output to L-SIMO or H-SIMO appropriately by utilizing a simple indirect sensing of the load current.

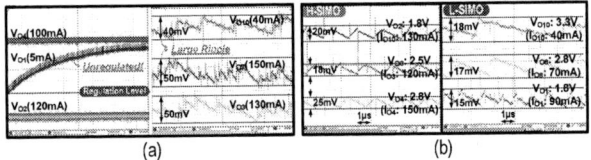

Fig. 7. (a) The regulation problems in conventional SIMO converters and (b) DBMO converter under diversified load condition.

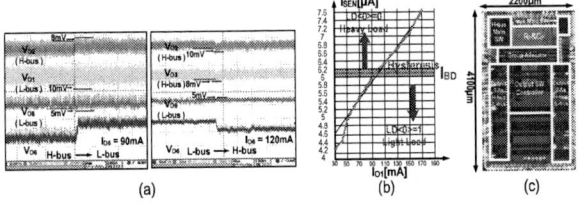

Fig. 8. (a) The bus change of V_{O6} from H-bus to L-bus and from L-bus to H-bus, (b) the plot of the sensed current by the GA (I_{SEN}) versus its target load current (I_{O1}) and (c) chip micrograph.

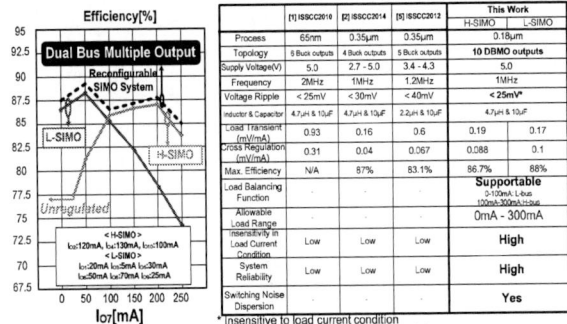

Fig. 9. (a) The measured efficiency plot of DBMO converter versus load current (I_{O7}) and (b) The comparison table with previous works.

Reference

[1] K-C. Lee, et al., ISSCC, pp. 200-201, Feb. 2010.
[2] Danzhu Lu, et al., ISSCC, pp. 82-83, Feb. 2014.
[3] Y-P. Su, et al., ASSCC, pp. 65-68, Nov. 2014.
[4] M-Y. Jung, et al., ISSCC, pp. 1-3, Feb. 2015.
[5] C-W. Kuanand, et al., ISSCC, pp. 274-276, Feb. 2012.

A Microcontroller with 96% Power-Conversion Efficiency using Stacked Voltage Domains

Kristof Blutman[1], Ajay Kapoor[1], Arjun Majumdar[1], Jacinto Garcia Martinez[1], Juan Echeverri[1],
Leo Sevat[1], Arnoud van der Wel[1], Hamed Fatemi[1], José Pineda de Gyvez[1] and Kofi Makinwa[2]

[1]NXP Semiconductors, Eindhoven, the Netherlands   [2]Delft University of Technology, Delft, the Netherlands

Abstract

This paper presents a CMOS 40nm microcontroller where for the first time, stacked voltage domains are used. The system features an ARM Cortex M0+ processor, 4kB ROM, 16kB SRAM, peripherals, and an on-chip switched-capacitor voltage regulator (SCVR). By using voltage stacking the test chip achieves state-of-the-art (96%) power-conversion efficiency and observed power savings run from 23% to 63% depending upon the payload current, while supply voltage variations are reduced from 5.6mV to 3.8mV (RMS).

Introduction

Voltage stacking [1] aims to reduce voltage conversion losses that are often a significant fraction of the power drawn by the load itself. The series connection of voltage domains allows operating current to be recycled, in contrast to conventional "flat" (parallel connected) voltage domains in which all current must pass through the system's power converter. Voltage stacking has been shown to confer substantial benefits for the case of independent disconnected systems [2]. But this is not representative of practical applications. Other works present only simulations [3], or the stacking of simple circuit blocks like multipliers [1]. In this paper, we present the first microcontroller with stacked voltage domains at the IP level.

Implementation

As shown in Figure 1, the system supports two operating modes: the conventional flat mode in which the logic and the memory are connected in parallel, and the stacked mode, in which the memory is connected in series with the logic. In stacked mode, the logic is between 1.1V and 0V, while the memory is between 2.2V and 1.1V. In both modes, the fully-integrated SCVR regulates the 1.1V node. In stacked mode, the current drawn by the memory is directly re-used by the logic in the bottom voltage domain, bypassing the SCVR. As a result, the re-used current is not subjected to power conversion losses, therefore improving power delivery efficiency [4]. Without loss of generality, we will assume that the input power rail is at $2V_{DD}$ simply to show the potential of voltage stacking. The down-conversion from a typical battery voltage to $2V_{DD}$ can be accomplished in several ways and falls outside the scope of this paper.

Figure 2 depicts the system block diagram. An external 2.2V supply powers the I/O pads and the SCVR, while the latter generates the 1.1V supply for the rest of the system. The block diagram also denotes the voltage domain partitioning. The bottom voltage domain includes the ARM Cortex M0+ core and the peripherals that run at the 80MHz external clock frequency. The 4kB ROM and the 16kB SRAM are placed in the top voltage domain. The two domains communicate via level shifters. These are custom cells that can establish connection between signals ranging from 1.1V to 0V and from 2.2V to 1.1V. To facilitate routing, their layout is compatible with standard digital cells. Despite their wide I/O voltage range, the level-shifters were realized with *standard* thin-oxide transistors, and their reliability was verified by extensive simulations. To allow its supply- and ground voltages to be arbitrarily set, the top voltage domain is located in a deep n-well.

Power savings from voltage stacking depend on how well the supply currents of the two stacked domains are balanced [5]. Figure 3 displays the post-synthesis power breakdown of the design for one typical application. Notice that memory and logic consume about the same power. This is the reason why we decided to place the Cortex-M0+ core and the rest of the peripherals in the bottom domain, while the ROM and the SRAM in the top domain. The choice of memory on top of logic is also the result of system architecture

considerations. Namely, in active mode the core always needs to read instructions from the memory. Consequently, memory and core power consumption are correlated. Our post-synthesis power analyses not only revealed a well-balanced current consumption, but also allowed less area overhead due to careful control of the level shifter instance count (13% core area).

Figure 4 shows the 2:1 SCVR. It consists of 16 non-interleaved converter instances. In flat mode, all instances are operated at a 10MHz switching frequency, while in stacked mode, only 6 instances are operated at a reduced 5MHz switching frequency, since less output power is required from SCVR. This is because most of the supply current is re-used and does not need to be delivered by the SCVR. Each instance employs a voltage-halving architecture [6] with a single 41pF flying capacitor, and is able to source or sink current from the 1.1V mid-node as required. Power switches are driven by non-overlapping clock signals derived from an external clock input. The SCVR re-uses the same level shifters as the core.

Measurement Results

The SCVR efficiency is presented in Figure 5. In flat mode, the SCVR load current is always positive as it is the sum of the logic- and memory currents, while in stacked mode, the SCVR load current can be either positive or negative, since it is the difference of these two currents. The SCVR has 81% peak efficiency for both the current sink and -source scenarios. Power conversion losses are mainly due to the switching losses associated with the periodic charging and discharging of the bottom-plate capacitances of the flying capacitors. This form of loss is relatively independent of the load and is mainly responsible for the reduced efficiency at low load currents.

Note that with voltage stacking a small load current imbalance yields a high system power efficiency (even though the SCVR itself becomes less efficient at low output currents), because most of the current in the top domain flows into the bottom one. This efficiency improvement can be seen in Figure 6a where the system's power efficiency in stacked mode is 96%, thus exceeding the SCVR's 81% maximum efficiency. This means that the benefit from voltage stacking is more significant than the reduced SCVR power efficiency at low load currents. The µA/MHz figure of merit is shown in Fig. 6b for both stacked- and flat modes as a function of the core's running frequency. The measurements were conducted on 10 samples running a selected matrix multiplication test program. The SCVR was most efficient at high loads (80MHz core frequency). It is evident that for low frequencies the stacked mode outperforms the flat mode by more than 2x, even after accounting for the power overhead from the level shifters. Running at the 80MHz nominal core frequency, the current drops by 23%: from 10.6µA/MHz in flat mode to 8.1µA/MHz in stacked mode. Figure 6c shows power efficiency as a function of core frequency. It can be seen that even when the flat mode efficiency is near 50%, the efficiency in stacked mode remains above 80%.

Over 400 test programs were run in both stacked- and flat modes to characterize all the possible current balancing scenarios. These consisted of common benchmark algorithms like FFTs, FIR filters, matrix multiplications and sorting, as well as stress tests that loaded memory and logic in an unequal way. Despite the variety and number of the programs, the current of the two voltage domains stays reasonably balanced, supporting the memory-on-logic partitioning choice, as indicated by the current profiles in Figure 7.

Figure 8 shows the supply voltage waveforms and the mid-node supply variation in both power modes while executing an FFT test program. In flat mode, 16 instances of the SCVR, switched at 10MHz, are used to process current $I_{BOT}+I_{TOP}$. This results in 1.042V mean supply voltage and 5.6mV RMS variation. In stacked mode, only 6 SCVR instances are turned on at 5MHz to process current I_{BOT}-I_{TOP}. This yields 1.064V mean mid-node voltage at 3.8mV RMS variation. Hence in stacked mode, the load requirements for the SCVR are relaxed. Figure 9 shows a microphotograph of our chip.

Acknowledgements

The authors would like to thank Rina Lim for her help in realizing the level shifters.

References

[1] S. Rajapandian *et al.*, ISSCC, pp. 298-299, Feb., 2005.
[2] S. Lee *et al.*, VLSI Symp., pp. 318-319, June, 2015.
[3] E. Ardestani *et al.*, TACO Vol. 12 Issue 4, December 2015
[4] Y. Liu *et al.*, ISSCC, pp. 400-401, Feb., 2013.
[5] K. Ueda *et al.*, JSSC, vol. 48, pp. 2608-2617, Nov. 2013.
[6] L. Chang *et al.*, VLSI Symp., pp. 55-56, June, 2010.
[7] B. Zimmer *et al.*, VLSI Symp., pp. 316-317, June, 2015.

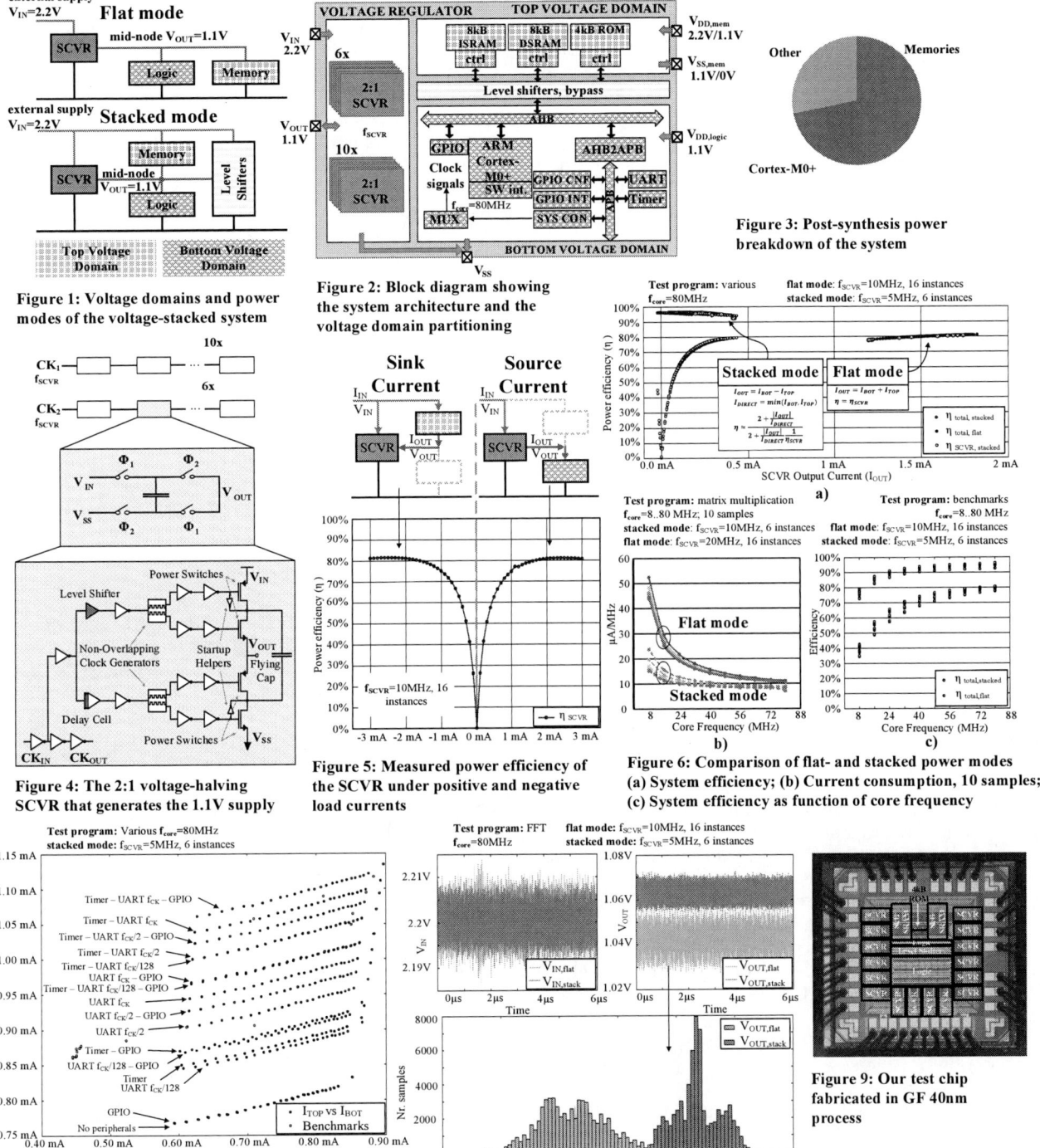

Figure 1: Voltage domains and power modes of the voltage-stacked system

Figure 2: Block diagram showing the system architecture and the voltage domain partitioning

Figure 3: Post-synthesis power breakdown of the system

Figure 4: The 2:1 voltage-halving SCVR that generates the 1.1V supply

Figure 5: Measured power efficiency of the SCVR under positive and negative load currents

Figure 6: Comparison of flat- and stacked power modes (a) System efficiency; (b) Current consumption, 10 samples; (c) System efficiency as function of core frequency

Figure 7: Measured current balance between flat- and stacked power modes running various benchmark programs and stress tests

Figure 8: Measured supply noise in flat- and stacked mode

Figure 9: Our test chip fabricated in GF 40nm process

A Fast, Flexible, Positive and Negative Adaptive Body-Bias Generator in 28nm FDSOI

Milovan Blagojević[1,2,3], Martin Cochet[1,2], Ben Keller[2], Philippe Flatresse[1], Andrei Vladimirescu[2,3], and Borivoje Nikolić[2]

[1]STMicroelectronics, Crolles [2]Dept. of EECS, University of California, Berkeley [3]Institut Supérieur d'Électronique de Paris

Abstract

This work demonstrates a fully-integrated, compact body-bias generator (BBG) with a fine voltage step and sub-100ns response time for use in process and voltage compensation as well as dynamic energy optimization. The generator is implemented in 28nm UTBB FDSOI, using only 1.0V core and 1.8V IO voltage inputs. A modular design enables easy integration into target mobile SoCs, scalable to power domains of any size. The fine resolution (5mV V_{th}), 100ns full-scale and 5ns incremental step response, low power (<10μW), and 1.2% area overhead enable fine-grained adaptive body-biasing (ABB). The ability to dynamically track a target frequency within 1% for 200mV of V_{CORE} change is demonstrated experimentally.

Introduction

The ultra-thin body and box (UTBB) FDSOI technology features a high sensitivity (>70mV/V) of transistor threshold voltages to body bias. In contrast to adaptive voltage scaling (AVS), enabling adaptive body-bias (ABB) require only two additional power grid lines without substantial consideration of IR drop. The load for body biasing is almost purely capacitive (see Figure 1), with low static current – under 5μA/mm² worst case. Optimal body bias can be used for run-time compensation of process, aging, temperature and supply-voltage variations, or to improve energy efficiency [1], but a compact, low power, fast BBG is required. This work implements a fully-integrated switched-capacitor solution that can be integrated on any SoC with low overhead to enable fast, fine-granularity threshold control.

IP Implementation

Figure 2 shows the two main building blocks of the BBG. The driver unit receives four 1-bit digital commands from the control unit to toggle on or off the charging and discharging of the transistor wells (see Figure 3). The nwell drivers are pMOS and nMOS power switches that provide 0V-1.8V range and high slew rates. Figure 4 shows the pwell negative voltage generation, which is based on a switched-capacitor (SC) charge-pump in contrast to current-source-based solutions in bulk CMOS [2]. The driver implements a 1:-1 SC charge pump with the negative bootstrap circuit for the negative gate drive signal G_{bot} [3]. The flying capacitor consists of a MOS/MOM stack and occupies the lowest five metal layers to achieve 8fF/μm² density at 1.8V. The driver unit design enables simple distribution across large body bias domains for a fast and uniform charging profile.

The control unit integrates a body-bias sensor and decision logic, as well as a digital interface for communication with a power management unit and debug-and-trace registers (see Figure 5). It operates between two main modes: transition or ON-mode, and keep-the-value or STEADY-mode. In the ON-mode body bias sensor is a direct voltage sample-and-compare circuit clocked at 1or 2GHz, thus enabling sub-ns closed loop control over the driver units. A pwell sampler employs a -1:0.8 switched capacitor structure with 20% gain-error to compensate for the lower pMOS body factor. The outputs of the analog upper- and lower-bound rail-to-rail comparators are used to generate exclusive charge or discharge commands. Once the voltages settle inside the bounds, the control unit switches to low-power

STEADY-mode and a 0.5MHz clock. The analog reference voltages are generated with a 5-bit two-path resistive DAC implemented using poly resistance only in the slow path and a combination of MOS and poly resistance in the fast path (see Figure 6). This enables 2ns settling of the new upper-bound/lower-bound values and only 400nA of DAC static consumption in STEADY-mode. The size of upper-bound/lower-bound window is programmable with a default value of 58mV. When new digital target values are received, or when voltages leak out of the window, the fast clock unit is engaged within a cycle and the control unit switches into ON-mode to set new body-bias voltages.

System Integration and Measurement Results

Both the driver and the control unit occupy the lower five metal layers and are generated as hard macro IP instances with a programmable digital interface for system integration. The body bias generator was embedded in a RISC-V processor SoC, implemented in a 28nm UTBB FDSOI process with LVT transistors. Figure 7 depicts the placement of two drivers and the control unit to supply the total body-bias area of 1mm². Figure 8 shows measurement results of the static levels of pwell and nwell voltage reachable with the BBG IP. The nwell has 58mV resolution which, according to simulation, translates to roughly a 5mV V_{th-n} step. Similarly, V_{th-p} achieves 5mV minimal step with a 72mV pwell resolution. Figures 9 and 10 display the dynamics of charging and discharging of the wells. The nwell reaches high slew rates of -80mV/ns and +65mV/ns during discharging and charging, respectively. The pwell voltage ramps down from 0V to -1.3V in 160ns with one driver unit, and in 90ns with two units, switching at 1GHz in both cases. The pwell discharges from -1.4V to 0V in 70ns with only one unit operating. During the ON-mode the BBG drives currents in the range of 40-200mA, while in STEADY-mode it sources only 3.3μA and 1μA from the 1.8V and 1V supplies respectively. High ON-mode currents are averaged over long periods of STEADY-mode and contribute less than 2μA in total average current. Figure 11 illustrates the maintenance of target nwell and pwell voltages with short recharge phases when the well voltages drift due to leakage. In this case, under 5ns ON-mode recharge of 5ns duration occurs every 1ms and 0.2ms for the nwell and pwell respectively. Figure 12 demonstrates the variation-compensation and energy-efficiency optimization capability of the BBG. On-chip critical path replicas (CPRs) and frequency counters are used to extract the switching frequency which is used to tune the next BB value. The design is able to maintain the target frequency of the CPR within 1% while dynamically changing V_{CORE} in the range of 760mV-970mV. Note that pwell and nwell may have asymmetric well voltages to achieve robust compensation. Table I compares our solution with other relevant BBG designs and summarizes the performance.

References

[1] J. Tschanz, et al, JSSC, vol. 37, no. 11, pp. 1396–1402, Nov 2002.
[2] T. Kuroda, et al, JSSC, vol. 31, no. 11, pp. 1770–1779, Nov 1996.
[3] Y. Tsukikawa, et al, in VLSI Circuits, pp.85–86, Jun 1993.
[4] N. Kamae, et al, in A-SSCC, pp. 53-56, Nov 2014.
[5] M. Meijer, et al, Trans. VLSI, vol. 20, no. 1, pp. 42–51, Jan 2012.
[6] D. Levacq, et al, JSSC, vol. 43, no. 11, pp. 2390–2395, Nov 2008.

978-1-5090-0636-6/16 $31.00 © 2016 IEEE

Acknowledgements. The authors would like to thank Stevo Bailey, Brian Zimmer, Vladimir Milovanović, and Stéphane Le Tual, for great help during design and integration. Fabrication was donated by STMicroelectronics.

TABLE I: Comparison and summary of BBG performances

Metric	This work	Kamae[4]	Meijer[5]	Levacq[6]
Technology	28 FDSOI	65nm bulk	90nm bulk	90nm bulk
BBG Area	0.012mm²	0.0052um²	0.03mm²	0.006mm²
Bias Area	1mm²	0.22mm²	1mm²	0.15mm²
Area overhead	**1.2%**	2.3%	3%	2%
Response time	**30/70ns**	1.6/2µs	4/5µs	-/70ns
V_{th} Resolution	5mV	4mV	2mV	>20mV
BBG Supply	1.0V, 1.8V	0.5V-1.2V	1.2V	2.5V, -1V,1V
BBG Power	**10µW**	600µW	177µW	1500µW
V-range: nwell, pwell	[0, 1.8] V, [0, -1.4] V	VDD±0.25V GND±0.25V	[0.7, 1.2] V [0, 0.5] V	/ [-0.85, 0.4] V

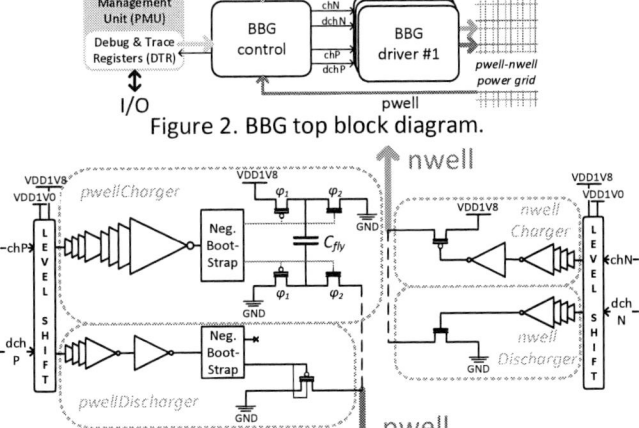

Figure 2. BBG top block diagram.

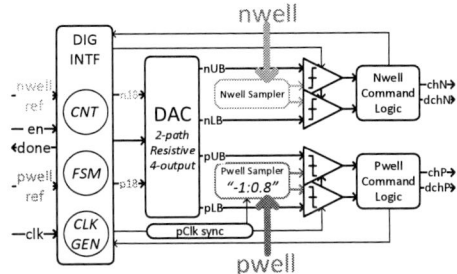

Figure 3. Driver unit - block diagram.

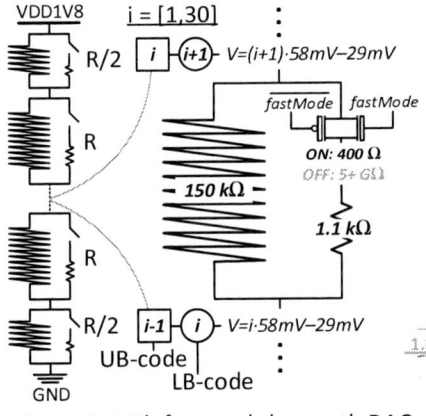

Figure 1. Load model.

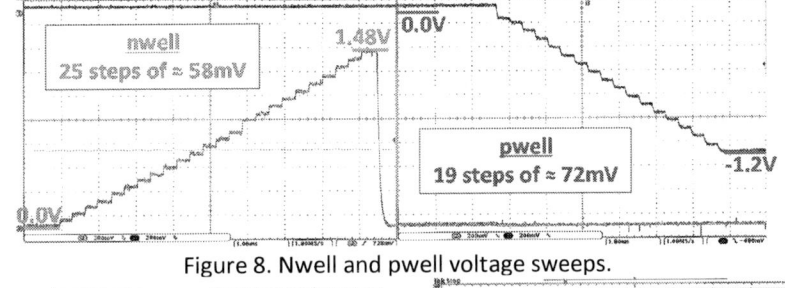

Figure 4. Negative voltage generation - "1:-1" switch-capacitor charge pump.

Figure 5. Control unit – block diagram.

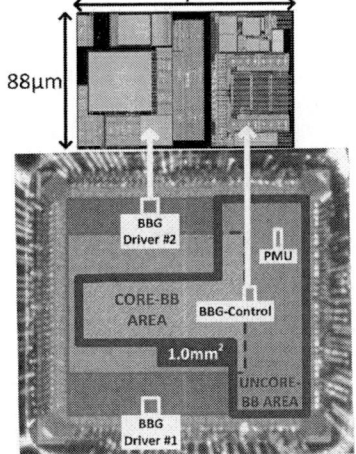

Figure 6. A 5b fast- and slow-path DAC.

Figure 8. Nwell and pwell voltage sweeps.

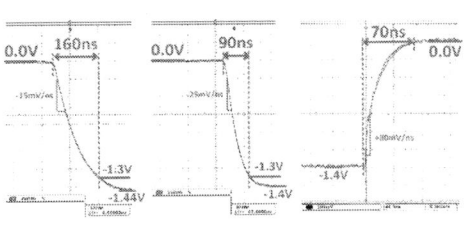

Figure 9. Nwell discharging and charging.

Figure 11. Maintaining the BB values.

Figure 7. Die-photo and BBG layout

Figure 10. Pwell charging and discharging.

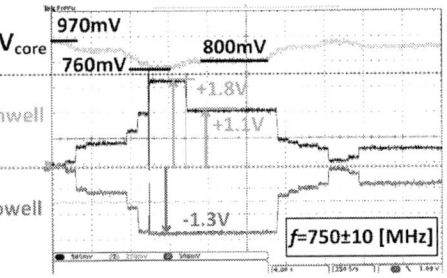

Figure 12. Adaptive BB for CPR frequency tracking on variable V_{CORE}.

A Bluetooth Low-Energy (BLE) Transceiver with TX/RX Switchable On-Chip Matching Network, 2.75mW High-IF Discrete-Time Receiver, and 3.6mW All-Digital Transmitter

Feng-Wei Kuo[1], Sandro Binsfeld Ferreira[2,3], Masoud Babaie[3], Ron Chen[1], Lan-chou Cho[1], Chewn-Pu Jou[1], Fu-Lung Hsueh[1], Guanzhong Huang[4], Iman Madadi[3], Massoud Tohidian[3], Robert Bogdan Staszewski[3,4]

[1]TSMC, Hsinchu, Taiwan.   [2]Federal University of Rio Grande do Sul, Porto Alegre, Brazil.
[3]Delft University of Technology, The Netherlands.   [4]University College Dublin, Ireland.   email: fwkuo@tsmc.com

Abstract

We present a new ultra-low-power (ULP) transceiver for Internet-of-Things (IoT) optimized for 28-nm CMOS. The receiver (RX) employs a high-rate (up to 10 GS/s) discrete-time (DT) architecture with intermediate frequency (IF) placed beyond the 1/f noise corner of MOS devices. New multi-stage multi-rate charge-sharing bandpass filters are adapted to achieve high out-of-band linearity, low noise and low power consumption. A transmitter (TX) employs an all-digital PLL (ADPLL) with switched-current-source digitally controlled oscillator (DCO) and switching PA. An integrated on-chip matching network serves both PA and LNTA, thus allowing a 1-pin direct antenna connection with no external antenna filters. The transceiver consumes 2.75 mW in RX and 3.6 mW in TX when delivering 0 dBm in Bluetooth LE.

Introduction

Ultra-low power (ULP) transceivers are key subsystems for wireless sensor networks and Internet-of-Things (IoT), which impose stringent requirements on cost (i.e., low silicon area and few or no external components) and power consumption. Recent ULP receivers (RX) achieve significant power reduction [1]-[4] typically using sliding intermediate frequency (IF) and low-IF continuous-time (CT) architectures, but with a sacrificed performance. To further decrease power consumption while maintaining adequate performance, we propose a fully discrete-time (DT) high-IF RX architecture with *complex-signaling* band-pass filters (BPF) and a progressively reduced sampling rate. The new approach exploits a DT filtering recently introduced for high-performance 4G RX [5], which is adapted here for ULP.

To eliminate the bulky and costly antenna filter, two techniques are employed. First, out-of-band (OOB) blockers are attenuated using a combination of a charge sampling mixer and a full-rate (~4×2.45 GS/s) charge-sharing (CS) BPF. Second, a dynamically switched fully integrated TX/RX matching network is proposed. The TX architecture is an improvement of [6] and consists of an ADPLL with switched-current-source digitally controlled oscillator (DCO), which maximally reduces its frequency pushing and 1/f phase noise such that it can operate with only a few kHz drift in *open-loop* in RX and during TX modulation. The digital PA (DPA) operates in class-E/F$_2$, at V$_{DD}$=0.5V supply. Fig. 1 shows the complete BLE transceiver.

Discrete-time Receiver with Direct Antenna Connection

The high-IF RX implements a multi-rate DT using three stages of cascaded quadrature charge-sharing (CS) BPFs. The circuitry after the low-noise transconductance amplifier (LNTA) comprises only gm stages, switches and capacitors, which are amenable to deep CMOS scaling. Gain is provided by LNTA and 3 highly linear programmable inverter-based gm-cells with 12 dB of maximum gain each. RX frontend is detailed in Fig. 2. The narrow-band LNTA with Q-factor ~10 is followed by a single-to-differential passive mixer and a 4/4 CS-BPF with 25%

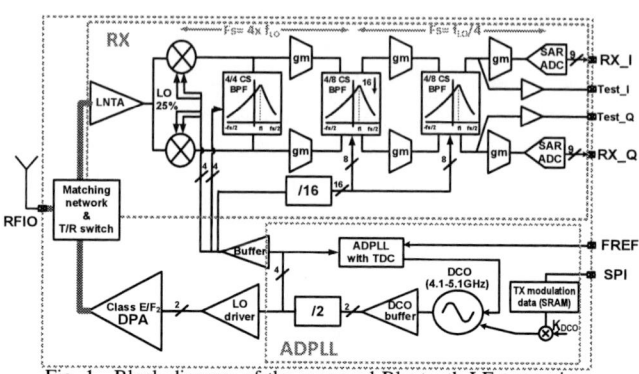

Fig. 1. Block diagram of the proposed Bluetooth LE transceiver.

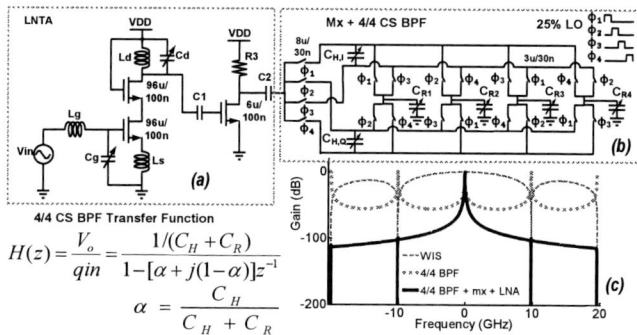

Fig. 2. LNTA and 1st-stage DT complex BPF using I/Q charge sharing.

duty-cycle clocks [5]. The mixer's charge-sampling operation at $4f_{LO}$ creates a window integration sampling (WIS) filter. Combined transfer functions of LNTA, mixer, and filter are shown in Fig. 2c and exhibit protection from OOB blockers (>100dB at $f_s = 4f_{LO}$). The three IF BPFs provide enough anti-alias filtering for the following 9-bit SAR ADCs, which consumes only 0.125 mW each for 20 MS/s.

The 5 MHz IF processing continues with two more stages employing a new 8-phase CS-BPF structure. The 4/8 CS-BPF has 8 rotating capacitors, sampled at 8 phases with 12.5% duty-cycle (Fig. 3) and works at 16x slower clock to further reduce power consumption. The 8-phase implementation improves selectivity, with an additional 5 dB of image rejection (Fig. 3b). Decimation is performed by reducing clock frequency and integrating the samples over the rotating capacitors, creating a sinc-like effect that works as an anti-aliasing filter for the decimated filters. The

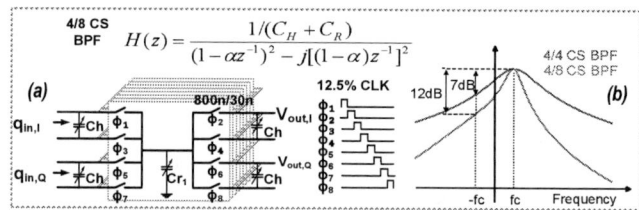

Fig. 3. 4/8 CS BPF (2nd and 3rd stages) (a); and comparison (b).

transfer function of the DT filters is accurate and controlled by the clock frequency and precisely programmable capacitor ratios to account for process, voltage and temperature (PVT) variations.

All-Digital TX and On-Chip Matching Network

The all-digital TX presented in Fig. 1 consists of a two-staggered-chains TDC topology, adopted here to double the TDC resolution to 12 ps. The V_{DD} of 0.5 V was chosen for the last PA stage to deliver low output power and to abide by the process reliability rules of gate-oxide breakdown. Furthermore, class-E/F$_2$ tuning exhibits the lowest systematic drain current and thus P_{out} at the same V_{DD} and load resistance among different flavors of switched-mode PAs, without sacrificing the power added efficiency (PAE). Consequently, this PA needs smaller impedance transformation ratio for $P_{out} < 3$ dBm, which results in a lower insertion loss for its matching network and thus higher system efficiency.

Fig. 4a shows the proposed implementation of the on-chip matching network with a 'soft' switch between the TX and RX paths (i.e., "T/R switch"). PA's transformer-based matching network (TXMN) acts as a second-order resonator in the RX mode. The input matching of LNTA is quite sensitive to the imaginary part of the impedance Z_{TX} seen from the output pad towards TX. Hence, the resonance of PA's matching network should be adjusted via C_1 and C_2 for the operating band. Furthermore, the equivalent Q-factor and thus the input parallel resistance, R_{TX}, of TXMN should be as high as possible to exhibit the lowest penalty on the RX NF. R_{TX} is about 250 Ω with less than 1 dB NF penalty in this design. In the TX mode, however, the resonant frequency of LNTA's matching network is pushed to higher frequencies via C_{gs}. Consequently, PA sees the RX path as a small capacitor in parallel with 1.2 kΩ modeling LNA matching network losses (see Fig. 4d). Compared to the 50 Ω load, they create a large impedance path for the TX signal, which leads to a negligible penalty (~4%) in the drain efficiency.

Measurement Results and Conclusion

Fig. 4 shows measured S$_{11}$ during LNTA and PA operations. The measured DT IF transfer function is presented in Fig. 5. The desired asymmetry between positive and negative frequency offsets with respect to LO is due to the presence of complex filters. The RX gain is 46 dB with a minimum of 24 dB image rejection obtained in the IF strip, thus allowing for the budget of 31 dB to be easily achieved when combined with a typical digital baseband. OOB blocking performance shows a large margin to the BLE specification, verifying the feasibility of the direct (i.e., filter-less) antenna connection. Sensitivity of -95 dBm is measured through packet error rate (PER) curve for PER = 30.8% [2]. RX core consumes 2.75 mW with a maximum gain of 46 dB, NF of 6.5 dB and peak IIP3 of -19 dB, as shown in Fig. 5. The TX consumes 3.6 mW at 0 dBm RF output. During TX and RX packets, the ADPLL is shut down immediately after settling [6]. The DCO tuning word is maintained on its update port while the second port is used to perform an open-loop modulation (see Fig. 1). This reduces the LO power from 1.4 mW to 0.6 mW. This is possible due to the very low frequency pushing (10–12 kHz/mV) and 1/f^3 corner frequency (<100 kHz) of the DCO. BLE micrograph shows an active area of only 1.4 mm^2 in TSMC 28 nm CMOS. Power breakdown is presented in Fig. 5.

Table I summarizes the proposed transceiver and compares it with state-of-the-art BLE designs. Besides being fully integrated with an on-chip T/R switch, as in [2] and [3], this work reaches similar (NF, linearity and sensitivity) or better (max P_{out}, PLL PN) RF performance but at a much lower power consumption, even better than [1] [4], which use an off-chip matching network and T/R switch.

REFERENCES

[1] Y. H. Liu, et al., *ISSCC Conf.*, pp. 446–447, 2013.
[2] J. Prummel, et al., *ISSCC Dig. Tech. Papers*, pp. 238–239, 2015.
[3] T. Sano, et al., *ISSCC Dig. Tech. Papers*, pp. 240–241, 2015.
[4] Y. H. Liu, et al., *ISSCC Dig. Tech. Papers*, pp. 236–237, 2015.
[5] I. Madadi, *et al.*, *VLSI Circ. Symp.*, pp. C308–C309, 2015.
[6] F.-W. Kuo, *et al.*, *ESSCIRC Conf.*, pp. 356–359, 2015.

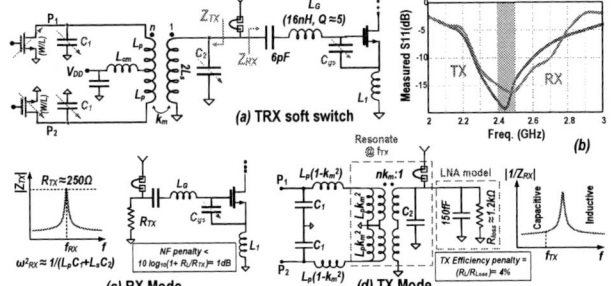

Fig. 4. On-chip antenna matching network schematics and performance.

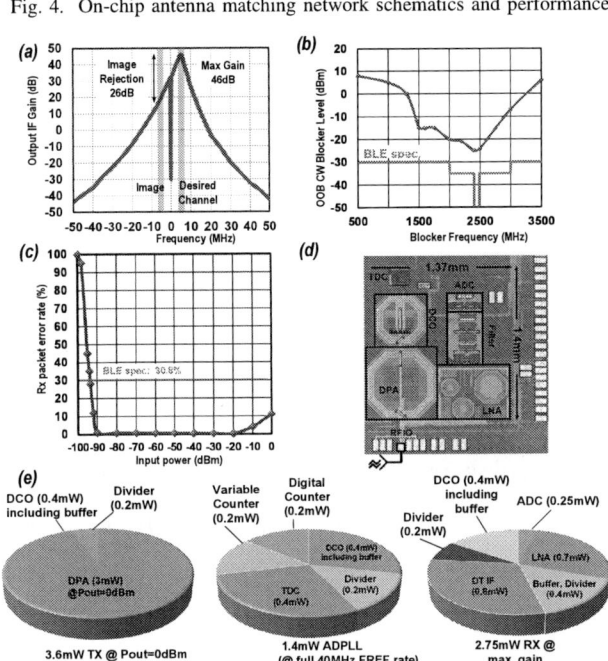

Fig. 5. Summary of chip measurements: gain, blocking test and PER measurement, power breakdown of main blocks.

TABLE I
PERFORMANCE SUMMARY AND COMPARISON WITH STATE-OF-THE-ART.

		This work	[1]	[2]	[3]	[4]
CMOS node		28nm	90nm	55nm	40nm	40nm
Data rate & modulation		1-Mbps GFSK	1-Mbps GFSK	1-Mbps GFSK	1-Mbps GFSK	1-Mbps GFSK
RX noise figure (dB)		6.5	6	N/A	6.5	N/A
RX sensitivity (dBm)		-95	-98	-94.5	-94.5	-94
RX IIP3 (dBm)		-19	-19	N/A	N/A	N/A
OSC FoM (dB)		188	183	N/A	N/A	N/A
PLL in-band PN (dBc/Hz)		-92**	-85	N/A	N/A	N/A
		-101***		N/A	N/A	N/A
Integrated PN (degree)		1.08**	2.3	N/A	N/A	N/A
		0.87***		N/A	N/A	N/A
TX max. Pout (dBm)		3	0	2.3	0	-2
Maximum PA PAE		41%	N/A	45%	<25%	N/A
T/R Switch		Yes	No	Yes	Yes	No
Supply voltage (V)		0.5# / 1	1.2	0.9-3.3	1.1	1
Power consumption (mW)	TX @ 0dBm	3.6	5.4	10.1	7.7	4.2*
	RX	2.75	3.3	11.2	6.3	3.3
TX efficiency (P$_{OUT}$/P$_{DC}$)		28%	18%	10%	13%	15%

*Measured @-2dBm **FREF=5MHz ***FREF=40MHz #Low-voltage DCO & DPA

A 380pW Dual Mode Optical Wake-up Receiver with Ambient Noise Cancellation

Wootaek Lim, Taekwang Jang, Inhee Lee, Hun-Seok Kim, Dennis Sylvester and David Blaauw

University of Michigan, Ann Arbor, MI, USA

imhotep@umich.edu

Abstract

We present a sub-nW optical wake-up receiver for wireless sensor nodes. The wake-up receiver supports dual mode operation for both ultra-low standby power and high data rates, while canceling ambient in-band noise. In 0.18μm CMOS the receiver consumes 380pW in always-on wake-up mode and 28.1μW in fast RX mode at 250kbps.

Introduction

A key component of a truly energy-autonomous wireless sensor node is an always-on wake-up receiver to enable asynchronous wake-up triggered by external interrupt signals at near-zero standby power. RF-based approaches have been widely adopted [1-3] to realize ultra-low power (ULP) wake-up receivers. However the RF frequency oscillator and amplifier limit their energy efficiency to several nJ/bit with high (10s of μW) standby power. An ULP optical wake-up receiver was proposed with only 695pW power consumption [4]. However, it has a low data rate of ~100 bps and lacks a critical capability in optical receivers, namely adaptation to highly variable lighting conditions, which can range from 100s of lux to 100 klux.

Wake-up Receiver Operation Including Noise Cancellation

We present an ULP wireless optical receiver to (re-)program/reset a wireless sensor that enables sub-nW asynchronous wake-up, high data rate data communication, and ambient background light tracking. In order to achieve both ultra-low standby power and high data rates, the proposed receiver employs dual mode operation: 1) Voltage mode for passcode verification at low bit rate: the photodiode voltage is used as the input signal and is directly sensed by a clocked comparator and digital demodulation logic. This approach enables ultra-low power operation by avoiding power hungry analog components. 2) Current mode for fast RX: the diode current is used as the input signal and is sensed by a trans-impedance amplifier to achieve a high bit rate since the effect of diode parasitic capacitance is eliminated. Both modes support a flexible data rate that is dynamically tracked by the clock recovery algorithm employed in the proposed receiver.

A visible light optical receiver is known to be vulnerable to in-band noise from various ambient light sources such as sunlight, incandescent, and fluorescent lighting, resulting in inferior bit rate and sensitivity. Addressing this critical challenge, we propose noise canceling circuity in both modes, improving bit rate and input sensitivity while enabling operation across 0.3 – 100 klux background light conditions.

Fig. 1 shows the system block diagram of the proposed ULP optical receiver. A 100×100um parasitic photodiode serves as a signal receptor, providing much smaller size compared to a typical inductor/antenna for RF receivers. This enables miniaturization of the entire optical receiver system to the sub-millimeter scale. The fast RX current mode block is power-gated until a valid passcode is successfully verified by matching the on-off keying (OOK) Manchester coded signal to the expected 16-bit passcode. In wakeup mode, the input signal is detected by comparing the diode voltage to a reference voltage (V_{REF}) (Fig. 2). To adapt to different ambient light levels and avoid saturation of the diode voltage, the diode is loaded with a tunable unary-coded resistor bank that consists of 36 off-state medium-V_{TH} transistors with geometric growth. This guarantees a monotonic increase of the resistor value with the selection code (SEL) while covering a background light level from 250 lux to 93 klux (Fig. 3). The use of off-state transistors for diode loading enables a small layout footprint and was previously shown to provide steep light/voltage response [4], thereby improving sensitivity.

Ambient light tracking and pass-code verification operate in parallel using two comparators that each compare the diode voltage to a reference voltage. The ambient light tracking logic searches for the SEL code that has a 50% probability of $D_{OUT\_AMB} = 1$ (Fig. 2) and biases the diode output voltage exactly at V_{REF}. Since the underlying modulation is OOK, the light threshold for data detection must reside above the ambient light level. Hence, after the ambient light tracker has tested the light level and updated the SEL code, the selection code is increased by a fixed value (3 in our implementation) for data signal detection. Fig. 3 shows how the ambient comparison is interspersed between data comparisons. This allows the resistor bank to be reused for both comparators and avoids mismatch issues that would arise if different banks were used for data detection and ambient tracking. At the same time a constant data sampling rate is maintained, which is critical for clock/data recovery. The voltage reference circuit uses two zero-V_{TH} NMOS transistor in series with a SVT PMOS [5] and consumes 50pW. The 50Hz clock is generated using a leakage-based differential thyristor oscillator [6] that also consumes 50pW. The passcode verification block logic uses 3V I/O transistors to achieve ultra-low leakage and hence low standby power.

After successfully verifying the 16-bit passcode, three signals (*power_gate_off*, *oscillator_on*, and *digital_reset*, Fig. 1) are enabled sequentially to power up and initialize the fast RX mode logic. The fast RX mode block consists of analog front-end (AFE) circuitry and digital clock-data-recovery (CDR) logic. The AFE employs a trans-impedance amplifier (TIA) with a DC noise canceling feedback circuit that amplifies the current signal and converts it to a rail-to-rail voltage. The digital CDR logic is synthesized with SVT transistors to support fast decoding speeds. The CDR logic extracts the data as well as the associated clock from the Manchester coded input signal. The AFE contains three amplifiers: 1) the light diode regulation amplifier regulates V_{DIODE} to V_{REF} from the passcode verification block, independent of the light level, 2) the ambient cancelation amplifier ensures the DC value of the V_{SIG} node is the same as the $V_{SIG-REF}$ and generates an adequate gate voltage to sink the DC ambient current to ground, and 3) the post amplifier consists of self-biased cascaded amplifiers to compare V_{SIG} with $V_{SIG-REF}$ and amplify their difference to full rail. A push-pull transistor stage follows the post amplifier to sharpen the data output edge. The fast RX mode CDR circuit is clocked by a current-starved ring oscillator at 2.5 MHz. This CDR circuit takes the signal data stream from the TIA block and generates the decoded *Data* and *CLK* with a flexible data rate (i.e., *CLK* rate) ranging from 0.98 kHz to 250 kHz.

The diode regulation and ambient cancelation amplifiers have an identical two-input cascaded structure, which uses the same three biases generated by the on-chip voltage reference generator (Fig. 4). However, these two amplifiers are sized differently to meet their gain and bandwidth requirements. Decoupling capacitors (C_1 & C_2 = 3.2pF) are inserted for noise-canceling feedback loop stability and also for controlling the bandwidth of the noise cancellation. During fast RX mode, the feedback loop cancels out ambient light noise if the noise spectrum is narrower than the loop bandwidth, which is designed to be 400Hz.

Measurement Results

The proposed design is fabricated in 0.18μm CMOS (die photo in Fig. 7) with total area of 0.85mm². Fig. 5 shows the measured waveform of *SEL[35:0]* and V_{CTRL} signals as the ambient light intensity changes. As expected, *SEL[35:0]* toggles in steady state and V_{CTRL} tracks the ambient light to cancel the DC noise current. Measured bit error rates (BER) are shown for the fast RX mode using a 3W LED source operating across an ambient light intensity ranging from low office light (500 lux) to full sun (100 klux). The DC noise canceling feedback circuit limits the BER increase to 3.4× across 500lux to 100klux at 0.5m TX-RX distance and adapts wide ambient lighting conditions automatically. Minimum required incident power on the 0.01mm² PV diode is 900nW at 850nm wavelength. This enables transmit distance of 25m with a standard 3W LED. The proposed receiver achieves its maximum energy efficiency of 112.5pJ/bit at the highest achieved bit rate, 250kbps. Table. 1 compares the proposed free-space optical receiver to prior work, showing 1.8× improvement in standby power, 2700× faster bit rate, and 1.25× higher energy efficiency.

References

[1] N. Pletcher, et al., ISSCC, 2008. [2] X. Huang, et al., ISSCC, 2010. [3] D. Y. Yoon, et al., JSSC, 2013. [4] G. Kim, et al., CICC, 2012. [5] M. Seok et al., CICC, 2009. [6] G. Chen et al., ISSCC, 2010.

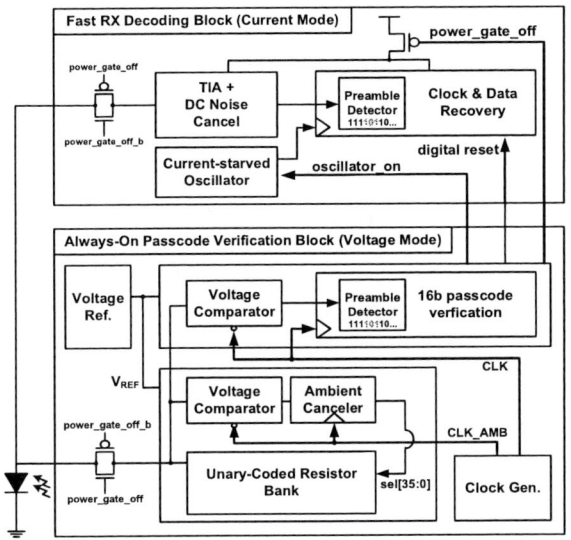

Fig.1 System block diagram

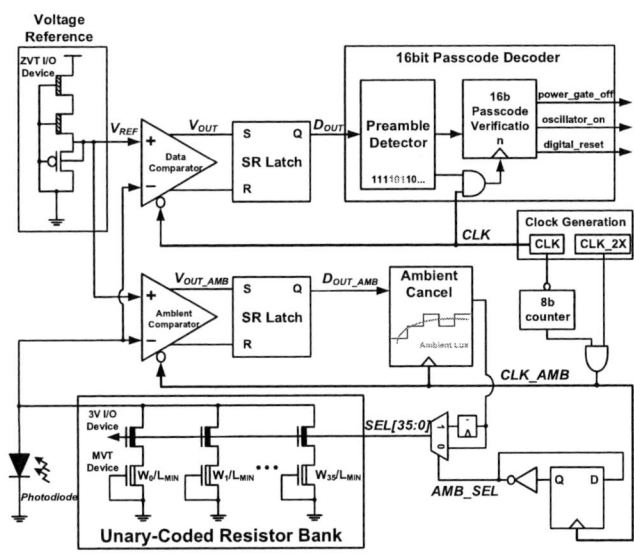

Fig.2 Voltage-mode passcode-verification block diagram

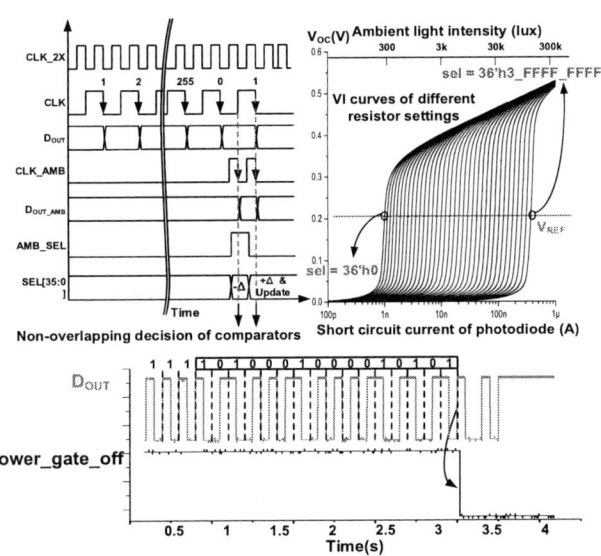

Fig.3 Operation of the passcode-verification block

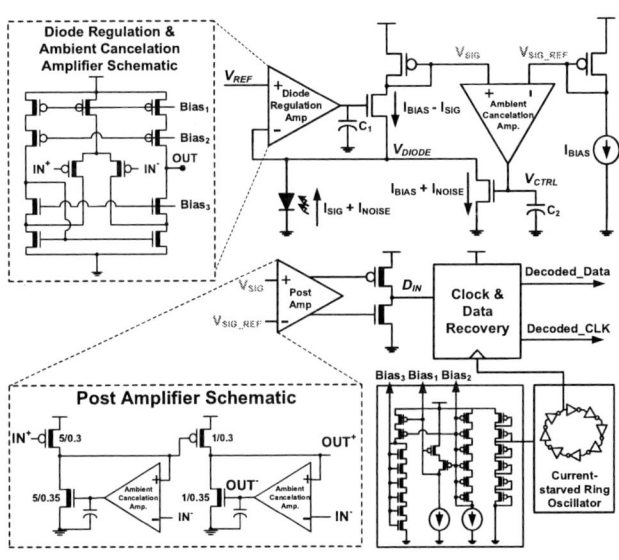

Fig.4 Current-mode fast-RX block diagram

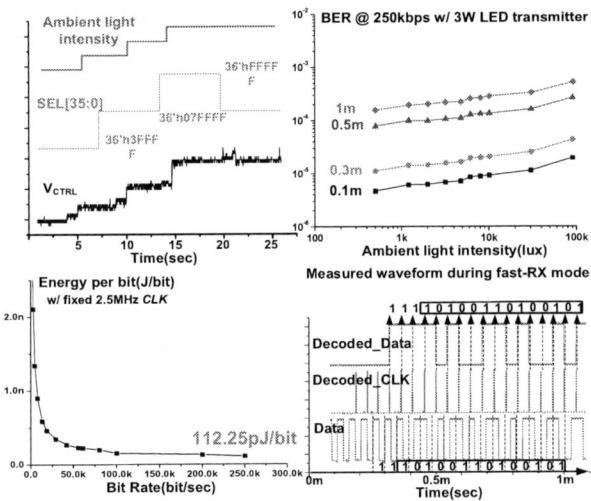

Fig.5 Measured results of fast-RX mode

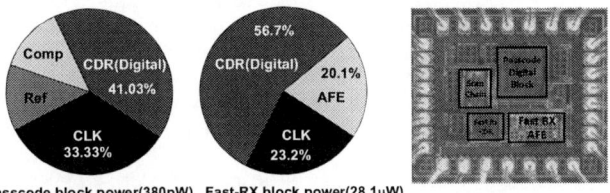

Fig.6 Measured power **Fig.7 Die Photo**

Table 1. Performance summary & comparison

	[1]	[2]	[3]	[4]	This Work
Transmit Method	RF	RF	RF	Optical	Optical
Technology	90nm	180nm	130nm	180nm	180nm
Supply Voltage	0.5V	1.8V	1.2V	1.2V	1.2V
Power	52μW	8.5μW/1078μW	16.4μW/22.9μW	695pW	380pW/28.1μW
Max. Data Rate	100kpbs	1k/200kbps	10k/200kbps	91bps	5bps/250kbps
Energy/Bit	0.52nJ/bit	8.5/5.39nJ/bit	1.64/0.11nJ/bit	140pJ/b	80/112.25pJ/b
BER	@-72dBm <10^{-3}	<10^{-6}	<10^{-5}	<10^{-5}	<10^{-5}

SleepTalker: a 28nm FDSOI ULV 802.15.4a IR-UWB Transmitter SoC achieving 14pJ/bit at 27Mb/s with Adaptive-FBB-based Channel Selection and Programmable Pulse Shape

Guerric de Streel[1], François Stas[1], Thibaut Gurné[1,2], François Durant [1,3], Charlotte Frenkel[1] and David Bol[1]

[1]ICTEAM Institute, Université catholique de Louvain, Louvain-la-Neuve, Belgium. [2]Now with Nokia Bell Labs, Antwerpen, Belgium. [3]Now with IMEC, Heverlee, Belgium. {guerric.destreel,david.bol}@uclouvain.be

Abstract

We propose a UWB transmitter (TX) SoC designed for ultra-low voltage (ULV) in 28nm FDSOI CMOS. Operated at 0.55V, it achieves a record energy efficiency of 14pJ/bit with embedded power management (PM), highly duty cycled digital baseband, programmable pulse shaping and wide-range on-chip adaptive forward back biasing (FBB) for V_T reduction, PVT compensation and tuning of both the carrier frequency (CF) and the output power.

Introduction

The high power consumption of RF frontend limits the datarate of Internet-of-Things (IoT) wireless sensor nodes (WSNs) and connected objects. UWB Impulse Radio (IR-UWB) shows potential for implementing medium-datarate links. In this work, a versatile TX SoC compliant with 802.15.4a is proposed.

FBB-based 802.15.4a IR-UWB transmitter at ULV

The 802.15.4a TX supports the 3 low-band channels (3.5, 4 and 4.5GHz) and 7 datarates (0.11-27.4Mb/s) with burst durations between 2 and 64 ns. To maximize the energy efficiency, the transmitter is operated at 0.55V with a mostly-digital architecture. This leads to significant challenges linked to the reduced speed of logic gates and the sensitivity to PVT variations at ULV. These challenges are addressed in the high-frequency TX blocks with wide-range FBB up to 1.8V enabled by FDSOI low-V_T (LVT) devices. Indeed, FBB is used for lowering the V_T of the transistors, adapted to compensate PVT variations and tune both the CF of the local oscillator (LO) to the selected 802.15.4a channel and the output power of the power amplifier.

The LO is free-running to avoid the power overhead of a PLL but as IEEE 802.15.4a imposes BPSK modulation for the chips inside every burst, limitation of LO jitter accumulation is critical. LO duty-cycling [1] can be used to limit jitter accumulation by limiting the LO run time to the ¼ of the symbol duration. Fig. 1 shows that the accumulated jitter of a differential ring oscillator (RO) is prohibitive in the worst symbol-duration case corresponding to the 0.11Mb/s datarate of 802.15.4a. We thus optimize jitter accumulation by 1) avoiding current starving in the LO thanks to FBB-based frequency tuning, 2) using a fast multiple-pass RO [8] and 3) aggressively duty cycling the LO within a symbol while meeting the 802.15.4a burst position scrambling required to avoid spectral lines in the output spectrum. As shown in Fig. 2, the coarse burst position inside the next symbol is precomputed at the symbol rate from the 802.15.4a LFSR output. The packet and symbol-level basebands implement coarse-position duty-cycling are clocked by the 31.25MHz crystal and start the LO for a short run time (16-64ns, depending on the selected datarate).A fine-position duty-cycling is performed by the pulse-level baseband clocked at a 500-MHz frequency obtained from the prescaling of the LO clock. Fig. 1 shows that the 3 jitter optimizations lead to an accumulated jitter divided by 500×.

For 0.55V operation, the transmitter uses a digital power amplifier (PA) with on-chip capacitive filtering to meet the FCC regulation (Fig. 3) and a high slew rate (HSR) amplifier setting the PA output bias at $V_{DD}/2$ between pulses. However, this does not lead to a sufficient attenuation in the stringent 960-1610MHz GPS band due to the single-order roll-off. To avoid using bulky off-chip filters, programmable pulse shaping is implemented. The digital PA is divided in 32 parallel drivers (thermometer code) that can be

independently enabled during the 2-ns pulse duration. Their activation sequence is controlled by the content of the pulse shape registers and non-overlapping enabling signals generated from a ring counter clocked at 2×CF. Depending on the 802.15.4a channel, the 2-ns pulse duration is thus divided into 14 (3.5GHz) to 18 (4.5GHz) time steps with programmable amplitude allowing the user to finely tune the pulse shape both in time and in amplitude for maximum spectral efficiency. Fig. 3 illustrates the impact of pulse shaping on output PSD. The roll-off improvement directly results in an increase of the maximum output power while meeting the FCC regulation. Adaptive FBB in the PA can further be used to tune the RF output power while compensating PVT variations.

SoC integration

The SoC architecture is presented in Fig. 4, showing the baseband partitioning. The packet-level baseband including forward error correction and FBB calibration is synthesized from standard cells along with SPI and configuration registers. It uses LVT library with gate length upsized by 4nm without FBB to reduce leakage. The rest of the TX is implemented at the full-custom level with LVT transistors and adaptive FBB. Embedded PM features an on-chip switched-capacitor DC/DC converter [9] to generate the 0.55V supply voltage from the 1.2V nominal voltage and dual Dickson charge pumps to generate the +1.8/-1.8V supplies for the FBB generator. In sleep mode these blocks perform power gating and only a regular-V_T sleep controller operated at 1.2V stays on. The measured efficiency of the DC/DC is 84% at 380µW.

The back biases (BBN) for the NMOS of the full-custom TX and PA are generated by two capacitive DACs with respectively 10 and 5 bits. The back biases for the PMOS (BBP) are generated with current matching loops similar to [10]. The measured BBN/BBP ranges are [0;1.6V] and [0;-1.7V], respectively. In the full-custom TX, the adaptive FBB is used to tune the 3.5-4.5GHz CF of the LO to the selected 802.15.4a channel. With the 10-bit DAC, a 5-MHz frequency resolution is obtained. The on-chip calibration loop, demonstrated in Fig. 4 is based on frequency measurement. It is performed before each data packet with a typical lock time of 2.2µs and a maximum lock time at startup of 573µs.

In order to fully benefit from the inherently duty-cycled nature of IR-UWB, 3 levels of duty-cycling are implemented. At system level, the sleep controller allows a transition from sleep mode to active in less than 1ms that can be done between two packet transmissions. Inside a symbol, the full-custom TX is duty-cycled by clock gating and is active only for 16 to 64ns depending on the data rate which correspond to between 1/4 and 1/128 of active time inside a symbol. Finally, the PA is duty-cycled at pulse-level and is only active for 2ns to 64ns when a pulse is emitted depending on the selected data rate. The active time is then limited to 1/32 or 1/128 of a symbol depending on the number of burst.

The 0.93-mm² SoC is fabricated in 28nm FDSOI CMOS, housed in a QFN40 package. Die microphotograph given in Fig. 5 shows the compact 0.55mm² core area with only 0.095mm² for the TX. The measured peak output PSD is -50dBm/MHz while respecting the FCC mask and the output power is -20dBm. The TX performances are summarized and compared with previous works in Fig. 6 along with the SoC power breakdown. At 27.24Mb/s, the complete SoC consumes 650µW with 380µW for the TX leading to a record 24pJ/bit and as low as 14pJ/bit for the TX only.

Conclusion

SleepTalker is, to the authors' knowledge, the first full ULV RF SoC in 28nm FDSOI. It demonstrates the interest of IEEE 802.15.4a TX with record energy efficiency and low area for IoT wireless sensor nodes.

References

[1] J. Ryckaert et al, "A 0.65-to-1.4nJ/burst 3-to-10GHz UWB Digital TX in 90nm CMOS for IEEE 802.15.4a", ISSCC, pp. 120-121, 2007.

[2] P. Mercier et al., "An energy-efficient all-digital UWB transmitter employing dual capacitively-coupled pulse-shaping drivers", IEEE J. Solid-State Circuits, vol. 44, pp. 1679-1688, 2009.

[3] Y. Zheng et al, "A 0.18μm CMOS dual-band UWB transceiver", ISSCC, pp. 114-115, 2007.

[4] S. Joo et al, "A fully integrated 802.15.4a IR-UWB transceiver in 0.13μm CMOS with digital RRC synthesis", ISSCC, pp. 228-229, 2010.

[5] J. Brown et al, "An ultra-low-power 9.8GHz crystal-less UWB transceiver with digital baseband integrated in 0.18μm BiCMOS", ISSCC, pp. 442-443, 2013.

[6] F. Padovan et al., "A 20Mb/s, 2.76pJ/b UWB impulse radio TX with 11.7% efficiency in 130nm CMOS", IEEE ESSCIRC, pp. 287-290, 2014.

[7] S. Geng et al, "A 13.3mW 500Mb/s IR-UWB transceiver with link-margin enhancement technique for meter-range communications", ISSCC, pp. 160-161, 2014.

[8] Y. A. Eken et al, "A 5.9-GHz voltage-controlled ring oscillator in 0.18-μm CMOS", IEEE J. Solid-State Circuits, vol. 39, pp.230-233, 2004.

[9] S. Clerc et al, "A 0.33V/-40℃ process/temperature closed-loop compensation SoC embedding all-digital clock multiplier and DC-DC converter exploiting FDSOI 28nm back-gate biasing" ISSCC, pp. 150-151, 2015.

[10] Couniot et al, "A 65 nm 0.5 V DPS CMOS Image Sensor With 17 pJ/Frame.Pixel and 42 dB Dynamic Range for Ultra-Low-Power SoCs", JSSC, pp. 2419-2430, 2015.

Acknowledgements

Funding was provided by the F.R.S.-F.N.R.S. of Belgium and FP7 MSP project. The authors thank STMicroelectronics for chip donation.

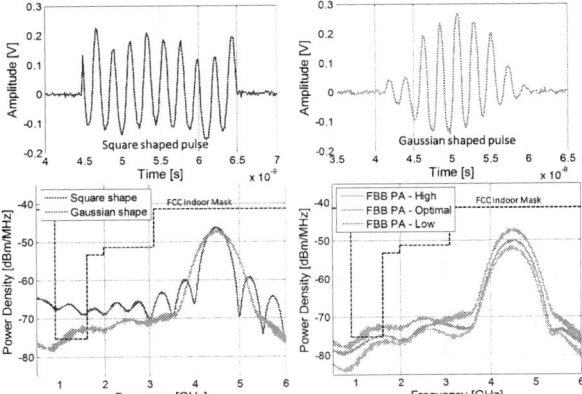

Fig. 3. Measured pulse waveform and spectrum for different shaping and output power tuning.

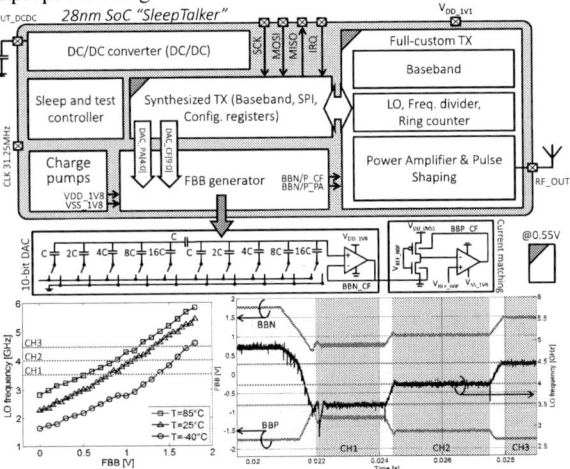

Fig. 4. SoC architecture with back bias generator details and functionality measurements of LO calibration at 0.55V.

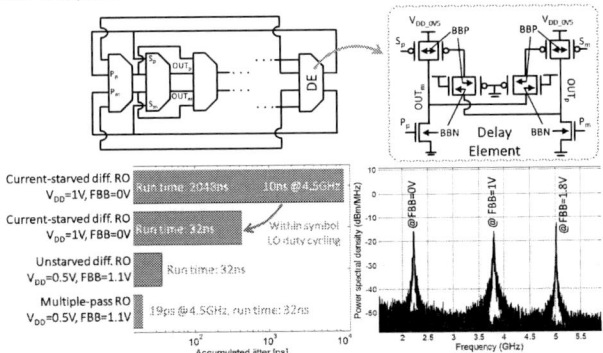

Fig. 1. LO architecture details with simulated accumulated jitter performances and measured frequency content.

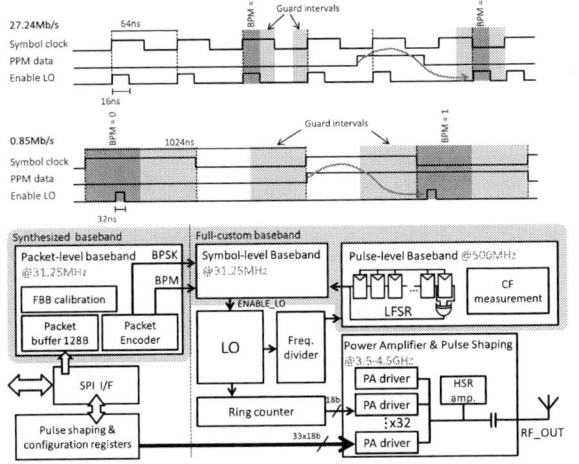

Fig.2. Aggressive LO duty-cycling to minimize jitter accumulation, and transmitter partitioning between crystal-frequency blocks synthesized from standard cells and high-frequency full custom blocks.

Ref.	This Work	[1]	[2]	[3]	[4]
Technology	28nm FDSOI	90nm	90nm	0.18μm	0.13μm
Data rate [Mb/s]	0.11, 0.85, 1.7, 6.81, 27.24	16	15.6	27.24	0.11, 0.85, 1.7, 6.81, 27.24
Power [μW]	TX: 380 SoC: 650	650	4360	N/A	5980
E_{bit} [pJ/b]	TX:14@27.24Mb/s SoC:24@27.24Mb/s	650@1Mb/s	280@15.6Mb/s	740@27.24Mb/s	219@27.24Mb/s
Die area [mm²]	TX: 0.095 SoC: 0.93	0.066	0.07	4.5	RFFE: 7.5 Dig. BB: 24.65
Supply voltage [V]	Aux: 1.2 Core: 0.55 * BB: +-1.8*	1	1	1.8	1.2
Output swing	Up to 350mV$_{PP}$	~100mV$_{PP}$	165-710mV$_{PP}$	N/A	Up to 720mV$_{PP}$
LO freq. range [GHz]	3.5-4.5	3-10	2.1-5.7	3-9	3.5-4.5

* Generated on-chip

Fig. 5. SoC microphotography and comparison to the state of the art.

A 2.4GHz Ternary Sequence Spread Spectrum OOK Transceiver with Harmonic Spur Suppression and Dual-Mode Detection Architecture for ULP Wearable Devices

Seong Joong Kim[1,2], Chang Soon Park[2], Youngkyu Kim[2], Seok-ju Yun[2], Young-Jun Hong[2], Sang-Gug Lee[1]

[1]KAIST, Daejeon, Korea, [2]Samsung Electronics, Suwon, Korea

Abstract

A novel 2.4-GHz Ternary Sequence Spread Spectrum (TSSS)-OOK Transceiver with spur suppression and dual-mode detection architecture is presented for the ULP wearable devices. A random bi-phase switching of a PA rejects the intrinsic spur of spread spectrum OOK TX by 22dB. A new TSSS-OOK TX supports spreading spectrum and dual reception of the coherent as well as non-coherent mode, implementing 12 dB SNR gain. The single-chip TSSS-OOK transceiver in 90nm CMOS occupies an active area of $2.1mm^2$ and measures 1 Mb/s, 22dB spur-suppressed output spectrum and 5.5% EVM at 2.17mW.

Introduction

With a growing interest of mobile health care and development of low-power sensor technology, there has been an increasing demand for wearable devices. For decades, On/Off Keying (OOK) transmitter has been popular for low-power, short-distance communications due to the structural simplicity. [1], [2] However, some improvement points had been found in applying it for a wearable device in a noisy environment. Fig. 1 shows the block diagram of conventional OOK TX along with our observations. The conventional OOK TX has a relatively clean output spectrum. However, the harmonic line spur occurs and deteriorate a TX spectrum mask when a spreading sequence is added to improve the reliability in a data connectivity.[3]

This work proposes a novel, TSSS-OOK transceiver architecture with spur suppression technique and additional transmission SNR gain, which provides the reliable data connectivity in a mobile wearable application.

TSSS-OOK Transceiver Architecture

The proposed transmitter architecture in Fig. 2 is composed of a TSSS-OOK TX, a reference RX for dual-mode detection, and a digital baseband. In TX, a bi-phased PA converts carrier phase between a bi-phase of 0 and 180 degree in a random order, which scrambles the patterned regularity of the ternary sequence and suppress the harmonic line spur. For Ternary Sequence Spreading Spectrum, TX binary data pass through Ternary Sequence Spreading Logic and is converted to ternary sequence of {1, 0, -1} composition according to pre-determined spreading ratio. Then, ternary sequence is applied to both Gaussian-pulse shaper logic and the phase randomizer logic of RF OOK carrier. To support the dual-mode detection capability of the TSSS-OOK TX, a dual-mode frequency synthesizer and a dual-mode reference receiver is integrated at a single VCO. A non-coherent RX is implemented in a super-regenerative oscillator (SRO) based architecture with an envelope detection and a coherent RX is implemented in a sliding IF(S-IF) architecture with a phase detection.

Harmonic Spur-suppression technique with Bi-phase PA

Fig. 3 shows the detailed block diagram of Bi-phase PA. The phase shifting signal is generated from Bi-phase controller at a ternary sequence chip rate. When it is an in-phase signal, M2 and M3 turns on and M6 and M7 operates. When PS signal is an Out-of-phase signal, M5 and M8 operate. On the other hand,

unary code TMPA[15:1] of Gaussian pulse shaping FIR filter is applied to the switching transistor M11[15:1] of the cascode amplifier array. The TMPA signal is updated at a 6 times Gaussian pulse rate ($f_{sample} = 1MHz \times 6 = 6MHz$) in order of 1,4,9,1,9,4,1 to make a single pulse shaping.

Dual-mode Carrier Generation for Dual-Mode Detection

The circuit schematics of the carrier generation blocks are shown in Fig. 4. The VCO has a current-reused complimentary type for low power consumption. The Dual-mode frequency synthesizer controls the capacitor bank of a single VCO. The PLL is a high-precision fractional-N type and FLL is a low power, digital feedback logic using counter and comparator. In a capacitor bank, BAND_SW<1:0> is commonly used between PLL and FLL, but then the PLL tunes continuously a varactor voltage to make a high-precision carrier during TX/RX frame. On the other hand, the FLL tunes a fine-resolution capacitor bank in a periodic time interval to lower power consumption, still providing sufficient frequency accuracy for a non-coherent mode.

Ternary Sequence Spreading Spectrum

The binary TX data are converted to ternary-sequence at the spreading logic as shown in Fig. 5. There are several spreading factors (SF): 1 bit to 4 chips (1/4), 3/8, and 5/32. At RX, TSSS-OOK carrier can be detected through both envelope detection and phase detection. Therefore, it has a spread spectrum effect and detection gain as well, providing an extended transmission gain of about from 3 dB to 12 dB in data link connectivity. The theory is partly introduced in the new low power Body Area Network (BAN) standard, IEEE 802.15.4q and this is the first prototype chip implementation based on it. [4]

The timing diagram for 1/4 ratio, as an example, is shown in Fig. 6. The Gaussian pulse shaping and random bi-phasing switching is done for the individual ternary sequenced chip. As marked, the phase inversion is happened whenever the polarity of the random bipolar controller is changed.

Measurement results and Conclusions

The single chip transceiver is fabricated in a 90nm CMOS and packaged in 8x8 64-pin QFN package. A 1.0V supply powers RF/analog circuits and digital circuits. Fig. 7(a) shows the measured TSSS-OOK TX output spectrum. When the spur suppression is OFF, 1MHz harmonic line spur at the spreading sequence rate is prevalent and degrade the spectrum mask characteristics. On the other hand, when the spur suppression is activated, the harmonic line spur is removed and meet the mask specification. Also, a Gaussian pulse shaping achieves about 20dB side lobe attenuation compared to the rectangular pulse. The carrier qualities of phase noise, EVM, and frequency error are also plotted with a dual-mode reception performance of reference receiver. In non-coherent mode, it has -81dBm sensitivity without ternary sequence spreading. However, it has about 12 dB SNR gain at coherent detection using a ternary sequence spreading factor of (5/32). In Fig. 8, the chip performances are summarized with prior works. Also, die

978-1-5090-0636-6/16 $31.00 © 2016 IEEE

micrograph is shown with the TX, RX and digital part marked.

In conclusion, a reliability-improved, novel OOK transceiver architecture with ternary sequence spread spectrum and spur-suppression technique is proposed for mobile wearable devices. Also, to verify the dual-reception capability and transmission SNR gain of the proposed architecture, a reference RX is implemented with a dual-mode frequency synthesizer.

Reference

[1] Denis C. Daly, et at., "An Energy-Efficient OOK Transceiver for Wireless Sensor Networks," in IEEE J. Solid-State Circuits, vol. 42, pp.1003-1011, Sep.2007.

[2] M. Vidojkovic, et al., "A 2.4GHz ULP OOK single-chip transceiver for healthcare applications," in ISSCC Digest Tech Papers, pp. 458-460, Feb. 2011.

[3] Chang Soon Park, et al., "Transmitter, Receiver, and Wireless Communication method thereof," U.S. Patent US14/204387, Mar., 11, 2014.

[4] Draft Standard for Local and metropolitan area networks, IEEE P802.15.4q/D3.0, Feb. 2015.

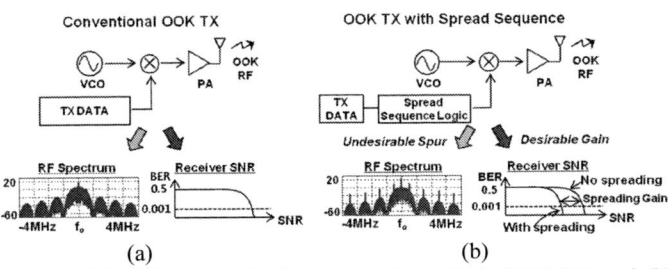

Fig. 1. OOK TX observation(a) Conventional ULP OOK TX and (b) OOK TX with Spreading Sequence.

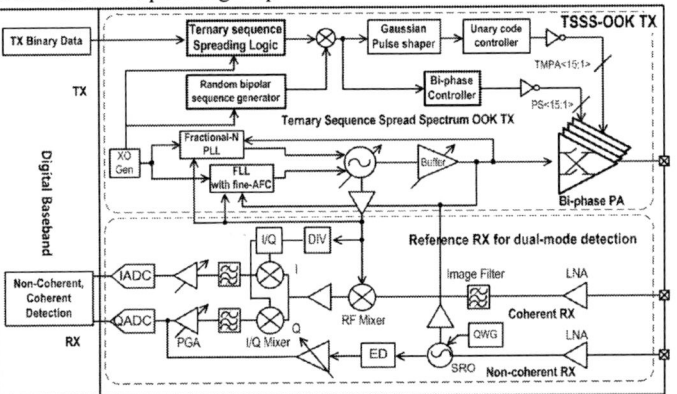

Fig. 2. Block diagram of Proposed TSSS-OOK Transceiver.

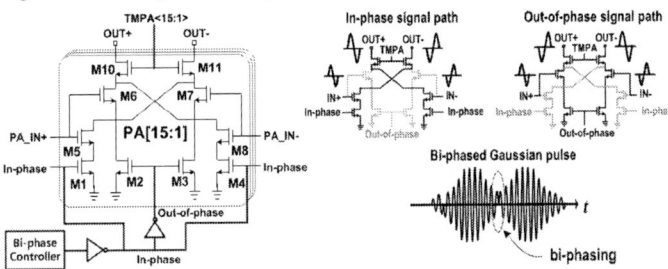

Fig. 3. Bi-phase, Gaussian Pulse Shaping PA.

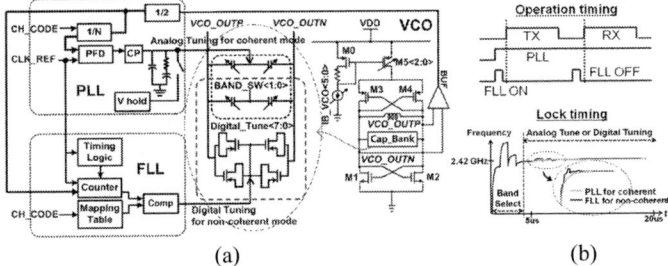

Fig. 4. Dual-mode frequency synthesizer for individual operation at coherent/non-coherent Mode (a) schematic and (b) timing diagram.

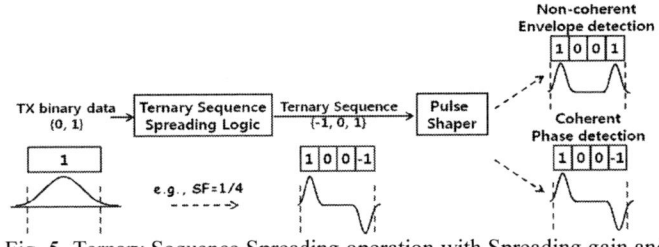

Fig. 5. Ternary Sequence Spreading operation with Spreading gain and Dual reception capability.

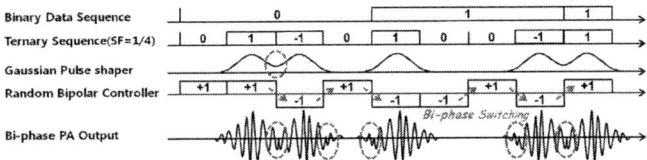

Fig. 6. Timing diagram of TSSS and spur-suppressing scheme.

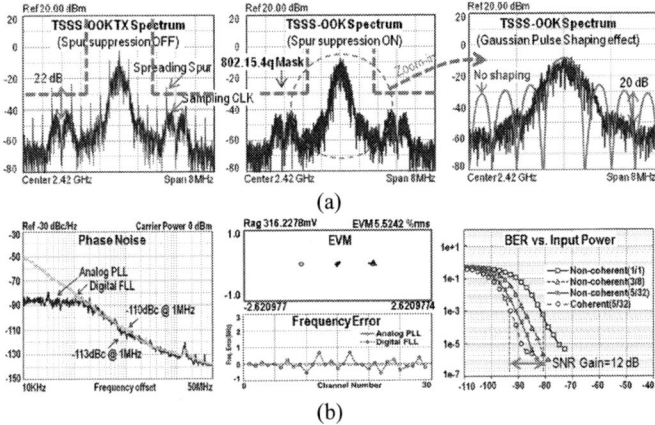

Fig. 7. Measured (a) TSSS-OOK output spectrum and (b) carrier performance including a dual-mode SNR gain using reference RX.

	JSSC2007[1]	ISSCC2011[2]	This Work	
Freq band	916.5 MHz	2.4GHz	2.4GHz	
TX Architecture	OOK	OOK	OOK/TSSS-OOK	
TSSS SNR Gain	0 dB	0 dB	12 dB	
Data rate	1Mcps	1,10Mcps	1Mcps	
Frequency Synthesizer	Fee running	PLL	Dual-mode of PLL/FLL	
Pulse shaping	Rectangular	Root-raised	Gaussian	
Spur rejection	NA	NA	22 dB	
Spur technique	NA	NA	Bi-phase PA	
PN @ 1MHz	NA	NA	-110dBc/Hz	
Output power	-11.4dBm	0dBm	0dBm,-10dBm	
EVM	NA	NA	5.50%	
Freq. tolerance	NA	NA	37 ppm	
RX Sensitivity	-65 dBm	-75dBm	Non-coherent(1/1):-81dBm Coherent(5/32):-93dBm	
Power consumption	TX:3.8mW RX:2.5mW	TX:2.53mW RX: 0.53mW	TX:2.17mW RX:0.714mW	
Process	180nm	90nm	90nm	

Fig. 8. Performance summary and die micrograph.

978-1-5090-0636-6/16 $31.00 © 2016 IEEE   57   2016 Symposium on VLSI Circuits Digest of Technical Papers

An 18 μW Spur Canceled Clock Generator for Recovering Receiver Sensitivity in Wireless SoCs

Yosuke Ogasawara, Hiroki Sakurai, Ryuichi Fujimoto, Kenichi Sami

Toshiba Corporation, Kawasaki, Japan
yosuke.ogasawara@toshiba.co.jp

Abstract

A novel spur canceled clock generator (SCCG) capable of recovering RX sensitivity degradations caused by digital clocks in wireless SoCs is presented. Clock spurs which degrade RX sensitivity are canceled by applying the SCCG to the digital circuits or ADCs. The SCCG is integrated into a Bluetooth® smart SoC fabricated in a 65 nm CMOS process. Measured clock spur reduction of over 35 dB and RX sensitivity recovery of 4 dB are achieved. The power consumption and occupied area of the SCCG are only 18 μW and 40 μm × 120 μm, respectively.

Introduction

RF receivers in wireless SoCs suffer from clock spurs which are digital clock harmonics that originate from ADCs or digital circuits [1]. Fig. 1(a) illustrates the propagation mechanism for clock spurs. Spurs are generated by the ADCs or digital circuits, and propagate to the low-noise amplifier (LNA) through power supply, ground lines, ESD protection diodes, and Si-substrate [2]. On-chip single-ended LNAs are particularly sensitive to clock spurs.

Fig. 1(b) shows the clock spectrum and the channel allocation of Bluetooth® smart transceivers. When 13 MHz clock is adopted, the clock harmonics from the 185th to 190th appear in the frequency range from 2405 to 2470 MHz. The harmonics at 2418, 2444, and 2470 MHz overlap with channels, and the RX sensitivities of these channels are significantly degraded. Meanwhile, the other harmonics at the frequency between two adjacent channels do not degrade the RX sensitivities.

In order to reduce the influence of clock spurs, deep N-well isolation [2] and ensuring a large distance between digital blocks and LNAs have been used. However, these methods offer only limited spur reduction capabilities. A spread spectrum clock generator [3] is proposed, but it is not sufficient. Its spur reduction is up to $-10 \log(1\text{MHz}/13\text{MHz}) = 11.1$ dB, where 1 MHz is channel bandwidth and 13 MHz is frequency spacing of clock spurs. Moreover, an active spur canceller which uses active feedback system [4] and an on-chip magnetic thin-film noise suppressor [1] are also proposed. However, the former requires huge power and the latter results in additional fabrication costs. In this paper, a novel SCCG with small power consumption and occupied area is presented.

Spur Canceled Clock Generator

Fig. 2 shows the block diagram of the proposed SCCG applied to an ADC clock. The SCCG is composed of inverter chains as delay elements, a selector, and a controller. The SCCG generates three kinds of pulse trains, namely, delayed, base, and advanced pulse trains. One of these pulse trains is selected in a specific order.

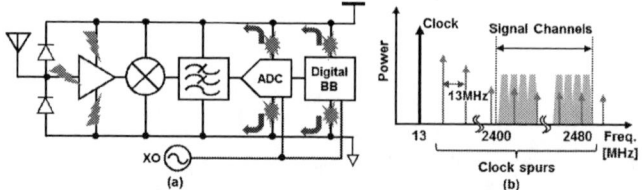

Fig. 1 (a) Propagation mechanism for clock spurs (b) Clock spectrum.

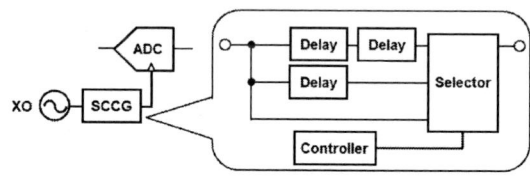

Fig. 2 Block diagram of SCCG and applied to ADC.

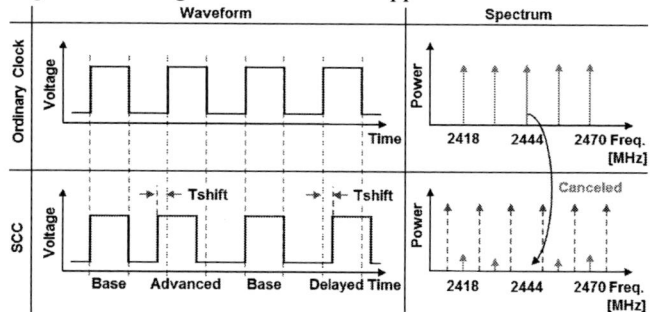

Fig. 3 Clock waveforms and spectra.

Fig. 3 shows waveforms of an ordinary clock and the spur canceled clock (SCC). Although the pulses in the ordinary clock form fixed intervals, the pulses in the SCC are shifted backward or forward by T_{shift} every second cycle. Thus, four states of the base, advanced, base and delayed are periodically repeated. Fig.3 also shows the clock spectra. In addition to the harmonics with 13 MHz interval, newly generated spurs indicated by dashed arrows appear in the SCC spectrum. It is notable that the clock spur at the desired channel at 2444 MHz is canceled by using the SCC with an appropriate value of T_{shift}. To clean up the clock spurs in all channels, the SCC needs to be used for the channels at 2418, 2444 and 2470 MHz, and the ordinary clock is used for the other channels.

Fig. 4 illustrates a principle of the spur canceling. The waveform of the SCC can be decomposed into the four pulse trains namely the base, advanced, base, and delayed pulses. Since these pulse trains are the same waveform except for their phases, their n-th harmonics are also the same waveform except for the phases. By setting T_{shift} to the half period of the n-th harmonic, the n-th harmonics of the advanced and delayed pulse trains have the reverse waveforms to those of the base pulse trains. Then, the n-th harmonics of the SCC are perfectly canceled each other.

978-1-5090-0636-6/16 $31.00 © 2016 IEEE   58   2016 Symposium on VLSI Circuits Digest of Technical Papers

The n-th harmonics of the SCC, *S(t)* can be represented as

$$S(t) = A_n\{\sin\theta(t) + \sin(\theta(t)+\varphi) + \sin\theta(t) + \sin(\theta(t)-\varphi)\} \quad (1)$$
$$= 2A_n(1+\cos\varphi)\sin\theta(t).$$

Here A_n is the amplitude of the n-th harmonics, $\theta(t) = 2\pi n f_{clock} t$ is the phase of the clock spurs, $\varphi = 2\pi n f_{clock} T_{shift}$ is the phase shift of the pulse, f_{clock} is the clock frequency that is 13 MHz. The amplitude of (1) is normalized to that of ordinary clock spur and is calculated as

$$A_{norm} = (1+\cos\varphi)/2. \quad (2)$$

Fig. 5 shows the relation between φ and normalized amplitude A_{norm}. A solid line labeled "both directions" in Fig. 5 shows the amplitude of the spur of the SCC shown in Fig. 3. When φ is set to π, the n-th harmonic is perfectly canceled. A dashed line labeled "single direction" in Fig. 5 shows amplitude of the spur for the SCC using the only two pulse trains of the base and advanced ones. Although the spur is also perfectly canceled, the "single direction" SCC is sensitive to variations of φ than the "both directions" SCC. It is because the variations of advanced and delayed pulses are canceled each other in the "both directions" SCC. The "both directions" SCCG can reduce the spur amplitude by over 20 dB within $\pm\pi/6$ of the phase variation. To cancel the clock spur at 2444 MHz, T_{shift} needs to be set to 205 ps. Although the T_{shift} causes a clock jitter, it is negligible small for the digital blocks using the 13 MHz clock whose period is 76.9 ns.

This clock jitter also limits SNR of the ADC. The clock jitter only has a negligible influence on the SNR for the case of Bluetooth® smart SoCs. T_{shift} of 205 ps corresponds to the clock jitter, J_{rms} =145 ps. The SNR can be calculated [5] as,

$$SNR = -20\log(2\pi f_{bw} J_{rms}) = 46.0 \text{ [dB]}. \quad (3)$$

Here f_{bw} is the bandwidth of the ADC that is 5.5 MHz. Since required SNR for our SoCs is sufficiently smaller than 46 dB, the RX sensitivities are not affected by the clock jitter.

Measured Results

Fig. 6 shows the measured spectrum at the ADC input. T_{shift} is set to the same value of 205 ps for the target channels at 2418, 2444, and 2470 MHz. When the ordinary clock was used, the spur power was -22 dBm. On the other hand, the clock spur in the target channels were canceled using the "both directions" SCC. Actually, the clock spur is suppressed by over 35 dB as shown in Fig. 6. The newly generated spurs do not affect the RX sensitivity because they are not in the target channel.

Fig. 7(a) shows the measured RX sensitivities of all channels. Without the SCC, the RX sensitivities are degraded at three spur-affected channels. By employing the "both directions" SCC, the RX sensitivities are recovered by 4 dB. Fig. 7(b) shows a measured packet error rate (PER) for the channel at 2470 MHz and 4 dB recovery is also observed.

A chip micrograph of the analog circuit blocks of the Bluetooth® smart SoC is shown in Fig. 8. It is fabricated using 65 nm CMOS process. The SCCG occupied only 40 μm × 120 μm, and it is absolutely small in the entire SoC. The power consumption of the SCCG is only 18 μW, and it is negligibly small comparing with entire receiver power consumption.

Conclusion

The RX sensitivities which are degraded by clock spurs are recovered by employing the proposed SCCG in the Bluetooth® smart SoC. The SCCG reduces over 35 dB of the clock spur, and recovers 4 dB of the RX sensitivity. The SCCG consumes only 18 μW and occupies very small area of 40 μm × 120 μm.

References

[1] M. Yamaguchi et al., "IC chip level low noise technology for high speed and high quality telecommunication systems," in APMC, pp. 540-542, 4-7 Nov. 2014.

[2] W. Yu-Chen et al., "Substrate Noise Coupling Reduction in LC Voltage-Controlled Oscillators," in IEEE Electron Device Letters, vol. 30, no. 4, pp.383-385, April 2009.

[3] M. Kokubo et al., "Spread-spectrum clock generator for serial ATA using fractional PLL controlled by ΔΣ modulator with level shifter," in IEEE ISSCC, vol. 1, pp. 160-590, 10 Feb. 2005.

[4] A. Trippe et al., "An adaptive broadband BiCMOS active spur canceller," in IEEE MTT-S, pp. 1-4, 5-10 June 2011.

[5] Shinagawa et al., "Jitter analysis of high-speed sampling systems," in IEEE JSSC, vol.25, no.1, pp.220-224, Feb 1990.

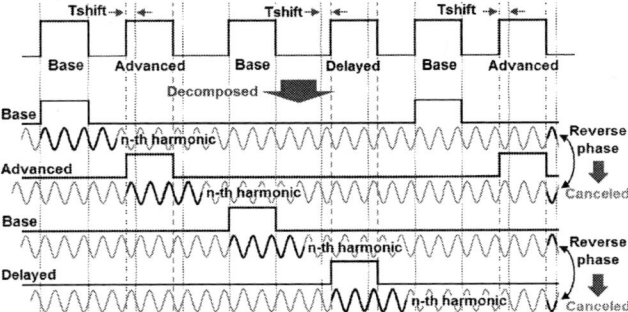

Fig. 4 Principle of spur canceling

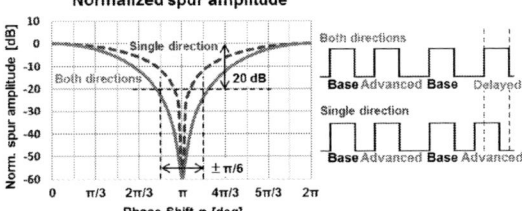

Fig. 5 Normalized spur amplitude for two kinds of SSC.

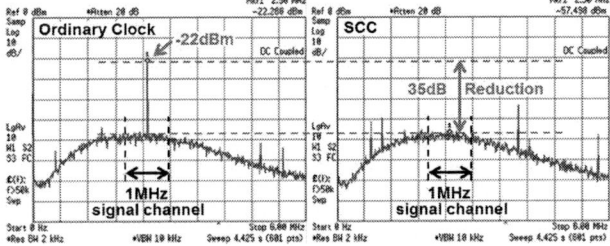

Fig. 6 Measured RX spectrum at ADC input.

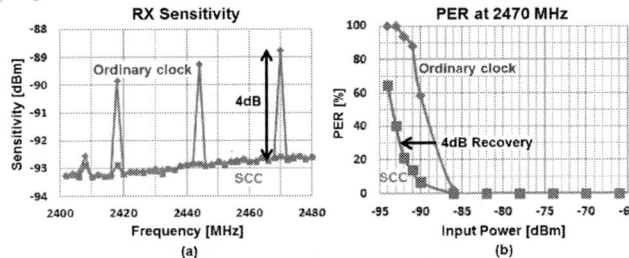

Fig. 7 (a) Measured RX sensitivity (b) Measured PER.

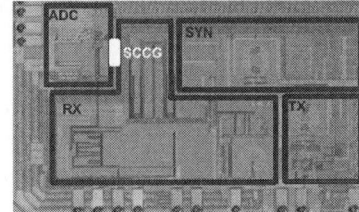

Fig. 8 Chip micrograph.

An Energy Harvesting Wireless Sensor Node for IoT Systems Featuring a Near-Threshold Voltage IA-32 Microcontroller in 14nm Tri-Gate CMOS

Somnath Paul, Vinayak Honkote, Ryan Kim, Turbo Majumder, Paolo Aseron[1], Vaughn Grossnickle[2], Robert Sankman[3], Debendra Mallik[3], Sandeep Jain[4], Sriram Vangal, James Tschanz and Vivek De

Circuit Research Lab, [1]Silicon Technology Prototyping Lab, [2]Platform Engineering Group, [3]Assembly & Test Technology Development, [4]Internet of Things Group, Intel Corporation, Hillsboro, OR, USA

Abstract

A wireless sensor node (WSN) integrates a 0.79mm^2 near-threshold voltage (NTV) 32-bit Intel Architecture (IA) microcontroller (MCU) in 14nm tri-gate CMOS, along with solar cell, energy harvester, flash memory, sensors and Bluetooth Low Energy (BLE) radio, to enable always-on always-sensing (AOAS) and advanced edge computing capabilities in Internet-of-Things (IoT) systems. The MCU features four independent voltage-frequency islands (VFI), a low-leakage SRAM array, an on-die oscillator clock source capable of operating at sub-threshold voltage, power gating and multiple active/sleep states, managed by an integrated power management unit (PMU). The MCU operates across a wide frequency (voltage) range of 297MHz (1V) to 0.5MHz (308mV), and achieves a peak energy efficiency of 17pJ/cycle at an optimum supply voltage (V_{OPT}) of 370mV, operating at 3.5MHz. The WSN, powered by a solar cell, demonstrates sustained MHz AOAS operation, consuming only 360μW.

Introduction

WSNs for IoT systems need to provide AOAS and advanced edge computing capabilities under stringent energy constraints, often supported mainly by harvested energy. Ultra-low power MCUs operating across a wide voltage-frequency range, including the NTV regime where energy per cycle is minimized, are key ingredients for enabling such energy-constrained WSNs [1-3]. Smart and fine-grained power management of different components of the WSN, including the MCU, are critical for realizing energy-neutral WSN systems. We demonstrate an energy-harvesting WSN (Fig. 1) that integrates a sub-mm^2, NTV and energy-optimized 32-bit IA MCU implemented in a 14nm tri-gate SoC process [4], a solar cell, a harvester, flash memory, sensors and a BLE radio.

NTV IA-32 MCU Architecture and Design

The MCU (Fig. 2) consists of a 32-bit IA core with 8KB I-cache (I\$) and 8KB data tightly-coupled memory (DTCM), with low latency and deterministic access. A 32-bit advanced high-performance bus (AHB) interconnect supports multiple masters including direct memory access (DMA) and test access port (TAP2AHB) for debug. The memory subsystem consists of 16KB of boot ROM and 64KB of shared memory (SMEM), used for both code and data. The SMEM can be power gated at 2KB granularity to minimize leakage from unused sections. The peripheral bus (APB) supports standard serial interfaces such as SPI and UART. The AON subsystem comprises of timer, GPIO and I^{2}C for communication with external sensors and a power management IC (PMIC). The power management and clock control unit (PMUCCU) supports management of multiple VFI and active/sleep states. A calibrated ring oscillator (CRO) is a frequency-locked loop that serves as a low-power, MHz on-chip clock source (Fig. 3). The 4mm^2 packaged MCU is fabricated in a 14nm tri-gate SoC process (Fig. 4). On-die logic and memory circuits are optimized for reliable NTV operation. The fully synthesized NTV design employs variation-aware pruning methods for the standard cell library to mitigate delay sensitivity to process variations [2]. The optimized cell library is characterized at 500mV for synthesis and timing convergence. To achieve low standby power, the on-die memory arrays use a custom 8T, 0.155μm^2 bitcell, built using 84nm gate pitch ULP transistors [4]. The leakage power of this bitcell is 26X lower than that of an 8T standard performance (SP) transistors bitcell, but area is 55% larger. The measured bitcell leakage is 8.3pA at 308mV. The chip timing convergence methodology is enhanced to provide variation-aware hold margin guard-bands for robust low voltage operation. The entire clock distribution network is designed using high performance (HP) devices to minimize variation-induced clock skew. 100% SP devices are used in logic to maintain sufficient speed in active mode. The IOs are designed using thick-gate (TG), high-voltage transistors.

MCU and WSN Measurements

The MCU is functional over a wide operating frequency (voltage) range (Fig. 5) of 297MHz (1V) to 0.5MHz (308mV). A 4.8X improvement in energy efficiency is achieved at V_{OPT} of 370mV and 3.5MHz operating frequency. The minimum energy per cycle is 17.18pJ, with the IA core continuously executing an AES encryption workload. At higher voltages ($V_{CORE}=V_{AON}=0.75V$), the core (x86+AHB) active power dominates MCU power, while at V_{OPT}, IO and CRO consume 63% of the MCU power (Fig. 6). The on-die CRO locks to a wide range of target frequencies from 1V down to 0.4V (Fig. 7). The CRO dissipates 60μW ($V_{CRO}=450mV$) while generating a 16MHz output to clock the MCU at V_{OPT} ($V_{CORE}=V_{AON}=370mV$). The CRO is functional down to deep sub-threshold voltage of 128mV (3.8μW, 7kHz). MCU active power can be reduced by up to 80% at 0.75V either by halting the IA-core or by clock gating idle functional blocks on the chip (Fig. 8). Energy consumption of the MCU improves by 40% for typical WSN workloads, when the on-die I\$ and DTCM are enabled (Fig. 9). For minimum energy operation, V_{CORE} must be set to V_{OPT}, which changes with active/sleep profiles of the workload (Fig. 10). The area, voltage/frequency and energy efficiency of the IA-32 MCU are compared with other 32-bit MCUs targeted for WSNs (Table I).

The MCU can cycle the WSN through four sleep modes (S0-S3) with varying wake-up latencies (Fig. 11). The PMU enables wide DVFS range by communicating clock frequency changes to the CRO and voltage change commands to the external PMIC through the I^{2}C bus. The PMU switches from a fully active state (S3) to a short sleep state (S2) with the IA core halted and clocks gated. In long sleep state (S1), the PMU switches the primary MCU clock from MHz to 32kHz RTC. In the lowest power deep sleep state (S0), V_{CORE} and V_{CRO} rails are power gated, with the AON subsystem clocked by RTC, resulting in 16X energy reduction. The WSN operates continuously using the energy harvested by a 1cm^2 solar cell from indoor light (1000 lux), with sensor data transmitted over BLE (Fig. 12). In the AOAS operating mode (BLE advertising + sensor polling), average power (P_{AVG}) for the entire WSN is 360μW, with the MCU contributing 290μW (13MHz, 0.45V), which further drops to 120μW in deep sleep state S0 (Fig. 11).

Acknowledgements

The authors thank S. Park, D. Kurian, S. Liff, M. Kumar, S. Jayaraman, T. Nguyen, S. Karpenko, J. Kulkarni, T. Wang, A. Srinivasan, Y. Hoskote, K. Caviasca, and M. Haycock.

References

[1] A.B. Warneke et al., *ISSCC Dig. Tech. Papers*, pp. 316-317, 2004.
[2] S. Jain, et al., *ISSCC Dig. Tech. Papers*, pp. 66-68, 2012.
[3] S. Paul et al., *Symposium on VLSI Circuits*, pp. C30-C31, 2013.
[4] C.-H Jan et al., *Symposium on VLSI Technology*, pp. T12-T13, 2015.
[5] M. Turnquist et al., *Symposium on VLSI Circuits*, pp. C320-C321, 2015.
[6] Y. Tsuji et al., *Symposium on VLSI Circuits*, pp. T86-T87, 2015.
[7] J. Myers, et al., *ISSCC Dig. Tech. Papers*, pp. 144-145, 2015.
[8] W. Lim et al., *ISSCC Dig. Tech. Papers*, pp. 146-147, 2015.

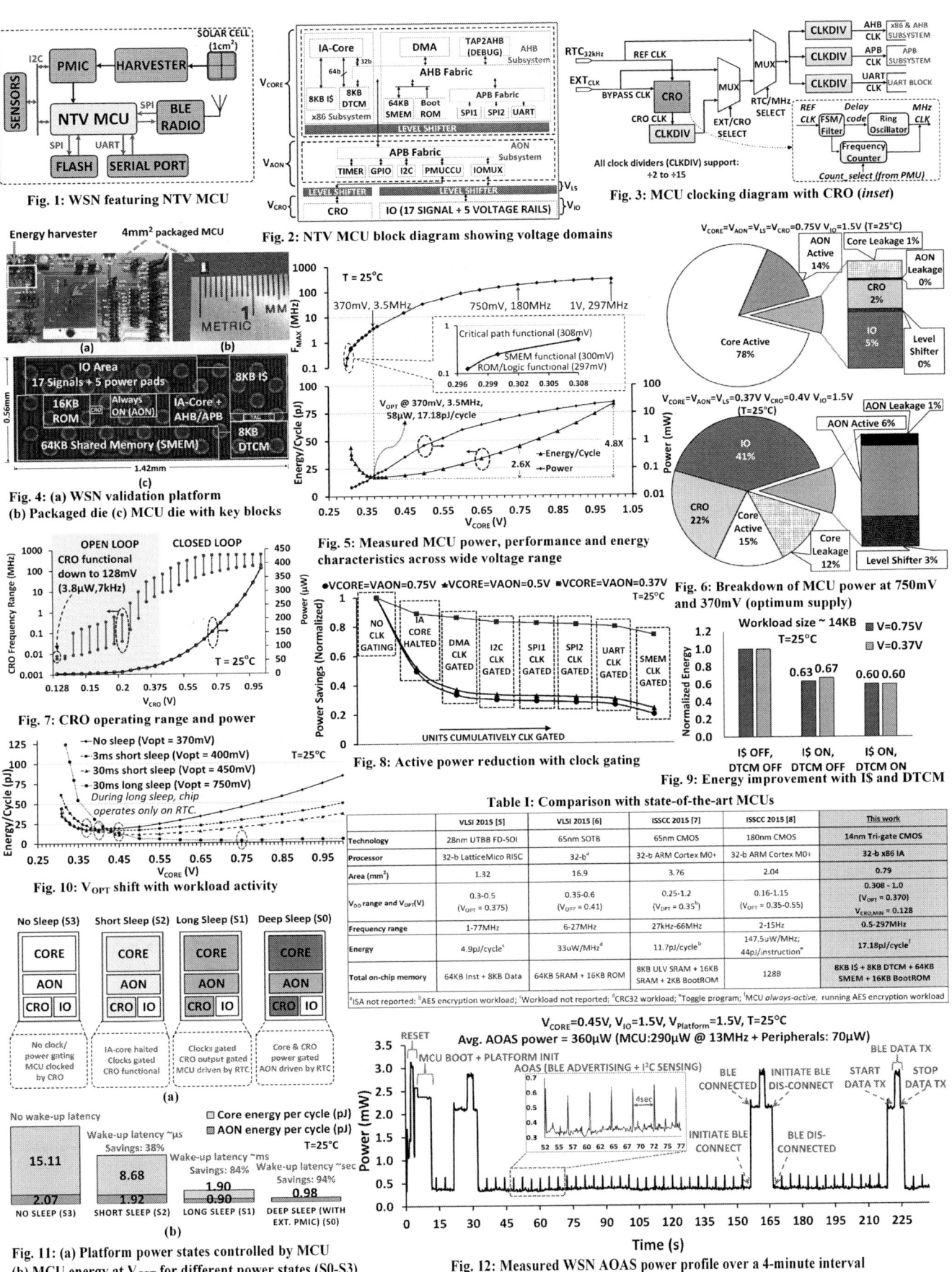

Fig. 1: WSN featuring NTV MCU

Fig. 2: NTV MCU block diagram showing voltage domains

Fig. 3: MCU clocking diagram with CRO (*inset*)

Fig. 4: (a) WSN validation platform (b) Packaged die (c) MCU die with key blocks

Fig. 5: Measured MCU power, performance and energy characteristics across wide voltage range

Fig. 6: Breakdown of MCU power at 750mV and 370mV (optimum supply)

Fig. 7: CRO operating range and power

Fig. 8: Active power reduction with clock gating

Fig. 9: Energy improvement with I$ and DTCM

Fig. 10: V_{OPT} shift with workload activity

Table I: Comparison with state-of-the-art MCUs

	VLSI 2015 [5]	VLSI 2015 [6]	ISSCC 2015 [7]	ISSCC 2015 [8]	This work
Technology	28nm UTBB FD-SOI	65nm SOTB	65nm CMOS	180nm CMOS	14nm Tri-gate CMOS
Processor	32-b LatticeMico RISC	32-b[a]	32-b ARM Cortex M0+	32-b ARM Cortex M0+	32-b x86 IA
Area (mm²)	1.32	16.9	3.76	2.04	0.79
V_{DD} range and V_{OPT}(V)	0.3-0.5 (V_{OPT} = 0.375)	0.35-0.6 (V_{OPT} = 0.41)	0.25-1.2 (V_{OPT} = 0.35[b])	0.16-1.15 (V_{OPT} = 0.35-0.55)	0.308 - 1.0 (V_{OPT} = 0.370) $V_{CRO,MIN}$ = 0.128
Frequency range	1-77MHz	6-27MHz	27kHz-66MHz	2-15Hz	0.5-297MHz
Energy	4.9pJ/cycle[c]	33uW/MHz[d]	11.7pJ/cycle[b]	147.5uW/MHz; 44pJ/instruction[e]	17.18pJ/cycle[f]
Total on-chip memory	64KB Inst + 8KB Data	64KB SRAM + 16KB ROM	8KB ULV SRAM + 16KB SRAM + 2KB BootROM	128B	8KB I$ + 8KB DTCM + 64KB SMEM + 16KB BootROM

[a]ISA not reported; [b]AES encryption workload; [c]Workload not reported; [d]CRC32 workload; [e]Toggle program; [f]MCU *always-active*, running AES encryption workload

Fig. 11: (a) Platform power states controlled by MCU (b) MCU energy at V_{OPT} for different power states (S0-S3)

Fig. 12: Measured WSN AOAS power profile over a 4-minute interval

Lensless Smart Sensors: Optical and Thermal Sensing for the Internet of Things

Patrick Gill and Thomas Vogelsang

Rambus Inc., Sunnyvale CA, USA
tvogelsang@rambus.com

Abstract

Lensless Smart Sensors (LSS) add optical and thermal sensing capabilities to the Internet of Things (IoT) in a form factor that cannot be achieved with traditional lensed systems. Different from lensed systems, LSS is based on diffraction instead of refraction, and different from other diffractive optical elements in that it can operate with a wide field of view (FOV) and over a wide wavelength band. LSS's use of computation to extract information from a captured scene makes LSS a good fit for applications where the goal is not to create an image for human consumption, but for machine viewing (e.g. to trigger actions in a connected device). Since the raw sensed image is encoded by the grating structure, LSS opens applications where the use of a camera would create privacy concerns. This paper describes the operational principle of LSS and discusses three examples in more detail.

Keywords: optical sensing, thermal sensing, diffractive optics, internet of things, computational optics

Introduction

An important part of the Internet of Things (IoT) is the multitude of sensors integrated with connected devices. According to a recent study [1], IoT devices in the market segments of building automation, healthcare and life sciences, industrial, environment, security & public safety and retail & logistics will all have sensors that operate in visible or infrared (IR)/thermal wavelengths. Such optical and thermal sensing provides access to rich information about the environment where the devices are deployed. Typically the output of an optical or thermal sensor in the IoT does not need to be an image optimized for human consumption, but instead processed information about the observed environment. Traditional lensed systems have been optimized to generate images for humans. Their size, performance, price, power consumption and privacy concerns may make integration into IoT devices not ideal. The diffractive optics utilized in Lensless Smart Sensors (LSS) are co-designed with computational algorithms to extract relevant information from a scene in a low cost and compact form factor. Fig. 1 shows as an example a $2 \cdot 2mm^2$ custom image sensor we built [2] that has ultra-low power image change detection circuitry on die.

image change detection logic (row and column logic and ADCs are hidden underneath the aperture)

opening with grating (red square shows the size of the pixel array underneath)

metal aperture

Fig. 1 Die photograph of image change detection sensor in 180nm technology with attached metal aperture and phase grating.

In the following sections we describe the operational principle of LSS and then discuss three sample applications – point range finding, eye tracking and thermal occupancy detection.

Operational Principle of LSS

LSS uses phase modulation to achieve an illumination pattern on the image sensor from which the observed scene can be reconstructed. This allows capturing and concentrating more light than approaches modulating the amplitude [3, 4]. To achieve modulation a phase grating is mounted on top of a conventional imaging array. The imaging array is sensitive to the wavelength LSS is designed for, i.e. a CMOS image sensor for visible & near IR and a micro-bolometer or thermopile array for thermal IR. The special property of the phase grating is its phase anti-symmetric structure. Due to this structure and the half-wavelength depth of the grating features, light from left and right of the anti-symmetric boundary cancels in a curtain under the boundary. These curtains are robust against depth and wavelength changes [5] different from the approach using the Talbot effect [6]. Fig. 2 shows a one-dimensional cut through the grating and the phase anti-symmetric features.

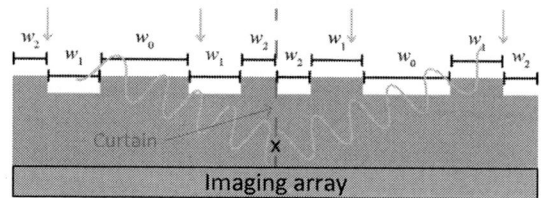

Fig. 2 LSS structure with phase anti-symmetric grating.

The grating material depends on the wavelength of LSS operation. For visible and near IR wavelengths, glass and, polycarbonates might be used, where at thermal wavelengths, silicon or germanium are possible materials. The materials are structured using either a stamping or an etching process.

Point Range Finding

Measuring the distance to an illuminated point with LSS does not require image reconstruction. Instead the stereoscopic shift of the Point Spread Functions (PSF) of a point light source viewed through two gratings is used to determine the distance. Since the center of the extended PSF can be determined with great accuracy, the ratio of measurable distance to baseline is much larger than for stereoscopy using lenses. The experimental setup is shown in Fig. 3a. We have used two identical phase gratings mounted in apertures on a shared pixel array. Each grating aperture has a diameter of 55μm and the distance between their centers is 1.86mm. With that setup we have been able to measure distances up to 50cm with an error of less than 8% as illustrated in Fig. 3b, a ratio of distance to baseline of 268:1.

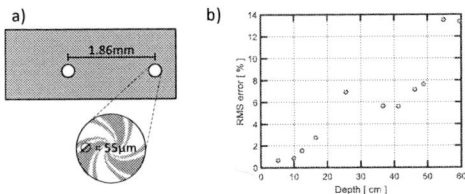

Fig. 3 Stereoscopic point range finding a) grating structure b) error as function of distance between point light source and LSS.

Eye-Tracking

Accurate eye-tracking is a task that is desirable for applications like smart glasses, virtual reality, and augmented reality systems. Eye-tracking hardware is difficult to design since the limited FOV and thickness of a lensed system makes unobtrusive integration into eyeglasses difficult. LSS has a wider FOV of up to 120° and is almost flat (distance between grating and imaging array 100μm - 500μm), so it can be mounted much closer to the eye. The eye-tracking method when using LSS follows a variant of the proposal by Zhu and Ji [7], using a binocular system to locate images of near-IR light sources reflected in the surface of the eye. Similar to the point tracking application above, no image reconstruction is required and the precision of the point location is improved by the extended PSF.

The geometrical arrangement is shown in Fig. 4.

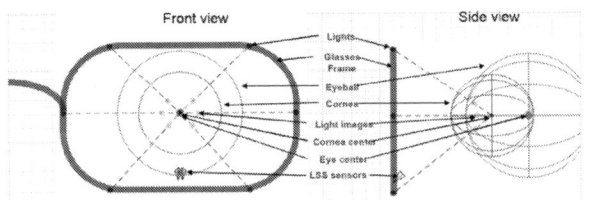

Fig. 4 Eye-tracking principle of operation: reflections in the eye of multiple LEDs are triangulated with a pair of LSS sensors allowing calculation of the gaze direction.

TABLE I
SIMULATED MEDIAN ERROR OF GAZE ANGLE

Eye azimuth	Eye elevation	1 set of 2 LSS	2 sets of 2 LSS
-40°	30°	0.61°	0.11°
-40°	0	0.55°	0.08°
-40°	-30°	1.17°	0.13°
0	30°	0.33°	0.06°
0	0	0.31°	0.05°
0	-30°	0.53°	0.05°
40°	30°	0.59°	0.11°
40°	0	0.55°	0.08°
40°	-30°	1.18°	0.12°

We have not built such an eye-tracking system yet. We have however simulated its expected performance varying different parameters. Table I shows the result of these simulations. Using two sets of binocular LSS pairs compared to one will improve the accuracy significantly.

Thermal IR Occupancy Detection

In the IoT, occupancy detection is an important function. For example, this could be detecting if and how many persons are in a room or if a person is sitting on the seat of a car. Thermal

IR wavelengths allow for easy distinction between people and the cooler environment around them and are therefore the ideal wavelengths to use for such functions. LSS's principle of operation works not only in visible or near IR, but also in thermal IR. The grating features have to be adapted to the longer wavelength While the operational principle remains the same in thermal as in visible wavelengths, the grating material needs to be transparent in thermal IR and the sensor needs to be a thermal IR sensitive pixel array [8]. Fig. 5 shows the experimental setup and a sample result.

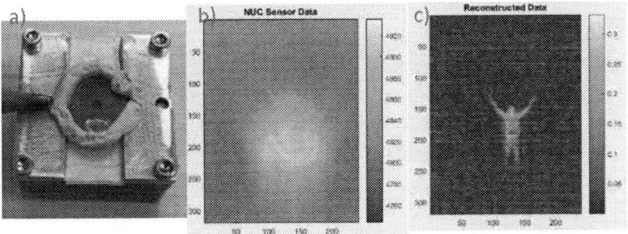

Fig. 5 LSS operating at thermal IR wavelength a) LSS mounted on micro-bolometer (the grating is in the circular opening of the aperture) b) sampled signal after non-uniformity correction c) reconstructed image.

The person in the experiment in Fig. 5 was standing about 2m away from the sensor. The field of view (FOV) is about 90°, making LSS an ideal solution to observe a large area from nearby..

Summary and Conclusions

LSS is a new technology for optical and thermal sensing that fits well to the requirements of the IoT. Its size and flat structure make it easy to integrate with devices produced in a semiconductor process. The quality of visual information that can be extracted from the phase pattern is good enough for many computer/machine vision applications, even when it would not be considered a beautiful image when viewed by a human.

References

[1] "Technologies and Sensors for the Internet of Things", online [http://www.yole.fr/iso_upload/Samples/Yole_IoT_June_2014_Sample.pdf], cited 1/22/16

[2] Gill, Patrick, et al., "Computational diffractive imager with low-power image change detection," *Computational Optical Sensing and Imaging*, Optical Society of America, 2015.

[3] Zomet, Assaf, and Shree K. Nayar. "Lensless Imaging with a Controllable Aperture," *IEEE Conference on Computer Vision and Pattern Recognition*, 2006

[4] Asif, M., et al. "FlatCam: Replacing Lenses with Masks and Computation," *IEEE International Conference on Computer Vision Workshops,* 2015.

[5] Gill, Patrick R. "Odd-symmetry phase gratings produce optical nulls uniquely insensitive to wavelength and depth." Optics letters 38.12 (2013): 2074-2076

[6] Wang, Albert et al., "Angle Sensitive Pixels in CMOS for Lensless 3D Imaging." *IEEE Custom Integrated Circuits Conference*, 2009

[7] Zhu, Zhiwei, and Ji, Qiang. "Eye gaze tracking under natural head movements." *Computer Vision and Pattern Recognition, 2005. CVPR 2005.* IEEE Computer Society Conference on. Vol. 1. IEEE, 2005.

[8] Erickson, Evan et al., "Miniature lensless computational infrared imager", *IS&T Electronic Imaging Conference 2016*, San Francisco, February 2016

Features of retinal prosthesis using suprachoroidal transretinal stimulation from an electrical circuit perspective

Yasuo Terasawa[1], Kenzo Shodo[1], Koji Osawa[1], Jun Ohta[2]

[1]Vision Institute, NIDEK Co Ltd, Gamagori, Aichi, Japan
[2]Graduate School of Materials Science, Nara Institute of Science & Technology, Ikoma, Nara, Japan
{Yasuo_Terasawa, Kenzo_Shodo, Kouji_Oosawa}@nidek.co.jp, ohta@ms.naist.jp

Abstract

Several research groups globally have been developing retinal prostheses since the 1990s. We have been developing a retinal prosthesis based on suprachoroidal transretinal stimulation (STS). In this paper, features of STS-based retinal prosthesis will be described in detail especially from an electrical circuit perspective.

Keywords: retinal prosthesis, STS, multiplexer, leakage current, wireless communication

Introduction

Retinal prostheses, which provide some visual information to patients clinically diagnosed with acquired blindness, have been under development since 1990s, and recently two devices are available in the market for commercial use [1], [2]. Suprachoroidal transretinal stimulation (STS) was proposed as a safe and effective retinal stimulation methodology at the Osaka University in Japan. We have been developing a retinal prosthesis based on STS as shown in Fig. 1. The advantage of STS is its safety because stimulating electrodes are placed outside the eyeball. In addition, STS has a potential to realize wider field of view since relatively large electrode array is implantable in the suprachoroidal area.

Overview of STS retinal prosthesis

In the STS retinal implant system, both electric power and the images captured by a camera are wirelessly transmitted from outside the human body to the implant (Fig. 2). The implant comprises of main unit, multiplexer (MUX), and a 49-channel electrode array. The main unit is implanted behind the ear of the patient. The MUX and electrode array are implanted on the eye. The main unit and MUX are connected by a subdermally-implanted flexible lead. The main unit receives both image information and electrical power, generates current pulses, and transmits the pulses to the MUX. The MUX connects one of 49 electrodes to the current source of the main unit.

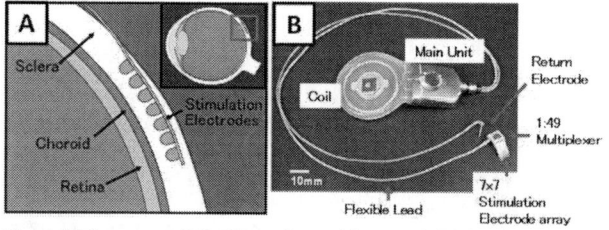

Fig. 1 STS system. (A) electrode positions and (B) implant unit.

In our previous version of device, all the electrodes and the main unit was directly connected with flexible lead because of the presence of only 9 electrodes [3]. If we apply the same

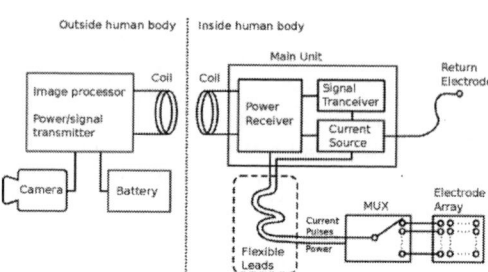

Fig. 2 Block diagram of STS retinal implant system

architecture to the current 49-channel device, the flexible lead (Fig.1B) will be too thick and rigid to implant. This is why we propose a novel architecture to place the MUX onto the eyeball. This architecture allows us to control several tens of electrodes using only a few conductive lines.

Features of STS retinal prosthesis

A. Fail-safe operation against water infiltration into non-hermetically sealed parts

It is impossible to completely avoid water infiltration into the non-hermetically sealed parts for extended periods (decades) in the human body. Applying a DC voltage to such parts, for example the "Flexible Lead" in Fig. 1B, is potentially dangerous because water infiltration could lead to metal corrosion and an irreversible electrochemical reaction. To avoid these problems, power from main unit to MUX is supplied in the form of AC voltage (6 V; 1 MHz) instead of DC voltage.

B. Charge balancing

If the net charge that passes through an electrode is not zero, an irreversible electrochemical reaction could occur. Most neural prostheses employ a biphasic current pulse to accomplish charge balancing; however, it is difficult to avoid charge imbalance given that there is no perfect current source and the accumulation of residual imbalance charge of stimulation pulses is inevitable. To solve this problem, all stimulation electrodes and the return electrode are periodically

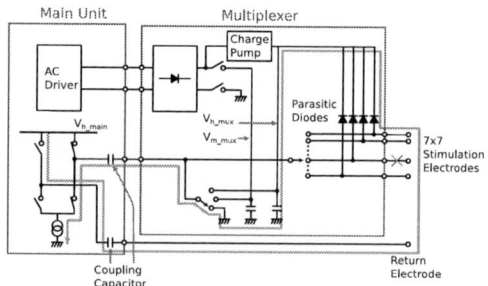

Fig. 3 Path of current leakage (indicated in orange) when selected electrode is broken.

shorted to discharge the imbalance charge accumulated into the coupling capacitors (Fig. 3).

C. Prevention of current leakage to unselected electrodes

If the impedance of an electrode increases abnormally owing to plausible issues such as breaking of wire (marked with red X in Fig. 3), current leakage via parasitic diode could occur (orange arrow in Fig. 3). This happens when the maximum stimulator voltage is applied between the unselected stimulation electrodes and the return electrode to attempt to induce a current pulse to a broken electrode. To circumvent this, the source voltage of MUX is set to a higher voltage than that of main unit, i.e. $V_{h\_mux} > V_{h\_main}$ in Fig. 3.

Similarly, if the potential of stimulation electrode exceeds the source voltage of MUX during electrical stimulation, unintended current leakage could occur. Therefore stimulation electrode is biased to the GND of MUX in the first phase of cathodic-first current pulse, and then biased to an intermediate potential (V_{m_mux} in Fig.3) in the second phase.

D. Communication between main unit and MUX with pulse period shift keying (PPSK)

The electrode selection signal is superimposed onto AC power supply voltage waves from main unit to MUX. High frequency is preferable for realizing high-speed communication; however low frequency is better for lowering noise emission. Therefore, we proposed the communication between the main unit and MUX by modulating the pulse period (Fig. 4), and demonstrated its functionality (Fig. 5). The advantage of utilizing PPSK is high-speed communication because in principle, assuming that carrier frequency is F (Hz), F bit/s communication is attainable. In addition, PPSK allows us to simplify the MUX receiver circuit because the detection of PPSK is asynchronous, which makes the reconstruction of reference carrier redundant.

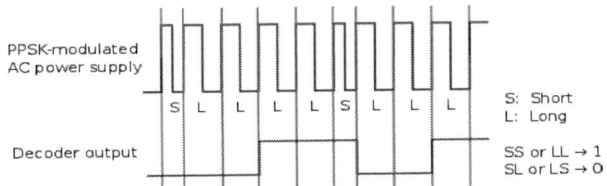

Fig. 4 Pulse period shift keying (PPSK) modulation

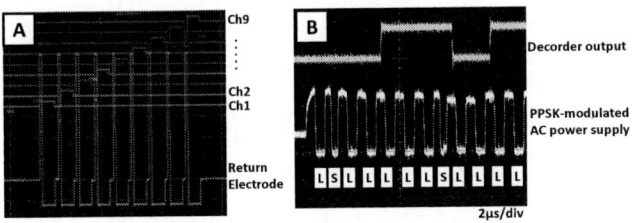

Fig. 5 (A) Example of current waveforms during electrical stimulation. Asymmetric biphasic pulses (1200μA 410μs + 240μA 2050μs) were used. (B) AC power supply waveform with PPSK modulation (yellow) and demodulated waveform (green).

The main unit and MUX chips were fabricated in 0.35μm HV CMOS. The photographs and their specifications are shown in Fig. 6 and Table I.

Fig. 6 Main unit chip (Left) and MUX chip (Right). Specifications are summarized in Table I.

TABLE I
Specifications of Main Unit chip and MUX chip

	Main Unit	MUX
Technology	0.35μm HV CMOS	0.35μm HV CMOS
V_{DD}	12V	6V / 15V (boosted)
Die size	5000 x 5000 μm	3400 x 2500 μm
Power Consumption	62mW	0.3mW
Data rate	93kbps	1Mbps

Results

The normal operation of the MUX was confirmed by monitoring the current waveforms during electrical stimulation (Fig. 6A). The AC power supply waveform from the main unit to MUX modulated with PPSK is shown in Fig. 6B. We confirmed that the signal was correctly demodulated by the MUX, and stable communication speeds up to 1Mbit/s were achieved.

By using these circuits, we fabricated a retinal prosthesis system for STS and successfully demonstrated its operation in vitro and in vivo.

Conclusions

We proposed a retinal implant system with an external MUX unit to control 49 electrodes using 5 conductive lines, and confirmed normal operation of the system. In addition, a clinical study of chronic implantation of this system to three patients clinically diagnosed with retinitis pigmentosa is under way [4]. In our subsequent work, we will attempt to develop a next-generation system that is able to apply pulses to several electrodes simultaneously using multiple current sources, and a system with one main unit and multiple MUX units.

References

[1] M.S. Humayun, et al., "Interim results from the international trial of Second Sight's visual prosthesis," Ophthalmology, 119, 779-788, 2012.

[2] K. Stingl, et al., "Functional outcome in subretinal electronic implants depends on foveal eccentricity," Invest. Ophthalmol. Vis. Sci., 54, 7658-7665, 2013.

[3] T. Fujikado, et al., "Testing of semichronically implanted retinal prosthesis by suprachoroidal-transretinal stimulation in patients with retinitis pigmentosa," Invest. Ophthalmol. Vis. Sci., 52, 4726-4733, 2011.

[4] T. Fujikado, et al., "Testing of Chronically Implanted 49-Channel Retinal Prosthesis by Suprachoroidal-Transretinal Stimulation (STS) in Patients with Advanced Retinitis Pigmentosa", Invest. Ophthalmol. Vis. Sci., 56, E-Abstract 3816, 2015.

Multi-modal Smart Bio-sensing SoC Platform with >80dB SNR 35µA PPG RX Chain

Ajit Sharma, Seung Bae Lee, Arup Polley, Sriram Narayanan, Wen Li, Terry Sculley, Srinath Ramaswamy

Kilby Labs, Texas Instruments Incorporated, Dallas, TX., USA

asharma@ti.com

Abstract

A multi-modal analog front end (AFE) and ultra-low energy bio-sensing CMOS SoC is presented. System/ circuit techniques enable signal path duty cycles as low as sub-1% and result in a 35µA Photo Plethysmography (PPG) RX Chain – 5X lower than published state of the art – while maintaining overall SNR > 80dBFS. The signal chain is adaptively synchronized by an ultra-low power FSM and includes a 1.3µW 14b 1kSPS SAR A/D. Input signal-aware, real-time data path adaptation is achieved by leveraging on-the-fly algorithms running on an external microcontroller (µC) to further reduce system energy. A programmable, asynchronous capacitive reset amplifier (PARCA) with NEF of 4.8 and dx/dt analog feature extractor demonstrate energy efficient ECG capture. A battery-powered, Bluetooth low energy (BLE) based, wearable platform with simultaneous ECG and PPG acquisition using this AFE has been demonstrated.

Motivation & Challenges

Optical heart-rate monitoring (HRM) using PPG [1] is widespread in wearable devices for health, fitness and mobile patient monitoring. The unique challenge is accurate, synchronous sensing of multiple biological parameters (PPG + ECG) over diverse subject types and use-cases – while operating from a limited energy budget (e.g. coin cell /paper battery). This paper presents a 130nm 100µW, energy efficient bio-sensing SoC for HRM using PPG and ECG. The PPG RX chain draws 35µA at pulse repetition frequency (PRF) of 100Hz, while maintaining a dynamic range (DR) > 80dBFS at duty-cycle of 1%. The multi-modal AFE includes a capacitive reset front-end for ECG and a separate instrumentation amplifier (INA). Energy efficiency without performance degradation is achieved through the combination of circuit and system-level techniques: (i) aggressive duty cycling of the analog signal chain, (ii) algorithmically assisted on-the-fly signal chain adaptation to relax block-level complexity and power, and (iii) feature extractor-based adaptive sampling [2].

An LED is pulsed at a fixed PRF onto the wrist (Fig 1). The AC/DC ratio or Perfusion Index (PI) is typically 0.05% − 1% (empirically obtained from ~30 human subjects) varying widely across subjects. PI is correlated to the modulation index (m) of the PPG waveform. Frequency of the AC component in the PPG signal yields HR. To achieve HR accuracy of ±1bpm, the SNR of the AC component must be ≥ 25dB. At a given PRF, the SNR is proportional to both, LED current (I_{TX}) and duty-cycle (Eq.in Fig 1). Both AC and DC components (I_{PLETH}) are proportional to the LED intensity through the current transfer ratio (CTR) – thus increasing I_{TX} to improve SNR creates the challenge of detecting a small AC signal in the presence of a large DC interferer. This implies at a low PI of 0.05%, the DR of the entire PPG RX chain must be > 80dB.

System- and Circuit-Level Architecture & Results

The AFE communicates to an external TI-MSP430 µC [3] (Fig 2) that runs multiple HR extraction algorithms and dedicated subroutines to adapt analog parameters on the fly. A transimpedance amplifier (TIA) with integrated DAC for static ambient cancellation senses the photocurrent. A 50mA (max), 8b LED driver capable of driving dual LED's enables both optical HRM and SpO_2 measurement. Dedicated switched-RC LPF's demodulate signals from the LED and ambient phases respectively. The ambient light measurement (measurement without LED) phase is inserted between LED phases to enable system-level CDS. The three front-end stages are time multiplexed into a DAC, PGA and ADC. A 14b duty-cycled, segmented capacitive SAR ADC with on-chip reference buffer and mismatch calibration is chosen for its energy efficiency, memory-less nature, and ease of time multiplexing without excessive cross-talk between time-interleaved measurements.

Aggressive duty cycling of analog and digital signal paths in the SoC enables LED on-times of < 40µs/sample, for energy efficiency. A reduced LED on time places stringent restrictions on settling time and bandwidth (BW) of the front-end TIA. To ensure the system is not RX SNR limited, duty-cycling of the TIA uses a two-step switching scheme to transition from bias-on to active via an intermediate low-power mode. This allows higher peak bias currents whist active, precluding the higher noise, low BW and PVT variations associated with sub-VT designs [4] – ensuring reliable operation across lots for volume manufacturing. Fig 3a plots measured time-averaged TIA current vs. duty-cycle. No duty-cycling induced settling artifacts are observed at the TIA output at the sampling instants (Fig 3b). The comparator in the SAR A/D is optimized to mitigate kick-back noise and allow duty-cycling – resulting in 75dB SNDR at 1kSPS and 1.3µW average power (Fig 3c).

Energy efficiency is further improved by reducing current requirement of the individual circuit blocks, through signal dependent, on-the-fly µC based feedback. To address PI of 0.5%, conventional solutions [1] use ΣΔ ADCs with DR > 20b to digitize both the DC and AC, extracting the AC component by digital post-processing. In addition to the front-end current DAC for static ambient interferer rejection [1, 8], a second switched capacitor DAC is used to reduce the dynamic DC component (I_{PLETH}) without impacting the HR-bearing AC component. A bang-bang control algorithm running on the µC sets the DAC value sample by sample to ensure the A/D output remains within programmed thresholds. The PGA amplifies the residual AC component prior to A/D conversion (Fig 4), effectively increasing the overall system DR from 14b to 19b at gain of 32 (Fig 4). Applying DAC correction every sample reduces in-band (0.5 − 5Hz) feedback interference. At duty-cycles > 2%, the entire PPG chain (TX + RX) SNR is measured using a loop-back method to be > 95dBFS (Fig 5) – sufficient to resolve HR in cases where PI ~ 0.05%. The corresponding time averaged current draw from the entire TX + RX chain (without LED) is 46µA.

The fully differential PARCA with capacitive gain of 200 senses ECG directly from electrodes without any off chip DC block (Fig 1). In contrast to analog DC servo loops [5], the µC controls when DC feedback is applied. Once the A/D output crosses programmed thresholds, the µC signals S1 to be closed asynchronously for ½ clock period, setting the DC bias at the summing node and ensuring linear operation. Absence of sampling precludes kT/C noise associated with switched cap AFEs. The reset instance can be algorithmically modified to prevent switching-induced charge injection. Compensation of PVT variations in pseudo-resistor topologies is avoided by varying the reset thresholds. The PARCA consumes 2.5µA with an NEF of 4.8. A dx/dt extractor based on [2] allows increasing the ADC sample frequency on-the-fly for accurate QRS peak detection while reducing system energy overhead.

SoC & Wireless Platform Summary

Table I tabulates the AFE current consumption in various operation modes. A battery powered BLE platform using this SoC synchronously captures ECG + PPG (Fig 6). The PPG and ECG HR's track to within ±5bpm and are transmitted by the TI-CC2541 every 2s to an iPhone® with BLE SIG HR profile. The platform lasts 3 days on a 250mAHr battery (Table II) with concurrent ECG and PPG acquisition (100Hz), adaptive µC algorithms enabled and BLE transmission every 2s. For comparison against published works, a figure-of-merit (FOM)

is derived based on the Walden-FOM commonly used for ADCs. The ENOB of the PPG RX chain is calculated from the loop-back SNR and the BW is set as the PRF – normalized in Table III to 100Hz for comparison. The AFE has optimized energy consumption by leveraging digital assists that exist in any sensing eco-system. Circuit techniques to enable aggressive duty cycling without performance degradation and the proposed cycle-to-cycle feedback to extend overall DR have resulted in the lowest FOM – 14.3pJ/sample – PPG-RX chain reported, with 5X lower current than state-of-the art.

References

[1] TI-AFE4404 datasheet (Texas Instruments Inc.) 2015.
[2] Yazicioglu, et al., JSSC Vol 46, No. 1, pp. 209 – 223, 2011.
[3] TI-MSP430 datasheet (Texas Instruments Inc.) 2015.
[4] Tavakoli et al., IEEE Trans. BioCAS Vol 4. No 1, pp 27 – 38, 2010.
[5] Harpe et al., ISSCC Digest 21.2, pp. 382 – 383, 2015.
[6] Glaros et al., IEEE Trans BioCAS Vol 7, No. 3, pp. 363 – 375, 2013.
[7] Winokur et al., IEEE Trans. BioCAS Vol 9, No. 4, pp 581-589, 2015.
[8] Muller et al., JSSC Vol 47, No. 1, pp. 232 – 243, 2012.

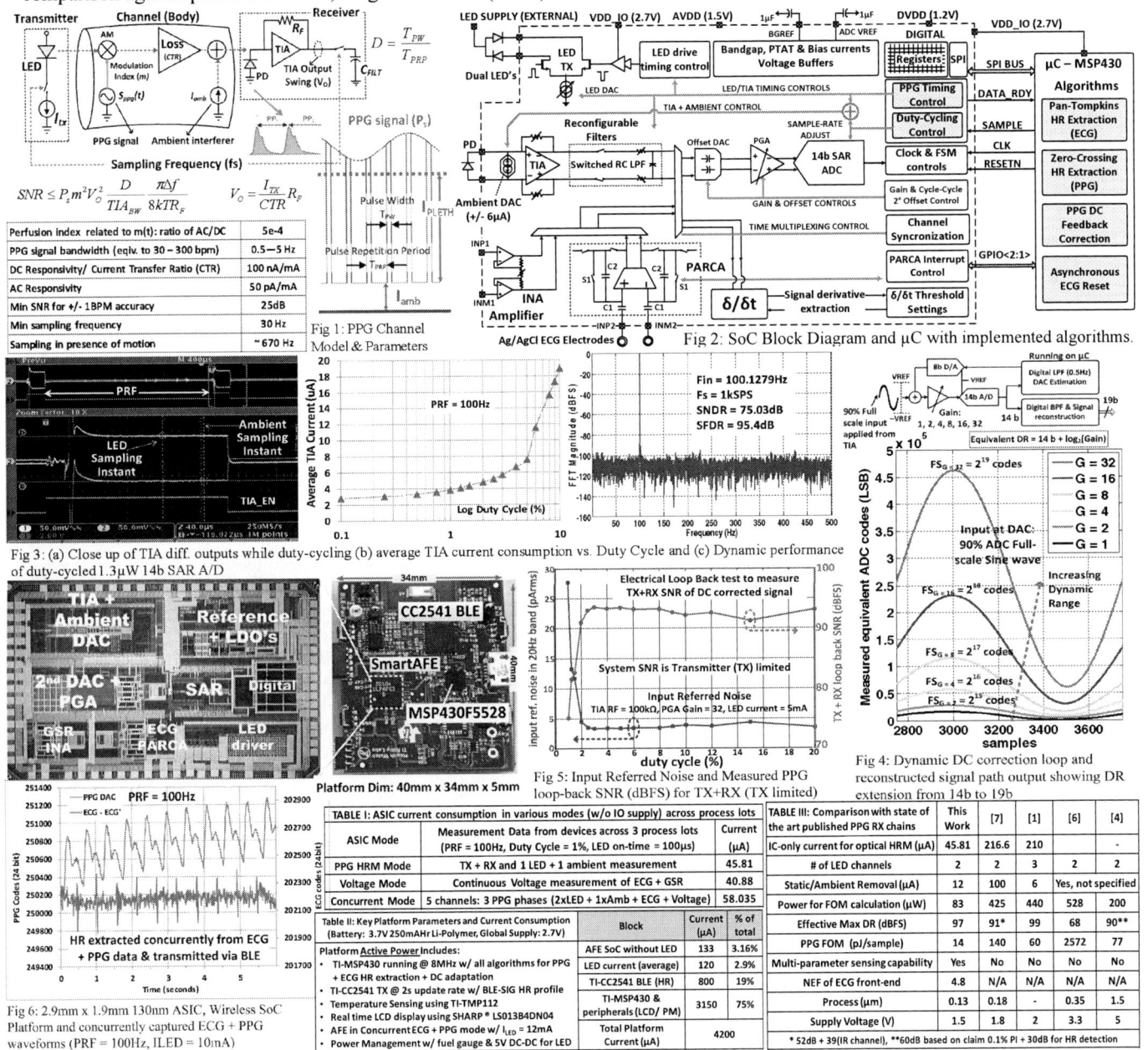

Fig 1: PPG Channel Model & Parameters

Fig 2: SoC Block Diagram and µC with implemented algorithms.

Fig 3: (a) Close up of TIA diff. outputs while duty-cycling (b) average TIA current consumption vs. Duty Cycle and (c) Dynamic performance of duty-cycled 1.3µW 14b SAR A/D

Fig 5: Input Referred Noise and Measured PPG loop-back SNR (dBFS) for TX+RX (TX limited)

Fig 4: Dynamic DC correction loop and reconstructed signal path output showing DR extension from 14b to 19b

Fig 6: 2.9mm x 1.9mm 130nm ASIC, Wireless SoC Platform and concurrently captured ECG + PPG waveforms (PRF = 100Hz, ILED = 10mA)

TABLE I: ASIC current consumption in various modes (w/o IO supply) across process lots

ASIC Mode	Measurement Data from devices across 3 process lots (PRF = 100Hz, Duty Cycle = 1%, LED on-time = 100µs)	Current (µA)
PPG HRM Mode	TX + RX and 1 LED + 1 ambient measurement	45.81
Voltage Mode	Continuous Voltage measurement of ECG + GSR	40.88
Concurrent Mode	5 channels: 3 PPG phases (2xLED + 1xAmb + ECG + Voltage)	58.035

Table II: Key Platform Parameters and Current Consumption (Battery: 3.7V 250mAHr Li-Polymer, Global Supply: 2.7V)

Block	Current (µA)	% of total
AFE SoC without LED	133	3.16%
LED current (average)	120	2.9%
TI-CC2541 BLE (HR)	800	19%
TI-MSP430 & peripherals (LCD/ PM)	3150	75%
Total Platform Current (µA)	4200	

Platform Active Power Includes:
- TI-MSP430 running @ 8MHz w/ all algorithms for PPG + ECG HR extraction + DC adaptation
- TI-CC2541 TX @ 2s update rate w/ BLE-SIG HR profile
- Temperature Sensing using TI-TMP112
- Real time LCD display using SHARP ® LS013B4DN04
- AFE in Concurrent ECG + PPG mode w/ I_LED = 12mA
- Power Management w/ fuel gauge & 5V DC-DC for LED

Platform Dim: 40mm x 34mm x 5mm

TABLE III: Comparison with state of the art published PPG RX chains

	This Work	[7]	[1]	[6]	[4]
IC-only current for optical HRM (µA)	45.81	216.6	210	-	-
# of LED channels	2	2	3	2	2
Static/Ambient Removal (µA)	12	100	6	Yes, not specified	
Power for FOM calculation (µW)	83	425	440	528	200
Effective Max DR (dBFS)	97	91*	99	68	90**
PPG FOM (pJ/sample)	14	140	60	2572	77
Multi-parameter sensing capability	Yes	No	No	No	No
NEF of ECG front-end	4.8	N/A	N/A	N/A	N/A
Process (µm)	0.13	0.18	-	0.35	1.5
Supply Voltage (V)	1.5	1.8	2	3.3	5

* 52dB + 39(IR channel), **60dB based on claim 0.1% PI + 30dB for HR detection

An FPGA-accelerated Partial Image Matching Engine for Massive Media Data Searching Systems

Takashi Shimizu, Yasumoto Tomita, Hidetoshi Matsumura, Masahiko Sugimura, Hironobu Yamasaki, David Thach, Takashi Miyoshi, Takayuki Baba, Yasuhiro Watanabe and Atsushi Ike

Fujitsu Laboratories LTD., 4-1-1, Kamiodanaka, Nakaharaku, Kawasaki, Japan
Phone/Fax: +81-44-754-2931, E-mail: shimizu.taka-01@jp.fujitsu.com

Abstract

We propose and demonstrate an FPGA-accelerated partial-image-matching engine for massive media-data searching systems. To take advantage of FPGA, a highly parallelized and pipelined architecture with an application-specific calculation was adopted. Our prototype system achieves 32 times better runtime performance than a CPU-based solution. *Keyword:* partial image matching, accelerator, FPGA.

Introduction

Today, huge amounts of media data such as digitized documents (Microsoft® PowerPoint, Word files) and CAD-designed data are stored on storage, and there is a high demand for fast data search to improve the working efficiency. Since most of these media data are widely occupied by figures copied or modified from original ones, a media data search system that finds partially duplicated images is required (Fig. 1). Since AI-based matching [1] requires massive training data to extract feature descriptors from each category, a brute-force matching of binary feature descriptor [2] was chosen for a partial image matching. However, it still requires a huge amount of computation resource, resulting in a long waiting time to retrieve matched documents.

Since the amount of media data is increasing while the transistor's scaling slows down, hardware acceleration using FPGA gathers strong attention [3], because FPGA can integrate application specific logic circuits with greater configuration flexibility compared to ASIC.

However, in order to achieve meaningful performance gain using FPGA, it is important to design an accelerator architecture that suits the given application to overcome FPGA's lower operating frequency, which is typically around several hundreds of MHz, roughly 1/10 of that of today's CPUs. We should utilize the following advantages that FPGA has over CPU: (1) an FPGA can be programmed to execute an application-specific calculation that may require several instructions on a CPU, (2) massively parallel configuration is possible within the circuit resource of an FPGA, and (3) utilization of processing units can be held high by carefully designing the data management architecture. In this paper, we present a partial-image-matching FPGA accelerator that achieves 32 times performance improvement by using FPGA-specific optimization for both hardware and algorithm. Unlike ref. [3], which achieved 2x performance improvement by using FPGAs while maintaining consistency with software.

Partial image matching algorithm

To perform the proposed partial-image matching algorithm [4], the query image is first scaled into 14 scaled images by using a simple bilinear filter. The feature extraction is then performed on each scaled image. Here, Canny edge detection [5] is used to generate keypoints from the images. A 128-bit feature descriptor is generated from 128 randomly selected point-pairs in the 48x48 region around each keypoint coordinate. Next, the feature extraction of the database images and matching process between the database and query

images are performed. The matching process consists of two steps, coarse matching and fine matching. In the coarse matching step, keypoints picked at every 12 pixels and all scaled query images are used for matching. In the fine matching step, the interval of keypoints is 6 pixels, and only coarse-matched query image and its neighboring-scale query images are used.

In both matching steps, the Hamming distance between the query-image and the database-image keypoint descriptors is calculated with changing the query-image keypoint. For each keypoint of database images, the keypoint in the query image that has the smallest Hamming distance is selected as the representative keypoint that casts a vote for a section in the voting space based on the target coordinates:

$$x = (xq - xd)$$
$$y = (yq - yd)$$

where (xq,yq) and (xd,yd) are the coordinates of the query image and the database image key-point, respectively. The entire voting space is divided into equal-sized rectangle sections and the section where the coordinate falls onto gets the vote. If partially duplicate parts with the query image exist in the database image, the key-point pairs of the duplicate parts will likely vote for the same section in the voting space (Fig. 1). When a section whose score exceeds a threshold is found, the ID of the database image is returned as a "matched" image.

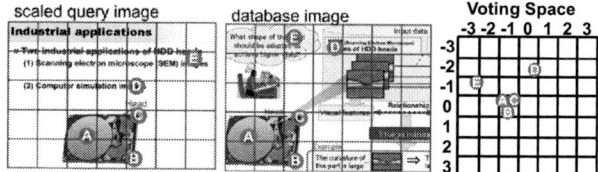

Fig. 1. Voting scheme for the proposed partial-image matching.

Accelerator architecture

The FPGA accelerator core consists of 14 feature extraction modules, two matching modules, and one control module, all of which are designed in Verilog-HDL (Fig. 2(a)). All modules are connected with the Avalon interface, which is the standard bus interface of IPs on Altera FPGAs.

The feature extraction module extracts feature descriptors. The module simply reads the pixels from the main memory, calculates 128-bit feature descriptor by using subtraction operators, and writes back the result to the memory. Input and Output buffers operate in a ping-pong fashion to enable the continuous data delivery to the processing circuits in the module, resulting in a high working rate of the processing circuits.

The matching module performs both the coarse and fine matching (Fig. 2(b)). Hamming distance calculations between 16 keypoints of a database image and 4 keypoints of a query image are performed simultaneously, resulting in the throughput of 64 keypoint-pairs/cycle. Then the voting and matched-image detection is performed. The operations in this module are integrated into a pipeline to enhance the working rate.

Figure 3 shows timing charts of the matching module. The coarse matching operation for each scaled query image is processed in a single stage of the pipeline, whereas the fine matching operation needs several cycles. When database images X, Y, and Z, for example, go through the coarse matching with no matching detected, a single stage is assigned to each image (Fig. 3 (a)). When a coarse match is detected for the database image X, the fine matching of image X is inserted between the coarse matching of images Y and Z to avoid an idle cycle of the pipeline stage (Fig. 3 (b)).

As described above, the task schedule of the matching module varies depending on the result of coarse matching whereas the feature extraction module has a fixed operation time. The control module has FIFOs corresponding to each feature extraction or matching module so that it controls the dispatch of the tasks depending on FIFO vacancy to keep a higher working rate of both feature extraction and matching modules.

In addition, not only the FPGA accelerator tasks but also the server application tasks that set the FPGA-accelerator parameters and displaying the results are pipelined to achieve multi-level pipeline structure. As a result, if the first database image is matched with the query image, it immediately appears on the display even though the query is in progress. This immediate response scheme improves the usability of the system.

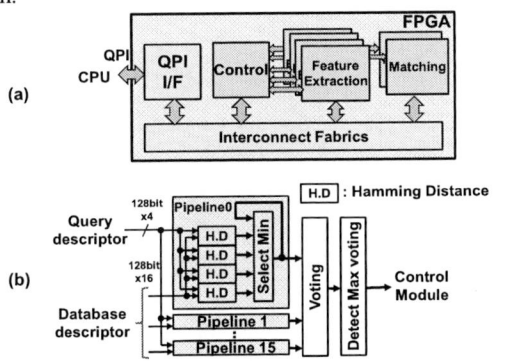

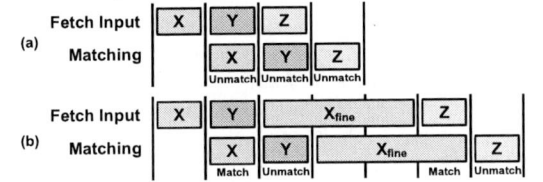

Fig. 2. Block diagram of (a) FPGA-accelerator, (b) Matching module.

Fig. 3. Timing diagram of matching module.

Experimental Results

In order to evaluate the runtime performance of the proposed FPGA-accelerated partial-image matching engine, we developed a document search system shown in Fig. 4. It consists of client computers for the user interface and a server computer that acts as the partial-image matching engine. The server computer communicates with multiple client computers using an Ethernet in order to share documents as a general file server.

Two windows appear on the display of a client computer. One is the Microsoft® PowerPoint window. The user selects a page of presentation materials as a query key. The other is a visualized document-search application window, which displays the searching results. The user's query request is sent from the proxy application to the server application through the Ethernet connection.

The server computer includes an application working on the CPU, an in-socket FPGA accelerator connected with CPU through an Intel® QuickPath Interconnect (QPI) interface, and storage for the image database. The combination of the FPGA accelerator and CPU provides all the computing resources needed in the document-search system, with computation intensive parts of the partial-image matching algorithm on the FPGA and the data flow control and accelerator control on the CPU, resulting in sufficient performance gain.

The task scheduling between CPU and FPGA accelerator is optimized so that time for the server application to retrieve a partial duplicate image from the image database is minimized. In this prototype system, database images are loaded in the main memory of the server computer in the boot process of the server application to avoid the inadequate transfer speeds of the HDD.

Table 1 summarizes the resource usage of the FPGA. We implemented the partial-image matching core on the Altera® Stratix®V A7 FPGA. The clock frequency of the accelerator is 200MHz. We evaluated the runtime performance of the FPGA-accelerated engine, and compared it with the engine implemented in the server application that runs on 1core of the Intel® Xeon® processor E5-2640 v3 (8core, 2.6GHz). The Hamming distance calculation is accelerated by a POPCNT instruction of Intel® Streaming SIMD Extensions (SSE) 4.2. In this evaluation, database contains 8,000 images generated from the presentation materials in our office, and the query images are randomly selected form the database. The FPGA accelerator significantly improves the runtime performance (Table 2). The speedup factor over a single core CPU-based solution is approximately 32.

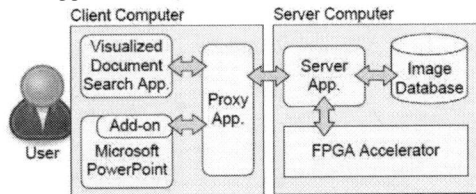

Fig. 4. Document search system prototype with FPGA-acceleration.

Table 1. FPGA usage.

Device	StratixV A7
Logic Utilization	196,399(84%)
Registers	218,552(23%)
DSPs	46(18%)
Block memory bits	23Mb(43%)

Table 2. Measured performance summary.

		Performance	Acceleration
CPU (1core)		38.4sec	x1
FPGA		1.17sec	x32.8

Conclusion

We proposed and demonstrated an FPGA-accelerated partial-image matching engine for massive media-data searching systems. By utilizing a highly parallelized and pipelined architecture with bit-level calculation, our prototype system achieves 32 times better runtime performance than a single core CPU-based solution.

References

[1] Z. Zhu et al., *IEEE ICCV2013*, pp.113-120, 2013.
[2] M. Calonder et al., *IEEE Trans. PAMI-34*, pp.1281-1298, 2012.
[3] A. Putnum et al., *ACM/IEEE ISCA2014*, pp.13-24, 2014.
[4] H. Matsumura et al., *IEEE WACV2016* (to be published).
[5] J. Canny, *IEEE Trans. PAMI-8*, pp. 679-698, 1986.

A 66pW Discontinuous Switch-Capacitor Energy Harvester for Self-Sustaining Sensor Applications

Xiao Wu, Yao Shi, Supreet Jeloka, Kaiyuan Yang, Inhee Lee, Dennis Sylvester and David Blaauw

University of Michigan, Ann Arbor, MI, lydiaxia@umich.edu

Abstract

We present a discontinuous harvesting approach for switch capacitor DC-DC converters that enables ultra-low power energy harvesting. By slowly accumulating charge on an input capacitor and then transferring it to a battery in burst-mode, switching and leakage losses in the DC-DC converter can be optimally traded-off with the loss due to non-ideal MPPT operation. The harvester uses a 15pW mode controller, an automatic conversion ratio modulator, and a moving sum charge pump for low startup energy upon a mode switch. In 180nm CMOS, the harvester achieves >40% end-to-end efficiency from 113pW to 1.5µW with 66pW minimum input power, marking a >10× improvement over prior ultra-low power harvesters.

Introduction

Energy harvesting from the ambient environment is essential for self-sustaining sensor nodes and there is a continuing need to harvest extremely small input power sources to enable new application fields. For example, a miniature 100×100um solar cell generates ~150 pW under low lighting conditions (32 lux). Efficient DC-DC upconversion from such a power source voltage to typical battery voltages is extremely difficult. Recent works using both boost and switched-capacitor (SC) DC-DC converters demonstrate various circuit techniques to reduce the minimum input power required for successful harvesting [1-3]. However, they are limited by the constant charge pump leakage and clock generation power and have demonstrated harvesting only down to 1.2nW.

Discontinuous Energy Harvesting

This work proposes a *discontinuous* harvesting approach based on the observation that at low power levels, charge pump efficiency plummets while the efficiency of the energy source remains high due to continuing operation at its maximum power point. Discontinuous harvesting operates in two modes, allowing it to achieve a balance between these two efficiencies and obtain higher overall end-to-end efficiency. In *harvest* mode, the charge pump is power gated, reducing its leakage to just a few pW while the power source charges a capacitor. In *transfer* mode, the charge pump is enabled and energy accumulated on the capacitor is transferred to the battery (Fig. 1). Since the capacitor voltage deviates from the MPPT point of the energy source, the transfer efficiency to the capacitor is reduced. However, since the charge pump operates at a much higher power level (µW) during transfer mode, its efficiency dramatically improves.

Using these two modes, the discontinuous harvester decouples the two main losses and allows us to optimally trade them off, enabling efficient operation across a very wide range of input power (23,000× in our implementation). An asynchronous mode controller with <15pW power consumption controls the mode switch. It maintains a constant power source voltage fluctuation (ΔV_{sol} in Fig. 1), thereby automatically increasing the duty cycle at low input power levels and maintaining optimal end-to-end efficiency. We further propose a moving-sum charge pump that was designed for low start-up energy, which reduces overhead during the harvest to transfer mode transition. In measurement, the harvester obtains 37% end-to-end efficiency at 66pW input power drawn from a 0.01mm^2 solar cell at 6 lux and has a maximum input power of 1.5µW.

Harvester Implementation

Fig. 2 shows the overall architecture of the proposed harvester, which consists of C_{buf}, an always-on mode controller, and a harvester in a gated power domain. During harvest mode, S1-3 are open to limit leakage from the battery and C_{buf} to 2.6pW (simulation). V_{sol} is monitored by an asynchronous mode controller; when V_{sol} crosses V_{ref_H} the mode controller switches to transfer mode. This closes S1-3, enabling power transfer from C_{buf} to the battery, and enters a *startup* phase where the charge pump conversion ratio is initialized while the clock and logic operate from the battery (4V). After the pump voltages stabilize, the system enters *operation* mode and switches to an

internally generated 1.2V supply to reduce switching power loss. Since the charge pump transfers charge from a capacitor and not a variable current source, the optimal pump frequency can be predetermined for both startup and operation modes, which significantly simplifies the charge pump design. The clock frequency change from startup to operation mode is performed by a glitch-free clock mux. As the charge pump drains C_{buf}, V_{sol} drops and an automatic conversion ratio modulator (ACRM) adjusts the conversion ratio to maintain optimum efficiency. When $V_{sol} < V_{ref\_L}$, the mode controller power gates the pump and changes to harvest mode.

A low-power mode controller is a key requirement that determines the lower bound of harvestable input power. An asynchronous design is used to save clock and logic power (Fig. 3). The mode is stored in flip-flop D1, which toggles based on comparators C1 and C2. The controller has < 100 gates, implemented in thick-oxide I/O devices, and consumes <15pW (measured). A diode stack is used to lower supply voltage from 4V to 1.6V, reducing the impact of GIDL.

In transfer mode, the conversion ratio is modulated based on $\Delta V = V_{in}*R - V_{out}$, where R is the conversion ratio and ΔV is an indicator of conduction loss. The ACRM (Fig. 4) approximates ΔV by multiplying input voltage V_{solar_pg} by $M*(R+1) = V_{mult}$, where M is a fixed weight and R is the current conversion ratio (Fig. 4). V_{mult} is then compared to a fixed threshold V_{ref_ACRM}. If $V_{mult} < V_{ref\_ACRM}$, a ratio counter increments, changing the conversion ratio to R+1. Multiplication is done by a switched-capacitor amplifier. Switch drivers for this amplifier are supplied by an auxiliary 2:1 DC-DC converter to reduce power consumption. The ACRM is duty cycled and only enabled every three SYSCLK cycles. After each modulation, ACRM shuts down its clock by itself.

As shown in Fig. 5 a 3-phase moving-sum charge pump is proposed to reduce startup energy. Charge in flying capacitors leak away during harvest mode and need to be restored during startup phase, presenting a power overhead. A traditional Dickson charge pump maintains high efficiency in operation mode, but startup energy is high due to the large number of flying caps and their high voltage potential. We design a "moving-sum" charge pump that consists of a Dickson charge pump with only 10 stages, followed by a modified series-parallel (S-P) charge pump with 4 flying caps to boost conversion ratio. In phases A and B, the Dickson stage operates conventionally except that four selected voltages are connected to the four flying caps in S-P stage. In phase C, the four S-P flying caps are connected in series to achieve 10-20× conversion ratios.

Measurement Results

The test chip is fabricated in 180nm CMOS. First examining moving-sum charge pump and ACRM performance, measured results in Fig. 6 show that the conversion ratio chosen by ACRM is within 2 of optimal, yielding < 10% efficiency degradation. The moving-sum charge pump achieves 60% peak efficiency at 256nW output power, and operates effectively across an output range of 4.2nW to 4µW (Fig. 7). We quantify the trade-off between startup energy and pump overhead (i.e., transfer phase efficiency) vs. solar cell efficiency by sweeping ΔV_{sol} in Fig. 8, with overall efficiency peaking at 120mV. Proposed harvester efficiency is measured with C_{buf} of 35.2µF, battery voltage of 3.8−4.2V, and a 0.01mm^2 solar cell as the energy source. The measured range of harvestable input power is 66pW to 1.5µW, marking a 23,000× range (Fig. 9). End-to-end efficiency of 37% is achieved at 6 lux with 66pW input power, and >30% efficiency is maintained up to a maximum power of 1.5µW at 43klux. Table 1 compares to prior work and shows 18× lower minimum harvestable power and 23× wider output power range.

References

[1] W. Jung et al., ISSCC, 2014.

[2] S. Bandyopadhyay et al. ISSCC, 2014.

[3] P. Chen, et al. ISSCC, 2015.   [4] K.Chew, et al. ISSCC 2013.

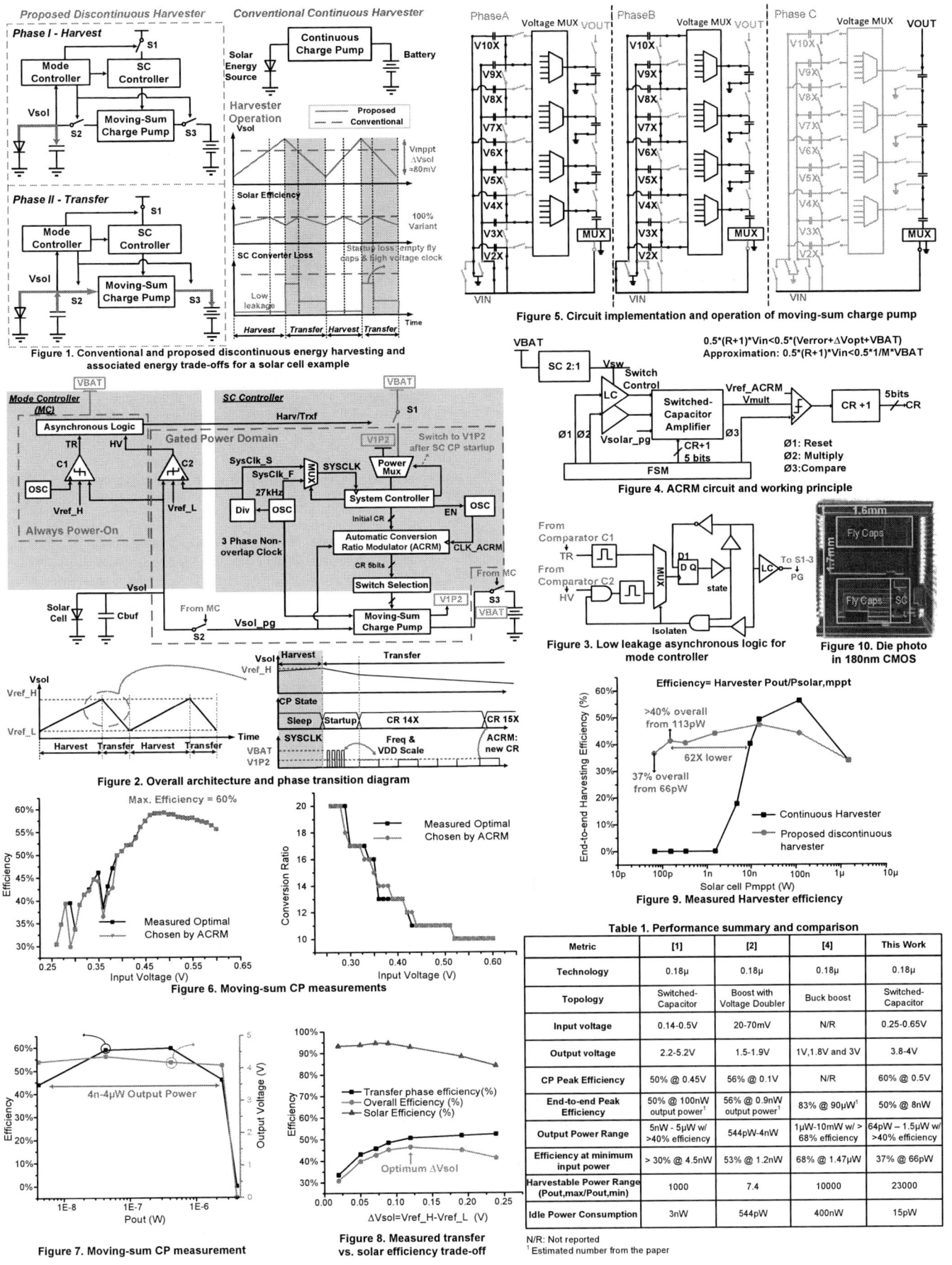

Figure 1. Conventional and proposed discontinuous energy harvesting and associated energy trade-offs for a solar cell example

Figure 2. Overall architecture and phase transition diagram

Figure 3. Low leakage asynchronous logic for mode controller

Figure 4. ACRM circuit and working principle

Figure 5. Circuit implementation and operation of moving-sum charge pump

$0.5*(R+1)*Vin<0.5*(Verror+\Delta Vopt+VBAT)$
Approximation: $0.5*(R+1)*Vin<0.5*1/M*VBAT$

Ø1: Reset
Ø2: Multiply
Ø3: Compare

Figure 6. Moving-sum CP measurements

Figure 7. Moving-sum CP measurement

Figure 8. Measured transfer vs. solar efficiency trade-off

Figure 9. Measured Harvester efficiency

Figure 10. Die photo in 180nm CMOS

Efficiency= Harvester Pout/Psolar,mppt

Table 1. Performance summary and comparison

Metric	[1]	[2]	[4]	This Work
Technology	0.18μ	0.18μ	0.18μ	0.18μ
Topology	Switched-Capacitor	Boost with Voltage Doubler	Buck boost	Switched-Capacitor
Input voltage	0.14-0.5V	20-70mV	N/R	0.25-0.65V
Output voltage	2.2-5.2V	1.5-1.9V	1V,1.8V and 3V	3.8-4V
CP Peak Efficiency	50% @ 0.45V	56% @ 0.1V	N/R	60% @ 0.5V
End-to-end Peak Efficiency	50% @ 100nW output power[1]	56% @ 0.9nW output power[1]	83% @ 90μW[1]	50% @ 8nW
Output Power Range	5nW - 5μW w/ >40% efficiency	544pW-4nW	1μW-10mW w/ > 68% efficiency	64pW – 1.5μW w/ >40% efficiency
Efficiency at minimum input power	> 30% @ 4.5nW	53% @ 1.2nW	68% @ 1.47μW	37% @ 66pW
Harvestable Power Range (Pout,max/Pout,min)	1000	7.4	10000	23000
Idle Power Consumption	3nW	544pW	400nW	15pW

N/R: Not reported
[1] Estimated number from the paper

A Wireless Power Transfer System with Enhanced Response and Efficiency by Fully-Integrated Fast-Tracking Wireless Constant-Idle-Time Control for Implants

Cheng Huang, Toru Kawajiri and Hiroki Ishikuro

Keio University, Yokohama, Japan; Email: doowtsewhuang@gmail.com, ishikuro@elec.keio.ac.jp

Abstract

In this paper, a 13.56 MHz fully-integrated wireless power transfer system with wireless constant-idle-time control is proposed. The massive off-chip components or wire required for transmitter (TX) voltage regulation in previous works are eliminated. Both wireless and local regulations are achieved with enhanced transient performance and total efficiency, and reduced circuitry and system design complexity. Thanks to the proposed wireless constant-idle-time control technique, an instant load-transient response, and a peak total efficiency of 67.6% with up to 13.7% improvement are observed in measurements with meat between coils at a distance of 6mm.

Introduction

Wireless power transfer (WPT) system is widely used as a battery-less solution to supply bio-implants such as retinal prostheses and cochlear implants typically in the power range of 10-100mW [1-3]. Because the received voltage and power at the receiver (RX) are highly dependent on the coupling and load conditions, output voltage regulation is necessary to ensure proper operation of biomedical functional circuits. Fast transient response is also important to avoid over-/under-shoot voltages that may induce failure operation or reliability issues.

In a complete WPT system, the total efficiency includes all the power losses in the transmitter (TX), wireless link and RX. Early designs only consider RX regulation, while TX is always transmitting the maximum power for worst cases [1-2], thus the total efficiency for most of the cases is degraded. Recently, several designs are reported with various control techniques to enable TX or both TX and RX voltage regulations to improve the total efficiency [2-6]. However, due to the complexity of these control techniques which require analog [3], duty-cycle-sensitive processing-required multi-bit digital [2], [5], or mixed [4] control signals to transfer wirelessly for TX regulation, a lot of additional off-chip components are used. For example, [2] requires an off-chip RX digital controller, TX pulse controller, decoder, and diodes; [3] requires an off-chip power inductor, DAC, DAC controller and data receiver; [4] requires a communication module and MCU. Moreover, due to the control complexity and limited system bandwidth, the transient performance is slow with wireless tracking time of 130μs [2] or 2ms [3]. Besides, wires are used in [5] and [6] for TX regulation feedback, which defeats the advantage of WPT.

Proposed Fully-Integrated Fast-Tracking WPT System

Fig. 1 shows the system block diagram of the proposed WPT system including two chips: RX and TX chips; and three coils: two for power transfer and one for current sensing, respectively. The proposed wireless constant-idle-time control includes the RX local regulation, which is realized inside the RX chip, and TX wireless regulation, which is realized based on the Load-Shift Keying (LSK) pulses generated in the RX chip, transferred through the wireless link, picked up by the sensing coil and recovered inside the TX chip. Since the control is comparator-based and only simple and strong pulses are transferred wirelessly for TX regulation, off-chip components in [2-4], such as decoder, DAC and MCU, are not required. The circuitry and system design complexity are significantly reduced with a faster transient performance.

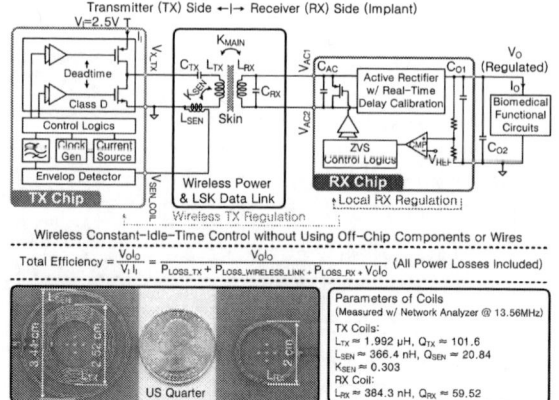

Fig.1. Proposed WPT system with two chips and three coils for implants.

Fig. 2 and Fig. 3 show the detailed circuitries and operation principles of the local RX and wireless TX regulations, respectively, for the constant-idle-time control. As shown in Fig. 2, a comparator (CMP) compares the V_O with V_{REF}. Once V_O is higher than V_{REF} indicating too much received power, the CMP will be triggered. A zero-voltage-switching (ZVS) signal V_{ZVS} is then generated when both V_{ACS} are high, indicating a rough zero voltage on the resonant capacitors considering circuit delays to avoid wasting the energy. A throttling signal V_{STOP_RX} is then generated with a period determined by R_1C_1 and the V_{CMP} to drive M_{STOP} shorting the two V_{ACS}. During this freewheel period, the rectifier stops and the coil current loops in the L_{RX}, and the input impedance of RX changes. This serves as the LSK signal for the later TX regulation. As shown in Fig. 3, the change of impedance reflects to the TX side through the wireless link, and the TX coil L_{TX} current increases immediately. It is then sensed by L_{SEN} and received by the TX chip. The AC coupling between L_{TX} and L_{SEN} are inductive instead of resonant to avoid extra power dissipation. The envelop signal is then recovered by the integrated half-wave envelop detector, filtered by a band-pass filter (BPF) with a given DC bias level, and compared with a DC threshold voltage by a CMP to generate V_{STOP_TX}. In this way, the LSK signals from the RX chip indicating too much transmitted power are wirelessly recovered inside the TX chip.

In the TX chip, a clock of 27.12MHz is generated and divided into a 50% duty-cycle (D) 13.56MHz clock by a 1/2 divider for the Class D driver, and further divided into a 1.695MHz slow clock by a 1/8 divider for TX regulation. Once V_{STOP_TX} goes high, the class D driver stops switching, and a '1' signal is sent to the DFFs for a constant delay. After 4 cycles of the 1.695MHz clock that are roughly 4 x 590ns, the stop-signal releases and the Class D driver switches again. In this way, the transmitted power of TX is reduced, and the proposed constant-idle-time control is realized wirelessly from RX to TX. Compared to the control techniques in [2] and [3] using wireless LSK signals for TX regulation, the system design complexity in this work is much lower without using on-chip ramp generators, error amplifiers, and the massive off-chip components. In terms of transient response, because the constant-idle-time control is CMP-based and does not require careful compensator design to ensure system stability in every

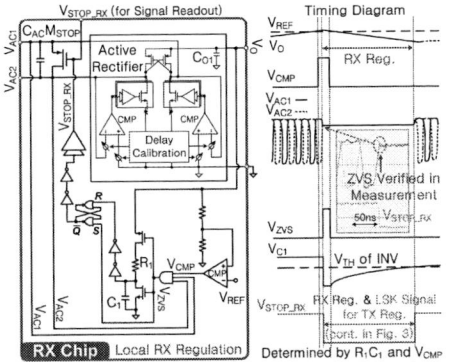

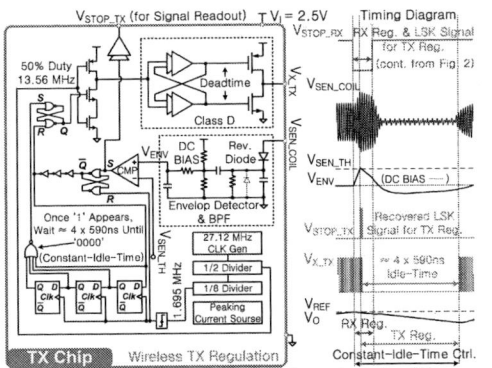

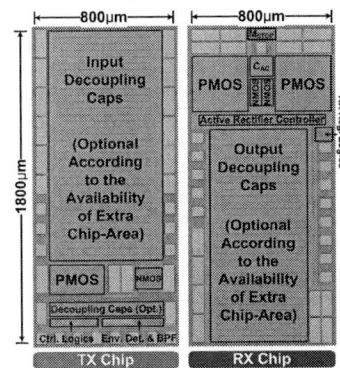

Fig.2. Detailed circuitry and operation principle of the local RX regulation in the RX chip.

Fig.3. Detailed circuitry and operation principle of the wireless TX regulation in the TX chip.

Fig.4. Chip photos of both the TX and RX chips.

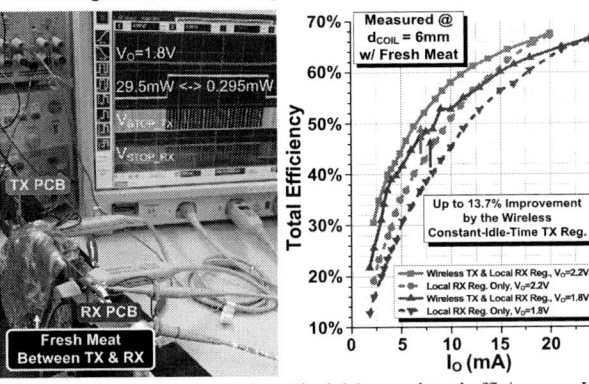

Fig.5. Measurement setups during a live transient measurement.

Fig.6. Measured total efficiency vs I_O w/ and w/o wireless TX regulation.

TABLE I. COMPARISON WITH PRIOR ART

	[1] ISSCC'13	[2] JSSC'15	[3] ISSCC'15	[4] ISSCC'15	[5] JSSC'13	[6] ISSCC'12	This Work
System Level on Silicon	RX Only	Both TX & RX	Both TX & RX	RX Only	RX Only	Both TX & RX	Both TX & RX
RX V_O	1.27-4V	3.6V	3.7V	4.2-5.8V	5V	15V	1.2-2.5V
P_{OUT} (MAX)	32mW	102mW	234mW	6W	6W	0.52W	49.4mW
Freq. (MHz)	13.56	13.56	13.56	6.78	6.78	13.56	13.56
Reg. Site	No Regulation	TX & RX (Rec.)	TX & RX (Rec.)	TX & RX (Buck/LDO)	TX	TX	TX & RX (Rec.)
TX Reg. Data Link	N/A	Wireless LSK	Wireless LSK	Wireless 2.4GHz	Wire	Wire	Wireless LSK
Loop Stability Compensation	N/A	Required (Dominant-Pole)	Required (Dominant-Pole)	Required (Pole-Zero)	Required (Dominant-Pole)	Required	Not Required (CMP-Based)
Off-Chip Components Required for TX Reg.	N/A	RX Digital Ctrlr., TX Pulse Ctrlr., decoder, Diodes	Inductor (for TX Buck), DAC & its Ctrlr., Data Receiver, BPF	Comm. Module, MCU	MCU, Wire	Wire	None (Fully-Integrated)
ΔV_O in Load Tran. ($\Delta V_O/V_O$)	N/A	112mV (3.1%)	162mV (4.38%)	N/A	162mV (3.2%)	400mV (2.7%)	Unnoticeable†
Load Reg.	N/A	7.2mV/mA	2.06mV/mA	N/A	N/A	N/A	0.9mV/mA
Tracking Time	N/A	130μs	2ms	N/A	500μs	30μs	Instant†
Peak Total Effi. (@ d_{COIL})	N/A	50%* (3mm)	62.4%* (3mm)	N/A	55%* (N/A)	50%* (5mm)	70.6%*# (6mm w/ Meat)# 67.6% (6mm w/ Meat)
Tech.	CMOS 0.35μm	CMOS 0.35μm	CMOS 0.35μm	BCD 0.35μm	BCD 0.18μm	HVCMOS 0.18μm	CMOS 65nm (I/O Devices)

* Power consumed by the off-chip active components, such as controllers, decoders, DAC and MCU, is not included.
† Please refer to the load-transient measurement in Fig. 7 for more details.
# All the previous works are measured without meat.

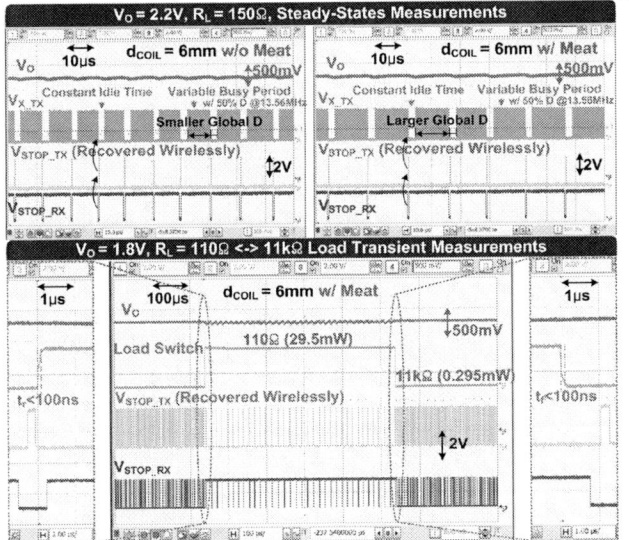

Fig.7. Measured waveforms in steady-states and during transients.

case, this work is much faster and easier to design. Real-time delay-calibrations in [7] are also applied for better efficiency.

Measurement Results

Both TX and RX chips are fabricated in TSMC 65nm process with standard I/O devices, occupying two 1.8×0.8mm^2 blocks, as shown in Fig. 4. The chip-areas are dominated by the optional decoupling capacitors due to the availability of extra chip-area. The controllers of TX and RX chips only occupy 0.021mm^2 and 0.03mm^2, respectively. To better emulate the operation conditions at real cases, a piece of fresh meat is inserted between TX and RX in the measurements, as shown in Fig. 5. Fig. 6 shows the measured total efficiency versus I_O with V_O regulated to 2.2V and 1.8V at d_{COIL}=6mm

with meat. A peak total efficiency of 67.6% is observed. With the help of the wireless constant-idle-time control, up to 13.7% efficiency improvement is achieved by enabling the wireless TX regulation. Fig. 7 shows the measured waveforms during steady-states and load-transients. As shown in the upper two figures, the V_{STOP_TX} is well recovered from V_{STOP_RX} and stops the V_{X_TX} for a constant idle time, while the busy periods varies according to different conditions. With meat inserted between coils at a distance d_{COIL}=6mm, the idle periods are less frequent and the busy periods occupy a larger portion of the total operation time, which indicates larger equivalent global D and more transmitted power from TX to RX due to weaker coupling. The transient performance is shown in the lower figure. With a load-transient between 110Ω and 11kΩ in less than 100-ns time-steps and d_{COIL}=6mm with meat, the over-/under-shoot voltages are unnoticeable and the tracking time is instant due to the CMP-based constant-idle-time control. As shown in Table I, compared to previous works, this work achieves both RX and TX voltage regulations without using any off-chip components or wires, a higher total efficiency and a much faster transient response under the measurements with fresh meat between TX and RX coils.

Acknowledgements

This work is supported by CREST/JST. We would like to thank Xun Liu from HKUST for her discussions.

References

[1] Y. Lu, et al., *ISSCC*, pp. 66-67. 2013.
[2] X. Li, et al., *IEEE JSSC*, vol. 50, no. 4, pp. 978-989, 2015.
[3] X. Li, et al., *ISSCC*, pp. 228-229, 2015.
[4] K.-G. Moh, et al., *ISSCC*, pp. 230-231, 2015.
[5] J.-H. Choi, et al., *IEEE JSSC*, vol. 48, no. 12, pp. 2989-3001, 2013.
[6] R. Shinoda, et al., *ISSCC*, pp. 288-290, 2012.
[7] C. Huang, et al., *IEEE CICC*, 2015, pp. 1-4, 2015.

A Fully Integrated 144 MHz Wireless-Power-Receiver-on-Chip with an Adaptive Buck-Boost Regulating Rectifier and Low-Loss H-Tree Signal Distribution

Chul Kim, Jiwoong Park, Abraham Akinin, Sohmyung Ha, Rajkumar Kubendran, Hui Wang, Patrick P. Mercier, Gert Cauwenberghs

University of California, San Diego, La Jolla, CA

Abstract

An adaptive buck-boost resonant regulating rectifier (B^2R^3) with an integrated on-chip coil and low-loss H-Tree power/signal distribution is presented for efficient and robust wireless power transfer (WPT) over a wide range of input and load conditions. The B^2R^3 integrated on a 9 mm^2 chip powers integrated neural interfacing circuits as a load, with a TX-load power conversion efficiency of 2.64 % at 10 mm distance, resulting in a WPT system efficiency FoM of 102.

Introduction

Emerging applications such as mm-sized modular neural interfacing devices [1] are now being enabled by fully integrating wireless power transfer (WPT) functionality on-chip. Conventional designs, which first rectify then regulate, suffer from cascaded losses that limit efficiencies [2-5], while prior-art regulating rectifiers require a large external coil. Although recent work has demonstrated complete on-chip integration of a regulating rectifier [6], robustness to link variations was limited. In addition, >100 MHz-range RF power delivered to an on-chip coil induces eddy currents in auxiliary on-chip wiring and metal planes, degrading WPT efficiency while also potentially disrupting the functionality of underlying circuitry. To avoid large metal planes, [3] reduced decoupling capacitance (decaps) to only 20 pF through inclusion of a high-performance high-power linear regulator.

Buck-Boost Resonant Regulating Rectifier (B^2R^3)

This paper introduces a fully-on-chip B^2R^3 architecture that robustly receives wireless power over a wide range of conditions by: 1) fully-integrating an on-chip coil with distributed perpendicular decoupling and loop-free H-tree geometry power line/signal distribution networks optimized for maximum RF power collection and minimum RF interference; 2) employing a regulating rectifier that dynamically adapts to a wide range (>14 dB) of RF input by switching between boost and buck modes, while providing a dual-rail supply from lower magnitude RF input signals than prior work [6]; and 3) including a 0.86 µW control feedback loop that offers fast load regulation performance while supporting both boost and buck regulating rectifiers.

The proposed B^2R^3 system is shown in Fig. 1. Here, RF energy received by an on-chip coil passes through the B^2R^3, to establish dual DC rails, VH and VL. Unfortunately, RF energy also couples to load circuits, and ad hoc routing of supply lines and placement of decaps can create many loops and metal planes, reducing the RX coil's Q by over 60 % (thereby reducing WPT efficiency), while also introducing noise to sensitive circuits. To remove loops and large planes from the layout, a fractal H-tree power and signal distribution network with 1 nF of distributed decaps is proposed. HFSS simulation and measurements show negligible loss in Q compared to an ideal, isolated coil. Furthermore, the same H-tree topology serves as a network backbone for cancellation of differential-mode interference in sensitive analog differential signals.

Although 2-step rectification and regulation can operate at wide input range, regulation is increasingly inefficient at larger RF input voltage as illustrated in Fig. 2. The proposed B^2R^3 accomplishes regulated rectification over wide input range through a mode arbiter that adapts to the sensed RF envelope. BOOST mode converts low RF voltage to larger regulated DC voltage, while $BUCK_{1,2,3}$ modes efficiently convert larger RF voltage down. For smooth transition between modes, a combined BUCK-BOOST mode operates at an intermediate region. Shown in Fig. 3, VHS tracks VH with DC offset defining a target VH. To retain the fast settling of conventional integrator-less bang-bang control, VHS is directly fed to the latch (path 2) while a parallel integration path (path 1) performs PID control to additionally remove static error at VH. The boost regulating rectifier employs a feedback-controlled V_{TH} cancellation scheme for regulation. The feedback loop dynamically determines the amount of V_{TH} cancellation through V_{OTA}. Conversely, the buck regulating rectifier is activated by V_{COMP} from the feedback, turning the rectifier on/off according to VHS and V_{OTA}.

In Fig. 4, the switched capacitor circuits implement floating voltage sources by transferring V_{OTA} to the NMOS devices in the boost rectifier. A zero short-circuit-current level-shifter consumes 190 nW at 1 MHz, a 35x improvement over a conventional design. In buck mode, the regulating rectifier dynamically determines the duty-cycle of the active switches (i.e., t_d and t_{pw}) with 1 MHz clock for maximizing t_{pw}. Instead of a high-speed comparator [6], low-power delay controlled inverter chains operating with a local feedback are employed. To prevent undesired energy transfer from large RF input, the power PMOSs are gated with a voltage (VHH) larger than the RF envelope. Three power switches are implemented in parallel ($BUCK_{1,2,3}$) to optimize PCE according to RF input.

Measurement Results

The B^2R^3 is tested with a 470 mm^2 TX coil located 6-16 mm away from the chip. In Fig. 5, the B^2R^3 produces regulated output voltages even under 50 % amplitude variation in PA input by adapting its modes dynamically. Load regulation is measured in Fig. 6. Owing to its fast loop response, even with only 0.25 nF decaps (two 0.5 nF in series), no sharp peak in V_{OUT} is shown, unlike typical >100 mV overshoot otherwise. Thanks to mode switching and high voltage gating, a link dynamic range of at least 14dB at 10mm distance is demonstrated in Fig. 7. The TX-RX link efficiency is -12.6 dB at 6.35 mm. The TX-load WPT efficiency is measured as 3.66 %, indicating a regulating rectifier PCE of 66.5 % (within 1.5 % of simulation). RF powering to integrated load circuits is validated by demonstrating regulated supplies during data transfer from the TX coil to the on-chip ASK demodulator through the integrated coil, shown in Fig. 8. Table I shows comparison with the state-of-the art. Standardized comparison for overall PCE (TX-regulated DC) is provided by the WPT system efficiency (WSE) FoM, defined here as overall PCE times cube of distance between TX and RX coils divided by cube square root of the RX coil area [3].

References

[1] S. Ha, et al., *Symp. on VLSI Circuits*, pp. 106-107, 2015
[2] R. Muller, et al., *ISSCC Dig. Tech. Papers*, pp. 412-413, 2014
[3] M. Zargham, et al., TBioCAS, pp. 259-271, April 2015.
[4] M. Mark, et al., *Symp. on VLSI Circuits*, pp. 168-169, 2011.
[5] S. O'Driscoll, et al., *ISSCC Dig. Tech. Papers* pp. 294-295, 2009.
[6] C. Kim, et al., *Symp. on VLSI Circuits*, pp. 284-285, 2015.

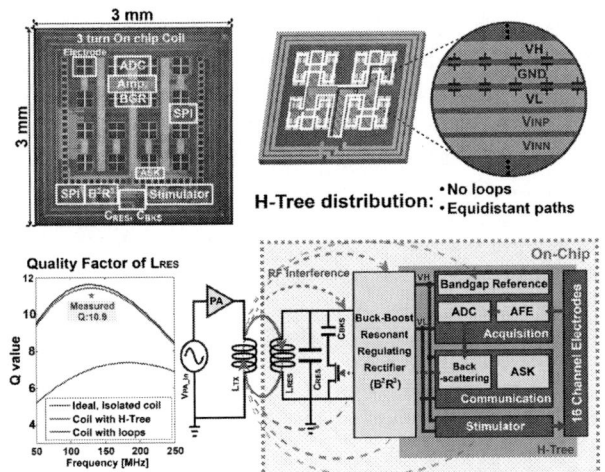

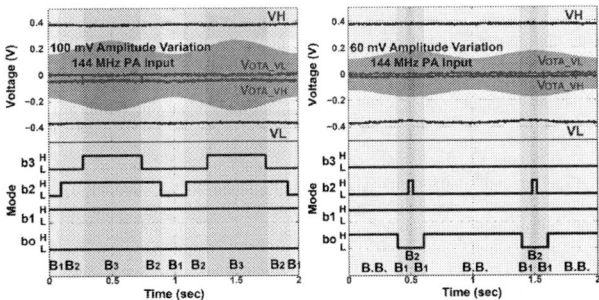

Fig. 1 Depiction of the on chip coil and the fractal H-Tree network. Significant routing is accomplished systematically and with minimal decrease in on-chip coil quality factor, and minimal RF interference to sensitive analog components.

Fig. 5 Measurement shows robustness of B^2R^3 to changes in RF input power dynamically thanks to the mode-change by the proposed mode arbiter. RF input power can vary due to a number of reasons such as link distance, alignment, and impedance matching.

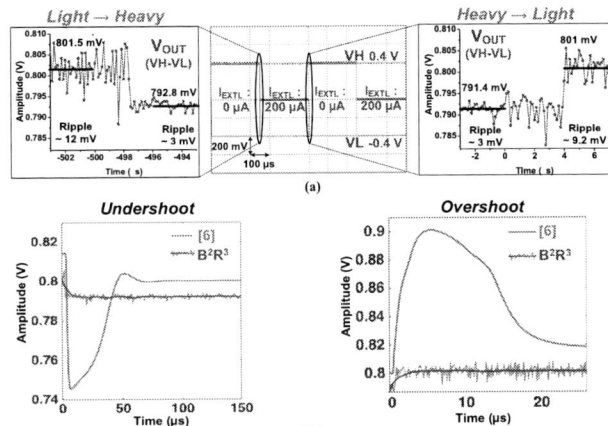

Fig. 6 (a) Measured load regulation response to external load perturbation, from open to 200 µA. Light to heavy load transition in left, and heavy to light load transition in right, showing 9.6 mV static ΔV_{OUT} (VH – VL). (b) Owing to the fast feedback loop, negligible over/undershoot is observed while a conventional design shows >100 mV overshoot.

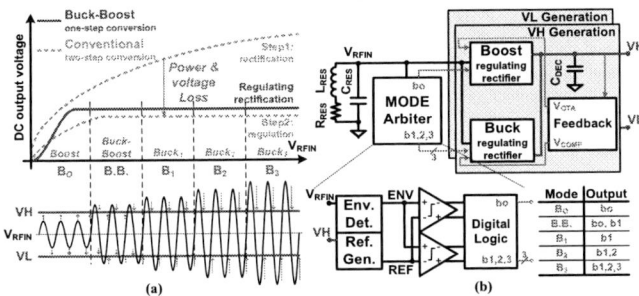

Fig. 2 (a) Conceptual operation of the conventional 2-step power conversion and the B^2R^3 showing the 5 regulating rectification modes that efficiently generate a dual supply according to RF input. (b) Overview schematics of the adaptive B^2R^3 mechanism with mode arbiter and its block diagram.

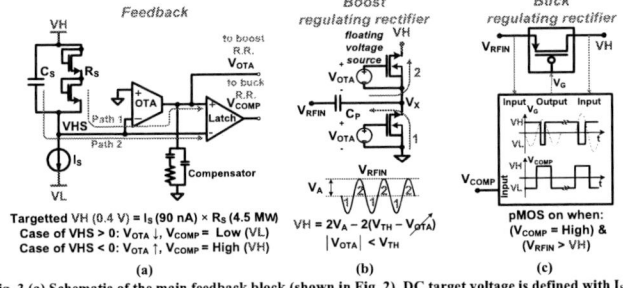

Fig. 3 (a) Schematic of the main feedback block (shown in Fig. 2). DC target voltage is defined with I_S and R_S as the main feedback loop forcing VHS to be close to GND. (b) With the floating voltage source, V_{TH} of NMOS in the boost regulating rectifier is cancelled dynamically by V_{OTA} from the feedback block for regulation fucntionality. (c) V_{COMP} from the feedback turns on/off the buck regulating rectifier to regulate VH.

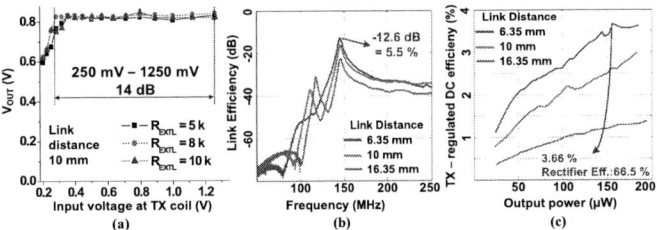

Fig. 7 (a) Measured transfer function showing wide input-range. At low RF inputs, boost mode can develop V_{OUT} while, at large RF input, high voltage gating helps regulation. (b) Link efficiency at various link distance and frequency, and (c) overall PCE (TX coil to regulated DC) are measured.

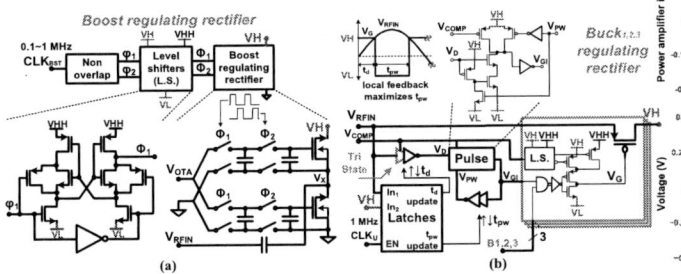

Fig. 4 Simplified schematic of the boost and buck regulating rectifiers. (a) Switched capacitors implement floating voltage sources with a no short-circuit-current level-shifter. (b) Local feedback loop defines t_d and t_{pw} with 1MHz updating CLK. PMOS is gated with higher voltage to avoid making a diode connection at large RF input.

Fig. 8 System level untethered operation is demonstrated by the ASK demodulation.

Table I State of the art comparison

	[2]	[3]	[4]	[5]	[6]	This Work
Res. Freq. (MHz)	300	160	535	915	144	144
RX Coil/ Value(nH)/ Area(mm²)	Off-chip/ 32/ 42.25	On-chip/ 130/ 4.36	Off-chip/ 5.73/ 1	Off-chip/ N/R/ 4	On-chip/ 23.7/ 8.64	On-chip/ 60.3/ 8.74
# of modes	1	1	1	1	5	5
Regulator	separate LDO	separate LDO	separate LDO	separate LDO	regulating rectifier	regulating rectifier
Regulation*(%)	N/R	N/R	N/R	N/R	1.87	1.12
Overshoot (mV)	N/R	N/R	N/R	N/R	100	< 1
Dec. Cap (nF)	4	0.02	1.39	N/R	1	0.25
Process	65 nm CMOS	0.13 µm CMOS	65 nm CMOS	0.13 µm CMOS	0.18 µm CMOS SOI	0.18 µm CMOS SOI
Overall PCE(%)	1.19ᵃ	0.62ᵇ	0.02 (-37 dB)	0.048 (-33.2 dB)	2.04	2.64
Distance (mm)	12.5	10	13	15	10	10
WSE FoM**	8.46	68.1	43.94	20.25	80.3	102.1

ᵃestimated from provided data, TX PW: 13mW, P_{DC_LOAD}: 0.160mW

ᵇestimated from provided data, estimated η_{LDO}: 68 % (V_{DD}: 3.1 V, V_{REC2}: 4.5 V), provided η from TX to output of the rectifier: 0.9 %

*Regulation = Static ΔV_{OUT} / nominal V_{OUT}

**WPT System Efficiency FoM = $\frac{\eta_{overall} \times D^3}{A^{1.5}}$, where $\eta_{overall}$ is TX-regulated DC efficiency; D is distance between TX-RX coils; and A is area of the RX coil.

A ±36A Integrated Current-Sensing System with 0.3% Gain Error and 400µA Offset from −55°C to +85°C

Saleh Heidary Shalmany[1], Dieter Draxelmayr[2], Kofi Makinwa[1]

[1]Delft University of Technology, Delft, The Netherlands, [2]Infineon Technologies, Villach, Austria

Abstract

This paper presents an integrated shunt-based current-sensing system (CSS) capable of handling ±36A currents, the highest ever reported. It also achieves 0.3% gain error and 400µA offset, which is significantly better than the state-of-the-art. The heart of the system is a robust 260µΩ shunt made from the lead-frame of a standard HVQFN plastic package. The resulting voltage drop is then digitized by a ΔΣ ADC and a bandgap reference (BGR). At the expense of current handling capability, a ±5A version of the CSS uses a 10mΩ on-chip metal shunt to achieve just 4µA offset. Both designs were realized in a standard 0.13µm CMOS process.

Introduction

Shunt-based CSSs are widely used in battery fuel gauges and in energy monitoring systems [1,2]. High current systems based on lead-frame shunts typically suffer from poor gain error, e.g. 5% over a ±15A range [3], mainly due to the shunt's temperature coefficient of resistance (TCR), ~0.3%/°C. At the expense of extra cost, better accuracy can be achieved with low-TCR in-package shunts, e.g. ±0.75% gain error and 50mA offset over a ±10A range [4]. The proposed CSS uses a temperature compensation scheme to achieve even better performance with a low-cost lead-frame shunt made from the heatsink of a standard package.

Proposed Design

As shown in Fig.1, the CSS consists of a chip on a lead-frame shunt, which senses the voltage drop V_{shunt} between the Kelvin-contacted points S1 and S2. The chip contains a bandgap reference (BGR) and two switched-capacitor 2nd-order ΔΣ ADCs; ADC_I digitizes V_{shunt} with respect to the BGR voltage V_{Ref}, and ADC_T uses the the BGR's PNPs to sense the shunt's temperature T. The shunt's spread is corrected by a room-temperature trim, while the effect of the shunt's TCR (>0.25%/°C), and the systematic non-linearity of the BGR and the temperature sensor are *all* digitally compensated by a *single* 2nd-order polynomial obtained by batch calibration [1].

A ±5A version of the CSS based on a 10mΩ on-chip metal shunt (Fig. 2) was also implemented. However, the insulating oxide between the metal shunt and the PNPs in the substrate gives rise to errors in the estimated T, which in [1,2] were corrected by an extra calibration step. In this work, thermal vias between the shunt and the gates of dummy PMOS devices improve the thermal coupling between the shunt and the substrate and reduce shunt self-heating, while preserving galvanic isolation. The result is better accuracy with simpler calibration.

By biasing a pair of PNPs at a 1:10 current-density ratio (Fig. 3), the BGR generates the voltages V_{BE} (CTAT) and ΔV_{BE} (PTAT). ΔV_{BE} is made accurate by dynamically matching both the current sources and the PNPs, while V_{BE} is corrected by a single PTAT-trim at room temperature.

At ADC_I's input, capacitor C_{S1} (=3pF) samples V_{shunt} using a low-leakage switching scheme [1,2], while capacitors C_{S2} (=3pF) and C_{S3} (=300fF), sample and accurately combine ΔV_{BE} and V_{BE}, respectively, to generate $V_{Ref}=\Delta V_{BE}+V_{BE}/10 \sim 120\text{mV}$. ADC_T digitizes T by charge-balancing ΔV_{BE} against $-V_{BE}/10$, resulting in an output bitstream with an average value of μ_T

$=\Delta V_{BE}/V_{Ref}$. To suppress the effect of the 1st integrators' offset and $1/f$ noise, CDS and low-frequency chopping (CHL) are used. Fig. 4 summarizes the digital backend and calibration process used in the CSS to accurately obtain the current I.

Compared to our previous CSS [1,2], the two ADCs used in this design facilitate the simultaneous measurement of current and temperature, which in turn, results in a more accurate response to transient currents. In addition, the ADCs employ current-reuse amplifiers (Fig. 3) and fringe capacitors to reduce power consumption and area by 4× and 2×, respectively. Further energy-efficiency is achieved by sharing the same PNP core between ADC_I and ADC_T.

Fabrication and Measurement

The CSSs were realized in a standard 0.13µm CMOS process (Fig. 5). They occupy 0.4mm² (CSS1 with heatsink shunt,) and 0.85mm² (CSS2 with on-chip shunt) and draw 13µA from a 1.5V supply. BGR, ADC_I and ADC_T consume 6.5µA, 4.3µA, and 2.2µA, respectively. For flexibility, the digital backend and decimation filter are implemented off-chip. At a clock frequency of 100kHz and for conversion rates up to 400S/s, ADC_I is kT/C-noise limited, achieving 1.4µV$_{rms}$ resolution in a conversion time T_{conv} of 18ms. ADC_T then achieves ±0.35°C inaccuracy and 10mK resolution. The measured spread of V_{Ref} is ±0.1%, dropping to ±0.035% after a digital PTAT trim.

Five samples of CSS1 were tested from −55 to +85°C. ADC_I's measured offset is less than 6µV (23mA), dropping below 110nV (400µA) after CHL (Fig. 6). Fig. 7 depicts the nonlinearities of V_{Ref}, the temperature sensor and R_{shunt}, which are digitally compensated by a *single* 2nd-order polynomial, as they are quite stable in the process used [1,6]. After calibrating the shunt (at +5A and ~25°C) and with temperature compensation, CSS1 achieves ±0.3% gain error from −55 to +85°C, and over a ±36A range (Fig. 8). It should be noted that this range is mainly limited by self-heating considerations (40°C at 36A) and not by the shunt.

Fifteen HVQFN-packaged samples of CSS2 were also characterized from −55 to +85°C. ADC_I's measured offset is then less than 5µV (500µA), dropping below 40nV (4µA) after CHL (Fig. 6). After calibrating the shunt (at +3A and ~25°C) and with temperature compensation, the CSS2 achieves a gain error of ±0.3% over a ±5A range (Fig. 9).

For the sake of comparison, the dynamic accuracy of both CSSs was evaluated with a 5A step, with T_{conv}=18ms (Fig. 10). This causes a temperature rise of ~1°C and ~20°C in CSS1 and CSS2, repectively. It can be seen that both systems maintain their accuracy throughout the current step.

A comparison with the state-of-the-art is shown in Fig. 11. Compared to [3–5], CSS1 represents a significant increase in current handling capability (>2×), accuracy (>2×) and dynamic range (>100×) despite the use of a *standard* lead-frame shunt.

Reference

[1] S. H. Shalmany *et al.*, *JSSC*, Apr. 2016.
[2] S. H. Shalmany *et al.*, *VLSI Circ. Symp.*, June 2015.
[3] UCC1926, http://www.ti.com.cn/cn/lit/ds/symlink/ucc2926.pdf
[4] INA250, http://www.ti.com/lit/ds/symlink/ina250.pdf
[5] LM3812/13, http://www.ti.com/lit/ds/snos028d/snos028d.pdf
[6] G. Maderbacher *et al.*, *ISSCC*, Feb. 2015.

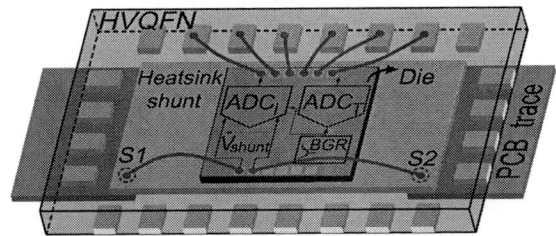

Fig. 1. The ±36A CSS based on a *heatsink shunt* in a standard HVQFN package.

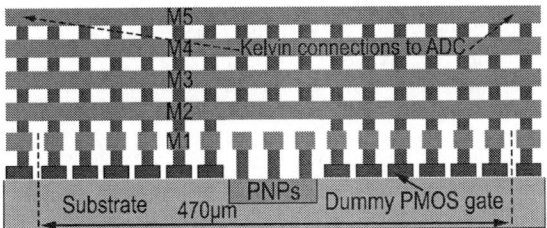

Fig. 2. Cross section of the *on-chip shunt* and the temperature-sensing PNPs underneath.

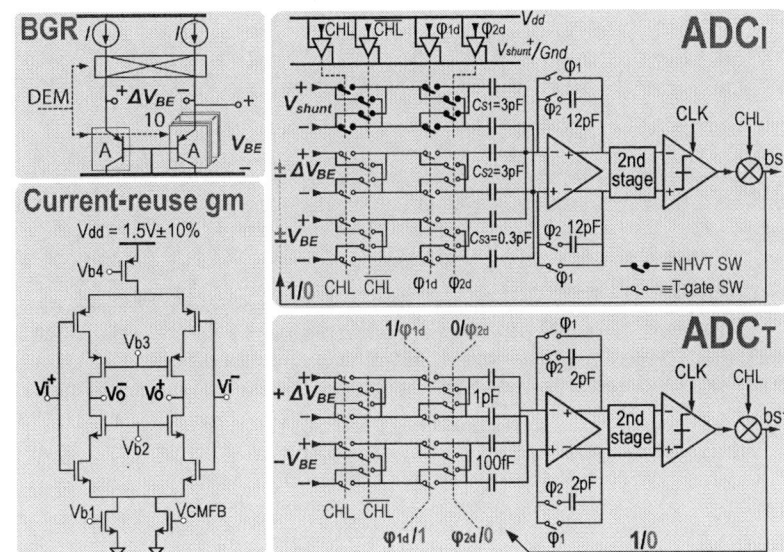

Fig. 3. A simplified block diagram of the readout electronics used in the CSS.

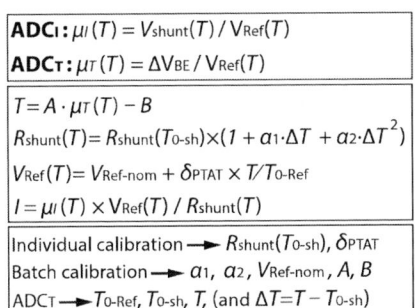

$$\text{ADC}_I: \mu_I(T) = V_{shunt}(T) / V_{Ref}(T)$$

$$\text{ADC}_T: \mu_T(T) = \Delta V_{BE} / V_{Ref}(T)$$

$$T = A \cdot \mu_T(T) - B$$

$$R_{shunt}(T) = R_{shunt}(T_{0\text{-}sh}) \times (1 + a_1 \cdot \Delta T + a_2 \cdot \Delta T^2)$$

$$V_{Ref}(T) = V_{Ref\text{-}nom} + \delta_{PTAT} \times T / T_{0\text{-}Ref}$$

$$I = \mu_I(T) \times V_{Ref}(T) / R_{shunt}(T)$$

Individual calibration	$R_{shunt}(T_{0\text{-}sh})$, δ_{PTAT}
Batch calibration	a_1, a_2, $V_{Ref\text{-}nom}$, A, B
ADC_T	$T_{0\text{-}Ref}$, $T_{0\text{-}sh}$, T, (and $\Delta T = T - T_{0\text{-}sh}$)

Fig. 4. Digital backend math & calibration.

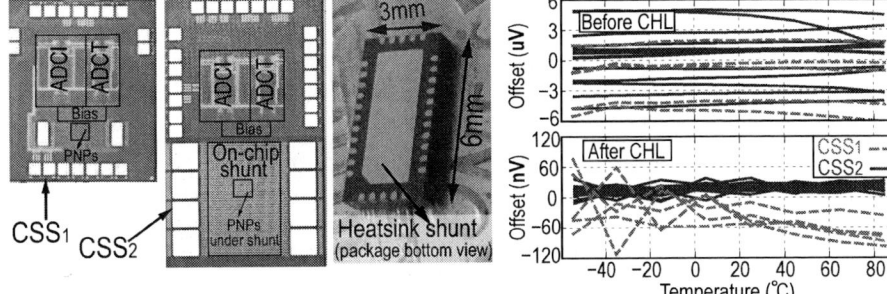

Fig. 5. Chip micrograph and HVQFN package.

Fig. 6. ADC$_I$'s offset over temperature.

Fig. 7. Nonlinearity: V_{Ref}, T-sensor, and R_{shunt}.

Fig. 8. CSS$_1$'s gain error.

Fig. 9. CSS$_2$'s gain error.

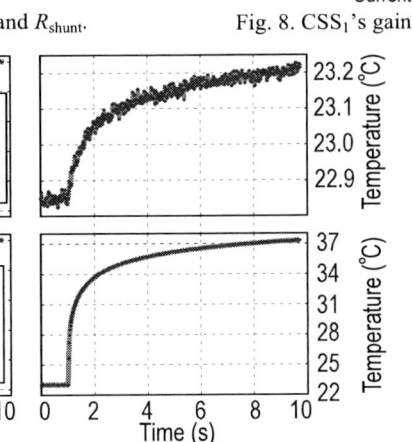

Fig. 10. Transient measurements for a 5A current step.

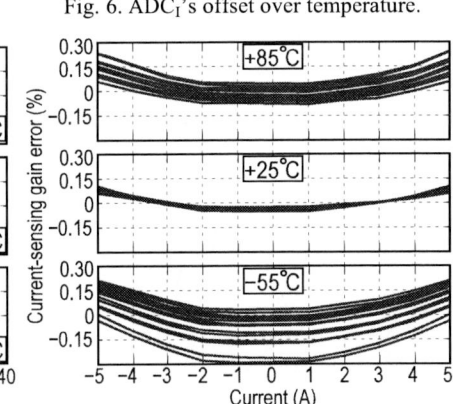

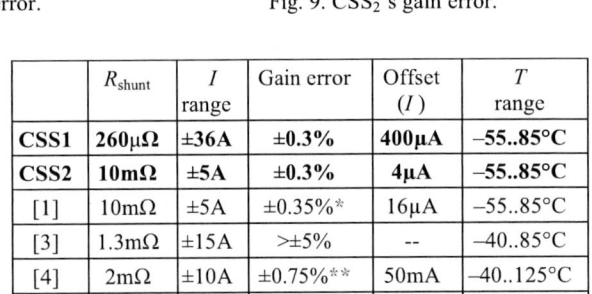

	R_{shunt}	I range	Gain error	Offset (I)	T range
CSS1	**260µΩ**	**±36A**	**±0.3%**	**400µA**	**−55..85°C**
CSS2	**10mΩ**	**±5A**	**±0.3%**	**4µA**	**−55..85°C**
[1]	10mΩ	±5A	±0.35%*	16µA	−55..85°C
[3]	1.3mΩ	±15A	>±5%	--	−40..85°C
[4]	2mΩ	±10A	±0.75%**	50mA	−40..125°C
[5]	4mΩ	±7A	±3%	11mA	−40..125°C

* Uses extra calibration

** Uses a non-standard low-TCR shunt

Fig. 11. Comparison table with the state-of-the-art.

A 114-pW PMOS-Only, Trim-Free Voltage Reference with 0.26% within-Wafer Inaccuracy for nW Systems

Qing Dong, Kaiyuan Yang, David Blaauw, and Dennis Sylvester

University of Michigan, Ann Arbor, MI, qingdong@umich.edu

Abstract

A sub-nW voltage reference is presented that uses only PMOS transistors, thereby providing inherently low process variation and enabling trim-free operation for LDOs and other applications in nW microsystems. Sixty chips from 3 different wafers in 180nm CMOS are measured, showing an untrimmed within-wafer σ/μ of 0.26% and wafer-to-wafer σ/μ of 1.9%. Measurement results also show a temperature coefficient of 48-124ppm/°C from −40°C to 85°C. Outputting a 0.986V reference voltage, the reference operates down to 1.2V and consumes 114pW at 25°C.

Introduction

Voltage references in LDOs, amplifiers, and ADCs for nW systems such as sensors and IoT devices can tolerate ~ 5% inaccuracy, but they require sub-nW power consumption [1]. Conventional bandgap voltage references achieve excellent uniformity across process variation and temperature, but their complexity leads to µW range power [2], which is unacceptable for emerging nW microsystems. To achieve low power, one approach is to use a V_{th}-based voltage reference with devices biased in the sub-threshold region [3, 4]. However, these sub-nW voltage references make use of native transistors, which are potentially at different corners than normal devices due to distinct doping processes, making them more sensitive to process variations. Also, native transistors are not provided by all fabrication technologies [3] and the output reference voltage is too low if an NMOS diode is used. Combining the native NMOS with stacked PMOS diodes can increase the reference voltage [4], but this further enlarges variation across corners.

For both bandgap references and the aforementioned sub-threshold references, post-fabrication trimming of each chip is required to alleviate the impact of process variations. However, this is a significant expense in cost-sensitive designs because of area overhead and testing complexity. In addition, non-volatile memory such as one-time-programmable (OTP) memory is required to store the trimming configuration information [3], requiring extra fabrication masks at increased cost. This paper proposes an ultra-low power PMOS-only voltage reference. By using only PMOS transistors, the reference has inherently low process variation. The untrimmed within-wafer σ/μ is 0.26%, and the untrimmed wafer-to-wafer σ/μ of 1.9%, which is sufficient for many applications in nW systems. With a 0.986V output reference voltage, the design can function down to 1.2V and consumes only 114pW.

Trim-free Voltage Reference Design

Fig.1 shows a simplified structure of the proposed trim-free voltage reference using only 4 PMOS devices (M1-M4). M1 is forward-biased and provides sub-threshold current flowing through the bottom PMOS diode M2. The current equations of M1 and M2 are expressed as in (1). By solving (1), V_{ref} can be expressed as (3). As M1 and M2 are the same type of PMOS, the difference between V_{th1} and V_{th2} comes solely from the body bias effect of M1. Random V_{th} mismatch is kept negligible by upsizing (> 20 µm²) of all 4 devices in this reference.

$$I_R = u_p C_{ox} \frac{W_1}{L_1} n V_T^2 exp\left(\frac{0-V_{th1}}{mV_T}\right) = u_p C_{ox} \frac{W_2}{L_2} n V_T^2 exp\left(\frac{0-V_{ref}-V_{th2}}{mV_T}\right) \quad (1)$$

$$I_L = u_p C_{ox} \frac{W_3}{L_3} n V_T^2 exp\left(\frac{V_{body}-V_{dd}-V_{th3}}{mV_T}\right) = u_p C_{ox} \frac{W_4}{L_4} n V_T^2 exp\left(\frac{0-V_{th4}}{mV_T}\right) \quad (2)$$

$$V_{ref} = V_{th1} - V_{th2} + mV_T ln\frac{W_1 L_2}{W_2 L_1} \quad (3)$$

$$= \gamma\left(\sqrt{2\phi_b - mV_T ln\frac{W_4 L_3}{W_3 L_4}} - \sqrt{2\phi_b}\right) + mV_T ln\frac{W_1 L_2}{W_2 L_1} \quad (4)$$

M3 and M4 generate the required body bias for M1. M4 is an off-state PMOS and M3 is a PMOS diode. The current equations of M3 and M4 are expressed in (2). As M3 and M4 are also the same type of PMOS, V_{th3} and V_{th4} are essentially identical. The combination of M3 and M4 provides a body-bias voltage V_{body} that tracks V_{dd} and creates

a constant V_{BS} (V_{body}-V_{dd}) for M1, as shown in Fig. 2. If the current through M3 (I_L) is much larger than the parasitic diode current (I_{dio}) from the source to the N-well of M1, V_{ref} can be expressed by (4). The left term of Equation (4) is complementary to temperature, whereas the right term is proportional to temperature (Fig. 1). With proper sizing of the four transistors, the first-order temperature dependency can be cancelled out. Moreover, V_{th} does not play a role in Equation (4) because each pair (M1/M2 and M3/M4) uses the same type of PMOS, thus significantly reducing process variation. Since I_{dio} is not well modeled, we designed I_L to be 3 orders of magnitude larger than I_{dio} to minimize the effect of I_{dio}. Proper sizing of these transistors can be determined using a global optimization tool.

As shown in Fig. 3, stacked PMOS diodes can replace M2 and M3 to generate a higher reference voltage, and multiple voltage levels can be generated in this manner. Three stages of PMOS diodes are used in our design to realize an approximately 1V output reference voltage. MIM capacitors C0 and C1 (both set to 1.78pF) are used to isolate the reference voltage from high-frequency power supply noise.

Fig. 4 compares simulated reference voltage distributions across corners for the proposed design as well as designs from [3] and [4]. The proposed design achieves < 4% inaccuracy across all corners, whereas [3] and [4] vary up to 10% and 19%, respectively.

Measurement Results

Sixty chips from 3 different wafers in 180 nm CMOS were tested. One wafer was in a typical corner with thin top-metal, another was found to be at a slow corner with ultra-thick top-metal, and the third was at a fast corner with ultra-thick top-metal. All measurements are reported without trimming.

Fig. 5 shows the measured reference voltage across temperature for all 60 chips. From −40°C to 85°C, the temperature coefficient of the typical wafer ranges from 48ppm/°C to 104ppm/°C, and those of the fast and slow wafers are 55.2−124ppm/°C and 56.1−117ppm/°C, respectively. The reference voltage distributions at 25°C of the 3 different wafers are shown in Fig. 6. Without trimming, the typical wafer shows a mean value of 986.2mV and standard deviation of 2.6mV. The average voltage difference between the fast and slow wafers is 3.6% (1.9% σ/μ), matching simulation and providing sufficient accuracy for many key circuit applications within nW systems.

Fig. 7 shows the measured sensitivity of reference voltage to power supply voltage. Line sensitivity is 0.38%/V from 1.2V to 2.2V. Fig. 8 shows the measured temperature coefficients at different supply voltages. Fig. 9 shows the measured power supply rejection ratio (PSRR) from 10Hz to 10MHz. High-frequency PSRR is −56dB, which can be further improved with larger loading caps C0 and C1.

Fig. 10 shows the measured power consumption across supply voltage and temperature. The output reference voltage is approximately 1V with 3 stages of stacked PMOS diodes. The power supply can be reduced to 1.2V while maintaining this approximately 1V reference voltage. To lower the minimum power supply, fewer stages of PMOS diodes can be used, but the output reference voltage will be lowered as well. At 25°C and 1.2V, the power consumption is 114pW, which is suitable for low-power sensor and IoT applications.

Fig. 11 shows the die photo; the proposed voltage reference occupies an area of 4880µm² (80µm x 61µm) with this area dominated by the two MIM capacitors, C0 and C1. Table I summarizes the results of the proposed sub-nW trim-free voltage reference and compares them with previous works.

References

[1] T. Jang et al., ESSCIRC, 2015. [2] G. Ge et al., JSSC, 2011.
[3] M. Seok et al., JSSC, 2012. [4] I. Lee et al., VLSI, 2014.
[5] A. Shrivastava et al., ISSCC, 2015. [6] Y. Osaki et al., JSSC, 2013.
[7] V. Ivanov et al, JSSC, 2012.

978-1-5090-0636-6/16 $31.00 © 2016 IEEE

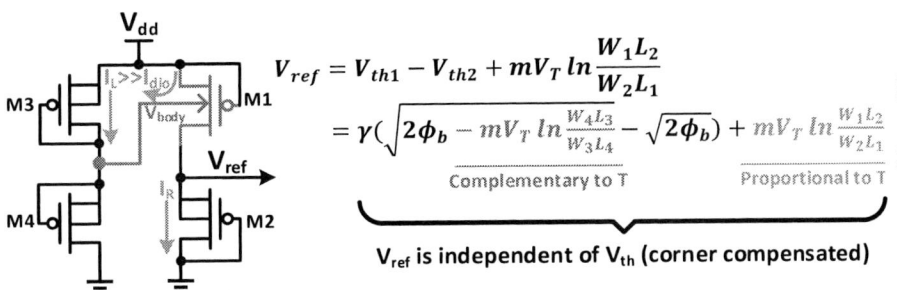

$$V_{ref} = V_{th1} - V_{th2} + mV_T \ln \frac{W_1 L_2}{W_2 L_1}$$

$$= \gamma \underbrace{\left(\sqrt{2\phi_b - mV_T \ln \frac{W_4 L_3}{W_3 L_4}} - \sqrt{2\phi_b} \right)}_{\text{Complementary to T}} + \underbrace{mV_T \ln \frac{W_1 L_2}{W_2 L_1}}_{\text{Proportional to T}}$$

V_{ref} is independent of V_{th} (corner compensated)

Fig.1 Simplified circuit of proposed voltage reference generator and its equations.

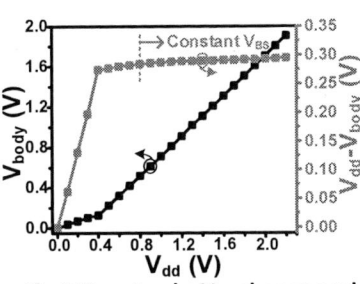

Fig.2 V_{body} tracks V_{dd} change and creates constant V_{BS} for M1.

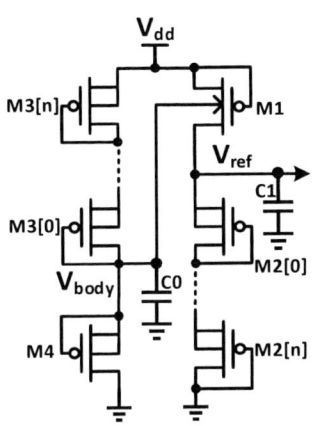

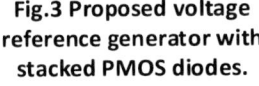

Fig.3 Proposed voltage reference generator with stacked PMOS diodes.

Fig.4 Comparison of V_{ref} simulation at all corners among the proposed design, [3], and [4].

Fig.5 Measured V_{ref} across temperature for 3 wafers in 3 different corners.

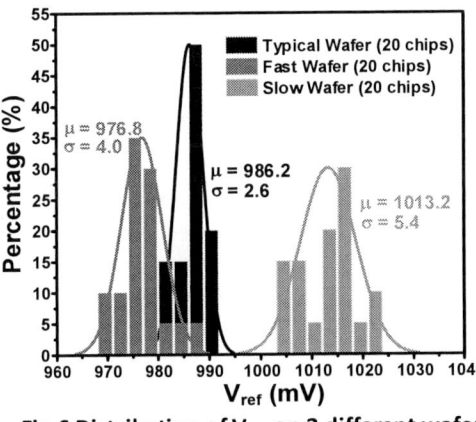

Fig.6 Distribution of V_{ref} on 3 different wafers.

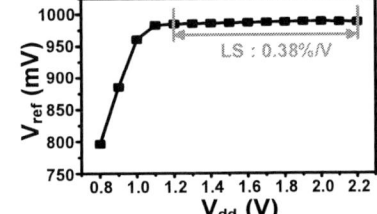

Fig.7 Measured line sensitivity.

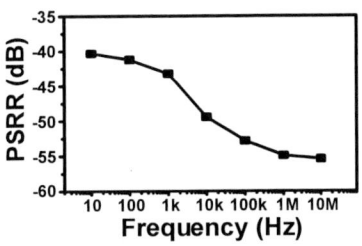

Fig.9 Measured PSRR.

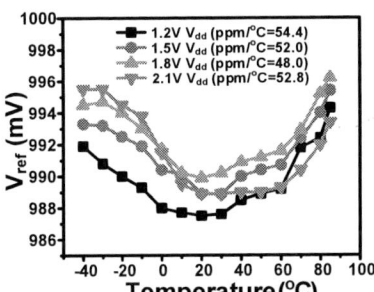

Fig.8 Measured temperature coefficients at different V_{dd}.

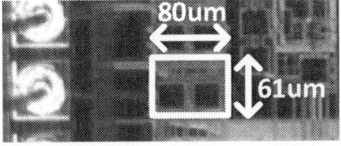

Fig.11 Die Photo in 180 nm CMOS.

Fig.10 Measured power across V_{dd} and temperature.

Table 1. Comparison table to related works.

Parameters	This Work	[3]	[5]	[6]	[7]	[2]
Process (nm)	180	180	130	180	130	160
Power (nw)	0.114	0.006	32	52.5	170	99000
Min. V_{dd} (V)	1.2	0.5	0.5	0.7	0.75	1.62
V_{ref} (V)	0.9862	0.3268	0.498	0.548	0.256	1.0875
Within-Wafer Untrimmed σ/μ (%)	0.26	0.8	0.67	1.05	1	0.5
Wafer-to-Wafer Untrimmed σ/μ (%)	1.9 (3 wafers)	NA	NA	NA	NA	~0.3 (2 wafers)
Temp. Range (°C)	-40 ~ 85	-20 ~ 80	0 ~ 80	-40 ~ 120	-20 ~ 85	-40 ~ 125
TC (ppm/°C)	48.0 ~ 124	54.1 ~ 176.4	75	114	40	5 ~ 12
LS (%/V)	0.38	0.044	2	NA	0.35	NA
PSRR (dB)	-42/-56 (100Hz/10MHz)	-49/-55 (100Hz/10MHz)	-40	-56 (100Hz)	-93	-76 (DC)
Area (um²)	4880	1425	26400	24600	70000	120000
Type	Vth	Vth	Bandgap	Bandgap	Bandgap	Bandgap
Chips Measured	60 chips in 3 wafers	14 chips	6 chips	9 chips	NA	61 chips in 2 wafers

Motor Control Used To Be Boring

Alexander Tessarolo

Texas Instruments Pty Ltd, Sydney, NSW, Australia

a-tessarolo@ti.com

Abstract

Motor Control may not be as trendy as say IoT, but it cannot be denied that the application areas of late have been much more interesting. The growing popularity of drones, electric vehicles such as the Tesla, mobility vehicles like electric bikes or the allure of hover-boards have seen an explosion of electric motors. But changes are also happening in the more traditional industrial sector of motor and servo drives. The demand for cost and energy efficiency is driving the need for new system topologies and greater integration levels. Devices now have to integrate communications with traditional control. Integrated Safety and Security are now part of the mix. The Motor Control device of today is vastly different then 10+ years ago and the challenges are much greater, requiring a broader set of design skills.

Introduction

Focusing on industrial motor control (motor control space is very broad with many common issues) Fig. 1 & 2 illustrates the basic components that make a motor/servo drive system:

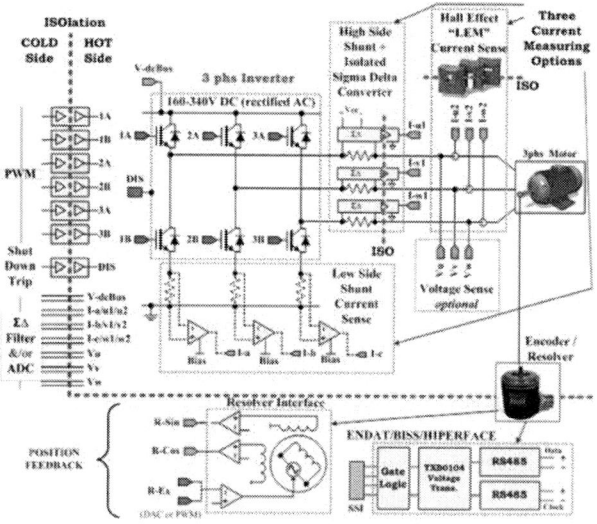

Fig. 1 Typical 3 Phase Power Stage

Pulse Width Modulation (PWM) peripherals drive power FETs/IGBTs (controlling the current driven into the motor coils), of each phase of a motor (most typically 3 phase). To control the speed and know the position of the motor, we measure current and optionally voltage on each phase. Sigma-Delta modulators are now more commonly used as they are easier to galvanic-ally isolate compared to an ADC, but suffer from higher latencies. For servo applications the exact position of the motor shaft needs to be known. This information is fed back to the controller via sensors on the motor shaft in various formats, either analog (Resolver)

quadrature clock signal (QEP) or digital bit stream in various formats (ENDAT/BISS/HIPERFACE/..).

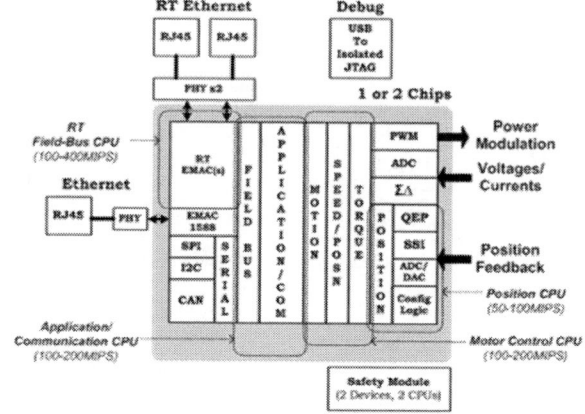

Fig. 2 Main Controller(s), Multi-CPU System

The main controller takes the sensor information (current, voltage, position) and applies the required control algorithm to drive the PWM signals. The control loop frequency can be anywhere from ~10kHz to ~40+kHz frequency without missing a beat (true real-time). This may increase in future as faster IGBTs using GaN technology come into the market.

There are various communication topologies and systems servicing single axis or multi-axis applications (such as robotic arms or CNC machines). The communication (termed Real-Time Field-Bus's) can be many and varied, some can be standard (EtherNET, EtherCAT, SERCOS III, ProfiNET, ProfiBUS etc.) or some can be custom.

The trend now is to try and integrate all of these components into a single device, using multiple CPU core sub-systems. It is very challenging to perform the real time control tasks and the communication and application level tasks in a single CPU. The availability of higher density process technologies is enabling the use of Multi-CPU systems. Also, the choice of memory sub-system is task dependent. For real-time control you try to avoid caching as you need deterministic and precise execution (tightly coupled memories). For communication and other tasks, a cached memory system is acceptable.

However, with the arrival of new high speed (100-200+MBits/sec) digital isolation technologies, it is now possible to move more of the control processing on the HOT side (see Fig. 1). This would improve overall system response, performance and enable greater levels of diagnostics. With fewer signals between the HOT and COLD side, cost of connectors and physical PCB area is reduced.

References [1] and [2] are providing more details.

Technology Challenges

From a semiconductor supplier perspective the choice of process technology is very critical. Some key factors affecting the choices are:

- Industrial products have very long life spans (20+years).
- Moving process nodes that integrate Flash and Analog require ~3x the effort and investment compared to a RAM only logic process node. So you try and improve return on investment by not changing process nodes as often (~10years) and also by skipping process nodes (i.e. 180nm to 65nm, skip 130nm and 90nm).
- Flash based processes have limited MHz capability compared to a high performance RAM process. You have to get very innovative in increasing performance (mainly via improving quality of MIPS with customized CPUs and application specific instructions).
- Process has to be reliable, meet safety requirements, and needs to support a high temperature range (-40 to 150C).

As we move into higher density nodes (65nm, 40nm) the following new challenges are faced.

- Developments are more expensive and take longer as devices are more complex which drives flexible and configurable solutions. Discipline in execution is required as mistakes are expensive.
- Leakage currents are much higher and can sometimes be greater than the active current. Typical low power techniques are ineffective as many industrial applications run at full speed 24/7. This puts pressure on packaging technology and efficient processing and peripherals (quality of MIPS is important).
- Flash and RAM need to be protected with ECC and Parity and integrated safety diagnostics are a must as soft error events are more prevalent.

In addition we have the following challenges from the industrial drive system:

- There is a plethora of communication (RT Field-Bus) and position sensing standards leading to a need to address all of them cost effectively. Adaptability to future revisions and support of custom interfaces becomes a must.
- New isolation technologies require new high speed I/F solutions (no standard exists). To date such interfaces are custom (Fast Serial Interface from TI coming 2016).
- The addition of Safety, Security, Real-Time Motor Diagnostics and higher speed control loops are increasing demands for MIPS. MIPS requirements will double in the next 5 years (TI TMS320F2837xD has 800MIPS total, we expect to need 1600 equivalent MIPS for next generation).

New System Architecture Approaches

With higher gate density process nodes we can create more complex devices, helping to address the above challenges:

- By using task optimized CPUs in a heterogeneous system we can maximize performance/task. For example, you may have a specialized bit processing CPU engine dedicated for processing of RT Field-Bus bit streams (i.e. PRU-ICSS system in TI AM4x/5x Sitara Processors). A generic CPU engine for application level/OS/communication (i.e. ARM Cortex-A/R class CPU). A math/DSP efficient CPU engine for real-time motor control algorithms (i.e. TI C28 CPU). A small and efficient CPU engine for handling position sensor information (i.e. TI CLA CPU) (optimized for Resolver signal processing algorithms).
- Hardware Floating-Point support is becoming the norm. 32-bit FPU is now common. 64-bit H/W FPU support will become common in the next 5 years.
- Fast integer division with resolutions beyond 32-bit (64/32 or 64/64) will also be required as position system accuracy increases.
- Customizing CPU instructions to improve the overall quality of MIPS. For example: adding instructions to handle SIN/COS/ATAN operations in a few cycles can significantly improve control algorithm performance.
- Configurable logic gates enable adaptation of new features or standards without having to spin a new device. This could be in the form of FPGA SOCs or as custom implementations (as found on TI TMS320F2837xD).
- Mixed technology and MCM packaging enabling the adoption of different process technologies (as found on TI TMS320M3x, ADI ADSP-CM40x).

An example of a monolithic device that incorporates some of the above solutions, targeting the Motor Control and Position sensing functions with integrated analog is shown in Fig 3.

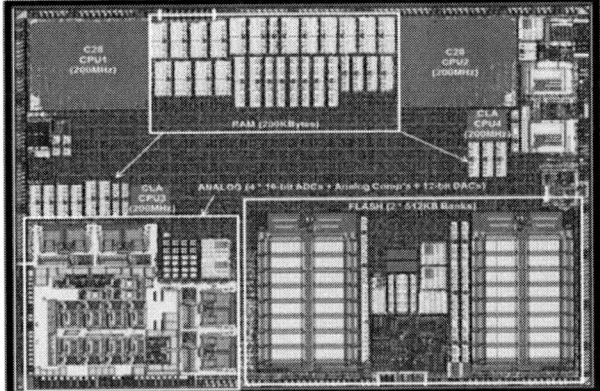

Fig. 3 TMS320F2837xD Controller, 35sqmm, 65nm FLASH

Motor Control Expanding In New Dimensions

We have barely scratched the surface of the changes happening in Motor Control. Even Wireless technology is being adopted for motor diagnostics (e.g. to predict faults before they cause damage). Not to mention that Power Control in i.e. Solar, Electric/Hybrid Vehicle Inverters leverage (re-use) much of the same technology, albeit at higher performance levels. Motor Control will continue to excite and challenge designers for years to come.

References

[1] Modern Power Electronics and AC Drives: Bimal K Bose
[2] Industrial Communication Systems, 2nd Edition: Bogdan M. Wilamowski, J. David Irwin

A Fully Integrated GaN-based Power IC Including Gate Drivers for High-Efficiency DC-DC Converters

Shinji Ujita, Yusuke Kinoshita, Hidekazu Umeda, Tatsuo Morita, Kazuhiro Kaibara, Satoshi Tamura, Masahiro Ishida and Tetsuzo Ueda

Engineering Division, Automotive & Industrial Systems Company, Panasonic Corporation, Nagaokakyo-shi, Kyoto, Japan
ujita.shinji@jp.panasonic.com

Abstract

In this paper, we present a state-of-the-art integrated GaN power IC capable of operating in a high frequency (MHz) regime. This realizes system size reduction, 60% maximum, of a power IC. The IC consists of two output power transistors (PT) and two gate drivers (GD). The key devices in the IC are normally-off gate injection transistors (GITs) for PT and GD and a normally-on hetero-junction field effect transistor (HFET) for GD. Novel local control of carrier concentration of an identical 2 dimensional electron gas (2DEG) at an AlGaN/GaN interface which made integration of the transistors with such a large threshold voltage difference possible is described. A specially developed post-passivation interconnection process giving low parasitic components is also described. The IC applied to a 12V-1.8V DC-DC converter shows high frequency switching operation well beyond the limit of Si pointing to future improvement in consumer electronics power supply systems.

Introduction

System size reduction is one of the most crucial issues of consumer electronics systems. In this regard, gallium nitride (GaN) is an excellent candidate for power transistors because of its excellent material properties such as high mobility and high breakdown voltage all of which contribute to superiority of the GaN based devices to Si ones as reflected in the higher Baliga's high frequency figures of merit (BHFFOM) of the former [1].

In this work, we present a novel integrated GaN power IC with a half-bridge architecture capable of operating in a high frequency (over MHz) regime. The IC enables one to improve an efficiency of DC-DC converters at higher frequency resulting in reduction of the system size as shown in Fig. 1. An efficiency of DC-DC converter using the IC can be realized about 90% at higher frequency of 3MHz beyond the limit of Si against an efficiency of that using Si-based discrete IC is 88% at 1MHz. The 3MHz operation can reduce a system size from 500mm² to 200mm² as compared to the 1MHz operation by reduction of values of passive components in inversely proportion to the frequency, i.e. an inductance of a coil is reduced from 350nH at 1MHz to 120nH at 3MHz. The IC comprises two output power transistors (PT) and two gate drivers (GD). The key devices in the IC are normally-off gate injection transistors (GITs) for PT and GD and a normally-on hetero-junction field effect transistor (HFET) for GD. In order to integrate the transistors with such a large threshold voltage difference, we realized a novel method to locally control the carrier concentration of an identical 2 dimensional electron gas (2DEG) at an AlGaN/GaN interface by local polarity tuning of the top surface of the AlGaN barrier layer [2,3]. Also, for realizing low parasitic components, a new post-passivation interconnection process is developed [4]. The IC applied to a 12V-1.8V DC-DC converter demonstrates high frequency, 3MHz maximum, switching operation well beyond the limit of Si, which, in turn, enables 60% reduction of system size. The result points to future system improvement in consumer power supply systems such as point-of-load (POL) devices, the commonly used ones in power supply systems of PCs.

Circuit Design

We aim at a 12V-1.8V DC-DC converter with a half-bridge architecture of which schematic circuit diagram is shown in Fig. 2. Instead of diodes, we employ GaN-GITs as the output PTs in order to avoid the junction voltage drop inherent to diodes. The PTs are driven by the GDs. By integrating PTs with GDs, the effect of parasitic

inductors L1-L4 can be reduced. This, in turn, contributes to reduction of operation loss as shown in Fig. 2(b).

The circuit diagram and the schematic cross section of the IC consisting the PTs and GDs are shown in Fig. 3(a) and Fig. 3(b),

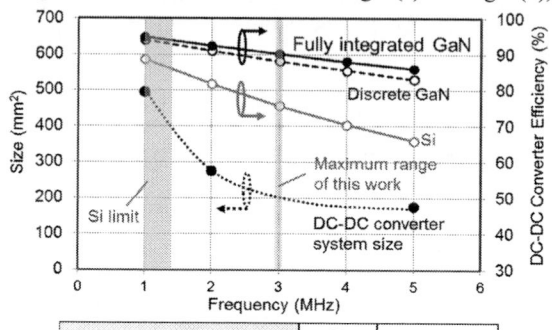

material	Si	GaN
Architecture	Discrete	Fully integrated
Operation frequency (MHz)	1	3
System size (mm²)	500	200
Efficiency of DC-DC converter (%)	88	90

Fig.1 Typical performance parameters of DC-DC converters predicted from materials properties.

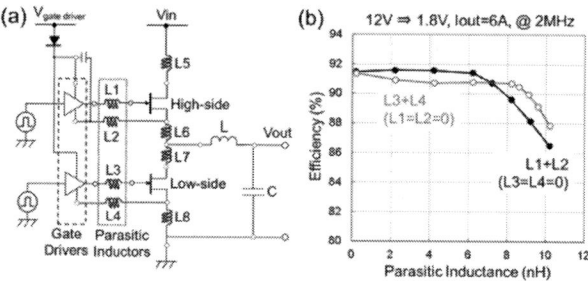

Fig.2 (a) Schematic circuit diagram containing parasitic inductors for DC-DC converters. (b) Simulated conversion loss as functions of parasitic inductance between PTs and gate GDs.

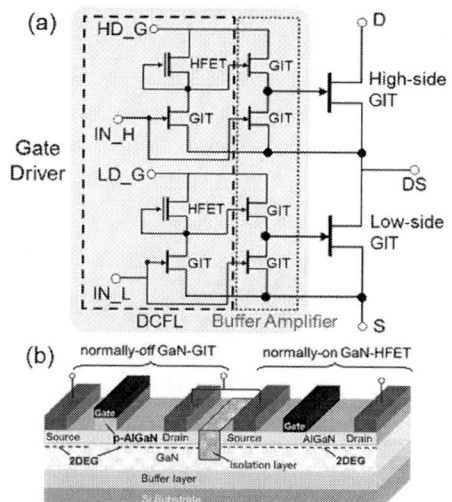

Fig.3 Circuit diagram and schematic cross section of GaN-based IC.

respectively. The GD has a direct coupled field effect transistor logic (DCFL) in which an inverter is formed with the normally-off GIT loaded by a normally-on HFET and the gate of the HFET is connected directly to the drain of the GIT. This is simply the depletion NMOS logic configuration of Si.

We have developed the normally-off GIT utilizing the 2DEG which has been difficult to be fabricated by introduction of the p-AlGaN layer between the gate electrode and the AlGaN/GaN hetero-junction. The Schottky gate electrode for the normally-on HFET is configured on the surface where the p-AlGaN layer is removed, resulting in integration of the GIT and the HFET without any additional process steps. The isolation between the GIT and the HFET is realized by ion implantation.

To reduce the effect of the large current flow through the channel during the off-state, we propose to add a buffer amplifier to the DCFL. A circuit diagram of the DCFL with the buffer amplifier consists of the GaN-GITs. Thus, the DCFL with the buffer amplifier acquires a momentary output current, enabling reduction of the gate width of the HFET. As a result, the power consumption of the GD is reduced.

Fabrication and Results

Post-passivation interconnection process realizes the integration of the GaN-based PTs with the GaN-based GDs. The post-passivation interconnection process has the two thick copper layers and the thick low-k dielectric layers. Fig. 4(a) shows the schematic cross section of the post-passivation interconnection. The reverse transfer capacitances (Crss) of the high and low sides of the PTs are plotted as functions of the dielectric layer thickness (A) as shown in Fig. 4(b). Also, the gate resistances of the high and low sides of the PTs are plotted as functions of the copper layer thickness (B) (Fig. 4(c)). ThicknessA and ThicknessB are set to be 8μm and 5μm, realizing the low parasitic reverse transfer capacitances and gate resistances. Due to this process, all of components in the IC can be connected with each other and the pads electrodes could be configured on the active region of the IC by the suppression of the parasitic capacitance.

A chip microphotograph of the fabricated GaN power IC is shown in Fig. 5. The size of the IC is as small as 5.1mm x 2.3mm. The chip includes all of components in the circuit of Fig. 3(a). Measured switching waveform of the fabricated GaN-based DCFL with the buffer amplifier is shown in Fig. 6(a). The GD discharges and charges a 1500pF capacitor. The total time of the rise time (tr) and the fall time (tf) is 12ns, which is faster by about 40% than that of the conventional Si-based GD with the same maximum output current. It is noted that the speed limitation is set not by GaN material properties but by external circuit parameters. Furthermore, the DCFL with the buffer amplifier realizes reduction of the power consumption by about 98.5% as compared to the DCFL without the buffer amplifier. Fig. 6(b) summarizes the measured operation efficiencies of the DC-DC

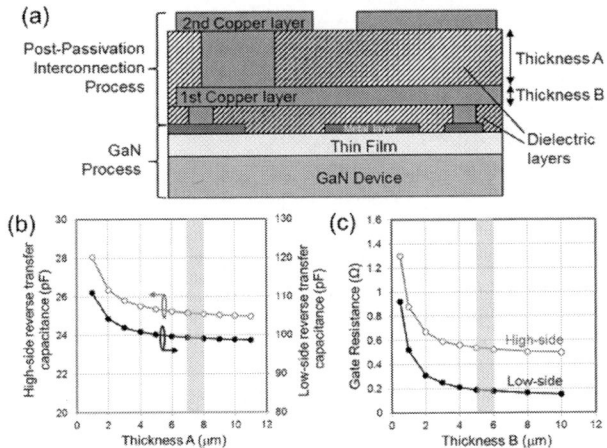

Fig. 4 (a) Schematic cross section of post-passivation interconnection. (b) Calculated reverse transfer capacitance and (c) calculated gate resistance of the PTs.

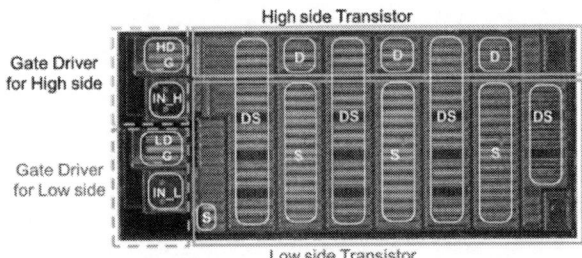

Fig.5 Chip microphotograph of the fabricated GaN-based power IC. Electrode Pads are shown surrounded by white lines. Symbols on the pads indicate terminals of the circuit as shown in Fig. 3(a).

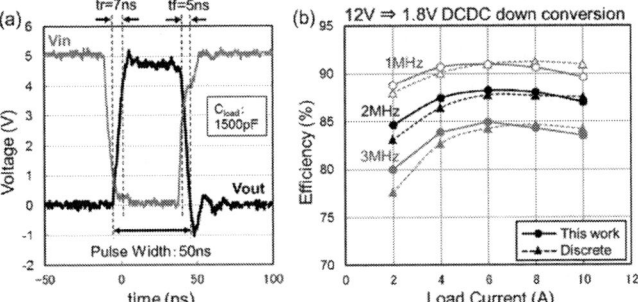

Fig.6 (a) Measured switching waveform of the GaN-based DCFL with the buffer amplifier. (b) Operation efficiencies of the fabricated 12V-1.8V DC-DC converter using the GaN-based power IC.

converter using the GaN-based power IC at the operation frequency up to 3MHz as a function of the output current. The peak efficiency as high as 88.2% and 84.9% is achieved with 12V- 1.8V DC-DC down conversion at 2MHz and 3MHz. The 3MHz operation beyond the limit of Si can reduce the system size of the DC-DC converter by about 60% as compared to the 1MHz operation, as indicated in Fig. 1.

The switching loss of the DC-DC converter at 2 MHz using the GaN-based power IC is calculated 0.68W at the output current of 6A flowing, which is lower by about 15% than that of the DC-DC converter implemented GaN-based GDs and PTs discretely. Based on an analysis of loss components such as output capacitance loss (Coss) and conduction loss etc., currently, efforts to reduce the loss are being made for higher frequency operation.

Conclusion

We demonstrated a high frequency operation, up to 3MHz, of an integrated GaN power IC exceeding the Si limit. We expect that low voltage DC-DC converters of consumer electronics systems such as power supply circuits for PCs can be reduced in size more than 50% in the near future by further improvement in the current efficiency of the present devices.

Acknowledgement

We would like to thank Managing Officer Eiji Fujii and Director Tsuyoshi Tanaka from Panasonic Corporation for their help and technical advice in this work. This work is partially supported by NEDO in Japan, under the Energy Saving Innovative Technology Development Project.

References

[1] T. P. Chow et al, "Wide Bandgap Compound Semiconductors for Superior High-Voltage Unipolar Power Devices", *IEEE Trans. Electron Devises*, vol. 41, No. 8, 1481(1994)

[2] T. Ueda et al., "GaN transistors on Si for switching and high - frequency applications", *Jpn. J. Appl. Phys.*, vol. 53, no. 10, 100214 (2014).

[3] M. Ishida et al., "GaN on Si Technologies for Power Switching Devices", *IEEE Trans. Electron Devises*, vol. 60, pp. 3053 (2013)

[4] S. Ujita et al., "A 26GHz Transceiver Chipset for Short Range Radar Using Post-Passivation Interconnection", *Jpn. J. Appl. Phys.*, vol.50 (2011) 04DE04.

A Transformer-based Digital Isolator With $20kV_{PK}$ Surge Capability and $> 200kV/\mu S$ Common Mode Transient Immunity

Ruida Yun, James Sun, Eric Gaalaas, Baoxing Chen
Analog Devices Inc, Wilmington, MA, 01887, USA

Abstract

This paper presents a transformer-based digital isolator that achieves best-in-class robustness with $20kV_{PK}$ surge capability and $> 200kV/\mu S$ Common Mode Transient Immunity (CMTI). Using polyimide as insulation material, the isolation barrier is more than $30\mu m$ thick and enables the exceptional surge performance. The transformer is fully differential with symmetric layout to improve noise immunity. The OOK Transmitter (TX) is based on the negative-G_m oscillator and operates in the voltage-limited domain. It enable the TX to keep generating the differential On-Off Keying (OOK) carrier signal during disturbances from very fast Common Mode Transient (CMT) noise.

Keywords: CMOS, digital isolator, OOK, CMTI.

Introduction

Digital isolators are widely adopted in a range of applications such as industrial automation, solar inverters and more[1, 2]. There is now a growing demand to make the digital isolators robust in the most noisy harsh environment. For instance, the isolation barrier has to withstand $> 15kV_{PK}$ surges to protect patients in patient monitoring systems, and communication must be immune to $> 100kV/\mu S$ CMT noise to enable next generation gate driver technology. However, digital isolators using SiO_2 insulation can only provide surge capability up to $12.8kV_{PK}$. Capacitive isolators are limited to CMTI of $50kV/\mu S$ even with the OOK data architecture[3, 4]. This work presents a polyimide-based digital isolator using transformers and OOK architecture, achieving $20kV_{PK}$ surge capability and $> 200kV/\mu S$ CMTI.

Circuit Design

Fig.1 shows the block diagram of the proposed digital isolator with chip-scale transformers. The transformers are built with two planar spiral metal windings stacked atop one another, with $30\mu m$ polyimide insulation separating the primary and secondary windings. The digital isolator core signal chain is fully differential to reject noise. For one input logic level (e.g. logic high), the OOK TX directly resonates with the transformer to generate a high frequency ($700MHz$) carrier signal. For the other input logic level (e.g. logic low), the TX is turned off to save power. The logic level that turns on carrier generation can be programmed. An AC-coupled bias network is used in front of the receiver (RX) to attenuate the CMT noise without affecting the carrier signal. The RX demodulates the received signal and converts it back into the digital domain. Each channel has dedicated regulators for its TX and RX to minimize cross-talk and reject power supply noise.

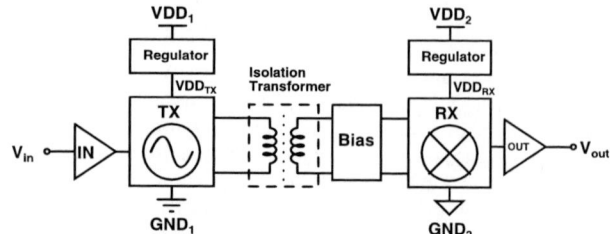

Fig. 1. Block diagram of the proposed digital isolator architecture.

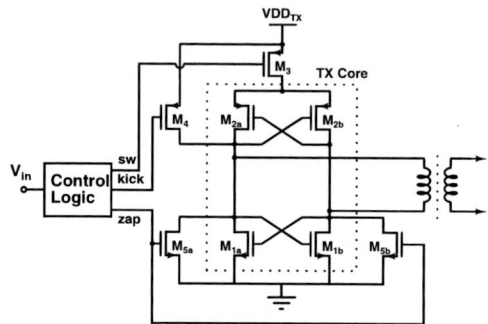

Fig. 2. Circuit diagram of the transmitter.

The TX is designed based on the negative-G_m oscillator. It has control logic to turn the oscillation on or off based on the input levels as shown in Fig.2. When the input is low, the control logic generates the zap signal to quench the LC tank and connect the transformer to GND. When the input is high, the carrier signal is generated by turning on the LC tank. In order to turn on the oscillation within $1nS$, a kick-start scheme is implemented by turning on M_4 first after an input rising edge [5]. The PMOS M_3 is then turned on to activate M_{1a}/M_{1b} and M_{2a}/M_{2b}, and the oscillator operates in the voltage-limited domain. VDD_{TX} is set at $1.25V$ and the TX core is biased in the weak inversion region. This not only reduces the TX power consumption, but also provides $20X$ more negative g_m than needed to sustain oscillation. The voltage-limited operation enables the TX to keep generating the (differential) carrier signal, even though the preferred TX bias condition is disrupted by the fast CMT noise. As the voltage difference between GND1 and GND2 changes because of the fast transient, a common mode current across the insulation is induced due to parasitic capacitance. During the positive CMT event, the TX is forced to source the current. As a result, the gate voltage of the PMOS pairs are driven close to their local ground and the NMOS pairs are effectively off. The PMOS pairs are biased in the strong inversion region to compensate the lost g_m from the NMOS pairs, such that the TX still oscillates and continues to generate the carrier . During the negative CMT event, the NMOS pairs contribute more g_m to sustain the oscillation.

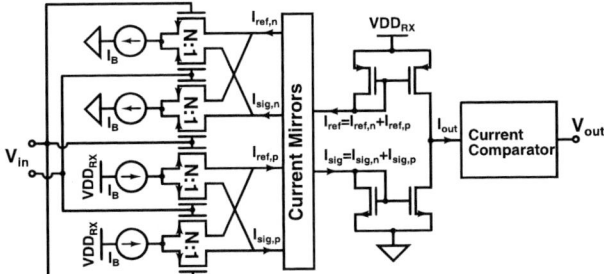

Fig. 3. Block diagram of the receiver.

The receiver (RX) uses asymmetric differential pairs as a rectifier to demodulate the signal in the current domain [6]. As shown in Fig.3, it consists of both NMOS pairs and PMOS pairs to implement rail-to-rail input range. Both pairs generate a corresponding signal current (I_{sig}) and reference current (I_{ref}), which are combined through current mirrors. When the oscillation is off, the weak $1X$ transistors generate the signal current I_{sig}, which is equal to $1/N$ of the tail bias current I_B. I_{ref} is very close to I_B by making N equal to 20. When the oscillation is on, the weak transistors fully turn on and I_{sig} can be approximated as I_B, such that I_{ref} is only a fraction of I_B. The resulting current, $I_{out} = I_{ref} - I_{sig}$ is then processed by a current comparator to differentiate between oscillation on and off.

Measurement Results and Conclusion

Fig.4 shows the packaged chip and cross-section view of the isolation barrier. Both dies are implemented in a $1.8V/5V$ $0.18\mu m$ CMOS technology. The chip operates in a wide supply range from $1.7V$ to $5.5V$ with maximum propagation delay of $11nS$ at $125°C$. It draws $2.8mA$ per channel while running at $1Mbps$. The maximum surge isolation voltage (V_{IOSM}) is tested according to IEC 60065. The test results show that the $30\mu m$ thick polyimide can withstand at least $20kV_{PK}$ surge voltage.

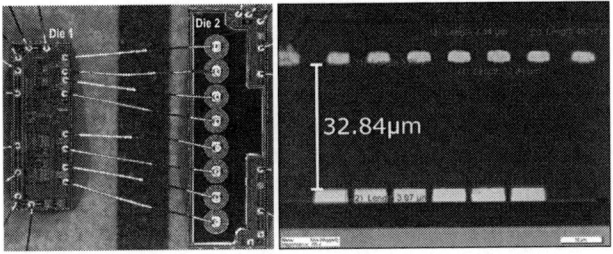

Fig. 4. Chip micrograph and cross-section view of the isolation barrier.

Fig.5 shows the measured CMTI performance for both positive and negative CMT pulses at the worst operating condition ($VDD_{1,2} = 1.7V$ and $tdegc = 125°C$). In the test setup, the high voltage pulse is injected into the floating side (VDD2/GND2) which is powered by a battery pack. The high voltage amplitude is gradually increased until the isolator outputs show glitches. The part eventually failed at $223kV/\mu S$ positive CMT, and even then the output glitch lasts only $4nS$. Table I summarizes the

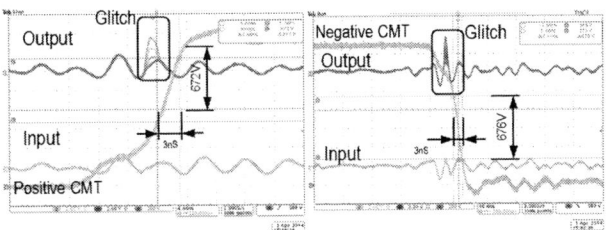

Fig. 5. CMTI test-to-fail results with oscillation on (input low). The part failed at $223kV/\mu S$ (positive CMT) and $225kV/\mu S$ (negative CMT) respectively with 4nS glitch.

TABLE I
Performance Summary.

	This Work	[1]	[2]	[3]	[4]
Insulation Material	Polyimide	Polyimide	SiO2	SiO2	SiO2
Isolation Element	Transformer	Transformer	Transformer	Capacitor	Capacitor
Data Architecture	OOK	Pulse	Pulse	OOK	OOK
Supply Range [V]	$1.7 \sim 5.5$	$3.0 \sim 5.5$	$3.3 \sim 5.5$	$2.25 \sim 5.5$	$2.5 \sim 5.5$
Max. Propagation Delay (t_p) [nS]	11	28	7	17.5	14
Idd per channel @ 5V, 1Mbps (idd/ch) [mA]	2.8	0.3	1.6	1.65	1.6
Max. Surge Voltage (V_{IOSM}) [kV_{PK}]	20	10	2	12.8	9.6
Min. CMTI [$kV/\mu S$]	200	25	35	50	35
Isolator FoM[1]	130	30	8	22	15

1. Isolator Figure of Merit (FoM) is defined as $V_{IOSM} \cdot CMTI/(t_p \cdot idd/ch)$.

performance of this work and compares it to other digital isolators. The minimal CMTI is $4X$ better and surge capability is $1.5X$ higher than [3], although the supply current per channel at $5V$ with $1Mbps$ input is $1.75X$ higher. This work achieves the best Isolator Figure of Merit.

References

[1] "ADuM1100, iCoupler Digital Isolator, Data Sheet, Rev. K, 07/2015," Analog Devices Inc.

[2] S. Kaeriyama, S. Uchida et al., "A 2.5 kV Isolation 35 kV/us CMR 250 Mbps Digital Isolator in Standard CMOS With a Small Transformer Driving Technique," IEEE J. Solid-State Circuits, vol. 47, no. 2, pp. 435–443, Feb 2012.

[3] "ISO7842, 8000 V_{PK} reinforced quad-channel digital isolator, Data Sheet, SLLSEJ0C, OCTOBER 2014, REVISED JULY 2015," Texas Instruments.

[4] "Si8642, Low power quad-channel digital isolator, Data Sheet, Rev. 1.7 6/15," Silicon Laboratories.

[5] A. T. Phan, J. Lee et al., "Energy-Efficient Low-Complexity CMOS Pulse Generator for Multiband UWB Impulse Radio," IEEE Trans. Circuits Syst. I, Reg. Papers, vol. 55, no. 11, pp. 3552–3563, 2008.

[6] A. Gerosa, S. Solda et al., "An Energy-Detector for Noncoherent Impulse-Radio UWB Receivers," IEEE Trans. Circuits Syst. I, Reg. Papers, vol. 56, no. 5, pp. 1030–1040, 2009.

INNOVATIVE SYSTEM ON CHIP PLATFORM FOR SMART GRIDS AND INTERNET OF ENERGY APPLICATIONS

Alessandro Moscatelli, STMicroelectronics, Milan Italy; alessandro.moscatelli@st.com

Abstract

Smart Metering and Smart Grid applications are booming worldwide, requiring innovative integrated circuit solutions to specifically meet multiple and evolving system requirements in terms of connectivity capabilities, system flexibility, sensing accuracy, security, power consumption, system miniaturization and cost of ownership. This paper introduces novel and future-proof system on chip solutions specifically developed to meet these challenging requirements of modern Smart Grid ecosystems.

Introduction

Smart Meters have been massively deployed worldwide replacing mechanical meters, initially just to provide more accurate and remote energy readings, then to enable additional energy services to Utilities and to final users including tariff flexibility, tamper detection, power quality monitoring, energy peaks management, remote loads connection/disconnection and many other services. Smart Grids are today evolving even beyond the smart metering infrastructure, by interconnecting the full power grid to Internet, including the integration of distributed and renewable energy sources, electric vehicles, street lighting systems, home energy gateways and so on, as shown in Fig.1, leading to a new technology revolution often identified as "Internet of Energy" (IoE) [1]. An IoE infrastructure can be seen as an interconnected and interoperable network of smart energy nodes, able to locally sense energy-related parameters, process data and actuations and able to securely receive and transmit data over the cloud to enable remote smart energy services. So, the basic functional blocks of an IoE system typically consist of: a programmable controller managing the main application, a metering sensing and processing unit, a communication modem processor and front end. Although multiple stand-alone technologies exist today to realize in principle each of these IoE functional blocks, some important technology challenges have to be properly addressed to effectively enable large scale industrial IoE implementations.

The first challenge is related to communication modem and front end sub-system. Multiple technologies commonly used in ICT world can be adopted also for IoE connectivity, but Power Line Communication (PLC) is becoming more and more attractive and actually already widely adopted in large smart metering deployments [2], being able to reliably transmit and receive data over long distances by exploiting as communication medium the same power cables which supply the IoE elements, so ensuring the needed ubiquity while avoiding the additional complexity and cost of installing dedicated wired communication bus or wireless infrastructures. Anyway, since power lines have not been conceived to convey communication data, Power Line Modems require demanding data processing coding and special silicon design, especially in the analog front end part which must be able to receive highly noisy and attenuated signals and amplify transmitted data with limited signal distortion. Furthermore, the recently increased PLC protocol standards fragmentation needs to be properly and flexibly managed to ensure interoperability among multiple suppliers. Another challenge of the IoE element is the sensing sub-system design which must provide highly accurate power line voltage and current values measurement over an extended power range and must also perform complex energy data post-processing.

The last but not least challenge is the integration of all above functional blocks along with the main application controller in a compact and cost effective IoE system to enable multiple, large scale implementations without affecting the needed system flexibility. Only a smart and innovative combination of mixed signal designs, multi-power silicon technologies and System on Chip (SoC) architecture can provide the needed technology breakthrough to overcome all above IoE challenges.

STCOMET: the IoE breakthrough SoC

STCOMET™ (STCOMET) platform [3] is the industry's first complete System-on-Chip (SoC) that combines a high precision metering sensing and processing unit with a flexible and performing power line modem and power line analog front end, along with a fully programmable application controller in a single device, so embedding all needed IoE node functionalities (Fig. 2), overcoming the limitations of today solutions, all based on multiple discrete components.

The chip exploits a multi-power 90nm CMOS silicon technology with embedded Flash and an advanced mixed signal design able to effectively combine different IoE application-specific IPs. Despite standard VLSI CMOS technologies, customized multiple gate oxides and dedicated HV transistors have been implemented in the silicon process to provide devices with voltage capabilities up to 10V for low noise, high power analog sections, while keeping high speed, low power 1.2V VLSI CMOS transistors for the digital cores and high performance logic.

The PLC modem architecture has been particularly designed to overcome today PLC performance challenges and meet all major emerging regulations and standards worldwide. It consists of a low-power, multi-core fully programmable DSP engine with embedded fast program and data RAM memories and PLC-specific instructions set, able to perform demanding PLC real-time functions. In particular, the PLC DSP is able to perform multiple modulation schemes to meet different protocol specification requirements (FSK, n-PSK, OFDM,...), advanced convolutional correction codings and Viterbi decodings to properly cope with power line noise environments, frame synchronization and many other PHY and MAC protocol layers real time functions. Different PLC DSP FW codes can be used to comply with all major PLC standards adopted worldwide such as ITU G.9903/PLC-G3, ITU G.9904/PRIME v1.3.6 and v1.4, IEEE 1901.2, METERS AND MORE, and their evolutions. The DSP interfaces a PLC Digital Front End (DFE) including dedicated programmable transmission/reception digital filter chains to fit the signal bandwidth in different PLC modulation cases. Furthermore the PLC DFE implements an automatic gain control (AGC) block, whose purpose is to adapt the received signal to the dynamic of the Analog Converter following the DFE block.

The PLC DFE also includes a current control (CC) block to safely limit the maximum output current delivered to the line.

The PLC DFE then interfaces with a PLC Analog Front End (AFE) including a Receiving section featuring a programmable gain amplifier (PGA) and a dedicated analog-to-digital converter (ADC) to achieve high RX sensitivity over a wide input range and a Transmission section featuring a dedicated Digital to Analog converter (DAC) which then feeds the transmitted signal to a buffering and amplifying Pre-Driver block.

The device PLC section is finally equipped with an integrated high-performance smart Power Line Driver (PLD). The PLD has been carefully designed to provide high output current - up to 1A(rms) - to drive very low power line impedances which may be found in real networks, while keeping a very low signal distortion to comply with most stringent EMC requirements such as EN50065 (fig. 3). Any over-temperature event will force the line driver to shut down, thus ensuring always safe operation. The entire PLC section is designed to support full 500kHz bandwidth in order to meet all available PLC regulations worldwide (CENELEC, ARIB, FCC…). The device also embeds an optional dedicated metrology sub-system designed for very high accuracy measurements of single-phase and split phase power and energy, interfacing whatever external sensors such as Rogowski coils, current transformers or shunt sensors. The metrology sub-system consists of specific analog and digital sections. The analog section is based on two programmable gain low-noise, low-offset amplifiers (PGA) and three high performance 24-bit 2nd order $\Sigma\Delta$ ADC converters. The digital section consists of a configurable hardwired Metering DSP and DFE. The metrology sub-system provides active and reactive power measurement with less than 0.1% error over 1:5000 range, so meeting most stringent metering accuracy classes worldwide (fig. 4) and a 3.6 kHz bandwidth at -3 dB, which makes it able to get up to 72nd harmonic content at 50 Hz. The hardwired DSP processing unit calculates instantaneous and RMS voltage and currents, active, active fundamental, reactive and apparent power and energy, but it can be even bypassed for direct access to the voltage and current values for further data processing by the application controller if needed. The metrology sub-system is fully configurable and allows fast digital calibration in a single point over the entire current dynamic range, while meeting all major metering regulations worldwide such as EN 50470-1, EN 50470-3, IEC 62053-21, IEC 62053-22 and IEC 62053-23 compliant class1, class0.5 and class0.2 AC metering applications. STCOMET also embeds a full IoE application controller sub-system based on state of the art ARM® Cortex™-M4F core and Floating Point Unit (FPU) with operating frequencies up to 96MHz and embedded Flash memory up to 1MB size. That application controller is able to run not-real-time upper layers of the protocol communication stack (MAC, Network layer…), so complementing the PLC DSP sub-system services. It can also run in parallel RF communication protocols so making possible to interface with external basic RF Transceivers to increase device connectivity capabilities. Furthermore, leveraging on internal processing and memory resources and on the rich set of integrated peripherals, the platform can manage the entire IoE node application keeping FW codes separation to meet the needed performance and legal constraints. This safe FW code management is possible thanks to the specific system architecture where each critical function is independently managed by dedicated cores and where also a Memory Protection Unit is used to divide the memory in several independent regions with defined location, size, access permissions, and memory attributes. This allows for example to modify the IoE node application FW without impacting the embedded certified protocol stacks and legal metrology code. The device also supports advanced security features by embedding True random number generator (TRNG) block and hardwired AES encryption peripheral processing 128-bit data blocks using a key with 128, 192, 256 bits possible size. The encryption block also supports multiple security modes useful for IoE secure communication, such as "Electronic Code Book" (ECB), "Cipher Block Chaining" (CBC), "Counter "mode (CTR), "Galois/Counter Mode" (GCM), GMAC and CCM mode. To increase IoE node system security features, the present platform can be even interfaced with an external secure element through the embedded ISO 7816 compliant interface.

Conclusions

Leveraging on a unique System on Chip architecture and modular and integrated mixed-signal design, STCOMET platform offers the highest system flexibility and programmability along with the highest performance for specialized IoE nodes functions in a single device, overcoming current technology limitations to a IoE wide scale industrial adoption.

References

1. Advanced Microsystems for Automotive Applications 2011: O. Vermesan et al." Internet of Energy – Connecting Energy Anywhere Anytime".
2. http://meter-on.eu/
3. www.st.com/powerline

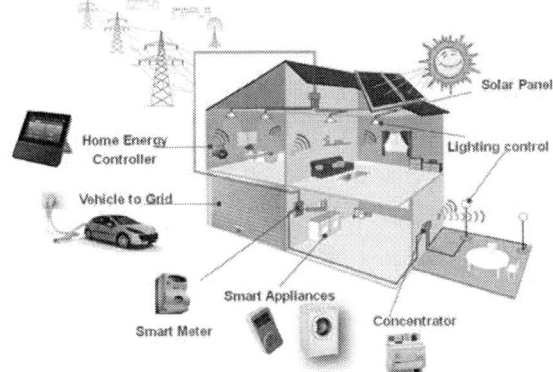

Fig.1: *The Internet of Energy (IoE) ecosystem*

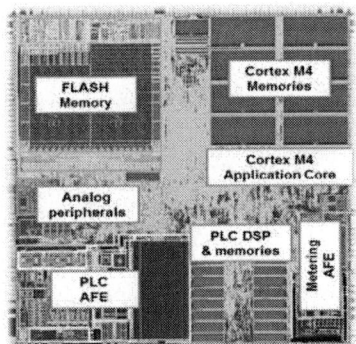

Fig.2: *IoE System on Chip die picture & functionalities*

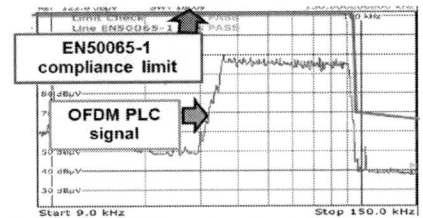

Fig.3: *STCOMET PLC conducted emission immunity*

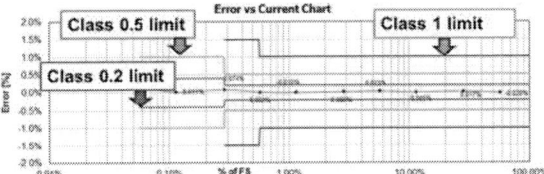

Fig.4: *STCOMET power metering accuracy*

A 65nm CMOS Transceiver with Integrated Active Cancellation Supporting FDD from 1GHz to 1.8GHz at +12.6dBm TX Power Leakage

Sameet Ramakrishnan, Lucas Calderin, Antonio Puglielli, Elad Alon, Ali Niknejad, Borivoje Nikolić

Berkeley Wireless Research Center, University of California, Berkeley, CA 94704, USA

Abstract

This paper presents an active transmitter (TX) cancellation scheme for FDD that synthesizes a replica of the TX current in shunt with the receiver (RX), virtually shorting out the TX signal for the RX while having minimal impact on TX insertion loss. The prototype in 65nm CMOS demonstrates >50dB cancellation of a +12.6dBm peak 20MHz modulated TX signal. A receiver integrated on the same prototype is able to down-convert the RX signal at 40MHz offset with <4.3dB noise figure (NF) degradation in the presence of the residual TX.

Introduction

Continuing growth in wireless connectivity has resulted in a proliferation of frequency bands that use frequency-division duplexing (FDD), requiring a large number of bulky and expensive off-chip duplexers. Analog subtraction of the transmit (TX) signal at the receive (RX) port [1-4] has been explored as an alternative for integration. However, the maximum cancelled leakage power has been limited to <+2dBm; accordingly, such systems still require 30dB of TX-RX isolation from off-chip filters. Furthermore, many cancellation schemes passively couple a portion of the TX signal into the RX to perform cancellation, degrading TX efficiency. Matching the time-varying, frequency selective TX-RX leakage channel over a wide TX bandwidth presents another challenge for electronic subtraction systems.

This paper presents a fully integrated transceiver with a single antenna interface providing >50dB of isolation for 20MHz modulated TX signal from 1.0-1.8GHz up to +12.6dBm. This is accomplished by connecting the antenna to the TX and RX via a series stacked transformer (Fig. 1.1). An RF current DAC is placed in shunt with the RX input, cancelling the TX current and creating a virtual ground across the RX for the TX signal (Fig. 1.2). The virtual ground shields the TX from RX port loading, preserving high TX efficiency. As the TX is designed with low output impedance, and the DAC has high output impedance relative to the RX, the RX signal experiences minimal loss (Fig 1.3). In order to track nonlinearity and time variance in the leakage signal over wide bandwidth, the DAC is driven with a digitally processed version of the TX data (Fig. 9).

Circuit Implementation

The chip top level (Fig. 2) includes an integrated TX, RX chain, cancellation current-steering DAC, and the TX/RX matching networks. The TX (Fig. 3.4) is a 500MS/s, 8-bit switched-capacitor power amplifier, [5-6] which provides a low, amplitude-independent output impedance to minimize RX insertion loss and amplitude-dependent TX/RX mixing products. A 25% duty cycle I/Q cell-sharing architecture [5], in which the TX unit cells are time-interleaved between I and Q phases, is used to maximize I/Q combination efficiency. This is implemented through data-dependent NAND gates (Fig. 3.1), which drive each unit cell in one of the ±I, ±Q, or ±I±Q phases. The maximum measured TX output power is +18.8dBm at 1.2GHz.

The 10-bit 500MS/s RF current DAC is segmented with 5 thermometer and 5 binary bits. It is important to note that the DAC cancels the TX signal in the current domain, rather than the power domain, substantially lowering its power consumption relative to the TX. The DAC can cancel up to 20dBm TX power, using an RX transformer with 2:1 turns ratio and a full-scale current of 60mA from a 1V supply. A single balanced current-switching DAC is used (Fig. 3.3), with Class-A operation for good dynamic linearity. DAC and TX phase noise skirts can fall in the RX band and desensitize the RX. However, phase noise correlated between DAC and TX LOs is cancelled [4]. Accordingly, the DAC is implemented in a Cartesian architecture with similar 25% duty-cycle I/Q cell sharing as the PA. This allows sharing of DAC and TX LO paths, to maximize correlation and minimize RX noise figure (NF) degradation.

In order to characterize the proposed cancellation architecture in a realistic environment, a receiver is realized on the same die. To enable multi-mode broadband testing, the receiver was designed for maximal linearity and re-configurability over NF performance. The RX is comprised of a low-noise trans-conductance amplifier (LNTA) (Fig. 3.2), current-mode passive mixers, and baseband trans-impedance amplifiers with programmable first-order filtering. The measured RX performance is summarized in Fig. 8. The chip, shown in Fig. 12, was designed in TSMC65nm process and measures 2.5mm x 2.5mm.

Measurement Results

Single tone cancellation is measured over TX output power ranging from -21.7dBm to +12.6dBm (Fig. 5), and varying TX center frequency from 1GHz to 1.8GHz at fixed 0dBm TX power (Fig. 6). In all cases, the post-cancellation residual TX is limited by the DAC LSB current and is independent of TX power. A cancellation of >50dB is achieved at a maximum TX output power of +12.6dBm, an order of magnitude higher power over the state of the art [1].

An off-chip adaptive digital filter on the DAC data (Fig. 4) is used to track the frequency-selective TX-RX leakage channel for modulated TX signals (Fig. 7). Over 50dB of isolation is achieved for a 20MHz modulated bandwidth TX signal with average power 6.6dBm and 6dB peak to average power ratio, a >20dB improvement over prior work [2,3].

Antenna load was varied to create a VSWR from 1:1 to 5:1 (Fig. 6). By adapting the filter coefficients, the residual leakage remains flat and is limited by DAC LSB over the entire range. The tested impedance region represents the measurement set up, rather than any cancellation limitations.

The RX NF degradation is measured as a function of TX output power for a TX-RX center frequency spacing of 40MHz (Fig. 10). For low TX power (below -5dBm), the NF is dominated by RX noise. For high TX power, the NF is dominated by *uncorrelated* phase noise between the TX and DAC. Cancellation of correlated phase noise by sharing a common TX/DAC LO path is measured at 19dB at 40MHz duplex offset through injection of spurs and white noise into the shared LO input port (Fig. 11).

When cancellation is enabled, the NF is improved by over 15dB, demonstrating the proposed scheme's ability to cancel the TX's RX-band noise. The RX NF is degraded by 1.1dB at the highest previously reported TX output power of +2dBm, and by 4.3dB at +10.6dBm.

Note that while the nominal RX NF is 7.6dB, 2.9dB is due to matching network loss (Fig. 8). Improvement of the matching network would leave NF degradation (Fig. 10) unchanged, as it is independent of this loss.

Acknowledgements

The authors acknowledge the students, faculty, and sponsors of Berkeley Wireless Research Center, chip donation by TSMC University Shuttle program, and Integrand tools. This work is funded in part by DARPA RF-FPGA, HR0011-12-9-0013, Nokia, Intel, and Qualcomm.

References

[1] J. Zhou et al, ISSCC, 2014. [2] J. Zhou et al, ISSCC, 2015. [3] D.-J. van den Broek et al, ISSCC, 2015. [4] D.-J. van den Broek et al, RFIC, 2015. [5] H. Jin et al, ISSCC, 2015. [6] S.-M. Yoo et al, JSSC, 2011.

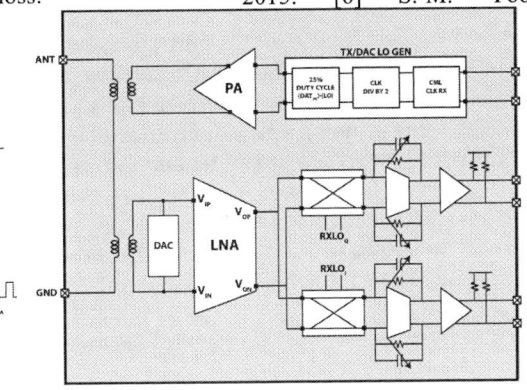

Fig. 1 (1) Proposed interface (2) TX equivalent (3) RX equivalent (4) Simultaneous operation of RX/TX

Fig. 2 Top level transceiver schematic

Fig. 3 Block schematics

Fig. 4 Test setup

Fig. 5 TX signal at RX input

Fig 6. Residual vs. frequency and load

Fig. 7 RX output in TX band, 20MHz TX

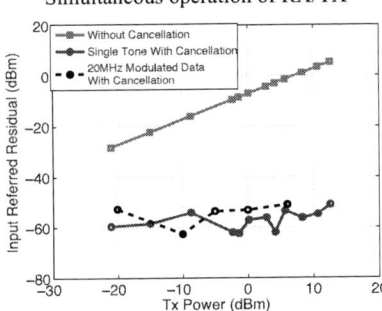

Gain (dB)	6-18
Baseband BW (MHz)	15-140
RF 3dB BW (MHz)	800
(OOB) IIP3 (dBm)	10.2
NF, No Matching Network	4.7dB
Matching Network Loss	2.9dB

Fig. 8 RX performance summary

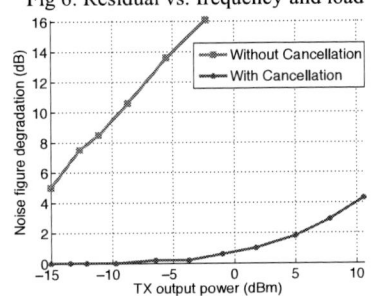

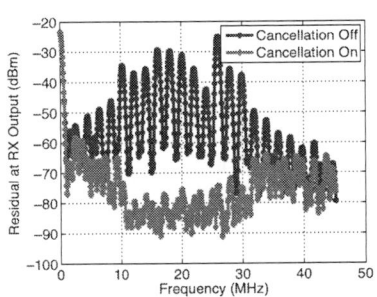

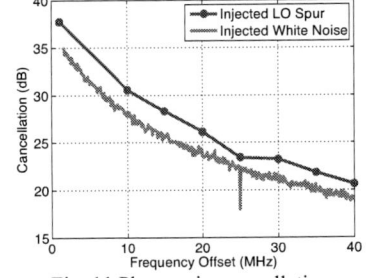

Fig. 9 TX and DAC after adaptation

Fig. 10 NF degradation vs. TX power

Fig. 11 Phase noise cancellation

Fig. 12 Die photo

	This Work	[1]	[2]	[3]
Technology	65nm	65nm	65nm	65nm
Frequency (GHz)	1.0-1.8	.3-1.7	.8-1.4	.15-3.5
Max TX Power Leakage (dBm)	+12.6	+2	-4	+1.5
Cancellation at Max TX Power (dB)	>50	>30	33	>27
Cancellation 20MHz Modulated Data (dB)	>50	-	20	27
Receive Noise Figure (dB)	7.6	4.2	7.5[1]	6.3
NF Degradation, +2dBm TX, 40MHz Offset (dB)	1.1	.8[2]	.9[3]	4[4]
RX Power (mW)	40	74.6-83.0	63-69	23-56
Single Antenna, No External Isolation	Yes	No	No	No
Fully-Integrated TX+RX	Yes	No	No	Yes[5]
Canceller Power (mW)	60	13-72	44-182	Unreported
Active Area (mm²)	3.9	1.2	4.8	2

1)Includes 2.7dB LC duplexer loss 2)TX power unreported(<+2dBm), 100MHz offset 3)TX power unreported(<-4dBm) 4)Full duplex 5)Unknown if measured with on-chip TX

Fig. 13 Comparison table

Digital PLL for Phase Noise Cancellation in Ring Oscillator-Based I/Q Receivers

Zuow-Zun Chen[1], Yilei Li[1], Yen-Cheng Kuan[1], Boyu Hu[1], Chien-Heng Wong[1], and Mau-Chung Frank Chang[1,2]

[1]University of California, Los Angeles, CA, USA, [2]National Chiao Tung University, Taiwan, ROC

Email: zzchen@g.ucla.edu

Abstract

A digital phase noise cancellation technique for ring oscillator-based I/Q receivers is presented. Ring oscillator phase noise, including supply-induced phase noise, is extracted from digital phase-locked loop (DPLL) and used to restore the randomly rotated baseband signal in digital domain. The receiver prototype fabricated in 65nm CMOS technology achieves phase noise reduction from -88 to -109dBc/Hz at 1MHz offset, and an integrated phase noise (IPN) reduction from -16.8 to -34.6dBc, when operating at 2.4GHz.

Introduction

Ring Oscillators (ROs) have gained increasing interest for applications in radio receivers due to their small area, wide tuning range, and multiphase output. However, their higher phase noise and higher sensitivity to supply noise may seriously deteriorate the received signal, specifically through two mechanisms: one is the reciprocal mixing of interference and the other is the close-in phase noise that falls inside the signal BW introduces random rotation of the constellation, as illustrated in Fig. 1. While the reciprocal mixing noise can be cancelled using the symmetrical property of phase noise [1], the alleviation of close-in phase noise effect remains challenging and is the focus of this work. Although increasing PLL BW can reduce RO phase noise, the BW is limited to $F_{REF}/10$ in conventional type-II PLL due to stability concerns. Larger BW also trades off higher spurious tones at PLL output. A delay-discriminator-based technique was proposed in [2] to cancel phase noise, but the approach showed limited jitter (or IPN) improvement and is conducted in analog domain. To circumvent the aforementioned constraints, this paper presents a digital phase noise cancellation technique capable of reducing both RO close-in and supply-induced phase noise for RO-based I/Q receivers. Fig. 1 shows the phase-noise-cancelling (PNC) receiver system diagram. The LO source is generated from a DPLL, which employs a RO as the digitally-controlled oscillator (DCO). The phase noise information is extracted from the DPLL and applied to the digital PNC circuit to restore the randomly rotated baseband signal.

Phase Noise Extraction

DPLLs are known for their compact digital-loop filter and re-configurability of loop parameters. In addition, the digital feedback signal contains useful information regarding the PLL dynamics. Here we take advantage of this feature to extract RO phase noise. Fig. 2(a) shows a simplified DPLL discrete-time phase-domain model. As indicated, RO phase noise component at PLL output (Φ_{LO}) equals to that of DLF_{IN} multiplied by ($-1/K_{TDC}$). This means RO phase noise in Φ_{LO} can be eliminated in principle by subtracting it with ($-DLF_{IN}/K_{TDC}$). The phase noise extraction concept also applies to RO supply noise. Fluctuations on RO supply alters the oscillation frequency and further transfers into phase noise at the RO output. Since supply-induced phase noise and RO intrinsic phase noise are indistinguishable, both of them are captured in DLF_{IN} and can be cancelled through the noise subtraction implemented in digital PNC circuit. On the other hand, time-to-digital converter (TDC) noise Q_{TDC} encounters a low-pass transfer function to Φ_{LO} but a high-pass function to DLF_{IN}, the calculated noise spectra are depicted in Fig. 2 (b) and (c). Therefore, while RO phase noise can be mutually cancelled by the subtraction, TDC noise still remains. For this reason, the sub-sampling TDC technique proposed in [3] with high time resolution and low TDC noise is employed in the DPLL. This helps to achieve a low-noise performance after cancellation.

Phase Noise Cancellation

Fig. 3(a) shows the block diagram of the digital PNC circuit. The input signal DLF_{IN} from DPLL is first applied to a digital LPF $F_n(z)$ to filter out the high frequency part of TDC noise (see Fig. 2(c)). A small low-pass corner reduces the TDC noise but increases the residue of RO phase noise after cancellation. In order to optimize the IPN after cancellation, $F_n(z)$ BW is set at the frequency where RO and TDC noise spectra intersect [4]. This renders both noise sources to have approximately equal contribution to IPN. Afterwards, digital filter output is scaled by ($-1/K_{TDC}$) and sent to the LUT to generate sine and cosine signals. The I/Q rotator then calculates the complex multiplication of ($SIGI+jSIGQ$) and $\exp(-j\Phi_{NC})$. On the baseband signal path, DC offsets are removed before the multiplication for correct rotation. After cancellation, the remaining noise that affects the baseband signal is minimized to ($\Phi_{LO}-\Phi_{NC}$), the calculated spectrum is depicted in Fig. 3(b). It should be noted that the latency introduced by the different delay time of the I/Q signal path and the phase noise extraction path affects the effectiveness of phase noise cancellation. To alleviate this issue, the delay of each path is carefully controlled using register arrays to insure their difference is within a single reference clock cycle.

Measurement Results

The PNC I/Q receiver prototype consist of two chips: a receiver chip and a PNC chip. Both are fabricated in TSMC 65nm CMOS technology and together tested with off-chip anti-aliasing filters (AFFs) and ADCs. The measured DPLL output frequency is from 4.5 to 6.5GHz, and 2.25 to 3.25GHz through a divide-by-2 and multiplexing circuit at the DPLL output. The phase noise result measured from a standalone DPLL chip, along with the extracted (DLF_{IN}/K_{TDC}) spectrum are reported in Fig. 4. As shown, the two noise spectra match up to around 12MHz ($\approx F_{REF}/4$), and after that TDC high-pass noise dominates the high frequency spectrum of (DLF_{IN}/K_{TDC}).

The PNC I/Q receiver is first tested by receiving a 2.4GHz single-tone signal. Fig. 5 shows the power spectrum and SSB phase noise of the digitized baseband signal. With PNC the spot phase noise at 1MHz offset is reduced from -88 to -109dBc/Hz. The IPN integrated from 1KHz to 15MHz frequency offset is improved from -16.8 to -34.6dBc. It is also

978-1-5090-0636-6/16 $31.00 © 2016 IEEE

noted the SSB phase noise result with PNC is dominated by noise from the RF signal source at low frequency offset. Furthermore, to verify the PNC technique on supply noise reduction a 240KHz sinusoidal tone is injected onto RO supply, as shown in Fig. 6. With PNC the supply-induced phase noise is suppressed by 38dB, from -33 to -71dBc. Fig. 7 shows the measured constellation results of receiving BPSK and 64QAM signals with 10MHz symbol rate operating around 2.4GHz. The EVM improvement demonstrates the proposed PNC technique greatly reduces RO phase noise, achieving an EVM of -37.5dB for 64QAM signal. Fig. 8 shows the die photos. The supply voltage of each building block is 1V for the DPLL, LO buffer, and digital PNC circuit, and 1.2V for the receiver chain. The digital PNC circuit consumes 2mW of power and 0.04mm² of area. Table 1 summarizes the system performance.

Acknowledgment

The lead author wishes to thank MediaTek for fellowship support. Authors also thank Jin-Fu Yeh, Jaewook Shin, and Long Kong for technical discussions.

References

[1] H. Wu, et al., *ISSCC Dig.*, pp. 30-31, Feb. 2015.
[2] S. Min, et al., *IEEE JSSC*, pp. 1151-1160, May 2013.
[3] Z.-Z. Chen, et al., *ISSCC Dig.*, pp. 268-269, Feb. 2015.
[4] X. Gao, et al., *IEEE TCAS-II*, pp. 117-121, Feb. 2009.
[5] A. Elshazly, et al., *IEEE JSSC*, pp. 2759-2771, Dec. 2011.

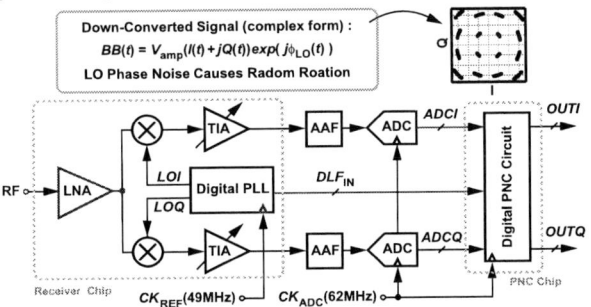

Fig. 1. The effect of LO phase noise (Φ_{LO}) on signal constellation and the proposed PNC I/Q receiver system diagram.

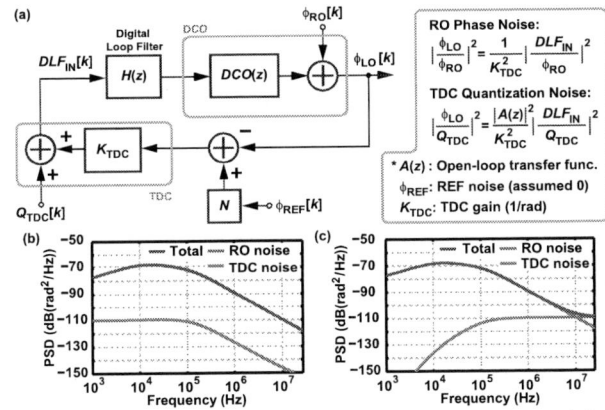

Fig. 2. (a) Simplified DPLL discrete-time phase-domain model. (b) Calculated PSD of Φ_{LO}. (c) Calculated PSD of (DLF_{IN}/K_{TDC}).

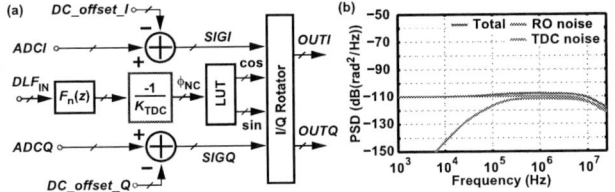

Fig. 3. (a) Digital PNC circuit. (b) Calculated PSD of (Φ_{LO}-Φ_{NC}).

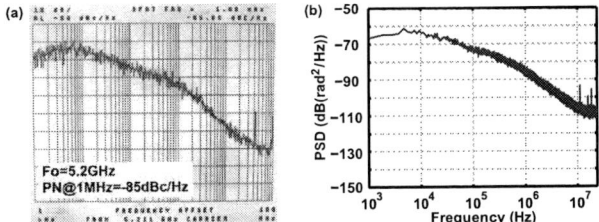

Fig. 4. Measurement results. (a) PLL output phase noise operating at 5.2GHz. (b) PSD of (DLF_{IN}/K_{TDC}).

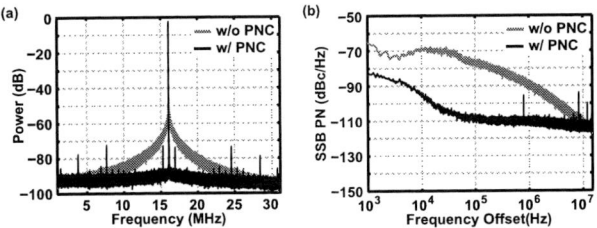

Fig. 5. 2.4GHz single-tone input signal test. (a) Power spectrum of baseband signal. (b) SSB phase noise of baseband signal.

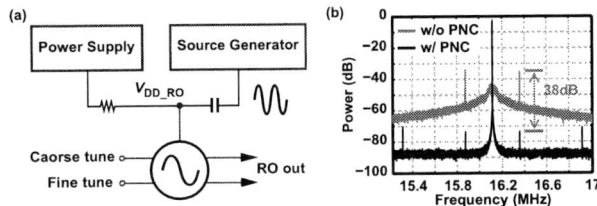

Fig. 6. 2.4GHz single-tone input signal test with supply noise injection. (a) Test setup. (b) Power spectrum of baseband signal.

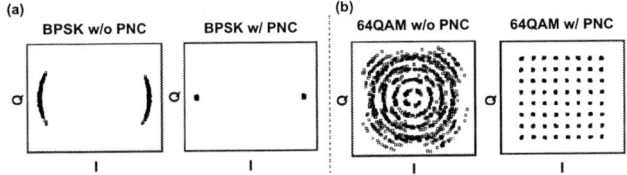

Fig. 7. Measured constellation with and without PNC. (a) BPSK signal. (b) 64QAM signal.

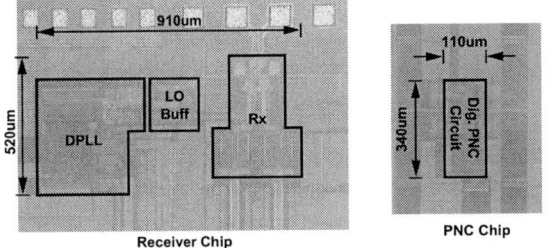

Fig. 8. Die photos of receiver and PNC chips.

TABLE1: Performance Summary

		This Work	[2]	[5]
Topology	Noise Extraction	Digital PLL	Delay Discriminator	Test-Signal Based
	Noise Cancellation	Digital PNC Circuit	NC Delay Cell	NC Transistors[5]
Operation Freq. (GHz)		2.25~3.25, 4.5~6.5	3.5~7.1	0.4~3
Spot PN w/o PNC (dBc/Hz)		-88@1MHz	-92.5@1MHz	N/A
Spot PN w/ PNC (dBc/Hz)		-109@1MHz	-105@1MHz	N/A
RMS Jitter w/o PNC (psec)		9.6 (1K~15M)[1]	5.2 (10K~10M)	8 (50K hits)
RMS Jitter w/ PNC (psec)		1.2 (1K~15M)[1]	3.8 (10K~10M)	4.8 (50K hits)
Core Area (mm²)		0.47[2]/ 0.17[3]/ 0.04[4]	0.12[3]	0.08[3]
Power (mW)		46.5[2]/ 14[3]/ 2[4]	29.6[3]	3.1[3]
Supply (V)		1 / 1.2	1.2	1
Technology (nm)		65	90	130

[1] Calculated from IPN [2] Off-chip AAFs and ADCs excluded [3] PLL circuit including noise extraction & cancellation [4] Dig. PNC circuit [5] Supply noise cancellation

A Chopping Switched-Capacitor RF Receiver with Integrated Blocker Detection, +31dBm OB-IIP3, and +15dBm OB-B1dB

Yang Xu, Peter R. Kinget
Columbia University, New York, NY, USA

Abstract

A 0.1-0.6GHz chopping switched-capacitor RF receiver with integrated blocker detector features tunable center frequency, programmable filter order, and very high out-of-band linearity. RF impedance matching, high-order OB interferer filtering, and flicker-noise chopping are achieved by passive SC circuits only. The 34-80mW 65nm receiver achieves 35dB gain, +31dBm OB-IIP3, +15dBm B1dB, and 4.6-9dB NF. The filter order is adapted to blocker power with a blocker detector with a 1us response time.

Introduction

Out-of-band (OB) linearity is a key challenge in the design of tunable RF receivers (RXs) for cognitive radio. They need to handle large continuous-wave (CW) close-by blockers as well as intermodulation and cross-modulation interferers in FDD or co-existence scenarios. N-path filters (NPF) [1] and mixer-first RXs [2] achieve high CW blocker tolerance, but offer only low-order filtering before the active baseband or RF circuits leading to a limited OB linearity for a close-by blocker. For co-existence applications, self interference (SI) cancellation [4] can reduce the close-by transmitter (TX) SI for a high cross-modulation linearity, but the canceller limits the power handling and has a limited bandwidth. A passive switched-capacitor (SC) RX [5] achieves programmable high-order and high-linearity filtering before the active circuits resulting in a high OB linearity even for a close-by blocker. Its noise figure (NF) is relatively high and it requires large baseband transconductors (Gms) to reduce the RX's in-channel flicker noise. Detecting the presence of the RF blocker also remains a challenge and is needed to configure the front-end filtering. In this work, a chopping SC RX with integrated blocker detector is proposed with high OB linearity at a lower supply and with an improved NF compared to earlier SC RXs. The integrated blocker detector detects the envelope of the unknown OB blocker, so that the filter order can be adapted to the blocker power with a fast response.

Chopping Switched-Capacitor Receiver

The chopping SC RX consists of 8 time-interleaved SC banks (Fig. 1). The RF impedance matching and sampling are achieved by capacitor Cs with switches s1 and s5 [5]. In each SC bank, the RF input signal is down-converted to baseband during sampling, then amplified by 8 Gms. The transconductances of the Gm cells are scaled as a sine wave to achieve 3rd and 5th order LO harmonic rejection. In [5] the switch s4 is a dominant contributor of noise; by moving it to the Gm output, its noise impact is reduced by the gain of the Gm. To reduce the parasitic Gm input capacitance, minimum length devices are used in the Gm; however, this results in large in-channel flicker noise. Adding input and output choppers reduces the flicker noise. High-order SC filtering is achieved with an RF NPF (switch s0, capacitor Ch0) and two discrete-time (DT) IIR filters (s2, Ch1 and s3, Ch2). The filter order can be programmed by enabling or disabling the switch clock signals. All the switches are driven by 8-phase non-overlapping clock signals p_1' to p_8'. The N-path filtering, sampling, IIR filtering and Gm amplification in one SC bank are achieved in sequential time intervals. The blocker detector

is attached to the last IIR filter capacitor Ch2 to measure the filtered Gm input voltage swing. The choppers before Cs up-convert the desired signal with chopping frequency f_{chop}, and the choppers after the Gms down-convert the desired signal back to baseband while up-converting the Gm flicker noise to f_{chop}. The choppers attached to Ch1 and Ch2 ensure the IIR filter transfer function is maintained while chopping. When driving the switches by 1/16 duty-cycle 16-phase non-overlapping clocks (Fig.1), the choppers can be merged with the SC circuits. For each bank the RF signal is sampled in phases p_i and p_{i+8} with opposite polarity to accomplish chopping. For a sampling frequency f_s, the RX LO frequency is $f_{lo}=f_s/8$ and $f_{chop}=f_s/16$.

Circuit Implementation

Cs capacitor pairs with sampling phase p_i and p_{i+4} share the same Chs in NPF and IIR filters to eliminate the DC and even-order LO harmonic responses (Fig. 2). The switches are implemented with CMOS transmission gates. The Gm with output switches acts as a switched Gm (Fig. 2) and the DC current of the Gm is cut off when the switches are off to save power. The on-chip LO divider generates the 16-phase non-overlapping clock signals.

The blocker envelope detector consists of 8 AC coupled common-source transistors operating in weak inversion with resistor and off-chip capacitor loads and a replica (Fig. 2).

Measurement Results

The 1.5x2mm² chip has been fabricated in a 65nm CMOS process (Fig. 2). Baseband supply is 1.1V and LO supply is 1.25V. Fig.3 shows the measured conversion gain, DSB NF and LO current for LO frequencies from 0.1 to 0.6GHz with different filter orders. The NF for a 0.1GHz LO is 4.6dB which is 2.2dB better compared with the earlier SCRX [5]. The measured and simulated NF versus IF frequency is shown in Fig. 3. The flicker noise corner is 100kHz with chopping while the simulated flicker noise corner without chopping is significantly higher. The measured wideband transfer function for a 100MHz LO is shown in Fig. 4. The spurious responses due to chopping are not higher than the responses from LO even-order harmonics.

Fig. 5 shows the measured OB linearity performance for a LO frequency of 0.2GHz and for different filter orders. The OB-IIP3 is measured with a two tone at 0.231GHz and 0.261GHz. The triple beat (TB) versus SI peak power is measured with a -30dBm adjacent-channel jammer and two-tone SI signals with a frequency offset of only -30MHz and a 5MHz frequency spacing. The high order filtering improves the B1dB, OB-IIP3 and TB. With NPF and 2nd-order IIR filters, the B1dB for a 30MHz blocker offset is 13dBm, OB-IIP3 is 31dBm, and TB for a -4dBm SI peak power is 62.5dB. In an FDD or co-existence application, the filter order can be tuned with the SI power level information available in the same device.

For an unknown CW OB blocker, the blocker detector can be used to adapt the filter order to the blocker power between two communication packets. Fig.6 shows the filter adaptation using the integrated blocker detector. The Gm input-referred blocker power is level of the blocker at the input of the (nonlinear) Gm filtering transfer function from the applied blocker input power. The detector output voltage increases with blocker power; when it is higher than a threshold (50mV), the filter order is increased to improve the gain compression. The blocker detector transient response (Fig. 6) settles in less than 1us; after increasing the filter order, the V_{det} stabilizes again in 1us. The detector only consumes 0.2mW (including 0.1mW from bias circuits).

When compared with the state of the art (Table I), this work achieves high OB linearity and power handling, better NF and power consumption than [5] and supports fast blocker detection. The frequency range and NF increase at higher LO frequencies are mainly limited by parasitic capacitance which will improve with process scaling.

Acknowledgments

We thank the Wei Family Foundation and DARPA CLASIC program for financial support.

References

[1] M. Darvishi, et al., *JSSC*, pp. 1370-1382, Jun. 2013.
[2] C. Andrews et al., *JSSC*, pp. 2696-2708, Dec. 2010.
[3] D. Murphy, et al., *JSSC*, pp. 2943-2963, Dec. 2012.
[4] J. Zhou, et al., *ISSCC*, pp. 342-343, 2015.
[5] Y. Xu et al., *RFIC*, pp. 39-42, 2015.

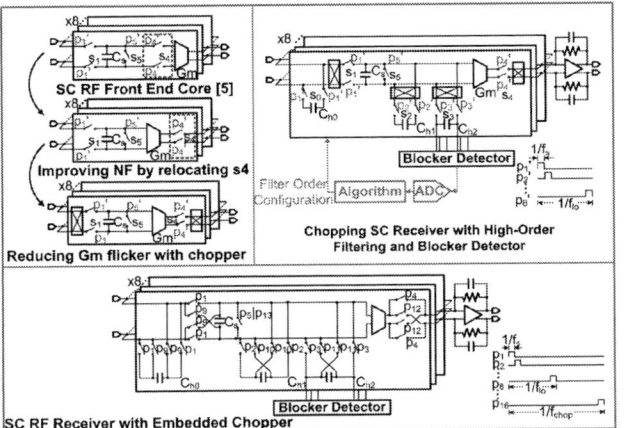

Fig. 1 Concept of the chopping switched-capacitor RF receiver with integrated blocker detector

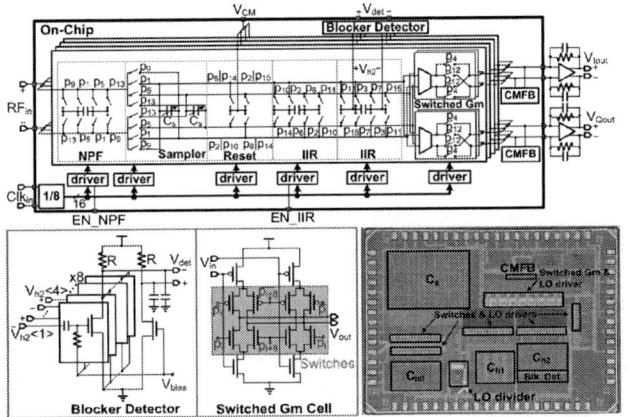

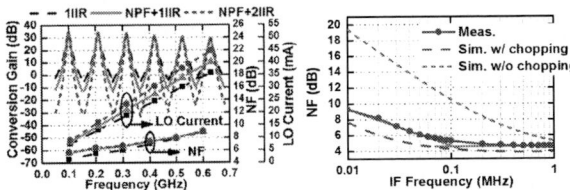

Fig. 2 Circuit schematic of the chopping SC RF receiver with integrated blocker detector and chip photo

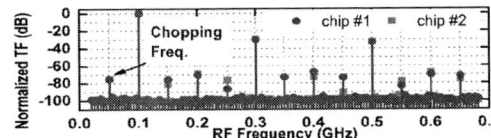

Fig. 4 Measured wideband transfer function for a 0.1GHz LO

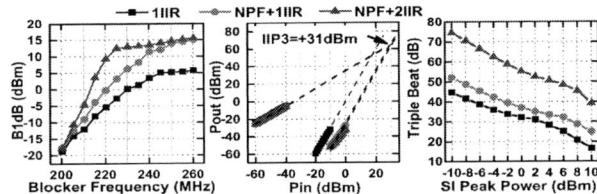

Fig. 5 Measured blocker 1dB compression point versus blocker frequency, out-of-band IIP3, and triple beat versus two tone SI peak power for an LO frequency of 0.2GHz.

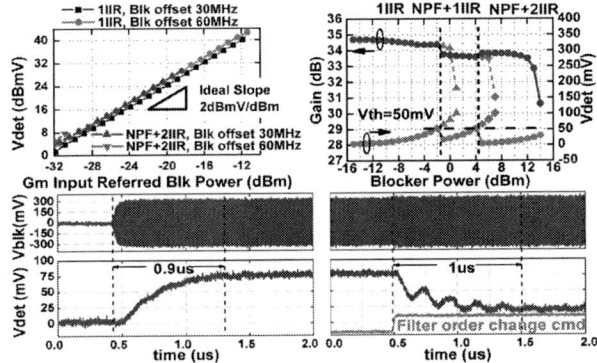

Fig. 6 Measured blocker detector output voltage versus Gm input referred blocker power, conversion gain versus blocker power with the adaptive filter order for 0.2GHz LO and 0.23GHz blocker, and blocker detector transient response.

Fig. 3 Measured conversion gain, double-sideband noise figure and LO current for a 0.1 to 0.6GHz LO and measured and simulated noise figure versus IF frequency for a 0.1GHz LO frequency with 1st-order IIR filtering.

TABLE I
Performance summary and comparison with the state-of-the-art

	This work	[5]	[3]	[4]	[1]
Technology	65nm	40nm	40nm	65nm	65nm
Architecture	Chop. SC	SC	FTNC	SI Canc.	NPF
Blocker detector	Yes	No	No	No	No
RF freq. (GHz)	0.1-0.6	0.1-0.7	0.08-2.7	0.8-1.4	0.1-1.2
NF (dB)	4.6-9	6.8-9.7	1.5-2.4[1] 3.5-5	4.8, 5.3[2]	2.8
OB-IIP3 (dBm)	**31**	24	13[1], 17	17	26
B1dB (dBm)	**15**	15	<0[1], <5	4	7
TB-4dBm[3] (dB)	**63**	NR	NR	48, 64[2]	NR
Max Handled Peak SI Power (dBm)	**>10**	NR	NR	NR, -4[2]	NR
OB-SFDR[4] (dB)	**92.9**	87.5	83.7[1], 85	84.1	91.5
Power (mW)	Ana: 24 LO: 9.5-55.8 Detector: 0.2	Ana: 52 LO: 7-53	35-78	63-69, 107-160[2]	18-57.4

NR = not reported; 1: with noise cancellation 2: with self interference cancellation, calibration is required 3: Triple beat at a -4dBm SI peak power 4: OB-SFDR = 2/3(OB_IIP3-(-174dBm/Hz)-10log (1MHz)-NF)

A 180 mW Multistandard TV Tuner in 28 nm CMOS

Jianhong Xiao, Weinan Gao, Xiaojing Xu, Dave Chang, Jiang Cao, Runhua Sun, Vijay Periasamy, Ning-Yi Wang, Xi Chen, Greg Unruh, Takayuki Hayashi, Tai-Hong Chih, Lakshminarasimhan Krishnan, Kuo-Ken Huang, Sunny Dommaraju, Guowen Wei, Bo Shen, Ardie Venes, Dongsoo Koh, James Y.C. Chang

Broadcom Corporation, 5300 California Ave, Irvine, CA 92617

Abstract

A 28 nm CMOS multistandard TV tuner is presented. A power-efficient RF front end and >80 dB dynamic range $\Delta\Sigma$ ADC, together with a smart AGC algorithm, enable this tuner to achieve 64 dB ATSC A/74 N+6 ACI while dissipating only 180 mW. A baseband resistor weighting harmonic rejection mixer clocked by a 7-13.6 GHz PLL and single-edge-triggered shift registers achieves >58 dB harmonic rejection ratio at frequencies up to 827 MHz.

Keywords: TV tuner, harmonic rejection, ATSC

Introduction

To support worldwide cable and terrestrial TV, multistandard TV tuners present many well-known design challenges, including a wide frequency range (50-1000 MHz), high sensitivity, >57 dBc blocker tolerance, and >70 dB harmonic rejection [1] [2]. Contemporary broadband communications SoCs, such as set-top box ICs, require fully integrated TV tuners to reduce cost, raising more hurdles. First, low-power design is critical in reducing the SoC thermal management cost. Second, this tuner shares the same F connector input with other integrated transceivers, such as MoCA and DVB-S (up to 2.4 GHz). This requires harmonic rejection capability at higher LO frequencies than a conventional TV tuner, e.g., 800 MHz vs. 330 MHz.

Tuner System

Fig. 1 shows the block diagram of the proposed direct conversion tuner. Variable-gain LNA and RFPGA amplify the incoming RF signal with 40 dB maximum gain. The mixer array includes three mixers: a 4-phase mixer covering 50-1000 MHz without odd-order harmonic rejection, an 8-phase HRM covering 50-827 MHz with 3rd and 5th harmonic rejection, and a 16-phase mixer covering 50-480 MHz with up to 13th harmonic rejection. A second-order RFLPF further improves the harmonic rejection by > 15 dB for signals below 500 MHz. Multiple switches are inserted to enable a RFLPF bypass mode and to minimize RFPGA parasitic loading from the mixers.

The $\Delta\Sigma$ analog fractional-N PLL generates frequencies within 7-13.6 GHz. This >1.5x frequency range, together with the post-divider chain, enables the 50-1000 MHz LO generation. The frequencies are much higher than previously reported tuner PLLs [1][2] to achieve precise phase matching of multiphase clock up to 827 MHz. Multiphase clock can be generated by shift registers that are either single-edge-triggered (SESR) or dual-edge-triggered (DESR). SESR requires twice the input frequency than DESR, but achieves much better phase matching because of its immunity from duty cycle variation [5].

A conventional tuner requires baseband VGA and a high-order channel selection filter before the ADC, which consume significant power and area. In this work, we propose a 432 MHz I/Q $\Delta\Sigma$ ADC with sufficient dynamic range (>80 dB) to eliminate these blocks. Furthermore, the ADC only consumes 7 mW due to a novel hybrid feedback feed-forward loop topology and a SAR quantizer, as we reported in [3]. The ADC outputs are then processed by a

demodulator block with embedded digital filter. A DC-offset-cancelation (DCOC) loop removes the baseband offset with a $\Delta\Sigma$ current DAC feedback. A digital calibration routine injects test tones to calibrate the RFLPF's bandwidth and mixer poles across PVT variation. A digital integer-N PLL generates clocks for ADC, AGC, DCOC, and calibration tones.

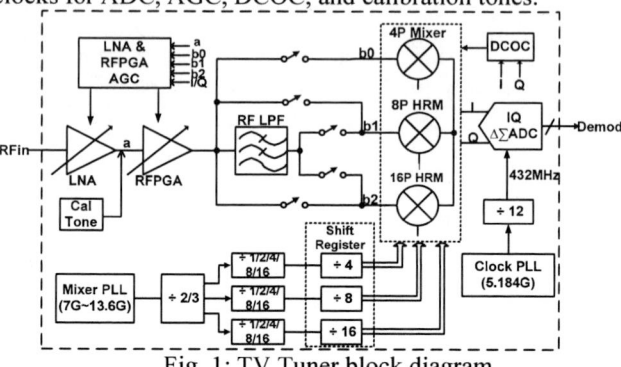

Fig. 1: TV Tuner block diagram

An AGC engine measures the signal amplitude of six nodes along the signal path via peak detectors. With this information, a smart AGC algorithm determines whether the far-out or close-in blockers dominate the composite RF input power. The AGC take-over points are then adjusted to optimize system noise and linearity. For a far-out blocker scenario, the system requires less gain back-off with good NF and moderate IIP3; for the close-in blocker case, the system requires more back-off with good IIP3 and moderate NF. This algorithm improves blocker tolerance by 4-5 dB without a power penalty.

Tuner Circuit Design

Compared to a typical WLAN/Cellular RF front end (RFFE), tuner RFFE requires complex gain control, including a 60 dB gain range and a <0.25dB gain step. In addition, IIP3 needs to be improved by 1 dB per dB gain reduction. Current-steering VGA is widely used to achieve a small gain step, but its IIP3 is constant during gain change. Significant power has to be consumed to overdesign the IIP3. To lower the power, we purposely implement gain control with passive stages and simplify the active stages to fixed-gain amplifiers, as shown in Fig. 2. The LNA π-type ladder covers a 40 dB gain range with 1 dB steps, and analog ramp signals drive the ladder switch for continuous gain change within each step. The LNA feedback resistor is also adjusted continuously to provide good S11 during gain change. The RFPGA's resistor ladder interpolates an overall 20 dB range to <0.25 dB steps, with multiple switches at adjacent tapping points (a2 to a9). Both passive stages achieve small gain step and good linearity with negligible power. To further lower the power, both amplifiers utilize a complementary PN structure. Flipped voltage follower (FVF)-based buffers are used to achieve high linearity with low power. RFLPF is a second-order Sallen-Key topology implemented by FVF buffers. By adjusting resistor and capacitor values, it achieves a wide programmable bandwidth of 100-500 MHz.

Fig. 3 shows the 8-phase HRM schematic, which uses a current-mode passive mixer topology for high linearity and low flicker noise. Similar to the LNA/RFPGA, the mixer Gm stage also uses a complementary PN structure. The switches are driven by nonoverlapping 8-phase clocks, which are derived from SESR. The mixer has two TIA stages and the second stage combines the first-stage outputs with proper resistor weighting to achieve harmonic rejection. Compared to RF Gm weighting in [1] [2], baseband resistor weighting can achieve much better matching regardless of LO frequencies. In 28 nm, resistors can achieve <0.3% mismatch with sufficient area, corresponding to >64 dB HRR. Without matching constraint, Gm-stage size is reduced to lower buffer power consumption. To further reduce mixer power, baseband I/Q signals are composed from the same first-stage outputs with proper polarity flip. Fig.4 (a) shows the measured 8-phase mixer HR3 histogram from 130 parts, achieving >58 dB rejection ratio at 827 MHz LO without calibration or filtering. It compares favorably to [4] for lower TIA count and power consumption. Compared to the HRM in [5], it does not require complex two-stage weighting but achieves similar performance. The 16-phase and 4-phase mixers use a similar architecture. Together with RFLPF, the tuner achieves > 75 dB harmonic rejection.

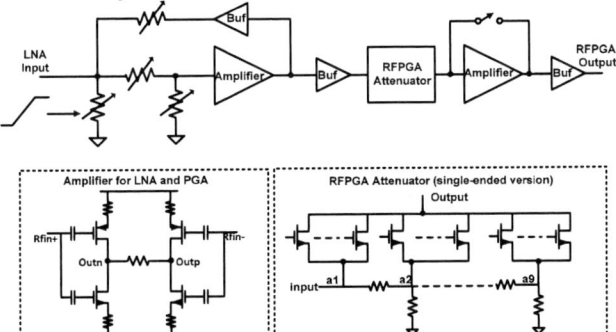

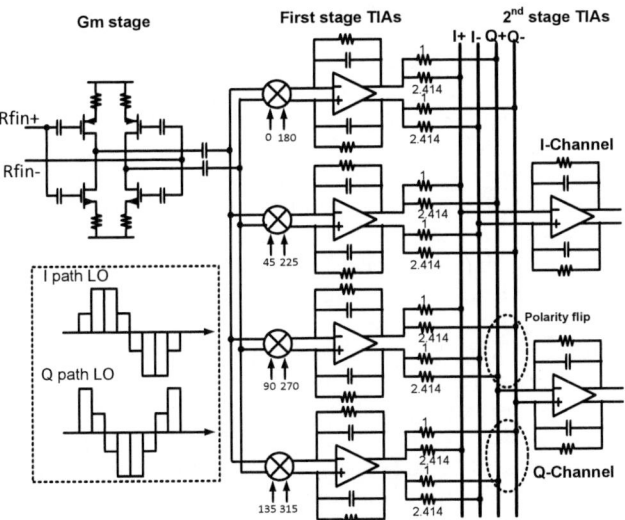

Fig. 2: LNA and RFPGA schematic

Fig. 3: 8-Phase harmonic rejection mixer schematic

Measurement Results

This tuner is fabricated in 28 nm CMOS technology with an overall area of 2.7 mm². Fig.4 (b) shows the die photo. It supports DVB-C, DVB-T/T2, ISDB-T, and ATSC standards with 4 dB NF and >12 dBm maximum single-channel input power. In DVB-T blocker ACI tests, it achieves 45 dB N+1 and 51 dB N+2 with an 18.6 dB threshold SNR. Fig.5 shows the ATSC +64 dB N+6 blocker test input spectrum and the overall ACI performance, which is >7 dB better than the A/74 specifications. With 1.8V supply, the tuner consumes only 180 mW, including 40 mW from two PLLs and 130 mW from the signal path. As shown in Table 1, this tuner achieves >5 dB better blocker tolerance at less than half the power of any work reported to date. In addition, the harmonic rejection mixer achieves >58 dB rejection ratio at frequencies up to 827 MHz, significantly higher than the 330 MHz reported in previous TV tuners.

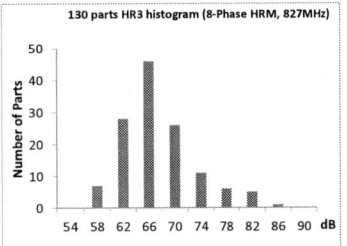
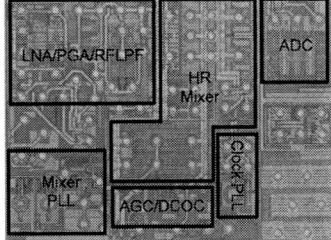

Fig. 4: (a) measured 8-phase mixer HR3 (b) Tuner Die Photo

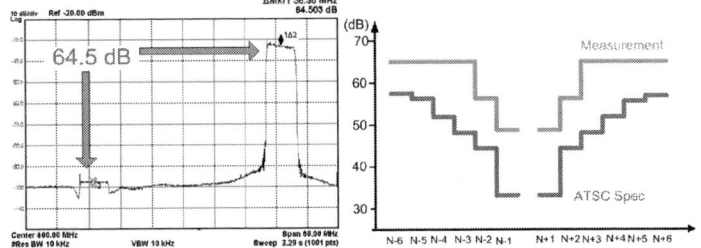

Fig. 5: (a) ATSC blocker test spectrum (b) test summary

	This work	[1]	[2]
NF @ max gain [dB]	4	4	3
LO PN @1MHz/855 MHz [dBc/Hz]	-124.5	-111	-122.2
Mixer HR3 @ 827 MHz [dBc]	>58	-	-
Mixer+RFLPF HR3 @ 330 MHz [dBc]	>75	72	>65
DVB-T N+1/N+2 ACI [dBc]	45/51	-	40/46
ATSC A/74 N+6 ACI [dBc]	64	59	59
Power [mW]	180	750	440
Area [mm2]	2.7	25	5.6
CMOS Technology	28 nm	0.18 μm	80 nm

Table 1: Performance comparison of state-of-the-art tuners

References

[1] S. Lerstaveesin, et al., "A 48-860 MHz CMOS low-IF direct-conversion DTV tuner," *IEEE J. Solid-State Circuits,* Vol.43, No.9, pp. 2013-2024, Sept. 2008.

[2] J. Greenberg, et al., "A 40 MHz-to-1 GHz Fully Integrated Multistandard Silicon Tuner in 80 nm CMOS," ISSCC Dig. Tech. Papers, pp. 162-164, Feb. 2012.

[3] G. Wei, et al., "A 13-ENOB, 5 MHz BW, 3.16 mW Multi-Bit Continuous-Time $\Delta\Sigma$ ADC in 28 nm CMOS with Excess-Loop-Delay Compensation Embedded in SAR Quantizer," 2015 VLSI, pp. 292-293.

[4] I. Lee, et al., "A Fully Integrated TV Tuner Front End with 3.1 dB NF, >+31 dBm OIP3, >83 dB HRR3/5 and >68 dB HRR7," ISSCC Dig. Tech. Papers, pp. 70-71, Feb. 2014

[5] Z. Ru, et al., "A Software-Defined Radio Receiver Architecture Robust to Out-of-Band Interference," *ISSCC Dig. Tech. Papers*, pp. 230-231, Feb. 2009.

Full Chip Integration of 3-D Cross-Point ReRAM with Leakage-Compensating Write Driver and Disturbance-Aware Sense Amplifier

[1)]Sangheon Lee, [1)]Jeonghwan Song, [1)]Changhyuk Seong, [1)]Jiyong Woo, [2)]Jong-Moon Choi, [2)]Soon-Chan Kwon, [2)]Ho-Joon Kim, [2)]Hyun-Suk Kang, [3)]Soo Gil Kim, [3)]Hoe Gwon Jung, [2)]Kee-Won Kwon, [1)]Hyunsang Hwang

[1)]Dept. of Mat. Sci. and Eng., POSTECH, [2)]Coll. of Info. and Comm. Eng., SKKU, and
[3)]SK Hynix Semiconductor Inc., Republic of Korea

Phone: +82-54-279-5123, Fax: +82-54-279-5122, e-mail: hwanghs@postech.ac.kr, keewkwon@skku.edu

Abstract

In this report, a fully integrated 3-D cross-point ReRAM is demonstrated with minimized disturbance and sneak current effect. HfO_X memory cells stacked on threshold-type selector exhibit superb leakage current suppression than cells with exponential selector. Remaining leakage current is diagnosed and compensated by leakage compensating write driver. Cells are prevented from disturbance by lowering read voltage at hot temperature, which sacrifices read margin. The read margin is recovered by cell current amplifier in read circuit.

Introduction

The 3-D cross-point array has been highlighted due to its high scalability, low power, and high speed [1-2]. Particularly, the vertical expansion of stacked cell arrays provides the sustainable density solution. However, increasing variability and uncertainty in 3-D cross-point array is not sufficiently investigated, and circuit solutions to supplement device variation are yet available. The selectivity targets of selection device have been set with weak justification [3].

In this research, sensing and write operation of 3-D cross-point ReRAM arrays are extensively investigated with fully integrated memory and assisting peripheral circuits. The study using a test vehicle has successfully extracted the intrinsic key parameter for reliable sensing and programming of 3-D cross-point ReRAM. In addition, temperature dependent device characteristics are compensated by dedicated circuit solution.

Test-Vehicle (TV) Information

A TV was devised to evaluate 3-D cross-point ReRAM as shown in Fig 1a. The TV has peripheral circuits and cell arrays such as 1T1R, cross-point, and 3-D cross-point types. The variable resistor is made of HfO_X. The rest of Fig 1 shows plan view and cross-section of the fabricated TV and measured I-V curves of 1R, 1S and 1S1R devices as reported in the previous works [4-5]. Two types of selectors are implemented in the test, exponential and threshold types of non-linear device. In order to exclude artifacts incorporated during the cell stacking process, 3-D cells are constructed by sophisticated wiring of two cells on the same plane as described in Fig 2.

Solutions for Large 3-D Cross-Point Array

Sensing limit for high-density ReRAM

The reliable sensing and sufficient read margin are essential in cross-point ReRAM [3] since the leakage current is inevitable. In order for reliable sensing in a large cross-point array, a decent selector is attached to the variable resistor in a memory cell unit.

For the flawless read, cell current incorporated with the leakage must be distinguishable. The minimum current of low resistance state (LRS) cell, I_{LOUT}, is detected when the LRS cell is merged in all high resistance state (HRS) cells. Likewise, maximum HRS cell current, I_{HOUT}, comes from LRS background as depicted in Fig.3. Fig.4 shows that the allowable array size, M^2, is limited by the ratio of read current difference of a cell to leakage current difference of a cell. As long as I_{HLKG} is similar to I_{LLKG} (ie. $\Delta I_{LKG}\approx 0$), the read margin is not deteriorated from that of unit cell. The absolute level of leakage current determines the power dissipation in cell array. With respect to the sensing margin, the ΔI_{LKG} is more decisive parameter than the selectivity is. Comparing the two types of selector, larger array size is allowed in T-1S1R cell although it has lower selectivity than the E-1S1R cell as shown in Fig.5.

Write disturbance (WD) suppression in cross-point array

Another remarkable advantage of the selector is write disturbance protection. When resetting a cell far from write driver, strong electric field is asserted unselected cell that resides close to driver. The amount of leakage current aggravates the disturbance due to IR drop. To reset the farthest cell in a cross-point array, high bias should be applied to the word-line, and it can induce a permanent breakdown stuck of near cells due to high voltage stress.

In this section, the effect of the selector on the WD is analyzed. The WD is occurred due to reaction between the switching oxide and the bottom electrode (BE). Once the device switches to the HRS, electric-field is applied to the gap between residual filament and the BE. To avoid high field influence, the selector is introduced as a voltage divider. In the case of the tunnel barrier, it requires continuous electric-field to maintain its on-state, and high voltage stress can be mitigated in the 1S1R (Fig. 6).

Disturbance-aware read and leakage-compensated write

At elevated temperatures, I_{LRS} decreases and I_{HRS} increases, resulting in sensing margin loss. The state flips back and forth by smaller electric field as marked in Fig.7. The leakage current increases due to the structural reason. It is desired to lower the read voltage at hot environment in order not to be disturbed. Fig. 8 suggests disturbance-aware sensing circuit. Temperature sensor adjusts read voltage to protect the memory from disturbance. The reduced sensing margin at lowered read voltage is recovered cell current amplifier as shown in Fig.9. Although absolute value of leakage does not influence the sensing performance, writing is fluctuated by the leakage as deduced in Fig.10. Prior to write operation, leakage is monitored and added to the write current so as to provide the desired current to the target cell using the leakage-compensation write driver in Fig. 11. Thanks to the leakage compensation, the set current to the specific cell is maintained constant regardless of the array size as shown in Fig. 12. Leakage-compensation must be effective way of reading 1S1R cells with small, but distinctive leakage in HRS and LRS states.

Summary

This study shows high read and write performances of ReRAM 3D X-point array. Not only selectivity of the device, but also the leakage properties of 1S1Rs and peripheral modules dramatically improve read and write disturbance immunity of 3D X-point array for high-density applications.

Acknowledgement

This work was supported by the R&D MOTIE/KEIT (10039191) and 8 ″ CMOS process of National NanoFab Center.

References

[1] P. Cappelletti et al., IEDM 2015 [2] S. Jo et al., IEDM 2014 [3] A. Chen et al., IEDM 2013 [4] J. Woo et al., VLSI 2013 [5] S. Kim et al., VLSI 2012

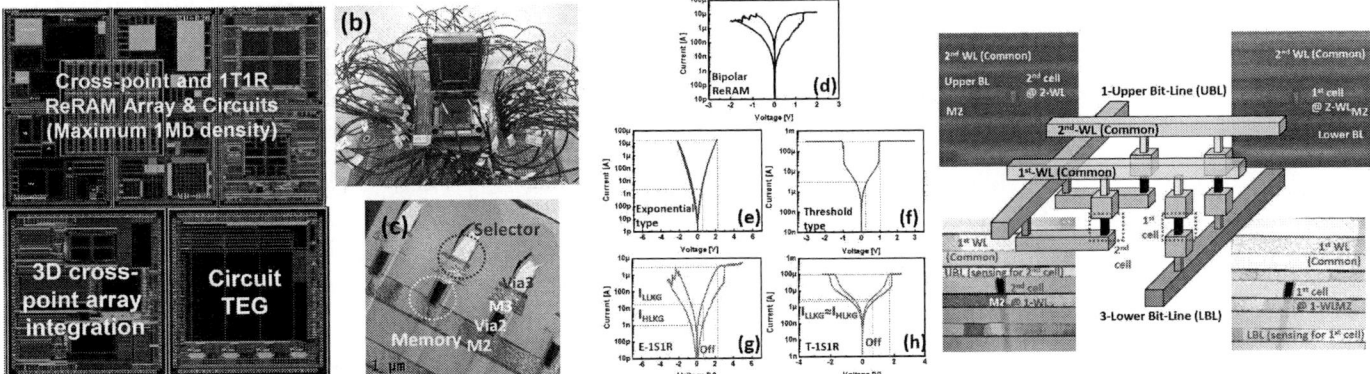

Fig. 1. (a) Layout of ReRAM TV having diverse types of cell arrays and peripheral circuits, (b) package of die (c) cross-section of fabricated TV, and measured I-V curves of (d) bipolar ReRAM, (e) exponential(E) and (f) threshold-type(T) selectors, (g) E-1S1R (h) T-1S1R

Fig. 2. Deployment of lower BL, upper BL, and shared WL in 3-D cross-point cell constructed by wiring two ReRAM cells on a plane

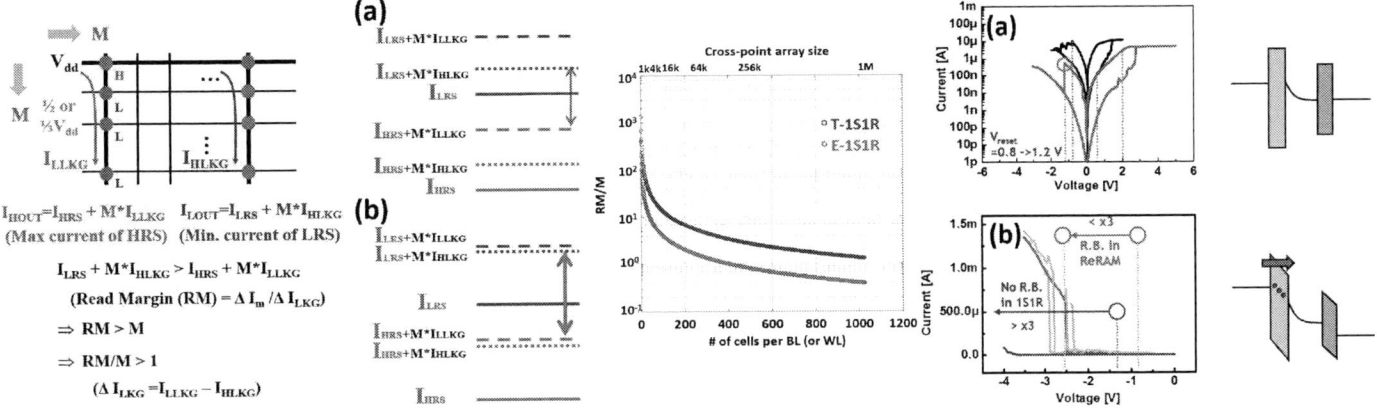

Fig. 3. Criterion for correct read with respect to cell current and leakage

Fig. 4. (a) Narrow and (b) wide read margins as $I_{LLKG} \gg I_{HLKG}$ and $\approx I_{HLKG}$, respectively

Fig. 5. RM/M as a function of array size for ReRAM cell in series with exponential or threshold selector

Fig. 6. Prevention of write disturbance in 1S1R cells

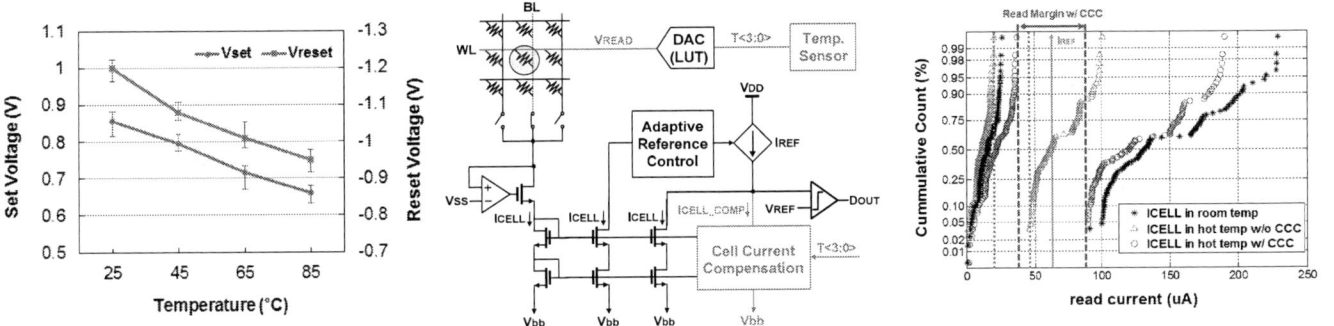

Fig. 7. Measured average set/reset voltage with temperature variation

Fig. 8. Proposed sensing circuit to adjust read voltage with temperature and to compensate reduced cell current

Fig. 9. Measured distribution of read current at hot temperature with cell current compensation (CCC) and without CCC

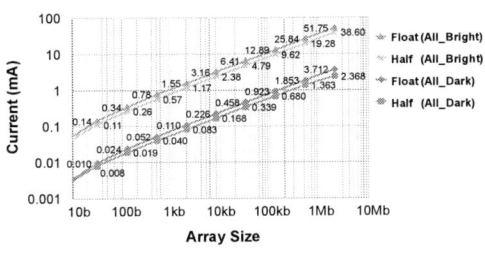

Fig. 10. Leakage of write current with bias condition of unselected WL/BL (all bright: cells are in LRS, all dark: cells are in HRS)

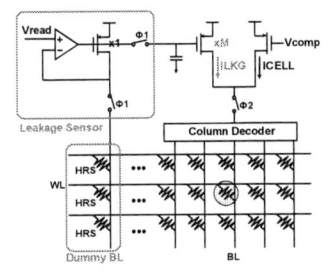

Fig. 11. Leakage current compensation circuit utilizing HRS dummy cells

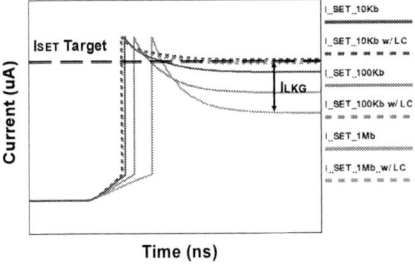

Fig. 12. Set current through the target cell in cross-point array with different array sizes when leakage current is compensated

Embedded Memory and ARM Cortex-M0 Core Using 60-nm C-Axis Aligned Crystalline Indium–Gallium–Zinc Oxide FET Integrated with 65-nm Si CMOS

Tatsuya Onuki, Wataru Uesugi, Hikaru Tamura, Atsuo Isobe, Yoshinori Ando, Satoru Okamoto,
Kiyoshi Kato, T R Yew[†], Chen Bin Lin[†], J Y Wu[†], Chi Chang Shuai[†], Shao Hui Wu[†], James Myers[††],
Klaus Doppler[†††], Masahiro Fujita[††††], and Shunpei Yamazaki

Semiconductor Energy Laboratory Co., Ltd. Atsugi-shi, Kanagawa, Japan
[†]United Microelectronics Corporation (UMC), Hsinchu, Taiwan
[††]ARM Ltd., Cambridge, United Kingdom
[†††]Nokia Technologies, San Francisco, CA
[††††]VLSI Design and Education Center (VDEC), The University of Tokyo, Bunkyo-ku, Tokyo, Japan
E-mail: to1153@sel.co.jp

Abstract

Low-power embedded memory and an ARM Cortex-M0 core that operate at 30 MHz were fabricated in combination with a 60-nm c-axis aligned crystalline indium–gallium–zinc oxide FET and a 65-nm Si CMOS. The embedded memory adopted a structure in which oxide semiconductor-based 1T1C cells are stacked on Si sense amplifiers. This memory achieved a standby power of 3 nW while retaining data and an active power of 11.7 μW/MHz by making each bitline as short as each sense amplifier. The M0 core adopted the flip-flop in which an oxide semiconductor-based 3T1C cell is stacked on the Si scan flip-flop cell without area overhead and achieved a standby power of 6 nW while retaining data. The combination of the embedded memory and the M0 core provided high-performance, low-power Internet of Things devices operating with a broad range of active standby power ratios.

Introduction

Microcontroller units (MCU) for Internet of Things (IoT) mostly require low power consumption to achieve long battery life. To reduce power consumption, both standby power and active power need to be reduced.

Power gating (PG) of memory and a CPU core while retaining data is an effective way to reduce standby power of the MCU. MRAM, ReRAM, FRAM, and a c-axis aligned crystalline indium–gallium–zinc oxide (CAAC-IGZO) FET have been studied for such an application [1–4]. However, MRAM and ReRAM have high active power, and FRAM has difficulties in scalability. On the other hand, the CAAC-IGZO FET has an extremely low off-state current that can be used to construct ultra-low-power memory. The characteristics of CAAC-IGZO FETs also have been demonstrated in a 20-nm node [5]. In terms of chip implementation, the results obtained using a 180-nm CMOS and a 60-nm CAAC-IGZO FET have been reported [4].

In this study, embedded oxide-semiconductor-based DRAM-type memory (DOSRAM) [6] and a Cortex®-M0 core were fabricated using a 60-nm CAAC-IGZO FET and a 65-nm Si CMOS. The chip did not require power supply for data retention; therefore, the significant reduction of standby power was achieved by PG. Moreover, in embedded memory, stacking of a cell array on a read circuit shortened bitlines and reduced the active power of the memory.

Design and Features

Fig. 1 shows the block diagram of the fabricated chip. The chip consists of 8KB DOSRAM as an embedded memory, an M0 core, a power management unit (PMU), and an AHB-Lite bus. Each of the DOSRAM and flip-flops (FF) in the M0 core includes a CAAC-IGZO FET and a storage capacitor (1OS1C),

which is the smallest unit of a CAAC-IGZO memory cell. PG of these two blocks is controlled by the PMU.

Fig. 2 shows the block diagram of the 8 KB DOSRAM which consists of four 2KB subarrays including 16 1Kb local arrays. The 1Kb local array has a stack structure in which a 1OS1C cell array of 8 wordlines × 256 bitlines is stacked on 128 sense amplifiers and a multiplexer (MUX). This stack structure enabled the reduction of active regions during memory access, as shown in Fig. 3. Fig. 3(a) shows a case without a stack structure in which long bitlines should be driven. Fig. 3(b) shows a stacked array that has short bitlines whose length is the same as that of the Si circuits under the cell array and additional global bitlines. Shortening of bitlines enables smaller bitline capacitance, which results in a reduction of storage capacitance. The MUX connects 64 of the 256 bitlines to global bitlines and reduces the number of long bitlines. Consequently, the drive load during memory access is reduced. Fig. 4 shows 2KB DOSRAM active energy simulation results, which indicate that the energy can be reduced by more than 70% in the stacked array. Fig. 5 shows the local array layout.

Fig. 6 shows the circuit diagram of an oxide semiconductor flip-flop (OS-FF). Three OS-FETs and one capacitor (3OS1C) are added to a scan FF, which is a standard cell. BK and RE signals transmitted from the PMU enable FF backup and recovery. As shown in Fig. 7, 3OS1C is added without any change in the original cell layout. This indicated that OS-FF layout was achieved without area overhead.

Fabrication and Measurement Results

Figs. 8 and 9 show the die photograph and cross-sectional image of the fabricated chip. A Si CMOS and wirings up to M2 metal were fabricated by a UMC 65 nm LL process, and the CAAC-IGZO FET and other wirings are stacked on the Si CMOS.

Fig. 10 shows 8KB DOSRAM retention characteristics. Even after 1 h at 85°C, 99.95% of the data were retained. This result indicated that the DOSRAM can achieve long-term PG without refresh operation.

The waveforms of chip backup and recovery are shown in Fig. 11. In the OS-FF operation at 30 MHz, backup time was two clock cycles (66 ns) and recovery time is three clock cycles (99 ns). The DOSRAM could retain data without power supply. There was no backup–recovery sequence but only a power on/off sequence is sufficient in DOSRAM PG.

A summary of the operating mode and power consumption of the chip is presented in Table 1. To measure the active power, a closed loop program was used in which in nine clock cycles, 32-bit data was read seven times from the DOSRAM, and 32-bit data was written twice to the DOSRAM at 30-MHz clock frequency and VDD = 1.1 V. The standby power of both the DOSRAM and the M0 core was reduced by PG.

Table 2 shows comparison between our chip and other low-power MCUs. The technology node, clock frequency, and active power of this work are superior to the other low-power MCUs. Regardless whether the active-standby ratio is high or low, this chip achieves the lowest power consumption.

Conclusion

Embedded memory and a Cortex-M0 core were fabricated by a CAAC-IGZO 60-nm process and a UMC 65-nm LL process. The CAAC-IGZO technology reduced both standby power and active power. This technology has proven to be effective as a low-power technology for IoT devices.

References

[1] S. Bartling *et al.*, *ISSCC Dig. Tech. Papers*, pp. 432–434, 2013.
[2] N. Sakimura *et al.*, *ISSCC Dig. Tech. Papers*, pp. 184–185, 2014.
[3] VK. Singhal *et al.*, *ISSCC Dig. Tech. Papers*, pp. 148–149, 2015.
[4] H. Tamura *et al.*, *IEEE Micro*, vol. 34, pp. 42–53, 2014.
[5] D. Matsubayashi *et al.*, *IEDM Tech. Dig.*, pp. 141-144, 2015
[6] T. Onuki *et al.*, *Jpn. J. Appl. Phys.*, vol. 54, 04DD07, 2015.

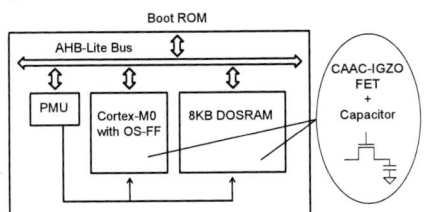

Fig. 1 Block diagram of fabricated chip.

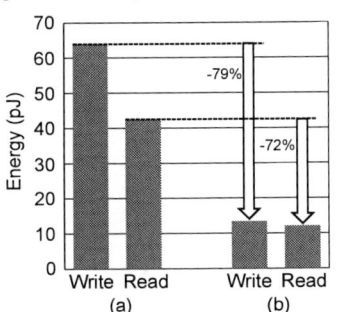

Fig. 4 Active energy simulation results (typical R.T.) of 2KB DOSRAM (a) without a stack structure: 256 bitlines, Cs = 30 fF and (b) with a stacked array: 256 short bitlines, Cs = 3.5 fF, 64 global bitlines.

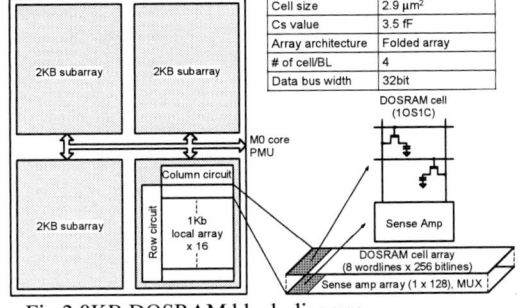

Fig.2 8KB DOSRAM block diagram.

Cell size	2.9 µm²
Cs value	3.5 fF
Array architecture	Folded array
# of cell/BL	4
Data bus width	32bit

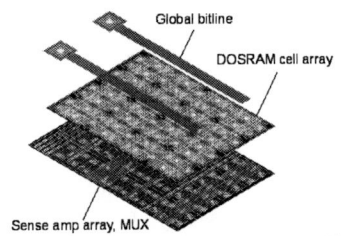

Fig.5 Layout of four sense amplifiers, DOSRAM cell array of 8 wordlines × 8 bitlines, and 2 global bitlines (Cs = 1.8 fF).

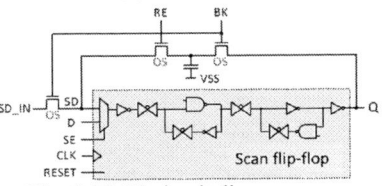

Fig. 3 Active regions during access (a) without stack structure and (b) with a stacked array.

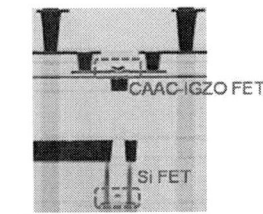

Fig. 6 OS-FF circuit diagram.

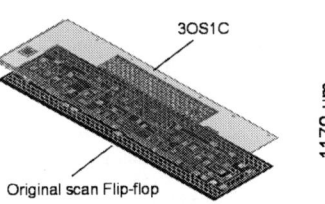

Fig. 7 OS-FF layout.

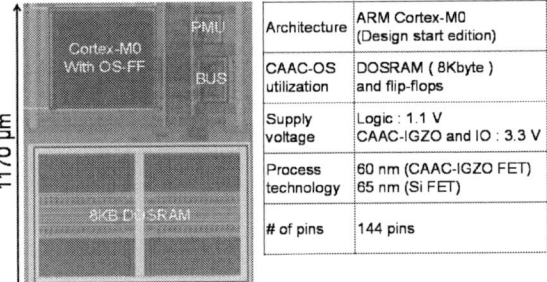

Fig. 8 Die photograph and specifications.

Architecture	ARM Cortex-M0 (Design start edition)
CAAC-OS utilization	DOSRAM (8Kbyte) and flip-flops
Supply voltage	Logic : 1.1 V CAAC-IGZO and IO : 3.3 V
Process technology	60 nm (CAAC-IGZO FET) 65 nm (Si FET)
# of pins	144 pins

Fig. 9 Cross-sectional image of fabricated chip.

Table 1 Operating mode and power consumption of chip. Measurement conditions are as follows: VDD = 1.1 V and temp. = 25°C.

	Active (µW/MHz)	Clock Gating (µW)	Power Gating (µW)
M0 core	16.5	11.6	0.006
8KB DOSRAM	11.7	1.43	0.003

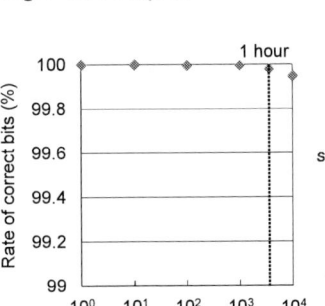

Fig. 10 8KB DOSRAM retention characteristics. Measurement conditions are as follows: VDD = 1.1 V and temp. = 85°C.

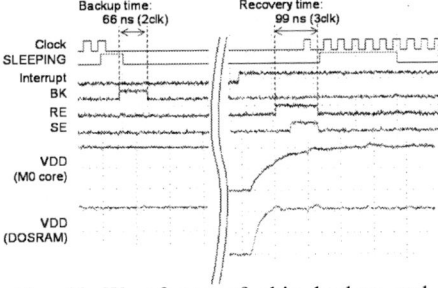

Fig. 11 Waveforms of chip backup and recovery operation. Measurement conditions are as follows: clock frequency = 30 MHz, VDD = 1.1 V, and temp. = 25°C.

Table 2 Characteristics of the chip developed in this study compared to other technologies

	ISSCC 2013 [1]	ISSCC 2014 [2]	ISSCC 2015 [3]	Micro 2014 [4]	This work
Technology (Si)	130nm HVT	90nm MVT	90nm SVT, HVT	180nm	65 nm LL
Memory device	FRAM	SpinRAM	N/A	CAAC-IGZO memory	CAAC-IGZO memory
Clock frequency (MHz)	8	20	16	30	30
Active Power (µW/MHz)	170	145	28.3	225.8	28.2
Standby Power (µW)	0	1.22	0.32	0.05	0.009
100K-cycle in 1sec (µW)	17.00	15.71	3.15	22.63	2.83
10K-cycle in 1sec (µW)	1.70	2.67	0.60	2.31	0.29
1K-cycle in 1sec (µW)	0.17	1.36	0.35	0.28	0.04

Versatile TLC NAND Flash Memory Control to Reduce Read Disturb Errors by 85% and Extend Read Cycles by 6.7-times of Read-Hot and Cold Data for Cloud Data Centers

Atsuro Kobayashi, Tsukasa Tokutomi and Ken Takeuchi

Chuo University, 1-13-27 Kasuga, Bunkyo-ku, Tokyo, 112-8551 Japan, E-mail: kobayashi@takeuchi-lab.org

Abstract

Versatile Triple-Level-Cell (TLC) NAND flash memory control with Read Hot/Cold Migration, Read Voltage Control and Edge Word Line Protection is proposed for data center application SSDs. Measured errors decrease by 85% and measured acceptable read cycles increase by 6.7-times.

Introduction

As smartphones mature, cloud data centers are expected as the emerging market of NAND flash-based SSDs. In smartphones, the user is mostly only one and the NAND flash is used as a data storage storing infrequently read-"cold" data (Fig. 1). The high reliability techniques [1-3] reduce data-retention errors of read-cold data. On the other hand, in data centers, many applications operate simultaneously and a lot of users often share the same data. These frequently read data is called read-"hot" data where both read-disturb and data-retention errors occur. To remove errors of both hot and cold data at data centers, this paper proposes the versatile control of 1Xnm TLC NAND flash. This paper first describes that the optimal read word-line voltages are different between hot and cold data. In the conventional SSD, read-hot and cold data are mixed in the same NAND flash block. Thus, different read voltages cannot be applied. To solve this problem, this paper proposes Read Hot/Cold Migration. The proposed SSD controller allocates hot or cold data in hot or cold regions in the TLC NAND flash. The second proposal, Read Voltage Control, optimizes the read voltages to decrease both read-disturb and data-retention errors. Finally, Edge Word Line Protection is proposed to improve the reliability of the worst edge word-lines.

New Observation of V_{TH}-down Shift by Read-disturb

In read-cold data, electron leakage from the floating gate decreases memory V_{TH} (Figs. 2-3). In contrast, this paper first reports that in read-hot data, read-disturb induces both V_{TH} increase and decrease, depending on the memory state of 1Xnm TLC NAND flash (Figs. 4-5). At lower V_{TH}, "Erase", "A" to "E", the electron injection to the floating gate increases V_{TH}. At higher V_{TH} "F" and "G" states, V_{TH} decreases probably by electron de-trapping from the inter-poly dielectric (IPD) as shown in Fig. 6 [4]. What's worse, even V_{TH}-down errors in hot data is 2.2-times larger than those in read-cold data (Fig. 7). These observations reveal that to minimize errors, the read reference voltages (V_{REF}) should be controlled differently for hot and cold data, "A" to "G" states, the read cycles and the data-retention time.

Proposal 1: Data Migration to Hot or Cold Region

To apply optimal read voltages for both read-hot and cold data, this paper proposes Read Hot/Cold Migration that separates data into read-hot or cold region by using LRU (Least Recently Used) table (Fig. 8). All data is initially stored in the read-cold region. With each read, the logical-block address (LBA) is added to an LRU table in the SSD controller. Next, by referring to the LRU table, hot data is detected and then moved to the read-hot region. As a result, read-hot data and read-cold data are separated in each region.

Proposal 2: Read Voltage Control (RVC)

Read Voltage Control (RVC) selects the optimal V_{REF} for both read-hot and cold region. As shown in Fig. 3, in read-cold region, RVC changes the optimal V_{REF} shift differently among each V_{TH} state when the data-retention time increases. The data-retention time is determined by the SSD controller [5]. On the other hand, in read-hot region, the optimal V_{REF} shift is changed among the V_{TH} states and by the read cycles, which is recorded in the SSD controller (Fig. 5). Moreover, in case of hot data, two edge word-lines in a NAND flash block, WL0 and 85, show the higher bit-error rate (BER) than middle word-lines by 9.6-times and 3.7-times, respectively (Fig. 10). In WL0 and 85, the higher electric field in the channel near the select gate accelerates the hot-electron injection from the channel into the floating gate [6]. The worst WL0 and 85 determine the overall reliability of

SSD. Even if only one word-line exceeds error correction capability of ECC, the entire block is judged as bad. To suppress errors of WL0 and 85, RVC shifts V_{REF} by 3.1-times compared with middle word-lines, WL1-84 (Fig. 11). In WL1-84, RVC selects V_{REF} in both positive and negative directions according to the location of word-lines and V_{TH} states. That is, RVC selects V_{REF} among 3 patterns of word-lines; two edge word-lines and middle word-lines (Fig. 12).

To apply RVC, the optimal V_{REF} is pre-recorded in a V_{REF} table in the SSD controller, for each write/erase cycle, read cycles and data-retention time before shipping. The conventional V_{REF} shift [1-3] decreases V_{REF} when ECC cannot correct errors. As shown in Fig. 13, when the conventional V_{REF} shift is applied, BER of hot data even increases by 3.5% compared with no V_{REF} shift. Therefore, no V_{REF} shift is assumed as the conventional method in the following analysis. Since, proposed RVC requires no extra read-retry [2, 3], the read-time decreases by 82% (Fig. 14). By using RVC, BER of WL0 and 85 decreases by 68% (Fig. 15). In addition, BERs of entire block in read-hot and cold region are reduced by 31% as shown in Figs. 15-16, respectively.

Proposal 3: Edge Word Line Protection (EWLP)

Even when RVC is applied, BER of WL0 and 85 of hot data is still higher than middle word-lines by 4.6-times and by 1.8-times, respectively (Fig. 15). To decrease errors of WL0 and 85 to the same level of middle word-lines, this paper also proposes Edge Word Line Protection (EWLP). In WL1-84, the number of "G" to "F" errors are similar to "Erase" to "A" errors (Fig. 17). This result means that the number of injected electrons to the floating gate is similar to electrons de-trapping from IPD. On the other hand, in WL0 and 85, the number of "Erase" to "A" errors is more than 10-times larger than other error patterns (Figs. 18-19). To reduce "Erase" to "A" errors of WL0 and 85, EWLP decreases the population of "Erase" state in WL0 and 85 (Fig. 20). Modifying populations of V_{TH} state was proposed in [7, 8], however, because encoding was applied to all word-lines, the flag bit increases the cell area as much as by 25%. In this work, modulation/encoding is applied only to WL0 and 85, and the flag overhead is only 0.58%. Using edge word-lines, WL0 and 85, as dummy word-lines [9] can remove errors of WL0 and 85 but will increase the cell area by 2.3%. By combining RVC and EWLP, measured errors of WL0 and 85 decrease by 94% and 93%, respectively (Fig. 21). In addition, BER of the entire block decreases by 36%. As the read cycle increases, the proposed methods become more effective and decrease more errors (Fig. 22). In addition, the acceptable read cycles increase by 6.7-times (Fig. 23). Because the proposals most effectively reduce both read-disturb and data-retention errors, it is suitable for data center application SSDs which store both hot and cold data.

Conclusion

Fig. 24 is a photograph of the measured SSD [1]. Table 1 summarizes this work. By combining RVC and EWLP, the worst BER decreases by 85% without read-time overhead or cell area penalty. The additional table size is only 9kBytes per 128Gbit TLC NAND flash. These tables can be stored in SSDs without memory capacity penalty. The proposals can be implemented in the SSD controller without chip size increase of the controller nor without changing the NAND flash chip, and thus are the ideal solution for the future SSDs.

References

[1] T. Tokutomi et al., ISSCC Dig. Tech. Papers, pp. 140-141, Feb. 2015. [2] T. Parnell, IMW Tutorial, May 2014. [3] Y. Cai et al, HPCA, pp. 551-563, Feb. 2015. [4] A. Serov et al., IRPS, pp. 887-890, April 2009. [5] S. Tanakamaru et al., JSSC, pp. 2920-2933, Nov. 2013. [6] J. D. Lee et al, NVSMW, pp. 31-33, Feb. 2006. [7] S. Tanakamaru et al., ISSCC Dig. Tech. Papers, pp. 204-205, Feb. 2011. [8] S. Yamazaki et al., Symp. VLSI Tech., Dig. Papers, pp. 112-113, June 2015. [9] K. T. Park et al, SSDM, pp.298-299, Sep. 2006.

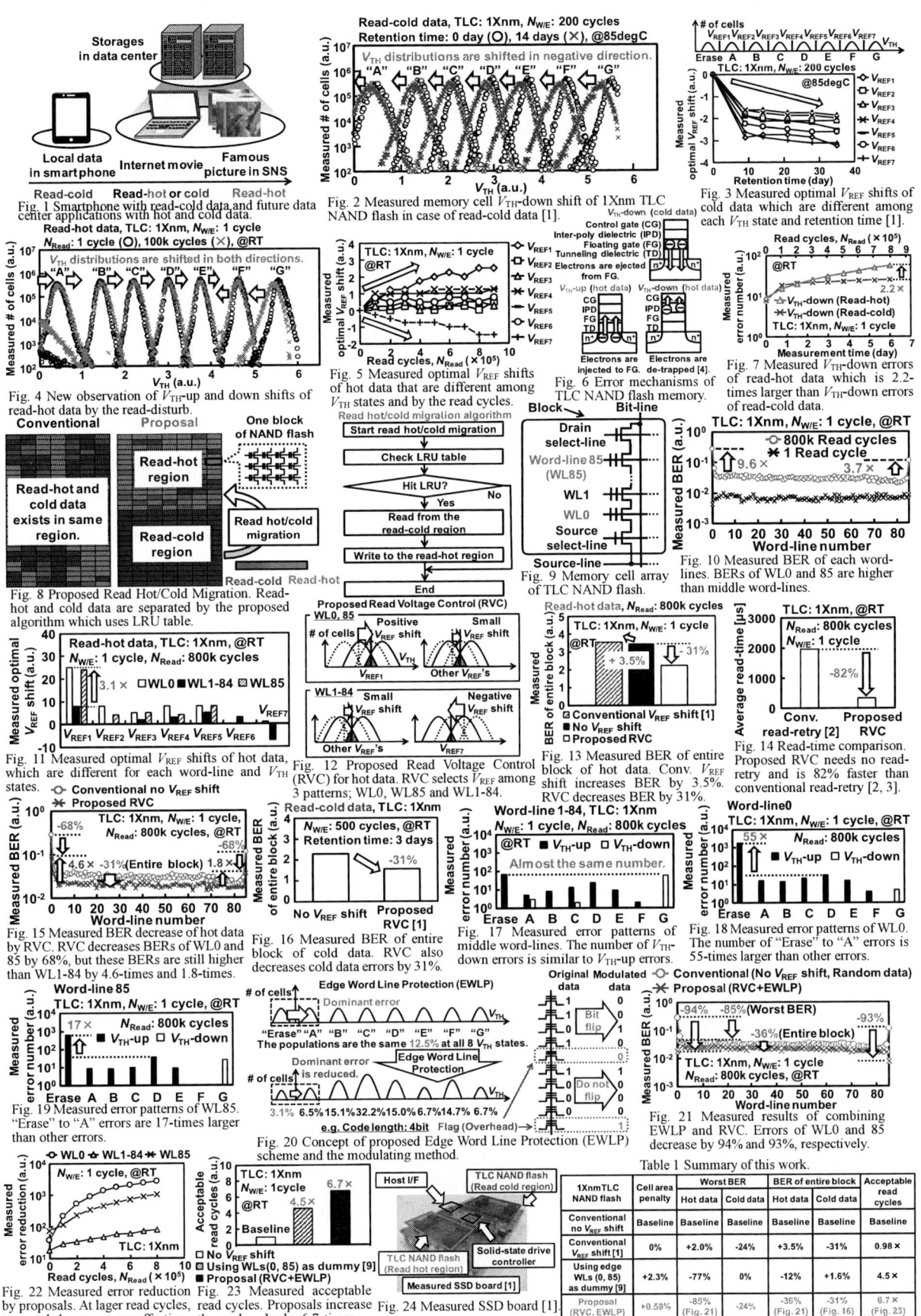

Fig. 1 Smartphone with read-cold data and future data center applications with hot and cold data.

Fig. 2 Measured memory cell V_{TH}-down shift of 1Xnm TLC NAND flash in case of read-cold data [1].

Fig. 3 Measured optimal V_{REF} shifts of cold data which are different among each V_{TH} state and retention time [1].

Fig. 4 New observation of V_{TH}-up and down shifts of read-hot data by the read-disturb.

Fig. 5 Measured optimal V_{REF} shifts of hot data that are different among V_{TH} states and by the read cycles.

Fig. 6 Error mechanisms of TLC NAND flash memory.

Fig. 7 Measured V_{TH}-down errors of read-hot data which is 2.2-times larger than V_{TH}-down errors of read-cold data.

Fig. 8 Proposed Read Hot/Cold Migration. Read-hot and cold data are separated by the proposed algorithm which uses LRU table.

Fig. 9 Memory cell array of TLC NAND flash.

Fig. 10 Measured BER of each word-lines. BERs of WL0 and 85 are higher than middle word-lines.

Fig. 11 Measured optimal V_{REF} shifts of hot data, which are different for each word-line and V_{TH} states.

Fig. 12 Proposed Read Voltage Control (RVC) for hot data. RVC selects V_{REF} among 3 patterns; WL0, WL85 and WL1-84.

Fig. 13 Measured BER of entire block of hot data. Conv. V_{REF} shift increases BER by 3.5%. RVC decreases BER by 31%.

Fig. 14 Read-time comparison. Proposed RVC needs no read-retry and is 82% faster than conventional read-retry [2, 3].

Fig. 15 Measured BER decrease of hot data by RVC. RVC decreases BERs of WL0 and 85 by 68%, but these BERs are still higher than WL1-84 by 4.6-times and 1.8-times.

Fig. 16 Measured BER of entire block of cold data. RVC also decreases cold data errors by 31%.

Fig. 17 Measured error patterns of middle word-lines. The number of V_{TH}-down errors is similar to V_{TH}-up errors.

Fig. 18 Measured error patterns of WL0. The number of "Erase" to "A" errors is 55-times larger than other errors.

Fig. 19 Measured error patterns of WL85. "Erase" to "A" errors are 17-times larger than other errors.

Fig. 20 Concept of proposed Edge Word Line Protection (EWLP) scheme and the modulating method.

Fig. 21 Measured results of combining EWLP and RVC. Errors of WL0 and 85 decrease by 94% and 93%, respectively.

Fig. 22 Measured error reduction by proposals. At larger read cycles, proposals become more effective.

Fig. 23 Measured acceptable read cycles. Proposals increase the read cycles by 6.7-times.

Fig. 24 Measured SSD board [1].

Table 1 Summary of this work.

1XnmTLC NAND flash	Cell area penalty	Worst BER		BER of entire block		Acceptable read cycles
		Hot data	Cold data	Hot data	Cold data	
Conventional no V_{REF} shift	Baseline	Baseline	Baseline	Baseline	Baseline	Baseline
Conventional V_{REF} shift [1]	0%	+2.0%	-24%	+3.5%	-31%	0.98×
Using edge WLs (0, 85) as dummy [9]	+2.3%	-77%	0%	-12%	+1.6%	4.5×
Proposal (RVC, EWLP)	+0.56%	-85% (Fig.21)	-24%	-36% (Fig.21)	-31% (Fig.16)	6.7× (Fig.23)

A 0.9um² 1T1R Bit Cell in 14nm SoC Process for Metal-Fuse OTP Array with Hierarchical Bitline, Bit Level Redundancy, and Power Gating

Z Chen, S H Kulkarni, V E Dorgan, U Bhattacharya, K Zhang

Logic Technology Development, Intel Corporation, Hillsboro, OR, USA

Phone: 503-613-2978, email: zhanping.chen@intel.com

Abstract

This work introduces the first high-volume manufacturable (HVM) metal-fuse technology in a 14nm tri-gate high-k metal-gate CMOS process. A high-density array featuring a 0.9μm² 1T1R bit cell and bit level redundancy is presented. An array efficiency of 50% is achieved with hierarchical bit line design to separate fuse programming from read/sense. A power gating scheme is adopted to reduce leakage current consumption and reduce high voltage exposure for reliability. Program conditions can be optimized for HVM and in-field programming (IFP) to achieve close to 100% bit level yield.

Keywords: High-density OTP-ROM, metal fuse, HVM, IFP

Introduction

On-chip one time programmable (OTP) fuse ROM enables the traditional functions of chip unit ID, cache repair and adaptive circuit tuning. While the industry transitioned to high-k metal-gate [1], salicided poly fuse technology [2-4] has been replaced by metal-fuse to maintain process compatibility for OTP-ROM [5]. Recent new applications such as code storage and IoT have been driving a dramatic increase in fuse bit count, necessitating the development of ever small bit cells and area efficient arrays. Additionally more fuse bit count directly leads to longer high voltage exposure, larger current consumption during fuse programming, higher leakage current during standby, and lower die level yield with iso-bit level yield. This paper presents the first metal-fuse memory technology in 14nm tri-gate high-k metal-gate CMOS with a 0.9um² bit cell, hierarchical bit line design to reduce overall area, bit level fuse repair scheme to improve die level fuse yield, and power gating scheme (PG) to minimize high voltage exposure and reduce current during fuse operation and standby for low cost and high volume products.

1T1R Bit Cell and Fuse Element Design

Fig. 1a shows a top-down SEM view of an unprogrammed and a programmed fuse bit cell with an industry leading 0.9μm² cell area in the 14nm technology. Each bit cell contains one PMOS transistor and one metal-fuse resistor (1T1R) which can be programmed based on thermally accelerated metal electromigration. Fig. 1b shows the temperature contour of the fuse element during programming where a local hot-spot is achieved by a momentary high current.

Lithography at advanced technology prefers uniformity in layout but this can be detrimental to yield as the collateral damage from void formation during programming affects the neighboring structures. Special layout enhancement in conjunction with lithography optimization was employed to balance the conflicting requirements. The resulting design is electrically more resilient to collateral damage, reducing the impact of such damage both within the element metal layer and among adjacent layers.

The use of metal-fuse enables 3-D integration with the fuse element (Metal4-Via3-Metal3) stacked over the tri-gate

transistors allowing continued area scaling, achieving a record small cell area of 0.9um², over 2x scaling over 22nm [6].

High-Density Array Design

The 0.9μm² bit cell is integrated into a 72-row (64 logic bits and 8 redundancy bits) by 32-column array (Fig. 2). Fuses are programmed by passing a high current through the selected cell. A short local bitline architecture (in red in Fig. 2) is chosen to minimize row dependence caused by bitline IR drop during fuse programming while a longer global bit line (in blue in Fig. 2) is adopted across all 64 logic fuse cells to share one sense amplifier and control logic to improve array efficiency to about 50% including the level shifters (LS), sense amplifiers (SA) and all other supporting logic shown in Fig.2.

On top of the fuse array, a power gating scheme using high voltage (HV) tolerance devices is developed (Fig. 3) to minimize HV exposure and reduce leakage current drawn by unselected cells by 6~10x depending on operation temperature.

During fuse programming, only one of those PG switches is on and the rest are off. Hence, HV exposure for unselected cells is minimized to prevent reliability concern. The leakage current from the unselected arrays is significantly reduced, allowing more fuse bits to be powered by one HV supply pin to save package cost as well as standby power consumption.

Bit Level Fuse Redundancy

The dramatic increase in fuse bit count per die necessitates efficient fuse repair schemes to maintain/improve die level yield. To repair a bit out of a 64x32 array, six/five bits are needed to address a row/column, respectively. Another bit is required to enable (en) a redundancy. Hence a total of 12 redundancy bits is needed to repair one random failing bit.

Each fuse column has 64 logic bits and 8 redundant fuse bits. Our 2k fuse array includes 256 redundancy bits to be used for repair. Fig. 4 illustrates a specific implementation where 16 random bits as well as serval rows and columns can be repaired. This scheme therefore has the flexibility to adapt to different process defect signatures.

With the efficient repair scheme, program voltages can be optimized to simplify design especially for in-field programming (IFP) when fuses are programmed by end users.

In-Field Programming

The 0.9um² fuse cell is designed for area efficiency in products with high bit counts. Die yield can be achieved by optimizing the two supply voltages: Vccfhv and Vccf. To reduce potential reliability concern due to HV exposure on Vccfhv, Vccf can be raised independently to over drive column NMOS while capping Vccfhv.

To proliferate such a high density fuse bit cell for IFP with minimal area overhead, only one voltage regulator (VR) on Vccfhv is required while Vccf can be tied to core supply (as low as 0.75V) during programming. For products with core supply lower than 0.75V, a simple voltage divider from

Vccfhv can be embedded at the point of use in each array.

Fig. 5 illustrates an IFP implementation using one VR with input at 3.3V, an output voltage ranging from 2.0V to 2.6V, and currant delivery capability up to 100mA.

When a fuse bit is asserted for programming, the VR supplies a specified voltage to the fuse arrays. During fuse read, the regulator is in bypass mode to power fuse arrays from a core supply which can be as low as 0.75V.

Measurements

Fig. 6a shows the die photo of a 14nm test-chip and Fig. 6b shows the magnified image of a 2k fuse array. Table I summarizes the key parameters of the new fuse technology.

Fig. 7 shows the measured bit level programming success rate with different program conditions, read voltages and temperatures. The right side of Fig. 7 shows almost 100% bit level yield down to Vccfhv/Vccf=2.1V/1.05V for HVM. All fuse bits can be read stably down to 0.75V. The left part indicates bit level yield with Vccf at 0.75V to allow SOC logic rail to drive Vccf. With Vccfhv raised to between 2.3V and 2.4V, close to 100% bit level yield can be achieved even at a low read voltage of 0.75V. The sample size for this data was over one million bits per programming condition.

All reported bit level yields are in the non-redundant read mode.

Conclusion

Third-generation metal-fuse technology is developed in a 14nm tri-gate high-k metal-gate CMOS process [7-8]. The technology features a 0.9um² fuse bit cell with hierarchical bitline design for efficient area. Close to100% pre-redundancy bit level yield can be achieved at Vccfhv/Vccf = 2.1V/1.05V and Vccfhv/Vccf = 2.4V/0.75V with read voltage as low as 0.75V for HVM and IFP, respectively. With 16 redundancy bits, program voltages can be further optimized for different applications to minimize design complexity while maintaining overall die level yield.

Acknowledgement

The authors are grateful to the technical staff at Intel Logic Technology Development for their contributions to this work.

References

[1] K. Mistry, et al., IEDM, pp. 247-250, 2007.
[2] M. Alavi, et al., IEDM, pp. 855-858, 1997.
[3] G. Uhlmann, et al., ISSCC, pp. 406-407, 2008.
[4] S. Chung, et al., VLSI, pp. 30-31, 2009.
[5] S. Kulkarni, et al., JSSC, vol. 45, pp. 863-868, 2010.
[6] S. Kulkarni, et al., JSSC, in press, 2016.
[7] S. Natarjan, et al., IEDM, pp. 3.7.1-3.7.3, 2014.
[8] C.-H. Jan, et al., VLSI, pp. T12-T13, 2015.

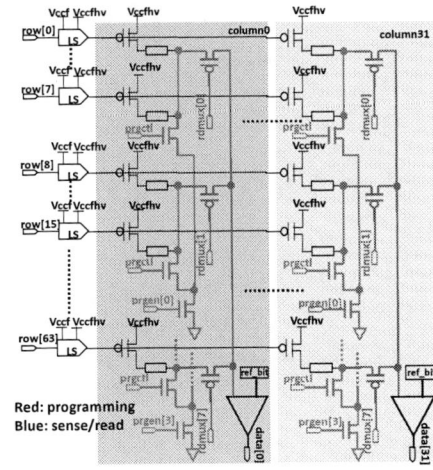

Fig. 2 High-density fuse array with hierarchical bit line design

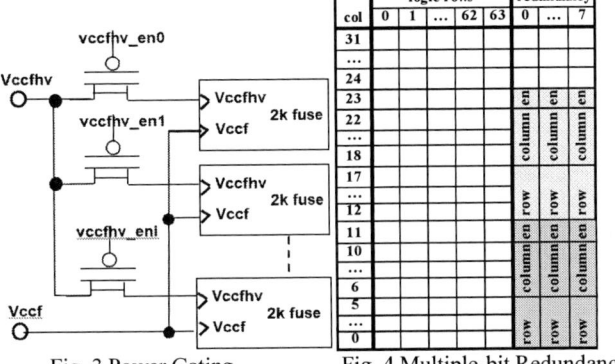

Fig. 3 Power Gating   Fig. 4 Multiple-bit Redundancy

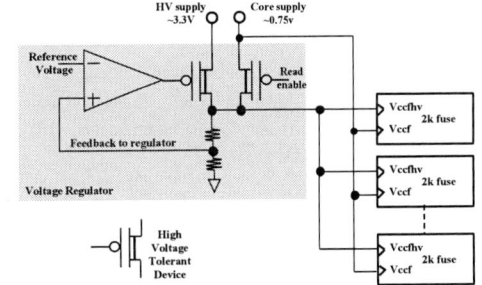

Fig. 5 In-field Programming

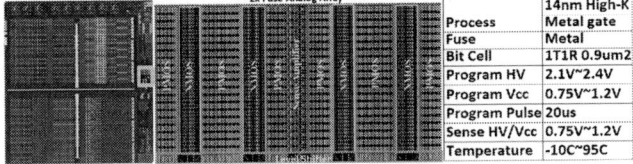

Fig. 6a Die photo   Fig. 6b 2K Array Floorplan   Table 1: Operation

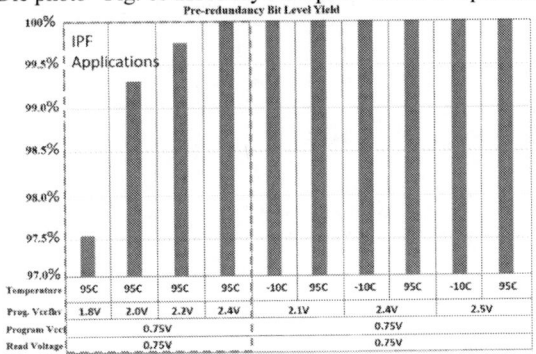

Fig. 7 Yield across Program Voltages

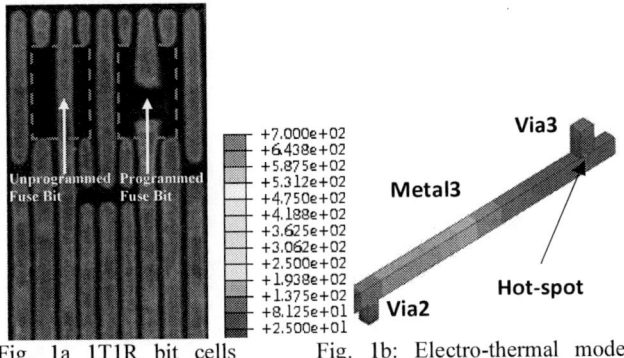

Fig. 1a 1T1R bit cells   Fig. 1b: Electro-thermal model

A 0.6mW 31MHz 4th-Order Low-Pass Filter with +29dBm IIP3 Using Self-Coupled Source Follower Based Biquads in 0.18μm CMOS

Yang Xu, Spencer Leuenberger, Praveen Kumar Venkatachala, and Un-Ku Moon

School of Electrical Engineering & Computer Science, Oregon State University, Corvallis, OR
xuyang@eecs.oregonstate.edu

Abstract

A highly compact low-pass filter (LPF) using self-coupled source follower based biquads is presented. The biquad cell synthesizes a 2nd-order low-pass transfer function in a single branch, using only two capacitors and a source follower with embedded local feedback for excellent linearity. A 4th-order Chebyshev LPF prototype is designed with a cascade of two biquads in 0.18μm CMOS, and occupies an active area of 0.1mm². A cut-off frequency of 31MHz is measured with a stop-band rejection of 76dB. The prototype filter draws only 0.46mA current from a 1.35V supply, and achieves an in-band IIP3 of +29dBm. The averaged in-band input-referred noise is 22.8nV/√Hz, resulting in a dynamic range of 71dB.

Introduction

Integrated low-pass filters (LPFs) are key building blocks for analog signal processing. They are typically implemented using either an active-RC [1] or a g_m-C [2] architecture. The performance of such filters depends on the operational transconductance amplifier (OTA) used to implement the integrator. However, implementation of a high-quality OTA is increasingly challenging. Consequently, large power penalty is incurred to satisfy performance requirements in OTA-based filter designs. In this paper, we propose a new OTA-free biquad LPF topology in a single branch, using only two capacitors and a self-coupled source follower with embedded local feedback. The proposed filter exhibits excellent linearity while consuming very little power.

Proposed Biquad Architecture

Fig. 1 shows the proposed highly compact biquad low-pass filter topology in a single branch, using two capacitors and a self-coupled source follower. The transistor M_2 is biased by the drain of M_1 via a level shifter, forming a negative feedback loop. The output impedance looking into the source of M_1 is significantly reduced by the loop gain. Therefore, inherently large linearity is obtained due to this embedded local feedback.

Transistors M_1 and M_2 re-use the same current, and are designed with the same size. In this way, they are identical and provide the same transconductance (i.e. $g_{m1} = g_{m2} = g_m$). Interestingly, for a fixed current, a larger g_m that is achieved with lower overdrive voltage (i.e. $V_{GS}-V_{TH}$) results in better linearity, which differs from g_m-C filters. Meanwhile, the small output impedance also allows low in-band input-referred noise, since the noise current from M_1 dominates and the noise contributions from M_2 and the current source I_1 are negligible. With the use of the level shifter, the input signal swing is enlarged, improving the linearity and signal-to-noise ratio. Two capacitors C_1 and C_2 are added to synthesize conjugate poles. Assuming the transistor's drain-source conductance is much smaller than its transconductance, the filter transfer function is

$$H(s) = \frac{1}{s^2 \frac{C_1 C_2}{g_{m1} g_{m2}} + s \frac{C_1}{g_{m1}} + 1}. \quad (1)$$

The proposed filter topology illustrates a 2nd-order low-pass transfer function. The pole frequency ω_0 and quality factor Q are respectively given by

$$\omega_0 = \frac{g_m}{\sqrt{C_1 C_2}}, \quad Q = \sqrt{\frac{C_2}{C_1}}. \quad (2)$$

A desired biquad can be synthesized by choosing proper values for C_1 and C_2 with a given g_m. As illustrated in (2), the Q is less sensitive to process, voltage and temperature variations. Moreover, pole frequency can be properly tuned via digital setting of capacitors array, enabling programmable bandwidth.

Design and Implementation

The proposed compact biquad filter is implemented in a pseudo-differential structure, as shown in Fig. 2. A supply voltage of 1.35V is chosen in this work, which satisfies the minimal supply requirement. All transistors use low-threshold devices, and are designed with large channel length to improve the transistor output impedance. The transistor M_3 along with a

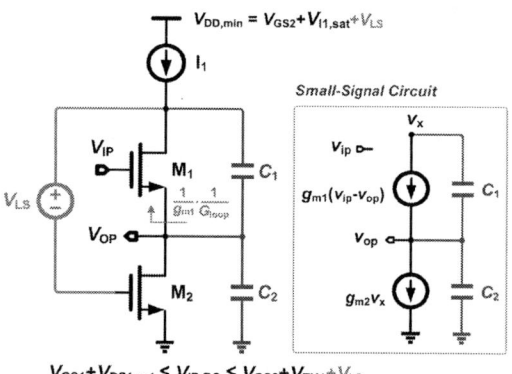

Fig. 1. Architecture of the proposed single-branch biquad low-pass filter using two capacitors and a self-coupled source follower.

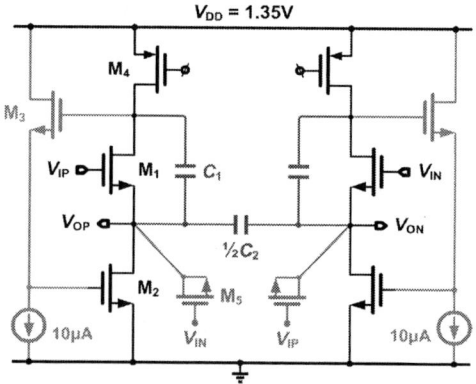

Fig. 2. Pseudo-differential implementation of the proposed biquad (all transistors employ low-threshold devices).

978-1-5090-0636-6/16 $31.00 © 2016 IEEE
2016 Symposium on VLSI Circuits Digest of Technical Papers

$10\mu A$ biasing current source realizes the level shifter (i.e. $V_{LS} = V_{GS,M3}$). MOS capacitors implemented by transistors M_5 (same type as M_1) with drain and source shorted are cross-coupled between the differential input and output nodes. As a result, they mitigate the effect of M_1's gate-source capacitance at higher frequency. In addition, no common-mode feedback circuit is needed, since the output voltage is set by the input common-mode voltage with a gate-source voltage drop of M_1.

A 4th-order filter prototype is implemented using a cascade of two biquads. The 1st biquad cell employs an NMOS-type source follower (as shown in Fig. 2), and the 2nd biquad uses a PMOS type. In this way, the input and output common-mode voltages of the entire filter are almost the same, compensating the gate-source level shift in each biquad. Fig. 3 presents the design parameters of each biquad. The low-Q biquad is placed in the first stage to provide attenuation of incoming interferers. The cascade of the two biquad cells synthesizes a 4th-order Chebyshev low-pass transfer function. The total capacitance for pseudo-differential implementation is 31.9pF.

Experimental Results

The prototype filter is fabricated in a $0.18\mu m$ CMOS process, and occupies an active area of $0.1mm^2$. Operating at a 1.35V supply, the filter consumes 0.62mW power and achieves a -3dB bandwidth of 31MHz. The measured magnitude response for two different biasing currents is depicted in Fig. 4. The feasibility of finely tuning the filter bandwidth by adjusting the biasing current is demonstrated. Moreover, the stop-band rejection is 76dB. The measured in-band 2nd- and 3rd-order intermodulation with two tones at 1.9MHz and 2.1MHz versus input power is shown in Fig. 5. The resulting IIP2 and IIP3 are +55.4dBm and +29.1dBm, respectively. Integrating the input-referred noise (IRN) density from 0.1MHz to 31MHz gives $126.4\mu V_{rms}$, and the averaged IRN over the bandwidth is $22.8nV/\sqrt{Hz}$. As measured with a 2MHz input tone, a +5.6dBm input signal ($1.21V_{pp}$ differential) creates -40dBc 3rd-order harmonic distortion (HD3) at the output, resulting in a dynamic range (DR) of 71dB. Moreover, the input 1dB compression point is +8.6dBm. A 6th-order Butterworth filter composed of three cascaded biquads is also implemented. Table I summarizes the filter performance and compares them with recent state-of-the-art LPF designs [1–5]. The presented self-coupled source follower based filters achieve the best power-per-pole/bandwidth efficiency and excellent linearity with very little power consumption.

Acknowledgement

The authors would like to thank Asahi Kasei Microdevices for chip fabrication and packaging. This work is supported in part by Analog Devices.

References

[1] S. Kousai, M. Hamada, R. Ito, and T. Itakura, "A 19.7MHz, fifth order active-RC Chebyshev LPF for draft IEEE802.11n with automatic quality-factor tuning scheme," *IEEE J. Solid-State Circuits*, vol. 42, no. 11, pp. 2326–2337, Nov. 2007.

[2] M. Oskooei, N. Masoumi, M. Kamarei, and H. Sjoland, "A CMOS 4.35-mW +22-dBm IIP3 continuously tunable channel select filter for WLAN/WiMAX receivers," *IEEE J. Solid-State Circuits*, vol. 46, no. 6, pp. 1382–1391, Jun. 2011.

[3] S. D'Amico, M. Conta, and A. Baschirotto, "A 4.1-mW 10-MHz fourth-order source-follower-based continuous-time filter with 79-dB DR," *IEEE J. Solid-State Circuits*, vol. 41, no. 12, pp.

2713–2719, Dec. 2006.

[4] B. Drost, M. Talegaonkar, and P. Hanumolu, "A 0.55V 61dB-SNR 67dB-SFDR 7MHz 4th-order Butterworth filter using ring-oscillator-based integrators in 90nm CMOS," *IEEE ISSCC Dig. Tech. Papers*, pp. 360–361, Feb. 2012.

[5] M. Tohidian, I. Madadi, and R. Staszewski, "A 2mW 800MS/s 7th-order discrete-time IIR filter with 400KHz-to-30MHz BW and 100dB stop-band rejection in 65nm CMOS," *IEEE ISSCC Dig. Tech. Papers*, pp. 174–175, Feb. 2013.

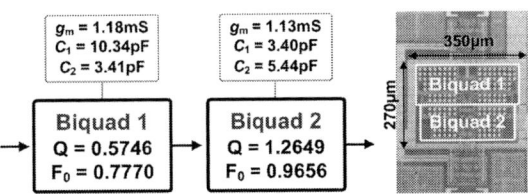

Fig. 3. Diagram of the 4th-order Chebyshev filter prototype and die micrograph.

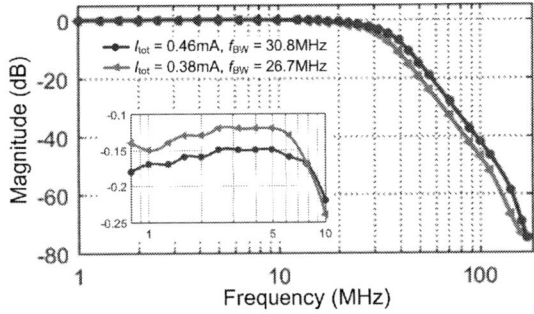

Fig. 4. Measured magnitude response of the 4th-order filter with two different biasing currents.

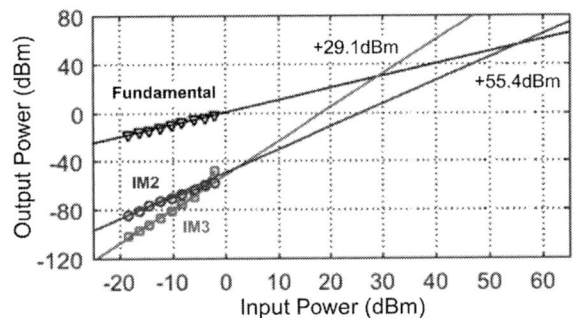

Fig. 5. In-band IIP2 and IIP3 measurement.

TABLE I. PERFORMANCE SUMMARY AND COMPARISON

	This work	[1]	[2]	[3]	[4]	[5]
Technology	$0.18\mu m$	$0.13\mu m$	90nm	$0.18\mu m$	90nm	65nm
V_{DD} (V)	1.35	1.5	1.0	1.8	0.9	1.2
Filter Order	4th	6th	5th	6th	4th	4th
Max. BW (MHz)	30.8	21.4	19.7	13.5	10	30	30
Power (mW)	0.62	0.93	11.25	4.35	4.1	19.1	1.96
Stop-band Rej. (dB)	76	70	N/A	85	N/A	N/A	45
$P_{1dB,in}$ (dBm)	8.6	10	5	7.6	5	N/A	0.7
In-band IIP3 (dBm)	29.1	28.2	18.3	22	17.5	16.7	16
IRN ($nV/\sqrt{Hz}$)	22.8	31.6	30	75	7.5	32.8	2.85
1% HD3 DR (dB)	71	68.5	69	63	79	65.5	81
Area (mm^2)	0.1	0.14	0.2	0.24	0.26	0.29	0.42

3.5mW 1MHz AM Detector and Digitally-Controlled Tuner in a-IGZO TFT for Wireless Communications in a Fully Integrated Flexible System for Audio Bag

T. Meister[1], K. Ishida[1], C. Carta[1], R. Shabanpour[1], B. K.-Boroujeni[1], N. Münzenrieder[2], L. Petti[2], G.A. Salvatore[2], G. Schmidt[3], P. Ghesquiere[4], S. Kiefl[4], G. De Toma[5], T. Faetti[5], A.C. Hübler[3], G. Tröster[2], F. Ellinger[1]

[1]Technische Universität Dresden, Dresden, Germany, [2]Swiss Federal Institute of Technology Zurich, Zurich, Switzerland, [3]Technische Universität Chemnitz, Chemnitz, Germany, [4]Siemens AG, Munich, Germany, [5]Smartex S.r.l., Pisa, Italy

Abstract

We developed a fully flexible AM (amplitude modulation) radio receiver suitable for integration in an "audio bag", by exploiting the heterogeneous integration of several fully flexible technologies. In this paper, we present a 2.9 mW 2-bit digitally-controlled tuner with a 576 kHz tuning range, a 3.5 mW 1 MHz AM detector and their integration in such a fully-flexible system. Their optimized power consumptions are essential because thin flexible batteries and organic solar cells serve as power supply. The circuits are fabricated in a low-temperature amorphous indium gallium zinc oxide (a-IGZO) technology. For the system integration textile techniques as well as flexible inkjet-printed packages and printed circuit boards (IPCBs) were used.

Introduction

Recently, great attention has been paid to research on flexible and wearable electronics. Many works exist that present circuits in cost-efficient flexible technologies based on organic and metal-oxide thin-film transistors (TFTs). Another branch of research focuses on advanced packaging technologies for flexible and wearable devices [1] around conventional rigid CMOS ICs. Either way, usually some rigid components remain in actual systems.

System Overview: The Audio Bag

The audio bag is a messenger bag augmented with fully-flexible components to enable radio reception of an AM signal and playback of its audio modulating signal. Fig. 1 shows the audio bag and its main modules. The flexible technologies that were combined include a-IGZO TFTs [2, 3], printed organic FETs (OFETs) [4], printed organic LEDs (OLEDs), textile antennas [2, 5], printed piezo-electric speakers [4], flexible rechargeable NiMH-batteries [6], organic photo-voltaic devices (OPVs) [6], button boards, inkjet-printed circuit boards (IPCBs), inkjet-printed passive components, and textile integration. The audio bag demonstrates wireless communication in a fully integrated system with flexible technologies, amongst which the a-IGZO TFTs provide the most prominent electronics functions.

Figs. 2 and 3 show the system architecture and features of the audio bag. The main modules combined in the textile frame are the user interface in the front of the flap, the electronics in the back of the flap, the solar power module [6] in the front of the bag, a piezo-electric printed speaker [4] on each side of the bag, and the textile antenna [2, 5] inside the bag. The control logic, receiver, and baseband amplifier are fabricated in a-IGZO TFT technology [2]. A variant of that technology [3] with a higher breakdown voltage is used to realize the amplifiers driving the speakers. The flexible packaging and integration of the TFT and OFET blocks are done on inkjet-printed circuit boards (IPCBs).

2-Bit Digitally-Controlled Tuner in a-IGZO TFT

The schematic and die photos of the tuner and related channel control logic are shown in Figs. 4 and 6. The tuner is a 2-bit switched capacitor (M1, M2, C1, and C2) that in conjunction with the textile loop antenna selects four different center frequencies. Compared to silicon devices, a-IGZO TFTs present higher leakage, slower switching speeds and larger on-resistance. This results in reactances and resonators of lower quality factors. For this reason, the tuner exploits the high quality factor of the textile loop antenna [5]. To select the center frequency f_0, a 2-bit control word {d1;d2} is provided by four buttons and the a-IGZO TFT channel control logic. The logic sub-blocks are realized in resistor-transistor logic using diode-connected TFTs as load. The supply voltage ranges from 4 V to 6 V. Fig. 5 shows measurements of the channel control logic.

Fig. 7 shows the real part of the measured impedance of the antenna and tuner as seen by the AM detector. The quality factor of the tuner depends on large voltage swings at nodes d1 and d2. In this regard, the output inverters are optimized.

Based on [7] we simulated the center frequencies f_0 for the four possible input words to be {230, 276, 375, 813} kHz. These predict well the measured f_0 = {226, 271, 340, 802} kHz. The measured real parts are roughly 2.5 times smaller than predicted, which is an acceptable accuracy at the f_0 of an LC-tank.

3.5mW 1MHz AM Detector in a-IGZO TFT

Figs. 9 and 12 show the schematic and the die photo of the proposed a-IGZO TFT AM detector. Design and biasing are optimized for the application defined frequency range of 100 kHz to 1 MHz (refer also to the f_0 tuning range above) and a low power consumption, taking the flexible battery and solar driven nature into account. The AM detection is achieved by charging and discharging capacitance C3 via the different impedances of transistors M3 and M4+M5. The frequency range of the detector allows its use with the tuner and large antenna (Fig. 10) in the audio bag and improves over a previously published flexible receiver [2], except in conversion gain. The power consumption (10-fold), area (4.7-fold), transistor count, and complexity are reduced (see Fig. 11). Thus, also an improvement in cost and yield is expected. The measured conversion gain vs. carrier frequency fc and baseband frequency fbb as well as the measured waveform are shown in Fig. 10.

System integration

Fig. 13 shows the integration of the a-IGZO TFT circuit blocks on an IPCB. The substrate material is a 120μm thick PET film. After printing, conductive polyurethane glue and an adhesive film are used to mount TFTs and interposers on the IPCB. Both are cured at room temperature. The cross-section and process flow for the IPCB are shown in Fig. 14. The IPCB process provides two metal layers (silver), printed capacitors (PVP), and resistors (carbon). We mount TFT circuits on an IPCB-interposer, which is then mounted on the main IPCB. The benefits of this modular prototyping approach are ease of testing and expandability.

Acknowledgement

This work was supported in part by the European Commission under project FLEXIBILITY under Grant 287568, in part by the German Research Foundation within the Cluster of Excellence "Center for Advancing Electronics Dresden-Organic Path", and in part by the DFG (Deutsche Forschungsgemeinschaft), within the project "Low-Voltage High-Frequency Vertical Organic Transistors" as well as WISDOM and the Coordination Funds of SPP 1796.

References

[1] H. Kim, et al., ISSCC, pp. 150–603, Feb. 2008.
[2] K. Ishida, et al., Symposium on VLSI Circuits, Jun. 2015.
[3] R. Shabanpour, et al., ISPACS, pp. 357–361, Nov. 2015.
[4] G. Schmidt, et al., J. Polym. Scie. Part B., pp. 1409–1415, 2015.
[5] T. Meister, et al., pp. 1–5, IMOC, Nov. 2015.
[6] T. Meister, et al., ECCTD, Aug. 2015.
[7] M. Shur, et al., J. Electrochem. Soc., 144(8):2833–2839, 1997.

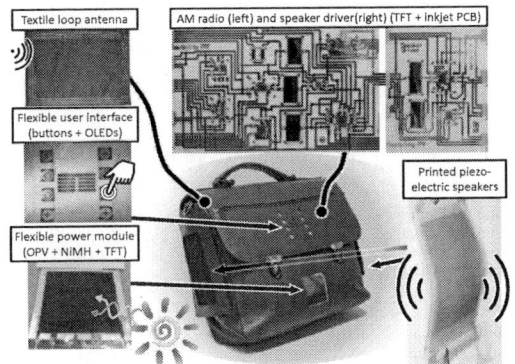

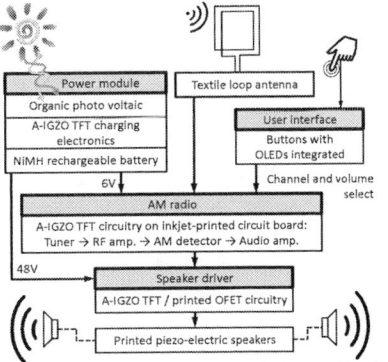

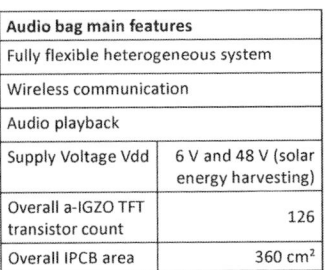

Audio bag main features	
Fully flexible heterogeneous system	
Wireless communication	
Audio playback	
Supply Voltage Vdd	6 V and 48 V (solar energy harvesting)
Overall a-IGZO TFT transistor count	126
Overall IPCB area	360 cm²

Fig. 1 Fully flexible modules of the heterogeneously integrated audio bag.

Fig. 2 System structure of the audio bag, where all components are flexible.

Fig. 3 Audio bag main features.

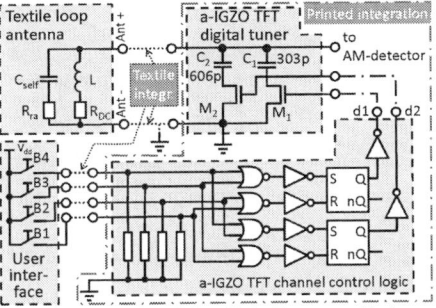

Fig. 4 Schematic of the digitally controlled tuner circuit and four-button channel control logic circuit.

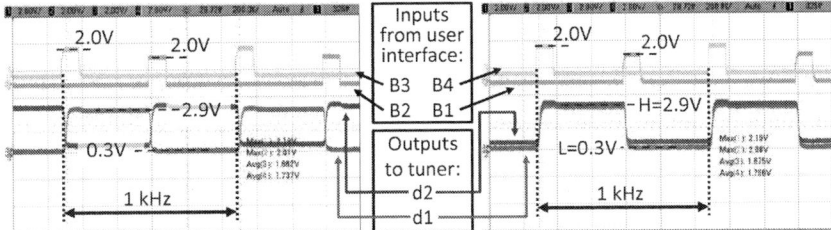

Fig. 5 Measurement results of the four-button channel control logic circuit.

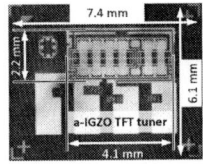

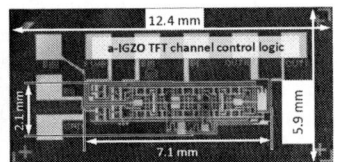

Fig. 6 Die photos of tuner and button control logic, showing core and cut frame dimensions.

{d1;d2}	{H;H}	{H;L}	{L;H}	{L;L}
f₀ [kHz]	226	271	340	802

Fig. 7 Measured characteristics of tuner with antenna.

a-IGZO TFT circuit	Tuner	Channel logic
Range / Channels	909 pF / 4 channels	2-bit / 4 channels
Supply voltage Vdd	—	4 .. 6 V
Power consumption	—	< 2.9 mW (@V_dd = 4V)
Transistor count	2	44
Core chip area	9.0 mm²	14.9 mm²
Total chip area	45.1 mm²	73.2 mm²
Bending radius	> 5 mm	> 5 mm

Fig. 8 Features of tuner and button control logic.

Fig. 9 Schematic of AM detector.

Bias network & Input coupling | RF amp. (Common source stage) | AM Detector w/ Source follower stage | Baseband amp. (Common source stage)

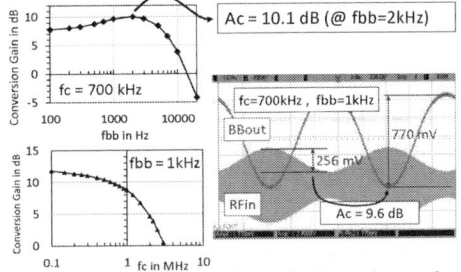

Fig. 10 Measured conversion gain Ac and waveform of AM detector (load condition: 1MΩ ∥ 15pF).

Comparison AM detector	This work	[2]	Factor of improvement
Technology	a-IGZO TFT	a-IGZO TFT	
Supply voltage Vdd	5 V	5 V	
Upper cutoff	ca. 1 MHz	> 20 MHz	Matched to audio bag antenna and tuner
Lower cutoff	< 100 kHz	ca. 1 MHz	
Conversion gain Ac	10.1 dB (3.2 x)	15.3 dB (5.8 x)	-5.2 dB (1/1.8)
Current at Vdd = 5 V	0.7 mA	7.2 mA	10.3
Transistors	7	24	3.4
Stages	2	5	2.5
Core chip area	5.7 mm²	26.85 mm²	4.7

Fig. 11 Comparison of AM detector to previous work.

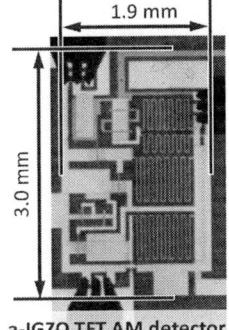

a-IGZO TFT AM detector

Fig. 12 Die photo of a-IGZO TFT AM detector (size with pads & cut-frame: 5.9 x 9.8 mm², not shown).

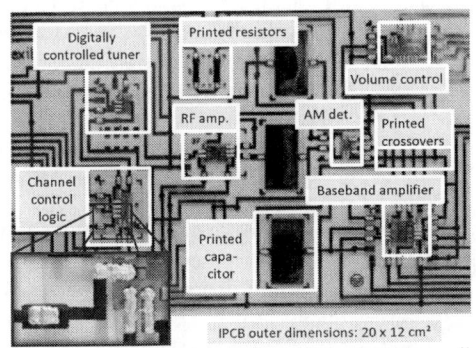

IPCB outer dimensions: 20 x 12 cm²

Fig. 13 Photograph of the fully flexible radio receiver, including digitally controlled tuner, channel control logic, AM detector, and baseband amplifier on an inkjet printed circuit board (IPCB).

Characteristic / Feature	Value
Resolution: min. lines and spaces	80 μm and 35 μm
Resistivity of interconnect (Cabot CCI-300 silver ink, cured at 150°C)	0.15 Ω/□
Resistivity of resistance layer (carbon)	25.8 kΩ/□
Relative permittivity of dielectric layer (cross-linked poly(4-vinylphenol), PVP)	4.5
Max. temperature during processing	150°C
Conductive glue	Polyurethane type

Substrates:
– Polyethylene terephthalate (PET), polyethylene naphthalate (PEN), polyimide, paper, etc.
– No preconditioning required

Fig. 14 Cross-section, process flow, and characteristics of inkjet printed circuit board (IPCB) integration technology.

A 16-channel Noise-Shaping Machine Learning Analog-Digital Interface

Fred N. Buhler[1], Adam E. Mendrela[1], Yong Lim[1,2], Jeffrey A. Fredenburg[3] and Michael P. Flynn[1]

[1]University of Michigan, Ann Arbor, MI, [2]Samsung Electronics, Yongin, Korea, [3]Movellus Circuits, Ann Arbor, MI

Abstract

A 16-channel machine learning digitizing interface embeds Inner-Product calculation within a Delta-Sigma Modulator (IPDSM) array canceling quantization noise and noise shaping the multiplicand. The prototype, with 16 independent IPDSM channels occupies a core area of 0.95mm^2 in 65 nm CMOS. Each channel performs up to 100M multiplications/s. The system is demonstrated with a standard machine learning scheme for image recognition. It achieves the same classification accuracy for the MNIST set of hand-written digits as with the same algorithm on floating point DSP.

Introduction

IoT devices are collecting and transmitting an ever-increasing amount of data to monitor health, the environment and manufacturing. For example, IoT devices can send vital data and real time monitoring from a patient to hospital staff but this requires large bandwidth and low power operation. Machine learning can overcome bandwidth and power limitations by decreasing the amount of transmitted data through feature extraction, or classification [1] at the sensor. A challenge is that these need energy intensive and accurate inner-product multiplication of the input signal with a pre-calculated basis vector. Traditionally, this involves the digitization of the sensor signal, communication of large amounts of raw ADC data, and then extensive DSP. A compelling approach is to embed machine learning functions within the digitization process. [2] uses a multiplying SAR ADC, however the multiplying SAR approach is limited to low accuracy of around 7 effective bits due to noise, nonlinearity and mismatch. Our new approach achieves classification accuracy equivalent to floating point DSP by embedding the Inner-Product calculation within a Delta-Sigma Modulator (IPDSM) array. The prototype achieves 1.6G 1 bit multiplications/s over 16 channels. Accurate digitization and an effective oversampled multiplicand resolution of 14 bits enables recognition accuracy similar to floating point DSP.

Machine Learning Noise-Shaping Architecture

Our new architecture embeds inner-product calculation, the key function in machine learning, within a $\Delta\Sigma$ modulator to achieve the effective accuracy of conventional DSP, at a fraction of the power, area and data bandwidth. This is done with little extra front-end power or area overhead. The inner-product, IP, is the vector multiplication of the signal sequence, X, with a basis vector φ:

$$P = \varphi X = \sum_{i=0}^{n-1} \varphi_i X$$

Conventionally, computing an inner-product on sensor data requires an ADC to digitize the sensor signal, followed by a digital multiply-and-accumulate (MAC) function. We combine accurate multiplication and digitization by introducing mixing within the modulator loop as shown the simplified first-order depiction in Fig. 1(a). This mixing is implemented either as a pass through or an inversion of the differential signal path – effectively multiplying by ±1. The

digitized inner-product is simply the accumulation of the sequence of digital outputs from the one-bit quantizer.

The new architecture has several unique and fundamental advantages over conventional digitization followed by DSP:

First, a unique advantage is that the IPDSM performs digitization and inner-product calculation with almost no quantization error. This is unlike a conventional $\Delta\Sigma$ modulator which simply shapes quantization noise for later DSP filtering. Simple time domain analysis (Fig. 1(a)) of a first order loop shows that almost all of the quantization error terms ($e_{(1)}$, $e_{(2)}$ etc.) cancel.

Second, the multiplicand in the IPDSM can achieve arbitrary resolution through a sequence of ±1 values and is noise-shaped for very high precision. The quantization noise of the multiplicand falls outside the signal transfer function (STF) of the main modulator and is rejected. The multiplicand is noise shaped to achieve an effective resolution of 14 bits. For comparison, [2] is limited to a 4 bit multiplicand.

Third, instead of the complicated filters and decimators that are required after a conventional $\Delta\Sigma$ modulator, the digital output of the modulator is processed by a simple accumulator. The IPDSM directly generates the 16 bit inner-product reducing the data output rate by four orders of magnitude over that of a traditional $\Delta\Sigma$ modulator.

Finally, this approach of combining digitization and inner-product calculation has several practical advantages. Implementation of ±1 mixing is inherently linear. Also, compared to a conventional $\Delta\Sigma$ modulator, the only extra cost is the control of the mixing sequence.

Implementation

In the prototype, each modulator is implemented as a second order structure with a 1 bit quantizer (Fig. 1(b). A second-order modulator with ring amplifiers [3] achieves the highest resolution and high bandwidth. A single-bit quantizer resolution ensures high accuracy of the feedback DAC without dynamic element matching (DEM). The amplifiers (Fig. 2) are three-stage ring-amplifiers for efficiency and wide output swing. Furthermore, the autozero of the ring amplifier suppresses 1/f noise which would otherwise contaminate the inner-product calculation. The mixing values for each modulator are stored in a dedicated 8K SRAM. The switch network, shown in Fig 3, implements mixing. Accumulation of the $\Delta\Sigma$ outputs, buffering, and time-interleaving of the output as well as synchronization are handled by a digital controller.

Measurements

The prototype (Fig. 4), with 16 independent IPDSMs is implemented in 65nm CMOS and occupies 0.95 mm^2. Each IPDSM performs up to 100M multiplications per second. The system was evaluated with a standard machine learning algorithm that recognizes handwritten digits from the MNIST database [4]. This algorithm requires extensive and accurate inner-product calculation.

The machine learning process is summarized in Fig 5. In this experiment, 3000 thresholded training images are used to generate a Principal Component Analysis (PCA) basis. The images expressed in the PCA basis are then used to train an Error Correction Output Code (ECOC) Machine Learner with one vs all coding. The IPDSM φ matrix is the product of the machine learner support vectors and the PCA basis.

As shown in Fig 6, IPDSM achieves the same 88% accuracy as floating point DSP on MNIST digit recognition. The plot compares the accuracy of IPDSM, with both fixed point DSP and floating point DSP. With a signal bandwidth of 800k multiplications/sec, the accuracy of IPDSM is equal to that of floating point DSP or 15 bit fixed point.

Fig 7 summarizes the measured performance and compares with the state of the art.

Acknowledgments

This work was supported in part by the National Science Foundation and Intel ISRA grant.

References

[1] T. Hastie, et al., The elements of statistical learning
[2] J. Zhang, et al., *ISSCC*, 2015
[3] Y. Lim, et al., *ISSCC*, 2015
[4] LeCun, et al. "The MNIST database of handwritten digits." 1998
[5] D. Gangopadhyay, et al., *JSSC*, F

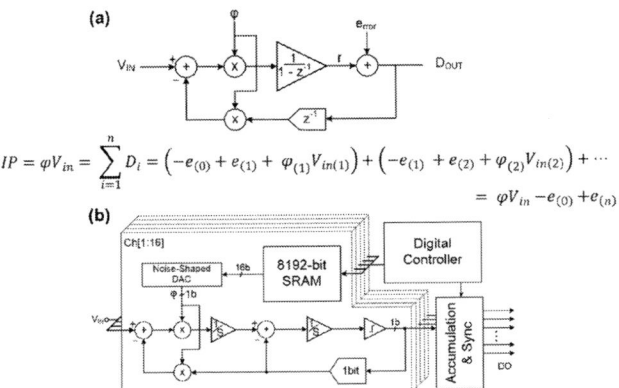

$$IP = \varphi V_{in} = \sum_{i=1}^{n} D_i = \left(-e_{(0)} + e_{(1)} + \varphi_{(1)} V_{in(1)}\right) + \left(-e_{(1)} + e_{(2)} + \varphi_{(2)} V_{in(2)}\right) + \cdots$$
$$= \varphi V_{in} - e_{(0)} + e_{(n)}$$

Fig. 1(a): A simplified first order modulator. 1(b) A block diagram of the IPDSM system.

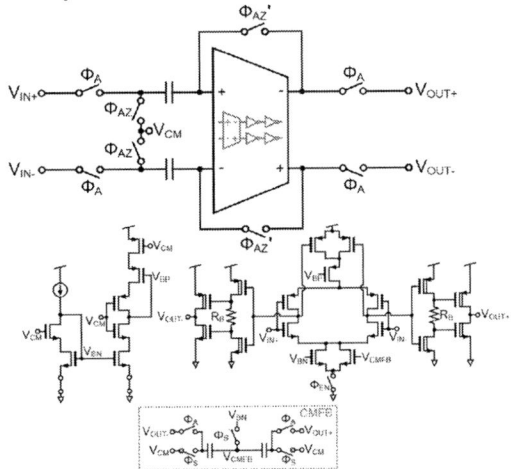

Fig. 2: Ring amplifier schematic with auto-zeroing switch configuration.

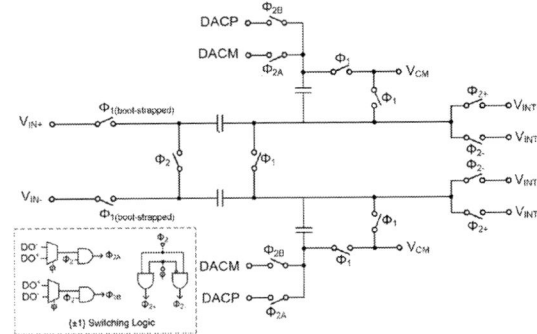

Fig 3: The switch network implementing ±1 multiplication.

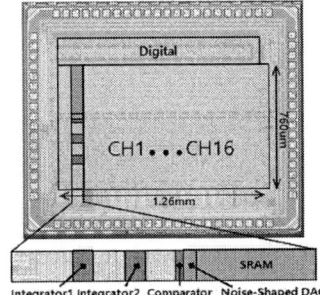

Fig. 4: Die microphotograph.

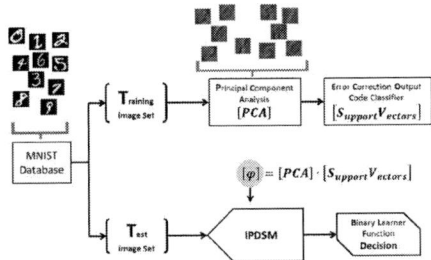

Fig. 5: Machine learning setup. IPDSM calculates the inner product and sends it to the BLF.

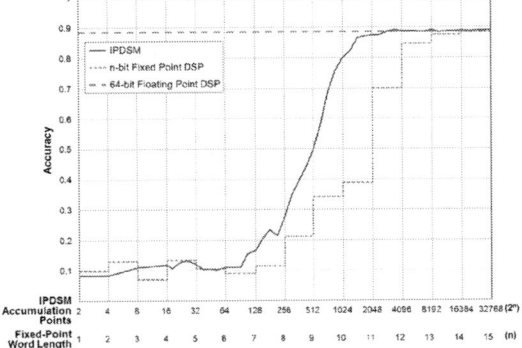

Fig. 6: Machine learning results of IPMDSM versus fixed and floating point DSP of MNIST digit classification.

Metric	[2]	[5]	This System
Technology	0.13 μm	0.13 μm	65 nm
Num. of Channels	1	64	16
Area (per channel)	0.106 mm²	0.0938 mm²	0.0594 mm²
Signal BW	20 kHz	1 kHz	100 kHz
SNDR	45.8 dB	40.6 dB	69.1 dB
ADC ENOB	7.31	6.5b	11.2b
Sampling Freq.	20 kHz	2 kHz	100 MHz
OSR	-	-	128
FOM_S	151.4	146.6	161.6
Power/ch. (analog)	286.8 nW	28 nW	241 uW
Power/ch. (digital)	376.8 nW		2.7 uW
Product Dynamic Range	40b	8b	16b
Multiplicand Resolution	4b	6b	14b
Application	Feature Extraction	Compressive Sensing	Feature Extraction

Fig 7: Performance summary and comparison with other published works.

A Field-Programmable Mixed-Signal IC with Time-Domain Configurable Analog Blocks

Yunju Choi, Yoontaek Lee, Seung-Heon Baek, Sung-Joon Lee, and Jaeha Kim

Department of Electrical and Computer Engineering, Seoul National University, Korea

{yunju, yoontaek, shbaek, sjlee, jaeha}@mics.snu.ac.kr

Abstract

A field-programmable mixed-signal IC for fast-prototyping and low-cost production of mixed-signal system is presented. The IC contains time-domain configurable analog blocks (TCABs) that can be programmed into a time-to-digital converter (TDC), digitally controlled oscillator (DCO), digitally controlled delay element, digital pulse-width modulator (DPWM), or phase interpolator (PI). The prototype IC fabricated in 65-nm CMOS demonstrates its versatile programmability with the successful operations as a 1-GHz PLL with 12.3-ps_{rms} integrated jitter, 50-MS/s ADC with 32.5-dB SNDR, and 1.2-to-0.7V DC-DC converter with 95.5% efficiency.

Introduction

The advent of IoT applications and their fast-changing market trends call for a variety of new analog/mixed-signal ICs and the ability to develop them with the least time and effort. While FPGAs are established solutions for fast-prototyping and low-cost production of digital IC products, the corresponding field-programmable devices for analog/mixed-signal ICs are still under development [1]-[3]. Particularly, a proper design of the configurable analog block (CAB), which is an analog-equivalent to the configurable logic block (CLB) in FPGAs, both realizing various analog functions and enabling versatile interconnection remains unsettled.

This paper presents a field-programmable mixed-signal IC that implements the CAB in time-domain. It is demonstrated that this time-domain CAB (TCAB) can emulate the functionalities of various mixed-signal circuits used in timing generation, data conversion, and power management ICs. Since the input and output signals of the TCABs are binary pulses with modulated pulse width, delay, or frequency, the signals can be easily routed to other TCABs or CLBs via digital programmable interconnects.

Architecture and Implementation

Fig. 1 illustrates the overall architecture of the proposed IC and the floorplan of the fabricated prototype IC. The IC contains a 2x8 array of TCABs along with a 5x4 array of arithmetic logic units (ALUs) and a 10x10 array of look-up table (LUT) based CLBs. The architecture can be programmed into various mixed-signal feedback systems such as PLLs, ADCs, and DC-DC converters by mapping their analog sensors (e.g. TDC) and producers (e.g. DCO) to the TCABs and digital controllers to the ALUs or CLBs. The IC also includes some gluing blocks that interface between the configurable units or with the external signals, such as the counters that convert the TCAB pulse frequency to a digital code, a global clock unit (GCLK) that distributes clock with optional frequency division, and voltage-to-time converter (VTC) and phase-frequency detector (PFD) that convert the external voltage and clock phase to pulse-width-modulated signals, respectively.

A noteworthy difference of the IC compared with the previous works in [1]-[3] is that the TCABs perform the analog functions in time-domain, receiving and producing digital pulses of which frequency, delay, or pulse width encodes the analog information. This time-domain, digital representation of analog information makes the TCAB array scalable without requiring power-hungry, noise-sensitive analog buffers between the TCABs. Instead, the switch/connection blocks connecting the TCABs in Fig. 1 can be implemented with the simple digital transmission gates and buffers. Moreover, in contrast to other works [2],[3] that provide a heterogeneous array of passive elements, transistors, and basic building blocks, each TCAB is a truly reconfigurable block that

can realize multiple time-domain analog functionalities via programming.

Fig. 2(a) shows the circuit implementation of the TCAB and its operation. Two pairs of digitally adjustable and complementary current sources of I_{1A}/I_{1B} and I_{2A}/I_{2B} selectively charge a pair of digitally adjustable capacitors C_1 and C_2 depending on the 8 input pulses, IN[7:0]. Whenever the voltage on the capacitor C_1 or C_2 exceeds a certain threshold, a pair of crossing detectors and an SR-latch assert the corresponding output pulse, OUT or OUTB, and reset the other. Only one of the two capacitors gets charged at a time in an alternating fashion. The 7-bit input digital codes $W_1[6:0]$ and $W_2[6:0]$ adjust the current levels while $C_{sw1}[6:0]$ and $C_{sw2}[6:0]$ control the capacitances, modulating the effective charging rates of the capacitor voltages. Interestingly, the basic TCAB operation is very similar to an integrate-and-fire neuron.

Fig. 2(b)-(e) depict how a single TCAB can perform various analog functionalities in time-domain when configured differently. First, a TCAB operates as a DCO if it is configured so that I_{1A} and I_{2A} charge C_1 and C_2, respectively (Fig. 2(b)). The oscillation frequency can be tuned by adjusting either the current levels or capacitances. This TCAB can also operate as a DPWM if C_1 and C_2 are adjusted in a complementary fashion to vary the pulse width while keeping the frequency fixed. Second, the TCAB can act as a gated oscillator when the input signals are connected to the MOS switches so that the TCAB oscillates only when the input is high (Fig. 2(c)). This TCAB can realize a TDC when it is followed by a counter that converts the pulse frequency to a digital code [4]. Third, the TCAB can serve as a digitally controlled delay element if a pair of complementary input signals steers I_{2A} between C_1 and C_2 (Fig. 2(d)). The outputs toggle only once whenever the input toggles after a delay set by I_{2A} and the capacitances of C_1 and C_2. Fourth, the TCAB operates as a phase interpolator when two pairs of complementary input signals steer I_{1A} and I_{1B}, respectively (Fig. 2(e)). It generates a clock with a phase which is a weighted sum between the two phases of input clocks [5].

Note that these TCABs mainly perform the conversion between the digital codes and timing properties of clocks or pulses, rather than amplification or filtering. In fact, it is what the analog circuit blocks actually do in many mixed-signal feedback systems such as digital PLLs, time-domain ADCs, and DC-DC converters, where the amplification and filtering are done mainly by the digital

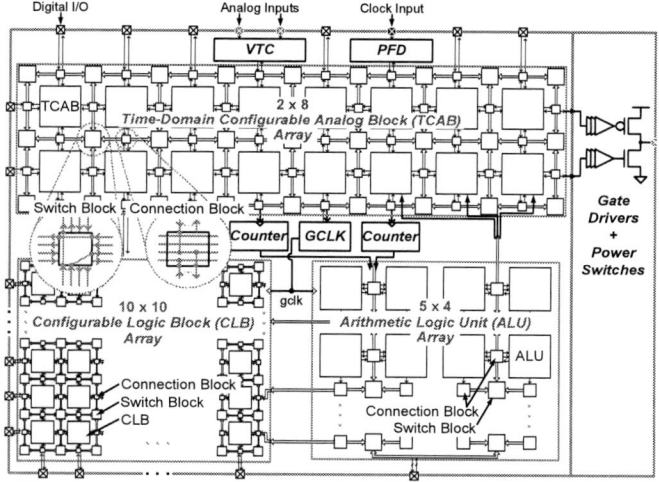

Fig. 1. Architecture and the floorplan of the proposed IC.

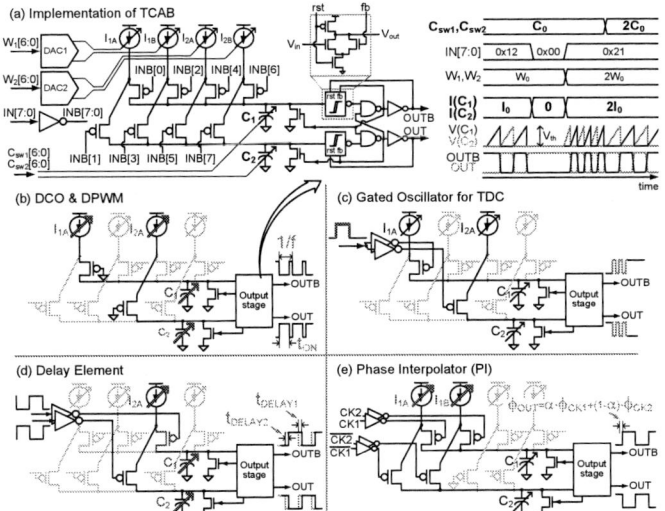

(a) Implementation of TCAB

(b) DCO & DPWM

(c) Gated Oscillator for TDC

(d) Delay Element

(e) Phase Interpolator (PI)

Fig. 2. Implementation and multiple functionalities of the TCAB.

controllers. The presented IC includes the arrays of 8-bit ALUs and 4-input CLBs, which can realize various digital filters and finite-state machines via field-programming.

Experimental Results

The prototype IC is fabricated in a 65-nm LP CMOS and operates with a 1.2-V supply. Its chip photograph and characteristics summary are listed in Fig. 3. Its versatile programmability is demonstrated by its operations and measured performances as a PLL, ADC, and DC-DC converter.

Fig. 4 illustrates the configuration and measured results of the proposed IC operating as a 1-GHz-output digital PLL consisting of a TDC, digital controller, DCO, and clock divider/buffers. As described earlier, the TDC is mapped to a PFD unit followed by two TCABs and two counters. The digital controller is realized by the ALU array. The DCO and clock dividers/buffers are mapped to a TCAB and GCLK units, respectively. An additional lock detector is realized by CLBs. The configured PLL successfully synthesizes a 1-GHz clock from a reference clock of 125MHz, dissipating 33.6mW. The integrated jitter is 12.3ps$_{rms}$ and phase noise is -102.91dBc/Hz at 10-MHz offset.

Fig. 5(a) describes the configuration and characteristics of the IC operating as a 7-bit, 50-MS/s time-domain ADC. First, a VTC unit converts the input differential voltage to a set of two pulse widths, which are then digitized by a pair of gated oscillators each mapped to a TCAB and counter unit. Since this open-loop architecture has poor linearity and dynamic range, it is augmented with a digital feedback loop realized by the ALUs that monitors the difference between the two gated oscillator outputs and adjusts their capacitances until the difference becomes zero. The programmed ADC achieves an SNDR of 32.5dB with a 100-kHz input, while dissipating 10.8mW. The measured INL and DNL are +0.84/-2.04 and +1.03/-0.59 LSB, respectively.

Fig. 5(b) shows the configuration and characteristics of the proposed IC as a DC-DC converter. It consists of the above-mentioned ADC, digital PID compensator mapped to ALUs, DPWM and delay elements mapped to TCABs for non-overlapping pulse generation, and buck converter stage with dedicated on-chip gate drivers, power switches, off-chip 3.3-μH inductor, and 0.22-μF capacitor. It achieves a peak efficiency of 95.5% with switching frequency of 2.2MHz, while converting a 1.2-V input to a 0.7-V output and supplying a current of 150mA. The reference tracking/load regulation waveforms and conversion efficiency under different load conditions are also shown in Fig. 5(b).

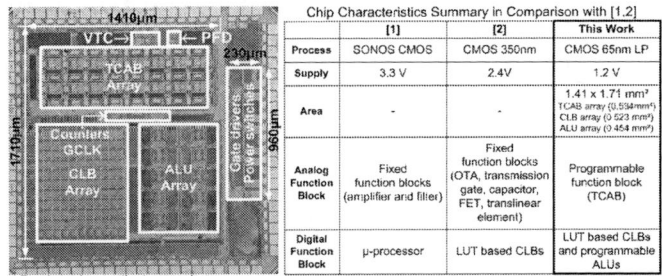

Fig. 3. Chip micrograph and characteristics summary.

	[1]	[2]	This Work
Process	SONOS CMOS	CMOS 350nm	CMOS 65nm LP
Supply	3.3 V	2.4V	1.2 V
Area	-	-	1.41 x 1.71 mm² TCAB array (0.534mm²) CLB array (0.523 mm²) ALU array (0.454 mm²)
Analog Function Block	Fixed function blocks (amplifier and filter)	Fixed function blocks (OTA, transmission gate, capacitor, FET, translinear element)	Programmable function block (TCAB)
Digital Function Block	μ-processor	LUT based CLBs	LUT based CLBs and programmable ALUs

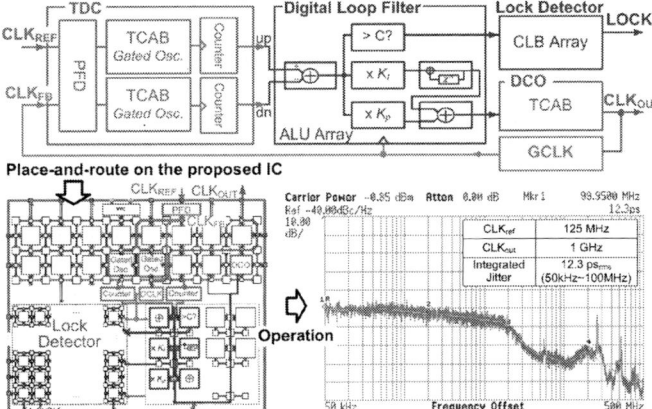

Fig. 4. Block diagram of the 1-GHz PLL programmed on the proposed IC and measured phase noise.

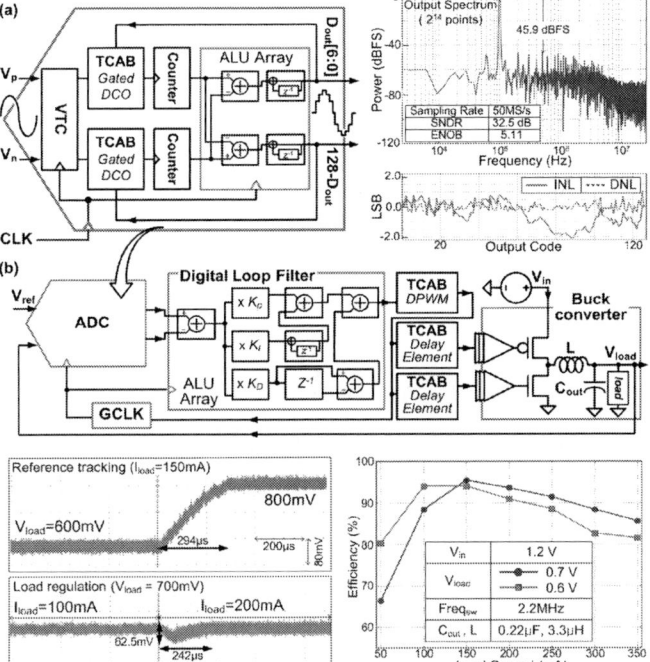

Fig. 5. Block diagram and measured results of the (a) ADC and (b) DC-DC converter programmed on the proposed IC.

Acknowledgements

This work was supported by Samsung Research Funding Center of Samsung Electronics under project number SRFC-IT1301-08.

References

[1] M. Mar, *et al.*, *J. Solid-State Circuits*, Mar. 2003.
[2] R. B. Wunderlich, *et al.*, *Trans. VLSI Systems*, Aug. 2013.
[3] C. Schlottmann, *et al.*, in *Proc. Custom ICs Conf.*, Sep. 2012.
[4] B. M. Helal, *et al.*, in *Proc. Symp. VLSI Circuits*, Jun. 2007.
[5] S. Ryu, *et al.*, *J. Solid-State Circuits*, Aug. 2014.

A 5.8 pJ/Op 115 Billion Ops/sec, to 1.78 Trillion Ops/sec 32nm 1000-Processor Array

Brent Bohnenstiehl, Aaron Stillmaker, Jon Pimentel, Timothy Andreas, Bin Liu, Anh Tran,
Emmanuel Adeagbo, Bevan Baas

University of California, Davis

Abstract

1000 programmable processors and 12 independent memory modules capable of simultaneously servicing both data and instruction requests are integrated onto a 32nm PD-SOI CMOS device. At 1.1 V, processors operate up to an average of 1.78 GHz yielding a maximum total chip computation rate of 1.78 trillion instructions/sec. At 0.84 V, 1000 cores execute 1 trillion instructions/sec while dissipating 13.1 W.

Introduction

Modern semiconductor fabrication technologies now enable the construction of integrated circuits containing over 1000 processors on a single chip [1]. However, for such systems to effectively compute workloads, new architectures are needed for the processors, the inter-processor interconnect, circuits that interact with larger memories, and the applications they execute [2–4]. KiloCore's 1000 MIMD processors are arrayed in 32 columns and 31 rows with 8 processors and 768 KB inside 12 independent memories in a 32nd row (Fig. 1). While caches are generally extremely effective, they are problematic for resources such as chip I/O and cache-coherency circuits when the number of processors is scaled into the 100s or 1000s and they also dissipate significant power [5]. In contrast, KiloCore processors do not contain explicit caches and instead store data and instructions inside i) local memory, ii) an arbitrary number of nearby processors, iii) on-chip independent memory modules, or iv) off-chip memory.

KiloCore Architecture

Each processor issues one in-order instruction per cycle into its 7-stage pipeline from either its 128 x 40-bit local instruction memory or an independent memory module. None of the 72 supported instruction types are algorithm-specific. Processor data memories are implemented as two 128 x 16-bit banks to sustain a throughput of one instruction per cycle for common instructions that require two source operands (Fig. 2). However, profiled code of five disparate applications (AES encryption, low-density parity-check (LDPC) decoder, 100-byte database record sorting, 802.11a/g receiver, and software single-precision floating-point arithmetic implementations) showed that only 0.34% of operands could not be mapped to an address in only one bank and thus needed to be written to both banks redundantly to avoid conflicts during subsequent reads. Our scheme permits conflict-free addressing with optimal memory space maximization.

Communication on-chip is accomplished by a high-throughput circuit-switched network and a complementary very-small-area packet-switched network (Fig. 3). The source-synchronous circuit-switched network supports communication between adjacent and distant processors, as resources allow, with each link supporting a maximum rate of 28.5 Gbps. Execution of an instruction with an input operand transferred by the circuit-switched network from an adjacent processor dissipates 16% less energy compared to if it were transferred from the local data memory. If an input operand comes from a processor ten processors away, only an additional 22% energy is required compared to a local access. Routers utilize wormhole routing, operate autonomously, and

contain a 4 x 18-bit buffer on each of the five input ports—one for each cardinal direction and one for the host processor [6]. Maximum throughput is 45.5 Gbps per router and 9.1 Gbps per port at 1.1 V. At 0.9 V, maximum throughput is 27.1 Gbps at 3.36 mW and at 0.67 V, it is 8.1 Gbps at 429 µW. Both network types contribute to an array bisection bandwidth of 4.2 Tbps.

Each of the 12 independent memory modules contains a 64KB SRAM, services two neighboring processors, and supports 28.4 Gbps of I/O bandwidth. When streaming instructions to a processor, a dedicated control module takes over program control and branch prediction control from the processor to more efficiently execute across branches (Fig. 4).

KiloCore's 1000 processors, 1000 packet routers, and 12 independent memories are clocked by local and completely-unconstrained (below the maximum operating frequency) clock oscillators that do not use PLLs and may change frequency, halt within 1-5 clock periods, and restart in less than one clock period to reduce power dissipation. Processors, routers, and memory modules with no work to do dissipate exactly zero active power (leakage only). At 0.9 V, idle processors leak 1.1% of their typical operating power. Information reliably crosses unrelated clock domains through dual-clock FIFO buffers [7].

Programming is accomplished by a multi-step process including a mapping step that assigns programs to processors (Fig. 5). New mappings may be computed during runtime for purposes such as: simultaneous execution of unrelated workloads, optimizing mappings with consideration of PVT variations, avoiding faulty or partially-functional processors, or for self-healing for failures due to wear-out effects.

Chip Design and Measured Results

The chip was fabricated in a 32nm PD-SOI technology. The entire array is 7.94 mm by 7.82 mm and contains 621 million transistors. Except for the 64 KB SRAMs inside the independent memory modules, all memories are built from clock-gated flip-flops with synthesized interfacing logic which greatly simplifies the physical design and likely lowers the minimum operating voltage. Each processor contains 575,000 transistors and occupies 239 µm by 232 µm; therefore 18 processors occupy 1 mm^2.

Processor maximum operating frequencies range from 1.70 GHz to 1.87 GHz with an average of 1.78 GHz at 1.10 V. The KiloCore chip is flip-chip mounted by 564 C4 solder bumps inside a stock BGA package that delivers full power to only the approximately 160 central processors; therefore, a maximum execution rate of 1.78 trillion MIMD operations per second per chip (Fig. 6) is achievable only with a custom-designed package. At a supply voltage of 0.56 V, processors dissipate 5.8 pJ per operation at 115 MHz, which enables a chip to process 115 billion operations per second while dissipating only 1.3 W (Table 1). Processors achieve their optimal energy times time of 11.1 (pJ x ns)/op at a voltage of 0.9 V. Independent memories operate from 1.77 GHz at 1.1 V down to 675 MHz at 760 mV. Routers operate from 1.49 GHz at 1.1 V down to 262 MHz at 665 mV.

Acknowledgments

This work was supported by DoD and ARL/ARO Grant W911NF-13-1-0090; NSF Grants 0903549, 1018972, 1321163, and CAREER Award 0546907; SRC GRC Grants 1971 and 2321, and CSR Grant 1659; and C2S2 Grant 2047.

References

[1] S. Borkar, *DAC,* 2007, pp. 746–749.
[2] S. Vangal et al., *ISSCC,* 2007, pp.98–589.
[3] D. Truong et al., *JSSC,* vol. 44, no. 4, pp. 1130–1144, 2009.
[4] M. Butts, *IEEE Micro,* vol.27, no.5, pp. 32–40, 2007.
[5] M. Horowitz, *ISSCC,* 2014, pp. 10–14.
[6] A. Tran, B. Baas, *TVLSI,* vol. 22, no. 6, pp. 1391–1403, 2013.
[7] R. Apperson, et al., *TVLSI,* vol. 15, no. 10, pp. 1125–1134, 2007.

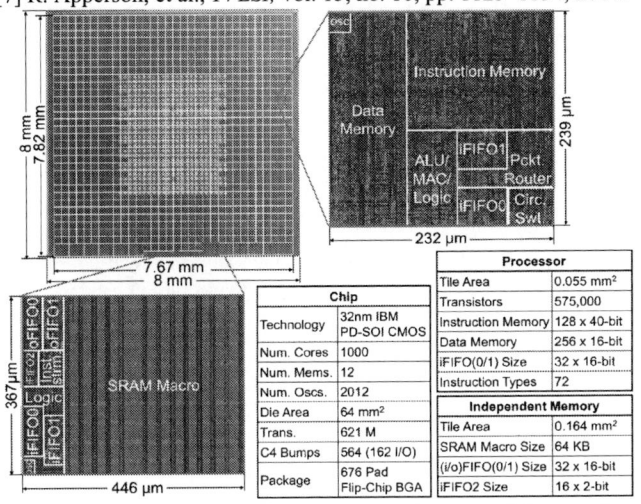

Fig. 1: Die photo of the KiloCore array, and annotated layout plots of a single processor tile and a single independent memory tile.

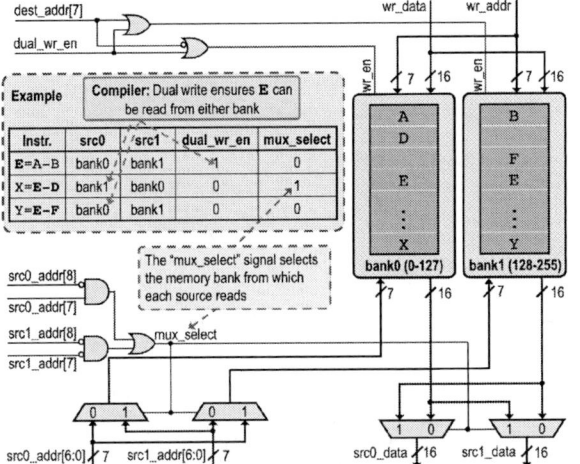

Fig. 2: Circuit diagram demonstrating dual memory bank reads and write backs.

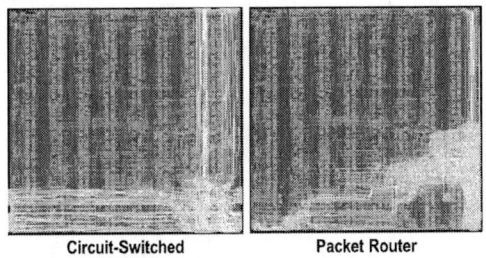

Fig. 3: Processor layout plots with inter-processor communication network circuits highlighted.

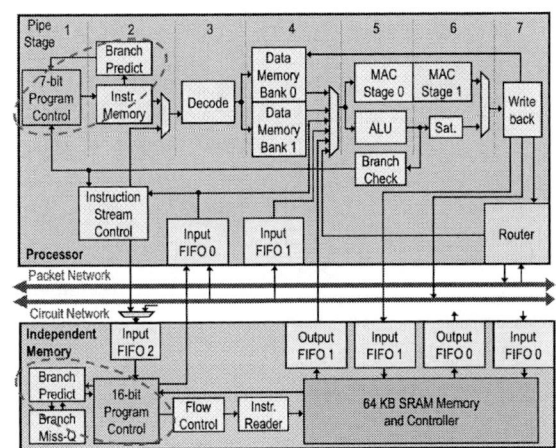

Fig. 4: Pipeline diagram of a KiloCore processor and an adjacent independent memory with program control highlighted in each block.

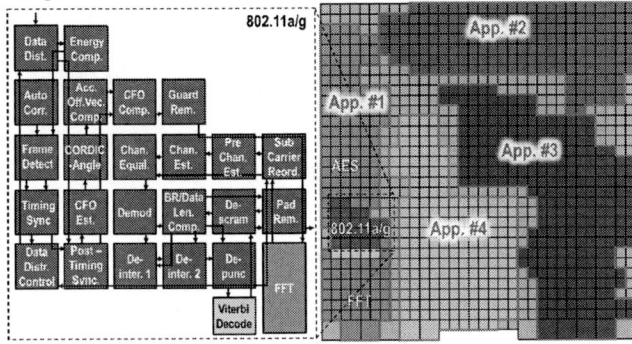

Fig. 5: Example multi-application mapping.

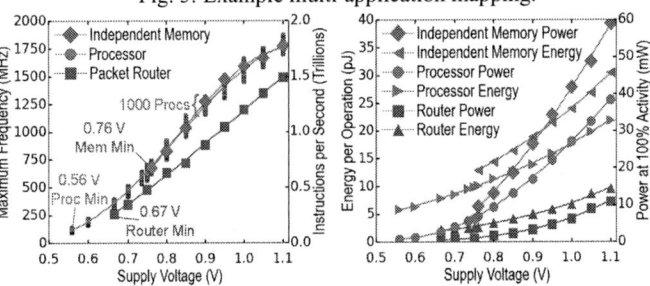

Fig. 6: Measured data at various supply voltages.

Table 1: Measured data and comparisons.

	Proc. Count	Clock Freq (MHz)	Supply Voltage (V)	Per Proc. Power (mW)	Energy /Op (pJ)	ExT (pJ x ns)	Max. Bisect. BW (Tb/s)
Sleepwalker Bol 65nm [JSSC'13]	1	25 / 23.6	0.4 / 0.375	0.065* / 0.052*	2.6 / 2.2	104 / 93.2	N/A
TeraFlops 65nm [2]	80	4000 / 3130	1.2 / 1.0	2260** / 1230**	70.6† / 49.1†	17.7 / 15.7	2.65
AsAP2 65nm [3]	167	1070 / 66	1.2 / 0.675	47.5 / 0.61	44 / 9.2	41.1 / 139	0.998
Am2045 130nm [4]	336	300	–	23.8*	79.4	265	0.713
KiloCore 32nm This work	1000	1782 / 1237 / 638 / 115	1.1 / 0.9 / 0.75 / 0.56	39.6 / 17.7 / 6.9 / 1.3	21.9 / 13.8 / 9.9 / 5.8	12.2 / 11.1 / 15.4 / 50.3	4.24

* Per processor power calculated by: energy per operation x clock frequency
** Per processor power calculated by: total power / number of cores
† Energy/Op calculated by: power / clock frequency / IPC (conservatively assumes processor is fully active every clock cycle)

28nm FDSOI Technology Sub-0.6V SRAM Vmin Assessment for Ultra Low Voltage Applications

R.Ranica, N.Planes, V.Huard, O. Weber*, D.Noblet, D.Croain, F.Giner, S.Naudet, P.Mergault, S.Ibars, A.Villaret, M.Parra, S.Haendler, M.Quoirin, F.Cacho, C.Julien, F.Terrier, L.Ciampolini, D.Turgis, C.Lecocq and F. Arnaud

STMicroelectronics, *CEA-LETI, 850 rue Jean Monnet, 38926 Crolles, France
Email : rossella.ranica@st.com ; Phone : +33 438 92 32 34

Abstract

Vmin measurements in 28nm FDSOI technology on 128Mb SRAM bitcells from -40°C to 125°C are reported in this paper. Adding the silicon ageing behavior and the process variability, we have developed a complete model and demonstrated end-of-life SRAM Vmin of 0.6V and 0.5V on 20Mb with $0.120\mu m^2$ and $0.152\mu m^2$ bitcells, respectively. This is the first report of a such extensive SRAM Vmin assessment at the 28nm node. The construction of write limited bitcells, combined with write assist design technique, was found to be the most efficient way to achieve ultra low Vmin in 28nm FDSOI technology. In addition, Vmin retention below 0.4V is demonstrated in $0.120\mu m^2$ bitcells, leading to the enablement of ultra-low leakage bitcells with 2pA/cell in retention mode.

Introduction

The market demand for ultra-low power and low cost technologies is relentless for System-on-Chip (SoC) and Internet of Things (IoT) products. The FDSOI technology at the 28nm node [1] has many assets to fully satisfy this demand. FDSOI transistors have an improved electrostatic behavior [2] as well as better Vth mismatch properties compared to planar bulk transistors [3], two pre-requisites for reducing the supply voltage (Vdd) and thus the power. Indeed, the Vth mismatch is responsible for SRAM fails that causes in turn the Vdd lowering limitation in SoC products. The benefit of the FDSOI technology for the SRAM Vmin associated with the improved Vth mismatch has been already demonstrated at the 28nm node [4].

In this paper, the full assessment of the SRAM Vmin for the 28nm FDSOI technology is presented for the first time. The path allowing the enablement of sub-0.6V end-of-life (EOL) Vmin for this technology is extensively described through measurements on several memory capacities up to 128Mb, on large temperature range, after ageing, and with various process centering.

Vmin Silicon measurements on 128Mb 6T-SRAM

A 128Mb 6T-SRAM chip has been implemented in 28nm FDSOI technology and this work is mostly focused on the single port high-density (SPHD) $0.120\mu m^2$ bitcell. SRAM device morphology is presented in Fig.1. The chip is made of 64 TopMem blocks containing each 2Mb of SPHD memory (Fig.2). A TopMem block includes 4 cuts of 0.5Mb SRAM and the control logic to drive and test the compilers. Chip architecture and its features are described in Fig2. The SEM topviews of a single TopMem block and of the memory array are shown in Fig.3.

Vmin has been measured on 128Mb fresh silicon for various temperatures in the range of -40/125°C on 100 dies (Fig.4). Values at 95% reported in the table show 120mV Vmin difference depending on temperature. Vmin is greatly improved with increasing temperature and Vmin down to 0.56V is measured at 125°C. This behavior is typical of write limited bitcells as we will explain in this paper.

Key parameters impacting SRAM Vmin

SRAM Vmin is function of large amount of parameters reported in the table of Fig.5: process centering, ageing, product usage (mission profile), operating conditions (temperature range) and memory cut size. A typical Vmin curve vs temperature on fresh silicon is presented in Fig.5, showing a V-shape behavior. Indeed, at low temperature, the higher Vth induces write margin (WM) decrease and this write limitation imposes a Vmin increase. At high temperature, the static noise margin (SNM) falls down and this read limitation degrades the Vmin too.

Using the simulation methodology described in [5], an accurate Vmin model has been developed and matches the fresh silicon behavior in all operating conditions, for all percentiles and for various SRAM capacity (Fig.6&7). Adding finally the contribution of ageing due to BTI stress on Vmin, the correlation between silicon and model is perfectly consistent, as demonstrated in Fig.8.

Vmin optimization in 28FDSOI technology

The impact of NMOS/PMOS process centering on Vmin is presented in Fig.9-10. SNM is efficiently increased by increasing Vth NMOS and decreasing Vth PMOS, thus changing the NP ratio (Fig.9). Vmin vs temperature curves in Fig.10 for those 3 NP process centering show the Vmin behavior from read limited (NP ratio 1, low SNM) to write limited (NP ratio 3, high SNM) bitcell centering. For write limited bitcells, Vmin clearly changes its shape: it is improved all along with temperature increasing and the strong Vmin degradation at high temperature, typical of read limited bitcells, is removed. 128Mb raw data of Fig.4 follows this write limited Vmin behavior (Fig.10b) and this is the process centering that we have chosen for minimizing the end-of-life Vmin.

Indeed, the main weakness of supporting a read limited bitcell is the ageing impact on Vmin [6]. During ageing, the |Vth| of Pull-Up transistors increases due to NBTI stress and cell stability is further reduced: higher Vmin is measured after stress at high temperature (Fig.11). And there is no way to determine at t=0 the bits that will fail after stress since Vmin drift after ageing show no correlation with fresh Vmin (Fig.11) [7].

Finally, we are able in Fig.12 to commit on Vmin values including process corner variability and ageing based on silicon results in Fig.10 and 11, respectively. Operating conditions are chosen according to targeted applications: -40/125°C for industrial, -20/105°C for networking and 0/85°C for consumer. These curves provide serious guidelines to guarantee EOL SRAM Vmin for a product considering its usage conditions and the memory capacity inside.

SRAM Vmin scaling on write limited bitcells

Another advantage of using write limited bitcells is that write assist (WA) design technique can be efficiently implemented to further lower the SRAM Vmin [8]. 20Mb aged Vmin is simulated in 0/85°C range as a function of enlarged bitcell area, with and without write assist (Fig.13). By adding WA, all bit failures due to write operation are removed. Bitcells become purely read limited and they benefit from their strong read stability, thus showing EOL Vmin of 0.6V for $0.120\mu m^2$ and 0.5V for $0.152\mu m^2$ bitcells respectively. Fig.14 explains the Vmin behavior vs bitcell area with and without write assist. In write limited bitcells, the WM decreases so abruptly at low Vdd that there is no benefit of lower σWM variability on larger bitcells. On the other side, after WA bitcells are read limited and their stability, illustrated with the SNM/σSNM metric in Fig.14, is clearly improved at low voltage with higher SRAM area.

The SRAM stability in retention mode, with bitline and wordline at 0V, is finally illustrated in Fig.15 with Vmin retention (Vmin ret) measurements. Vmin ret below 0.4V is demonstrated on $0.120\mu m^2$ bitcells allowing the enablement of ultra-low leakage bitcells with 2pA/cell in retention mode at room temperature.

Conclusion

We present for the first time an extensive study of key parameters impacting SRAM Vmin and provide guidelines for its efficient lowering in 28nm FDSOI technology. Silicon results on large memory cut sizes up to 128Mb are shown on extended operating range. Write limited bitcells combined with write assist are found to be the best way to reduce Vmin below 0.6V for $0.120\mu m^2$ SRAM. Finally, Vmin below 0.4V and extremely low leakage of 2pA/cell are demonstrated in retention mode. Supply voltage of entire SoC can thus be lowered down to 0.4V during off-mode resulting in extremely low power consumption, reinforcing the high potential of 28nm FDSOI technology to hit ultra low voltage market.

References

[1] N. Planes et al, p133, VLSI 2012 [2] F.Arnaud et al, p48, IEDM 2012 [3] O.Weber et al, p245, IEDM 2008 [4] R.Ranica et al, p210 VLSI 2013 [5] V.Huard et al, IRPS 2008 [6] J.C.Lin et al, p1, IEDM 2006 [7] C.Liu et al, p.277, IEDM 2015 [8] T.Song et al, p232, ISSCC 2014

Fig.1: 28nm FDSOI device architecture. TEM cross-section on NMOS SRAM 120µm².

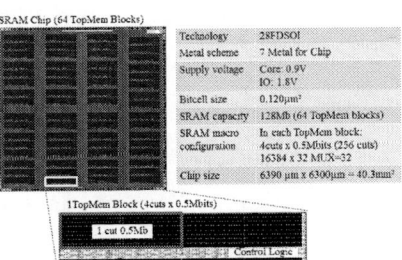

Fig.2: 128Mb 6T- SRAM chip architecture and its features. Zoom on one TopMem block of 2Mbits.

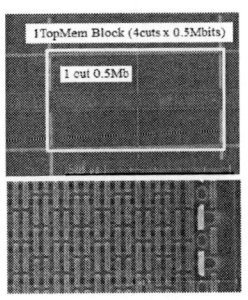

Fig.3: TopMem Block topview and SRAM array.

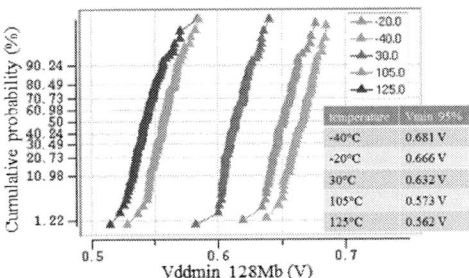

Fig.4: Vmin distributions measured on 128Mb chip at different temperatures (SPHD 0.120µm²). Vmin fresh values at 95% listed in the table.

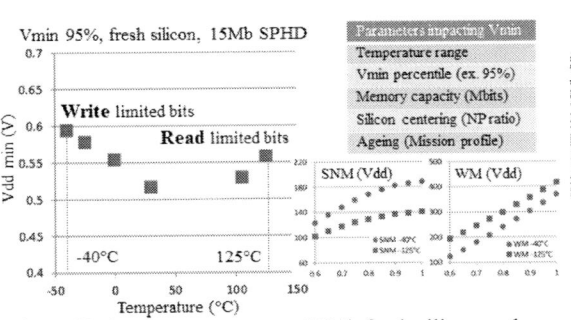

Fig.5 : Vmin measurements on 15Mb fresh silicon and SNM/WM behavior in -40/125°C temperature range. Key parameters impacting Vmin listed in the table.

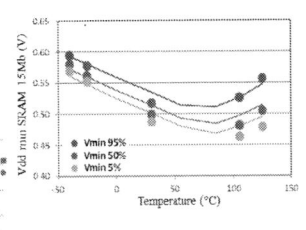

Fig.6: Si-model correlation on fresh silicon for different Vmin percentiles across temperature (15Mb SPHD bitcell).

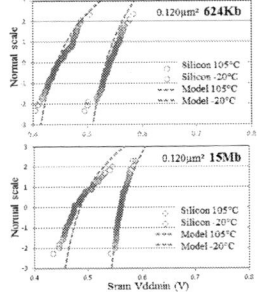

Fig.7: Si-model correlation on full Vmin distributions vs cut size.

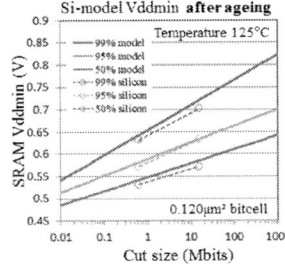

Fig.8: Model correlation with aged silicon for EOL behavior prediction at 125°C.

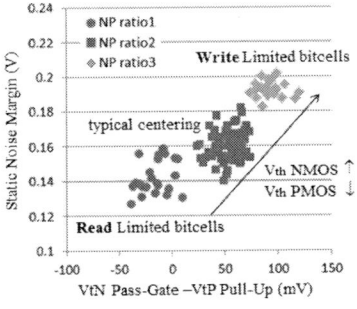

Fig.9: Static noise margin behavior vs VtN-VtP for different NP ratio process centering.

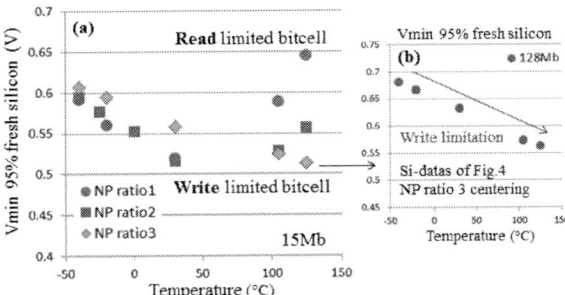

Fig.10: Impact of process centering on fresh Vmin behavior vs temperature. With NP ratio3, the bitcell becomes write limited, Vmin is lowered by 150mV at 125°C vs NP ratio1.

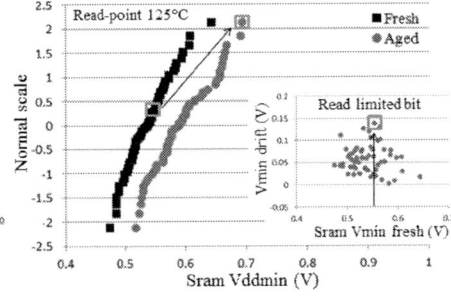

Fig.11: Vmin distributions on fresh and aged silicon (after 168h stress). 15Mb SPHD 0.120µm² with typical centering is measured.

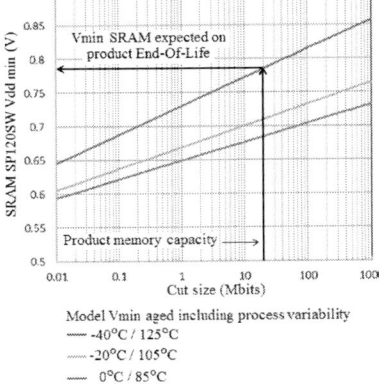

Fig. 12: Vmin simulations vs SRAM capacity (process centering and ageing included) to predict product behavior.

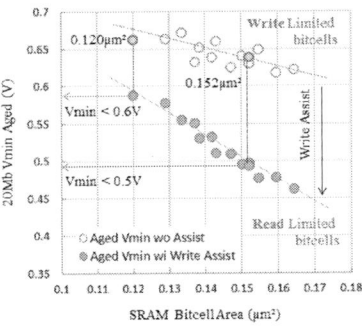

Fig. 13: 20Mb aged Vmin simulated in 0/85°C temperature range with and without WA vs increasing bitcell area, including process centering. When adding write assist, Vmin values are strongly correlated to bitcell area.

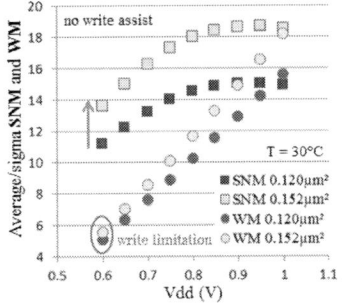

Fig.14: Better average/sigma SNM at low Vdd achieved on larger bitcells (0.152µm²). Better read stability explains lower Vmin when write assist is plugged and bitcells become read limited.

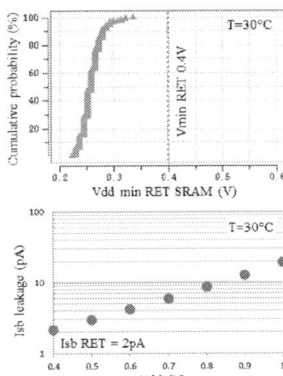

Fig.15 : Vmin retention and Stand-By leakage across Vdd on 0.120µm² SRAM.

A 400mV Active VMIN, 200mV Retention VMIN, 2.8 GHz 64Kb SRAM with a 0.09 um² 6T bitcell in a 16nm FinFET CMOS Process

Azeez Bhavnagarwala, Imran Iqbal, An Nguyen, David Ondricek, Vikas Chandra, Robert Aitken

ARM Research, San Jose CA 95134

azeez.bhavnagarwala@arm.com

Abstract

We propose and demonstrate in silicon simple, new circuit solutions using a standard 1-1-2 fin 0.09 um² 6T SRAM commercial bitcell in a 16nm FinFET CMOS process to enable a 400mV active VMIN, 200mV retention VMIN SRAM in a 64Kb CMOS array with 128b/BL. Active VMIN is enabled with a self-triggered feedback on an under-driven BL with faster and more robust signal development on the BL at lower voltages – providing dual read assist, and also a 2X tighter offset voltage distribution when compared to conventional differential voltage sensing. 200mV retention VMIN is enabled by reusing write assist circuit overhead while engaging two key observations: insensitivity of bitcell stability to systematic variations and sensitivity of bitcell data to noise on the power grid in the subthreshold/near threshold region. Average FMAX of 140MHz and 2.8GHz are measured across all chips for VDD at 0.4V and 0.9V respectively.

Introduction

SoCs in wireless devices require 6T SRAM arrays to scale performance 1.5X every 2 years, density by 2X every node [1] (Fig 1), with an expectation of much lower active and retention currents in the emerging IoT space. Wireless sensor nodes, Bluetooth LE 802.15.4 radio based IoT endpoint SoCs, for example, require a battery life of weeks to months. High end wearables such as smart watch or smart glass require GHz class performance to enable functionality and user experience while also demanding sub-pA retention current per bit. Arrays with 8T[2] and 10T[3] bitcells enable scaling operating voltages to near-threshold values. However, this translates into higher cost in SoC silicon area and does not leverage the same yield learning integrated into the 6T bitcell design during CMOS logic tech development or leverage the SoC flows for design, verification and test readily available for 6T arrays. This work shows near threshold operation of 6T arrays using new assist circuits without trading off performance or reliability.

Circuits

The circuit schematic of the self-triggered feedback on the under-driven BL is shown in Fig. 2a. Under-driving the BL pre-charge voltage to VBIAS before a read access improves the read disturb margin [4] and reduces the BL read voltage swing to the difference between VBIAS and the logic threshold of the NAND, improving performance by eliminating the need for larger swings of conventional single-ended sensing. The minimum signal voltage developed on the BL in this scheme is limited by the distribution of the trip point of the NAND (V_{LT} or logic threshold voltage) whose standard deviation, SD, is given by $\sigma V_{LT} = \sigma V_T / \sqrt{2}$ (Fig 3), corresponding to a 2X improvement over differential voltage sensing (DVS) where the minimum voltage swing is limited by the distribution of V_{os} whose SD is $\sigma V_{os} = \sqrt{2}\sigma V_T$ with σV_T equal to the SD of V_T in each of the pair of sensing devices. At lower VDD, the input offset SD of the DVS increases [5] by over 20% as VDD is lowered to the near-threshold region while the SD of V_{LT} in the NAND gate remains practically

unchanged (Table 1). The feedback path from the output of the NAND back to the BL increases the BL slew rate considerably past the 'knee' of the BL voltage transition (Fig. 2b) when the BL voltage equals V_{LT}, improving the sense delay at the output of the NAND while also providing a second read assist by 'writing back' data to the accessed bitcell. Read assist schemes for Write back implemented with DVS [6] cannot recover disturbs in bitcells from the fast end of the read current distribution since these typically have a lower pass gate (PG) VT and can potentially flip before the DVS is enabled to capture and write back the data. The uncertainty in the arrival delay of the Sense Enable with DVS (due to differences in logic propagation path for sense enable Vs data propagation path including bitcell) also adds overhead to the bitpath delay not present in our self-triggered sensing scheme.

Write Assist is implemented by lowering the cell voltage for the selected columns from VDD to VBIAS (Fig 4). While previous work [7] identifies data retention failure as limiting write assist by VDD lowering, the mechanism for data retention failure has not been characterized and is generally attributed to bitcell device variation imposed limitations. However, considering bitcell stability is insensitive to systematic variations (Fig 5) in the near or subthreshold region, that σVTsat improves in absolute value due to less DIBL at lower VDD [8] and that sensitivity of bitcell noise margins is very high to VDD variations (SNM model in Fig 11 using [10]), we observe barely 10-20mV of noise on the power supply terminal is sufficient to disturb data in the bitcells directly strapped to the power grid (Fig 6, 7).

Measurements

Write Assist headers with VBIAS<300mV have sufficiently high R_{ON} to be a noise shield for the bitcell array as a low-pass filter. Retention VMIN measurements of arrays with and without the Write Assist header (Fig 7) show average VMIN reductions of over 40mV to 170mV. The sub 200mV retention VMIN distribution permits VBIAS to be lowered without disturbing unaccessed bitcells during Write Assist enabling an average active VMIN of 400mV for over 90% of measured chips (sample of 50 chips) (Fig 8). The measured SD of active VMIN ~ 19mV with mean VMIN ~ 375 mV is projected to enable 6σ yield at VDD=0.49V. FMAX characterized for VDD and VBIAS shows improvements in FMAX by as much as 47% at 0.9V as VBIAS (Fig 9, Table 2) is lowered 250mV below VDD. VBIAS may be locally generated within the array [9] enabling true single VDD SRAM operation.

References

[1] https://en.wikipedia.org/wiki/Exynos, etc
[2] K-H Woo et al, Symp VLSI Ckts, 2015, pp 266-267
[3] J Myers et al, 2015 ISSCC Dig of Tech papers, pg 144
[4] A. Bhavnagarwala et al, IEDM Dig, Dec 2005, pg 658
[5] M H Abu Rahma et al, CICC 2011, pg 114
[6] H Pilo et al, 2006 Symp on VLSI Ckts, pg 15
[7] Y Wang et al, 2011 IEDM Dig, pg 32.1.1
[8] V.De et al, Proc. Symp. VLSI Tech., pp. 198-199, June 1996
[9] H. Pilo et al, IEEE JSSC Vol 47 No. 1 (2012), pp 97-106.
[10] R Swanson et al, IEEE JSSC Vol 7, No 2, April 1972, pp 146-153

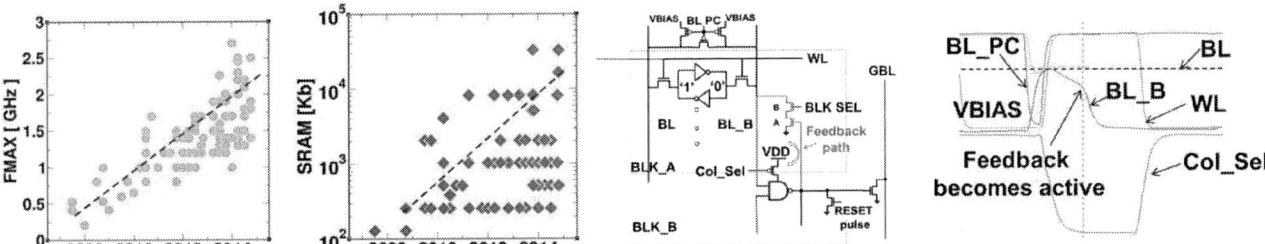

Fig. 1: SoC trends from leading wireless devices show markets demanding SRAMs to be faster, denser and operate at lower voltages to drain lower active and retention currents

Fig. 2a,b: Self-triggered feedback on an under-driven BL enables dual Read Assist and lowers BL delay. VBIAS < VDD enables lower active power for Read

- $K_n(V_{LT} - V_{TN})^\alpha = K_p(V_{DD} - V_{LT} - V_{TP})^\alpha$

- $\frac{V_{DD} - V_{LT} - V_{TP}}{V_{LT} - V_{TN}} = \left(\frac{K_n}{K_p}\right)^{\frac{1}{\alpha}} = \tau; \quad V_{LT} = \frac{V_{DD}}{1+\tau} + \frac{\tau V_{TN} - V_{TP}}{1+\tau}$

- $f(V_{LT}) = f(V_{LT}^{Tn}) * f(V_{LT}^{Tp})$ [joint distribution]

$f(V_{LT}^{Tn}) = \frac{f(V_{Tn})}{\frac{\partial V_{LT}}{\partial V_{TN}}} \sim 2f(V_{Tn})$;

$f(V_{LT}^{Tp}) = \frac{f(V_{Tp})}{\frac{\partial V_{LT}}{\partial V_{TP}}} \sim 2f(V_{Tp})$

For $\sigma_{VTp} \cong \sigma_{VTp}$, $\sigma V_{LT} = \sqrt{\frac{1}{4}\sigma_{Vtp}^2 + \frac{1}{4}\sigma_{Vtn}^2} \cong \frac{\sigma V_T}{\sqrt{2}}$

Fig. 3: Simplified analytical model of inverter logic threshold variance

Statistical simulations of NAND trip point (LT)				Mean	Sigma
VDD [V]	T [C]	Corner	VBIAS (V)	LT (mV)	LT (mV)
0.5V	125	SFG	0.4	261.6	6.11
0.5V	-40	FSG	0.4	205.4	6.67
0.8V	125	SFG	0.6	434.8	6.19
0.8V	-40	FSG	0.6	372.3	6.47

Table 1: Monte Carlo Sims show sigma of NAND logic threshold maintained as VDD lowered to near threshold region

Fig. 4: Column based Write Assist lowers column supply voltage to same voltage level as under-driven BL

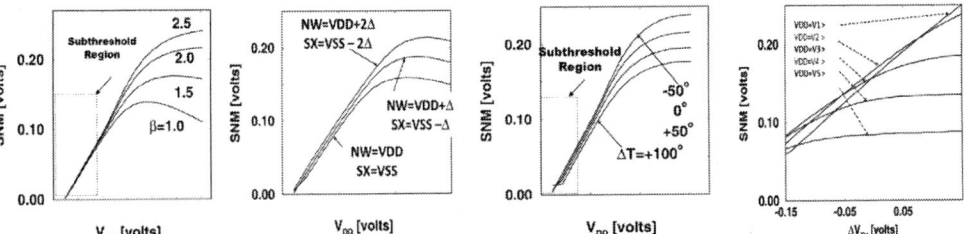
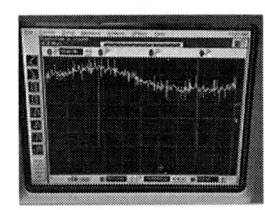

Fig. 5: Systematic variations (that affect all bitcell devices uniformly) in the near and sub threshold regions do not limit SRAM retention VMIN due to insensitivity of bitcell characteristics to these.

Fig. 6: Peak-peak noise<20mV to VDD shows retention VMIN dependence on WA header (Fig7)

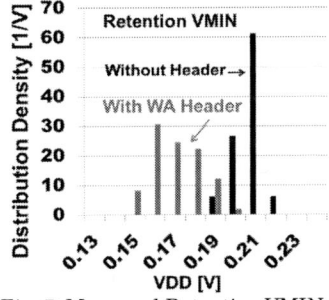

Fig. 7: Measured Retention VMIN dist. with and without WA header

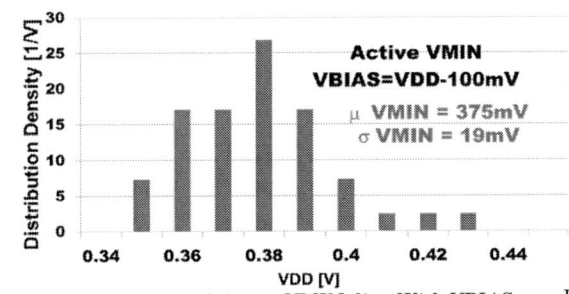

Fig. 8: Measured Active VMIN dist. With VBIAS = VDD-100mV, CLK=5MHz

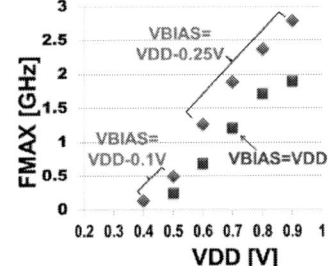

Fig. 9: Measured FMAX dependencies on VDD with improvements at VBIAS<VDD

Fig. 10 (at right): Voltage Schmoo measurements of VDD Vs VBIAS showing window of VBIAS enabling VDDmin down to 360mV

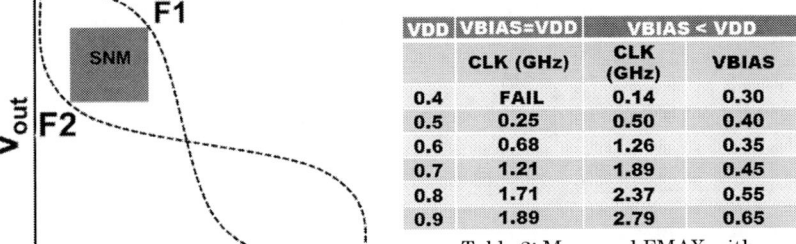

VDD	VBIAS=VDD	VBIAS < VDD	
	CLK (GHz)	CLK (GHz)	VBIAS
0.4	FAIL	0.14	0.30
0.5	0.25	0.50	0.40
0.6	0.68	1.26	0.35
0.7	1.21	1.89	0.45
0.8	1.71	2.37	0.55
0.9	1.89	2.79	0.65

Table 2: Measured FMAX with VBIAS=VDD and VBIAS < VDD

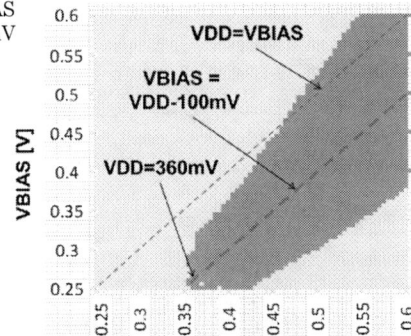

$Near\ F1:\ V_{out} = \frac{V_{dd}}{2} + \frac{V_{TN} - V_{TP}}{2} + \frac{\eta}{2\beta} ln\left|\frac{k_p}{k_n}\right| + \frac{\eta}{2\beta} ln\left|\frac{1 - e^{-\beta(V_{dd} - V_{out})}}{1 - e^{-\beta(V_{out})}}\right|$

$Near\ F2:\ V_{in} = V_{dd} - V_{out} + V_{TN} - V_{TA} + \frac{\eta}{2\beta} ln\left|\frac{k_a}{k_n}\right| + \frac{\eta}{\beta} ln\left|\frac{1 - e^{-\beta(V_{dd} - V_{out})}}{1 - e^{-\beta(V_{out})}}\right|$

Fig. 11: Sensitivity of bitcell retention VMIN to VDD in subthreshold

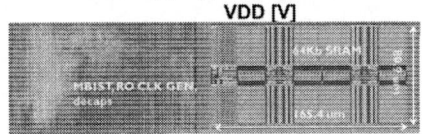

Fig. 12: Layout of SRAM with MBIST, CLK_GEN and decaps

978-1-5090-0636-6/16 $31.00 © 2016 IEEE   117   2016 Symposium on VLSI Circuits Digest of Technical Papers

A 350mV-900mV 2.1GHz 0.011mm² Regular Expression Matching Accelerator with Aging-Tolerant Low-V_{MIN} Circuits in 14nm Tri-Gate CMOS

Amit Agarwal, Steven Hsu, Mark Anders, Sanu Mathew, Gregory Chen, Himanshu Kaul, Sudhir Satpathy, Ram Krishnamurthy
Circuit Research Lab, Intel Corporation, Hillsboro, OR 97124, USA, amit1.agarwal@intel.com

Abstract

A regular expression matching on-die accelerator, consisting of a hybrid deterministic finite automata (HDFA) and Bloom filter (BLM), with measured 2.1GHz operation, is fabricated in 14nm tri-gate CMOS and occupies 0.011mm². HDFA integrates probability-based truncation, unique transition-state pairs isolation, parallel common transitions detection, and NFA empty transitions. BLM implements a fused 2-hash NOR match bit-line, 1bit read circuits, and sparse H3 hash. These techniques and aging-tolerant read/write register file circuits (120mV/180mV improved V_{MIN}) achieve 350mV-900mV wide dynamic voltage range with peak throughput of 15.2Gbps-17.1Gbps consuming 3.7mW-4.5mW measured at 750mV, 25°C and maximum energy-efficiency of 17.5Tbps/W-12.5Tbps/W measured at near-threshold 350mV.

Introduction

Energy-efficient complex regular expression (regex) search using large signature databases is the performance/power limiter for pattern matching applications such as anti-virus, big data analytics, and network intrusion detection [1-2]. Front-end regex matching filters using compressed signatures reduce the text search space, providing high-throughput pattern matching with an order of magnitude lower area (targeted for both client/server) compared to an exact match accelerator. A programmable regex matching on-die accelerator consisting of a HDFA and BLM filter supporting float and anchored regexes, respectively, with energy-efficient aging-tolerant low-V_{MIN} circuits, is fabricated in 14nm CMOS [3] (Fig. 1), while achieving 15.2-17.1Gbps peak throughput at 3.7-4.5mW power and a dense layout of 0.011mm².

HDFA and BLM Organization and Circuits

The 3-cycle HDFA with probability-based truncation (Fig. 2) is composed of 64 x 29b 1R/1W state address (STA) and 64 x 24b 4R/1W transition-state (TST) register files, followed by parallel transition (PTD) and transition-state (TSD) detection units with 12 and 4 range comparators, respectively (Fig. 3). Continuous prefix evaluation, to check for the start of DFA, results in identical transitions leading to early states (Fig. 4). Storing HDFA transitions as unique transition-state pairs in TST eliminates redundant storage for 57% lower area (Fig. 12). Common pairs across states (Fig. 5), stored in PTD for parallel scanning to reduce state count, enable 30% further area reduction and 3.6x improved throughput. During scan a state is read from STA entry to provide 12 enable bits which represent valid common pairs, 2 unique pair pointers, and an empty bit (Fig. 6). Setting this empty bit expands the number of transitions per state, using 4 additional unique pair pointers stored in the next entry. In the next cycle, the TST reads up to 4 transition-state vectors. The 4-way banked 2R/1W memory with 2-way 4:1 MUXs implements a 4R/1W register file, reducing TST area by 30% compared to a conventional 4R/1W.

The PTD and TSD compares 12 common and 4 unique transition-state vectors, respectively, against incoming text characters using range comparators (Fig. 7) to determine the next state. The range comparator, composed of two 8b logarithmic carry-tree circuits, detects transitions using i) a range of characters e.g. [A-C], ii) a class of characters e.g. [0-9], or iii) up to two characters with an integrated equality path, reducing STA entries by 5%. Case insensitivity circuits ignore the fifth input text bit without adding gate stages to the critical path. A single HDFA supports 50 float regexes and can be scaled to support more regexes by replication. The HDFA operates with 1.1 cycle per character throughput while achieving 0.9% positive probability (pp) resulting in 110x reduction in exact match search space (pp x number of signatures for exact match scan at filter match location) and 20x lower area compared to full exact match design.

The 2-cycle 4KB BLM memory, divided into four 256x32b arrays, scans for anchored regexes within incoming text, achieving 9.4% pp for 10K signatures (Fig. 8). Two 3-character prefix hashes using random 24x15 sparse H3 matrices (only four 1s per column, 57% lower gate count) produce two 9b row and one 6b column addresses for the 1R/1W register file with 1bit read

circuits. Anchored regex fixed strings are hashed to generate 1b BLM locations to store 1 during configuration phase. During scan, three 8b text characters are hashed to activate two word-lines within a single column of a fused 2-hash NOR match bit-line circuit to merge two complemented read data bits using a 1R/1W cell, indicating a match. These circuits reduce power and area by 25% and 28%, respectively, compared to 2R/1W (Fig. 12). The 1b read circuit, consisting of a 16-way local bitline (LBL) with column-select AOI followed by bit-select mid-level bitline, reduces redundant bitline switching, saving 40% power with 5% area increase. BLM read addresses are hashed using another H3 sparse matrix to generate a pattern set table, reducing exact match signatures to be scanned at BLM match location with overall 5300x reduction in exact match search space.

Aging-Tolerant Low-V_{MIN} Circuits

Robust register file operation at ultra-low voltage is limited by PVT variations [4] and is further degraded by aging. A shared-P circuit improves the write V_{MIN} contention [5] and is completion limited, which worsens with aging. Write completion is improved using a proposed gated-shared-P with diode clamp which reduces the aging impact (Fig. 9). A contention-free keeper delays the keeper activation to reduce read contention [4], improving read V_{MIN}, and is noise limited, which worsens with aging since it is only a partially gated keeper. Proposed fully gated contention-free keeper reduces aging in the entire keeper stack, improving noise V_{MIN}. Keeper stacks shared across LBLs reduce the select load by 50% with same area. Statistical simulations based on measured device data across 3σ systematic and 6σ random variations with aging, 0-110°C, show 120mV/180mV improved read/write V_{MIN}.

14nm Tri-gate CMOS Measurements

The HDFA operates at a maximum throughput of 15.2Gbps consuming 3.7mW (measured at nominal 750mV, 25°C) with an active leakage power of 20μW (Fig. 10,11). Throughput scales up to 19.1Gbps with 6.8mW power consumption at 900mV. HDFA circuit optimizations for ultra-low voltage operation enable robust functionality measured down to 350mV consuming 44μW at 770Mbps, with peak energy efficiency of 17.5Tbps/W (4.2x higher than nominal). The 4KB BLM throughput is 17.1Gbps (measured at 750mV, 25°C) with 4.5mW of total power. The BLM is functional across 350mV-900mV with scalable throughput of 860Mbps-21.5Gbps and total power consumption of 69μW-8.5mW. Peak energy efficiency of 12.5Tbps/W (3.3x higher than nominal) is measured at near-threshold voltage of 350mV. HDFA throughput and power from 15.2Gbps-13.4Gbps and 3.7mW-4.1mW, respectively, while scanning synthetic text with 1-15x higher pp, resulting in increased exact match search space (Fig. 13). Regex matching filter configured with reduced signature count increases the HDFA depth using lower transition probability-based truncation and decreases the BLM collisions, resulting in improved pp. HDFA and BLM pp is improved to 0.02% and 1.5%, respectively for 20% signature count, while HDFA power is reduced by 29% with 11% improved throughput.

Conclusion

A 2.1GHz regex matching on-die accelerator to reduce the exact match search space for large databases is fabricated in 14nm tri-gate CMOS, and occupies 0.011mm². HDFA and BLM with aging-tolerant low-V_{MIN} circuits enable a wide dynamic supply voltage range 350mV-900mV with peak energy efficiencies of 17.5Tbps/W and 12.5Tbps/W measured at 350mV, respectively.

Acknowledgements

The authors thank M. Haycock, M. Mayberry, S. Borkar, V. De, J. Tschanz, K. Vaidyanathan, H. Choday and B. Giridhar for encouragement. This research was, in part, funded by the U.S. Government. The views and conclusions contained in this document are those of the authors and should not be interpreted as representing the official policies, either expressed or implied, of the U.S. Government.

References

[1] C. Johnson, et al., ISSCC Dig. of Tech. Papers, pp. 104-105, Feb. 2010.
[2] C-C Wang, et al., ISSCC Dig. of Tech. Papers, pp. 390-391, Feb 2008.
[3] C-H Jan, et al., IEEE Symp. on VLSI Tech., pp. T12-T13, June 2015.
[4] A. Agarwal, et al., IEEE Symp. on VLSI Circuits, pp. 105-106, June 2010.
[5] M. Yuffe, et al., ISSCC Dig. of Tech. Papers, pp. 264-266, Feb. 2011.

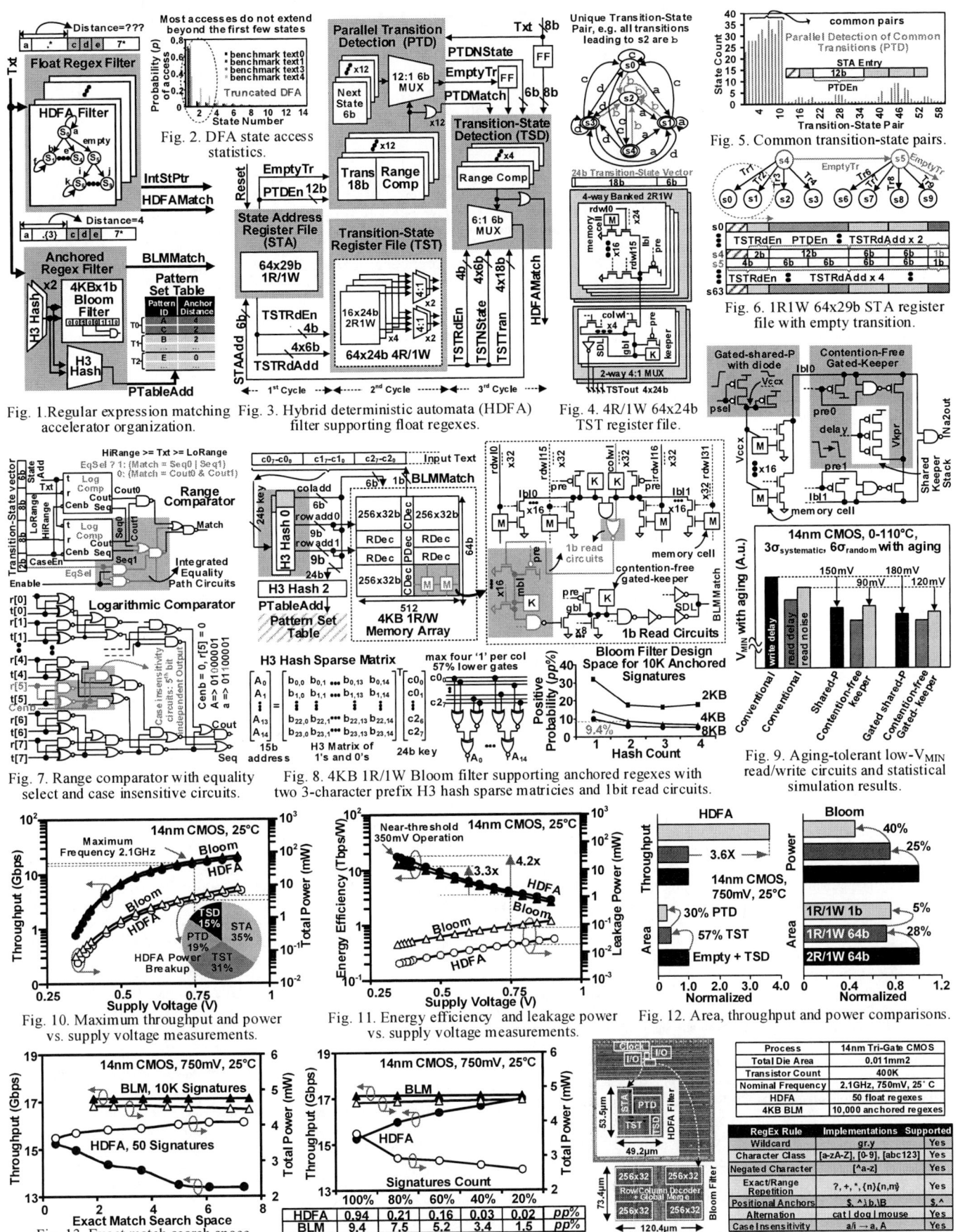

Fig. 2. DFA state access statistics.

Fig. 1. Regular expression matching accelerator organization.

Fig. 3. Hybrid deterministic automata (HDFA) filter supporting float regexes.

Fig. 4. 4R/1W 64x24b TST register file.

Fig. 5. Common transition-state pairs.

Fig. 6. 1R1W 64x29b STA register file with empty transition.

Fig. 7. Range comparator with equality select and case insensitive circuits.

Fig. 8. 4KB 1R/1W Bloom filter supporting anchored regexes with two 3-character prefix H3 hash sparse matricies and 1bit read circuits.

Fig. 9. Aging-tolerant low-V_{MIN} read/write circuits and statistical simulation results.

Fig. 10. Maximum throughput and power vs. supply voltage measurements.

Fig. 11. Energy efficiency and leakage power vs. supply voltage measurements.

Fig. 12. Area, throughput and power comparisons.

Fig. 13. Exact match search space measurements.

Fig. 14. Reduced signature count measurements.

	100%	80%	60%	40%	20%	
HDFA	0.94	0.21	0.16	0.03	0.02	$pp\%$
BLM	9.4	7.5	5.2	3.4	1.5	$pp\%$

Fig. 15. 14nm die micrograph and supported regex rules.

Process	14nm Tri-Gate CMOS
Total Die Area	0.011mm2
Transistor Count	400K
Nominal Frequency	2.1GHz, 750mV, 25°C
HDFA	50 float regexes
4KB BLM	10,000 anchored regexes

RegEx Rule	Implementations	Supported
Wildcard	gr.y	Yes
Character Class	[a-zA-Z], [0-9], [abc123]	Yes
Negated Character	[^a-z]	Yes
Exact/Range Repetition	?, +, *, {n}{n,m}	Yes
Positional Anchors	$, ^,\b,\B	$,^
Alternation	cat | dog | mouse	Yes
Case Insensitivity	a||A → a, A	Yes

Unified Technology Optimization Platform using Integrated Analysis (UTOPIA) for holistic technology, design and system co-optimization at <= 7nm nodes

S. C. Song, J. Xu, D. Yang, K. Rim, P. Feng, J. Bao, J. Zhu, J. Wang, G. Nallapati, M. Badaroglu,
P. Narayanasetti, B. Bucki, J. Fischer, and Geoffrey Yeap
Qualcomm Technologies Incorporated, 5775 Morehouse Drive, San Diego, CA. ssong@qti.qualcomm.com

Abstract - We propose complete technology-design-system co-optimization method in which power, performance, thermal, area and cost metrics are all simultaneously optimized from transistor to mobile SOC system level. This novel method, Unified Technology Optimization Platform using Integrated Analysis (UTOPIA), incorporates thermally limited performance, wafer process complexity and die area scaling model in addition to author's previous transistor-interconnect optimization method. Thermal model in UTOPIA evaluates/optimizes device and technology parameters not only for peak frequency but also for sustained performance after thermal throttling. Optimum N7 technology is selected using proposed UTOPIA method, showing significant overall gain over N10 technology.

Introduction - Previously, we presented transistor and interconnect optimization method, aka. "Critical Path Aware Binning & Mapping of statistical Path (CPA-BMP)' [1, 2]. As technology scales below N28, sustained performance limited by thermal throttling becomes one of critical factors determining ultimate mobile SOC performance. Thermally limited sustained performance optimization is, therefore, important in order to more accurately project system level power/performance and thus design transistor/BEOL technology accordingly. We have developed analytical thermal model working with CPA-BMP, in which peak burst mode frequency is throttled down depending on thermal requirement of system. In order to address area and cost portion of PPTAC (Power, Performance, Thermal, Area, Cost) metric [3], mobile SOC die area scaling and process complexity model have been developed and merged with CPA-BMP and thermal model. Area scaling model calculates layout scaling of routed logic, memory array (bitcell, periphery) and analog/IO based on key design rules and technology assumptions. Process complexity model is using step-by-step process flow with process complexity index (equivalent to process cost). This novel method, Unified Technology Optimization Platform using Integrated Analysis (UTOPIA), enables evaluation and optimization of power-performance and area-cost simultaneously, taking mobile SOC system/design and device/process technology all into consideration, which eases complex trade off among various technology details, enabling more cost effective and energy efficient scaling <= N7.

UTOPIA overview – Key sub-modules in UTOPIA method are depicted in Fig. 1, including thermal, cost and area scaling model. Key input and output parameters of each sub modules are feed back and forth each other for dynamic evaluation and optimization. Fig. 2 shows conceptual flow of UTOPIA method. Two parallel routes, Power-Performance-Thermal (PPT) route and Area-Cost (AC) routes, are defined. PPT route includes both electrical and thermal evaluation. Final metric of this route is battery consumption after full day of use. Area-Cost route is using wafer process complexity model and mobile SOC die area scaling model. Final metric from this route is process complexity normalized by die area assuming scaled die has same function (i.e., iso-functional) as previous node. In PP route, further improved CPA-BMP method is used employing 3 step binning and mapping. 1st step is to bin all critical path (~20k) into 6 bins depending on path delay. 2nd step is to sub-bin each 6 bin from 1st step to two sub-bins depending on total capacitance of path, making total 12 bins. 3rd and final step is to sub-bin 12 bins to final 24 bins depending on wire capacitance portion over total capacitance. 24 bins are mapped to 24 Ring Oscillators, capturing key path parameters.

Thermally limited sustained performance - Fig.3 shows Frequency-Total Power plot for each functional block in mobile SOC using advanced CPA-BMP method for N7. N10 device behaviors and process/device TCAD are used to predict N7 transistor behavior. Active leakage power is included in total power assuming same activity factor from N10. General concept of thermal limited performance model is shown in Fig.4. Thermal limited performance is especially important for high demand operations (e.g., Gaming and Video playback), where CPU and GPU tend to use peak

frequency to digest large workload. Thermal throttle occurs when junction temperature reaches 85°C in this model to prevent thermal run-away. Steady state power (and thus corresponding sustained frequency) after throttle is determined by required skin temperature (i.e., smartphone case temperature) at 40°C. Fig. 5 shows transistor Vt design using thermally limited preformation optimization for CPU core, accounting both single and multicore operation. Work load is fixed for all cases. Three different Vt devices (Vt1<Vt2 <Vt3) are considered. At single core case, higher Vt show better performance and lower battery consumption, attributed to much lower leakage power during steady state operation after thermal throttle, which allows larger room for dynamic power to reach required steady state power, and thus higher sustained frequency is achieved. This effect is more pronounced in quad core case, showing much less battery consumption at higher Vt. Due to parallel processing, time to complete operation is much shorter than single core operation, and thus starting frequency can be set well below steady state power (meaning, no thermal throttle) with greater energy saving. Energy-Performance trade-off is however needed at Vt3 as operation time becomes longer at quad core due to too high Vt.

Area and Process Complexity projection - Logic cell area estimation uses 27 representative cells (Combinatorial and Sequential), shown in Fig. 6. Relative usage % in array and CPP grid of each cells are used to calculate overall logic area. N7 cell height is calculated using required total number of Fin and required horizontal metal tracks. N7 CPP grid for each cell is estimated based on key design rule (e.g., tip to tip distance of horizontal metal) and active break scheme at cell boundary. Metal stack information is used for routing resources scaling and corresponding cell utilization rate for final routed logic area calculation. Fin depopulation scales cell height more than fin pitch scale, allowing more aggressive logic area scaling. Logic area scaling can be further accelerated by novel layout construct (e.g., self-aligned middle of line), offsetting rising wafer cost [4]. Wafer processing complexity model is built based on step-by-step process flow. Tool, consumable and labor costs are assigned to each step. Process complexity index is divided into each process modules (Fig.7), and thus changes in complexity is instantly reflected depending on technology option. This modular approach in complexity model allows mix and match of appropriate modules to build different process flows for various technology options. 1st order analytical equation is also used based on weighted sum of mask count taking cost equivalency of different litho steps.

N7 evaluation - Fig. 8 shows battery consumption using DOU (Day Of Use) model, applied to N7. Each usage in DOU uses multiple functional block in mobile SOC at the same time, and thus DOU evaluation allows optimization of various Lg and Vt device (used in different functional blocks in Mobile SOC) simultaneously. Results from two routes in UTOPIA, Power-Perf-Thermal and Area-Cost, are plotted together in Fig. 9. Optimized N7 tech flavor shows continuous scaling of PPTAC in line with N14 and N10 scaling trend. Fin depopulation with taller fin, parasitic R and C control using Wrap-Around Contact and Air spacer, and low R BEOL wire and via are the key elements of optimized N7 technology [5].

Conclusion – Complete technology-design-system co-optimization method is proposed unifying electrical behavior (Power, Performance), Thermal, and physical characteristic (Area, Cost), namely Unified Technology Optimization Platform using Integrated Analysis (UTOPIA). Thermally limited sustained performance method balances peak speed at burst mode and leakage power at steady state mode. Systematic layout scaling and modular process complexity model are added for complete PPTAC optimization. UTOPIA method allows fast turn-around of all aspect of technology evaluation, enabling complex technology trade-off at <= N7.

References [1] N. Mojumder et al., VLSI Tech Sym, p.10.2, 2014 [2] N. Mojumder et al., VLSI Tech Sym, p. JFS1-3, 2015, [3] G. Yeap, IEDM, p.1.3, 2013, [4] M. Bohr, IDF, 2014 [5] S.C. Song et al., VLSI Tech Sym, p. JFS3-4, 2015

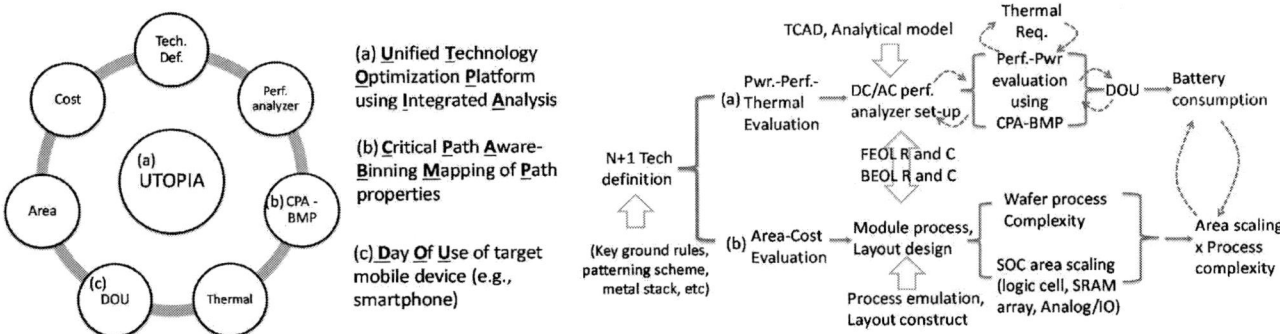

(a) **U**nified **T**echnology **O**ptimization **P**latform using **I**ntegrated **A**nalysis

(b) **C**ritical **P**ath **A**ware-**B**inning **M**apping of **P**ath properties

(c) **D**ay **O**f **U**se of target mobile device (e.g., smartphone)

Fig. 1 UTOPIA method has several sub modules including advanced CPA-BMP, thermal, cost and area scaling model. Key input and output parameters for each sub modules are feed back and forth each other for dynamic evaluation/optimization.

Fig.2 Conceptual flow of UTOPIA evaluation. (a) Power-Performance-Thermal evaluation route taking both electrical and thermal characteristics. Final metric is batter consumption after full day of use. (b) Area-Cost evaluation route using wafer process complexity model and mobile SOC die area scaling model (iso-functional). Final metric from this route is area-normalized-process complexity.

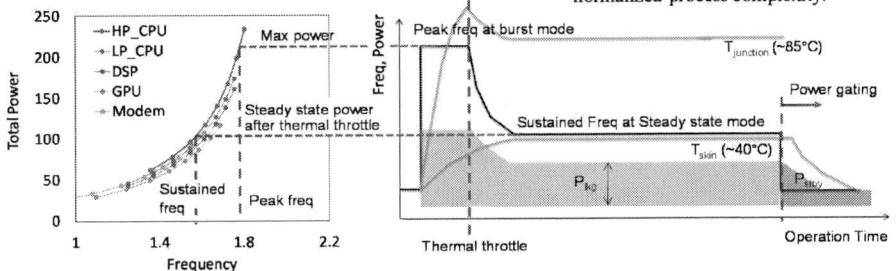

Fig.3 Freq-Power plot for each functional block in mobile SOC. Current node, N, device behavior and TCAD are used to predict next node, N+1, transistor behavior.

Fig.4 Thermal limited performance model is used for high demand operation in DOU model (e.g., Gaming and Video playback). Thermal throttle occurs when junction temperature reaches 85°C. Steady state freq is determined maintain skin temperature (i.e., smartphone case temperature) at 40°C.

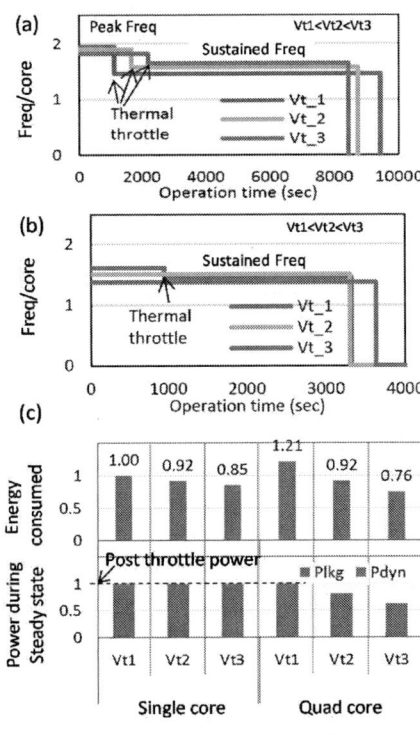

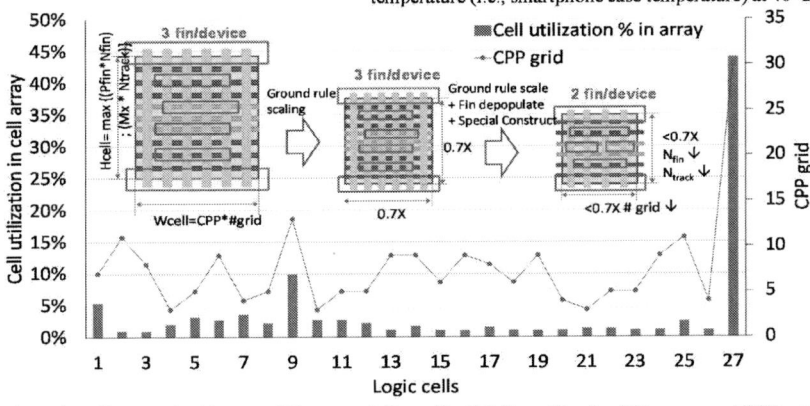

Fig.6 Logic cell area estimation uses 27 representative cells. Relative utilization % in array and CPP grid of each cells are used to calculate overall logic area. N+1 Cell height is calculated using required total number of Fin and required horizontal metal tracks. N+1 CPP grid for each cell is estimated based on key design rule (e.g., tip to tip distance of horizontal metal) and active break scheme at cell boundary.

Fig. 5. Sustained performance after thermal throttle improves at Vt2 than Vt1, shown in (a) Single core and (b) Quad core. Energy consumption is lower as Vt becomes higher shown in (c). Higher sustained freq at Vt2 is attributed to lower portion of leakage power. Too high Vt (Vt3) degrades quad core frequency.

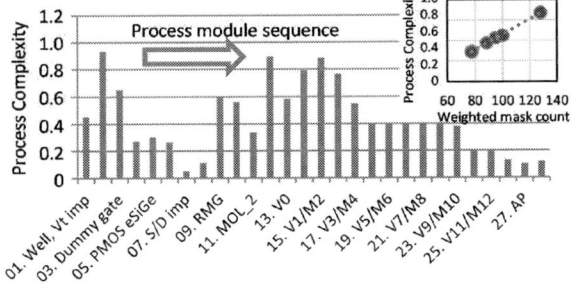

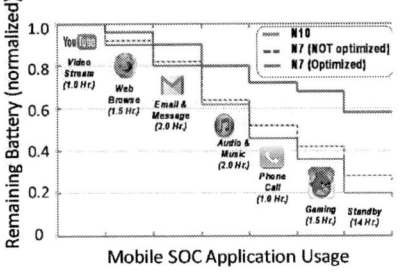

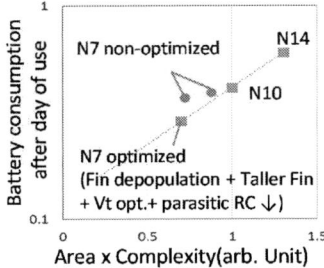

Fig. 7 Wafer process complexity model is built based on step-by-step process flow. Complexity index is assigned to each process modules, and thus changes in complexity is instantly reflected depending on choice of N+1 option. Analytical model also used based on normalized mask count taking different litho steps to weighted number of mask.

Fig. 8 DOU (Day Of Use) model based battery consumption evaluation allows various device/tech parameters (Lg, Vt, gate pitch, fin geometry, metal pitch, etc) in different functional block in mobile SOC simultaneously.

Fig. 9 Simultaneous electrical, thermal and physical evaluation using UTOPIA method to screen different technology flavor at N+1 node.

978-1-5090-0636-6/16 $31.00 © 2016 IEEE   121   2016 Symposium on VLSI Circuits Digest of Technical Papers

A 12-bit 1.6 GS/s Interleaved SAR ADC with Dual Reference Shifting and Interpolation Achieving 17.8 fJ/conv-step in 65nm CMOS

Jae-Won Nam, Mohsen Hassanpourghadi, Aoyang Zhang, and Mike Shuo-Wei Chen

University of Southern California, Los Angeles, CA, USA, Email: jaewon.nam@usc.edu

Abstract

A 12-bit SAR ADC architecture with dual reference shifting and interpolation technique has been proposed and implemented with 8-way time interleaving in 65nm CMOS. The proposed technique converts 4 bits per SAR conversion cycle with reduced overhead, which is a key to achieve both high speed and resolution while maintaining low power consumption. The measured peak SNDR is 72dB and remains above 65.3dB at 1-GHz input frequency at sample rate of 1.6 GS/s. It achieves a record power efficiency of 17.8fJ/conv-step among the recently published high-speed/resolution ADCs.

Keywords: SAR, ADC, time interleave, high speed, CMOS.

Introduction

High-speed (>GS/s) high-resolution (>10bit) ADCs with low power consumption are of increasing interest for many applications. This work aims to push the conversion speed of a single SAR ADC as high as possible to allow modest time interleaving and achieve an overall sample rate >GS/s. To enhance SAR conversion speed, a flash-based approach, i.e. M-bit/cycle scheme, has been used [1, 2] to reduce the required number of SAR conversion cycles; however, they typically require an excessive number of capacitive DAC (CDAC) and comparators, especially when M is large. In this work, we implement a cost-efficient 4-bit/cycle scheme that requires only two CDACs and shared pre-amplifiers via the proposed dual reference shifting and interpolation technique. Including redundancy, the system uses only 4 conversion cycles for a 12-bit resolution and achieves 200MS/s for a single ADC.

Proposed ADC Architecture

The concept of the proposed dual reference shifting and interpolation technique is illustrated in Fig. 1. The key idea is to shift the kth-cycle residue voltage $(V_{RES,k})$ up and down by a certain voltage, $V_{REF,k}$, and perform linear interpolation in between these two voltages, i.e. $V_{UP,k}$ and $V_{DW,k}$ as shown in Fig. 1. For example, the mathematical expression of a specific interpolation node can be written as: $\lambda \cdot (V_{RES,k} + V_{REF,k}) + (1 - \lambda) \cdot (V_{RES,k} - V_{REF,k}) = V_{RES,k} - (1 - 2\lambda) \cdot V_{REF,k}$, where λ is a certain interpolation ratio. As a result, a set of voltage differences, $(V_{RES,k} \pm \alpha_k \cdot V_{REF,k})$, can be created via an interpolation network, where $\alpha_k \in \{0, 1/8, 2/8 \dots 7/8\}$ in this implementation. By detecting their polarities, it generates a thermometer code and hence provides an efficient way to quantize $V_{RES,k}$.

Fig. 2 shows the simplified block diagram of the proposed ADC architecture. We implement the dual reference shifting via two separate CDACs together with digital control logic. The bootstrapped sampling switches are used to enhance linearity. The CDAC outputs are connected to two shared pre-amplifiers, i.e. transconductance (G_m) cells, which are followed by an interpolation network to generate all the required voltage differences. A bank of 15 latches then generates codes for the encoder, SAR control logic and bit

cache to ensure proper operation. Lastly, an asynchronous clock generator is used to generate all the required clock phases within a single ADC.

Fig. 3 shows the proposed CDAC implementation. During the sampling phase, the analog input is sampled into a subset of binary weighted capacitors. The extra capacitors ($C_{X0} \sim C_{X2}$) are used for conversion redundancy to tolerate incomplete DAC settling and other earlier decision errors. In the first SAR conversion cycle, most of the LSB capacitors are connected to $\pm V_{REF}$ to shift the sampled input up/down by $g_1 \cdot V_{REF}$ (as defined in Fig. 3) via charge redistribution. Note that, the first four MSB capacitors are intentionally left floating to provide signal attenuation and to prevent signal exceeding supply rail. In the following cycles, we connect the MSB capacitors to $\pm V_{REF}$, according to the decision results from prior cycles, and we connect certain numbers of LSB capacitors to $\pm V_{REF}$, as shown in Fig. 3.

The interpolation network is implemented across the differential output of two G_m-cells via resistor ladders, as shown in Fig. 4. The G_m-cells perform V-to-I conversion, sending AC currents to the resistive ladder, which results in different interpolation factor at the internal nodes. To enhance the settling speed, we configure the G_m-cells in high-speed and lower-gain mode for the first two conversion cycles; whereas high-gain and lower-speed mode is used for the remaining two cycles. This is achieved by switching the extra parallel resistors (R_L of each G_m-cell) at the G_m outputs on and off. Note that, there are built-in calibration routines to compensate the offset and gain mismatch from the latches and the G_m-cells. The calibration is activated upon start-up.

Fig. 5 shows the asynchronous clock generation scheme. It is mainly composed of a pulse generator with multiple feedback loops. By exploiting different time delays in different paths, various clock phases are created with proper durations for control different parts of the ADC. Note that, some of those delays are tunable for testing purposes.

In the final implementation, an 8-way time interleaving is used, as shown in Fig. 6. The global 1.6 GHz clock is routed in a tree structure adjacent to the signal trace to reduce the clock skews in between ADCs. This global clock is locally divided down into subsections of 200MHz each via a ring counter, which is carefully designed to minimize jitter. To increase the input bandwidth, the input traces are also terminated on chip. The final ADC outputs are decimated and multiplexed to off-chip.

Measurement Results

The ADC prototype occupies an active core area of 0.9 mm^2. It consumes 37.7mW through 1.1V analog and 1.2V digital/clock supplies, while the analog and digital/clock sections dissipate 20.6mW and 17.1mW/3.0mW, respectively. The measured dynamic performances of single and 8-channel ADCs are illustrated in Fig. 7. In the single channel spectrum plot, a SNDR of 72.8dB at a sub-sampled 749MHz input signal is measured with a 200MS/s conversion rate. The SNDR of the time-interleaved mode at 9.0MHz input frequency is 71.8dB; and maintains more than 65.3dB up to 1GHz, while SFDR measures from 83 to 69dB. The proposed ADC's FOM, is 7.9 fJ/conv.-step at a 9.0MHz input signal, and 17.8fJ/conv.-step at Nyquist frequency with 1.6GS/s. Fig. 8 shows the performance summary and chip micrograph of the proposed ADC. It achieves the best FOM among the published >GS/s, >10 ENOB state-of-the-art ADCs [3] (Fig. 9).

Acknowledgements

The authors thank ONR for funding support; Cheng-Ru Ho, Shiyu Su, and Jong Park for fabrication assistance.

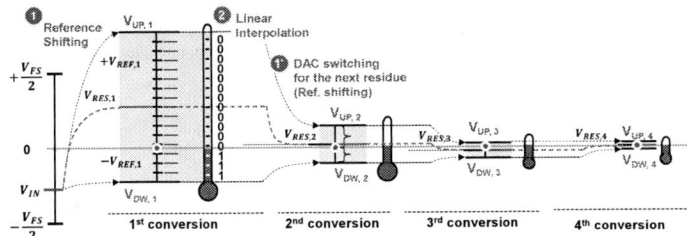

Fig. 1 Proposed reference shifting and interpolation technique.

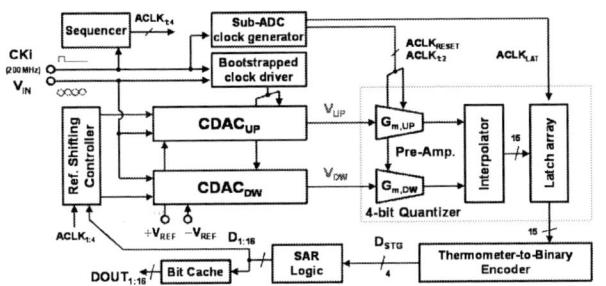

Fig. 2 Proposed sub-ADC block diagram.

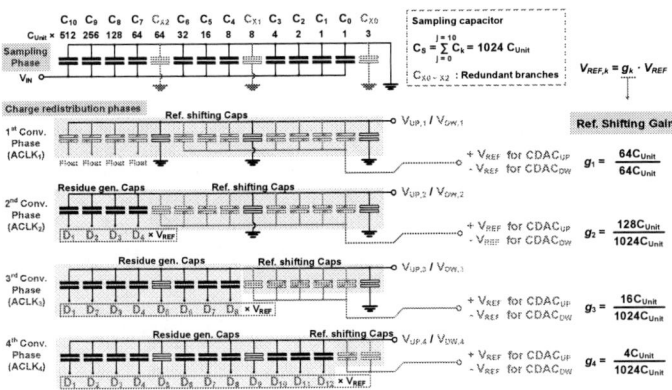

Fig. 3 Proposed capacitor DAC implementation (CDAC$_{UP}$ / CDAC$_{DW}$).

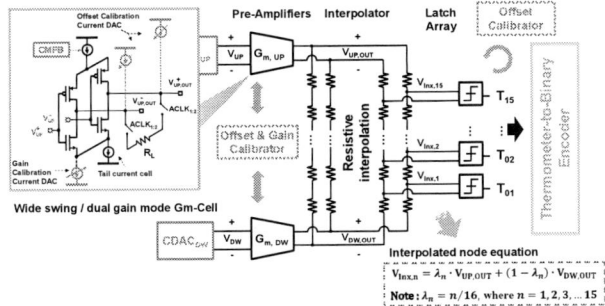

Fig. 4 Proposed interpolation network.

References

[1] H.-K. Hong *et al.*, *ISSCC*, 2015.
[2] C.-H. Chan *et al.*, *ISSCC*, 2015.
[3] B. Murmann, "ADC Performance Survey 1997-2015," [Online].

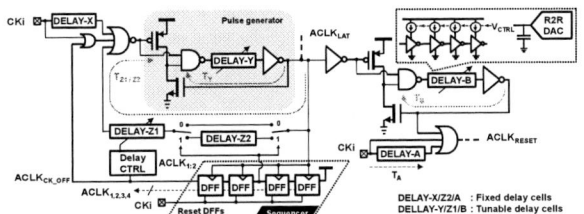

Fig. 5 Asynchronous clock generator for sub-ADC.

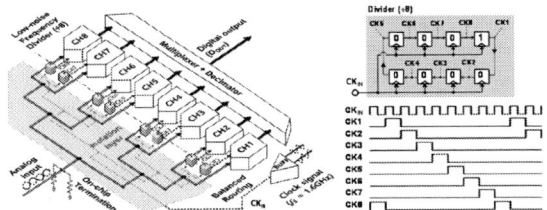

Fig. 6 Block diagram of the TI-ADC and clocking scheme.

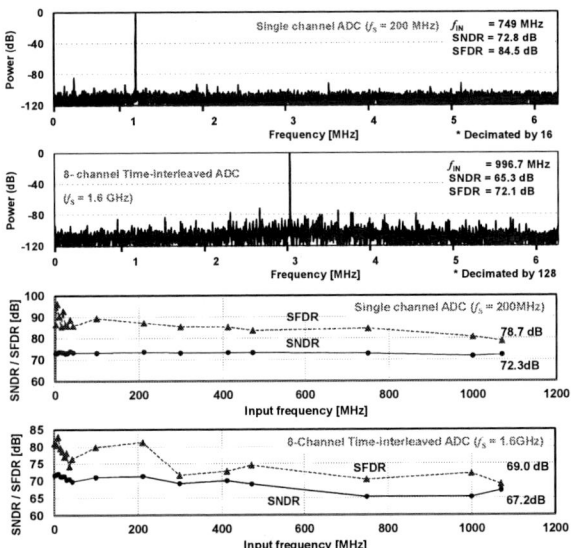

Fig. 7 Measured dynamic performance.

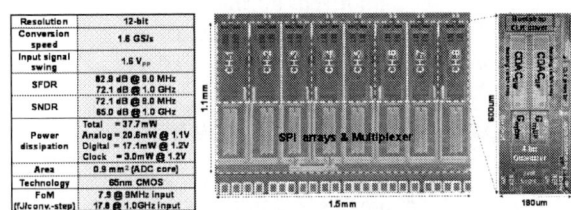

Fig. 8 Performance summary and chip micrograph.

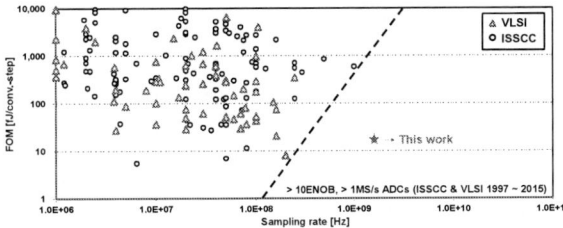

Fig. 9 FOM comparison with state-of-the-art ADCs [3].

A 14.6mW 12b 800MS/s 4×Time-Interleaved Pipelined SAR ADC achieving 60.8dB SNDR with Nyquist input and sampling timing skew of 60fs$_{rms}$ without calibration

Yuan-Ching Lien[1,2]

[1]MediaTek Inc.
[2]University of Twente, Enschede, The Netherlands
yuanching.lien@mediatek.com

Abstract

A 12b time-interleaved pipelined SAR ADC is presented. The proposed sampling circuit makes timing skew immune to mismatch of control circuit for time interleaving and reduces the main mismatch source to only sampling switch to achieve very low sampling skew of 60fs$_{rms}$ without calibration. MDAC transfer curve of pipeline stage is folded and OP output is kept half without degrading its gain and bandwidth by the proposed MDAC. The proposed OP loading reset scheme also enhances the settling speed without sacrificing ADC conversion time. Operating at 800MS/s, this ADC consumes 14.6mW from 1V supply and achieves SNDR of 60.8dB with Nyquist input.

Introduction

Time-interleaved(TI) ADC is an excellent way to achieve high conversion rate with good power efficiency. But offset, gain mismatch and sampling timing skew introduce errors that greatly degrade the SNDR of TI ADCs. The offset and gain mismatches are less input signal dependent and easier to calibrate. However, timing skew that increases with input frequency is difficult to deal with and calibration is often a burden. Master clock sampling is a widely used technique to reduce the sampling error. Nevertheless the achievable sampling skew is usually limited to several hundreds of fs$_{rms}$ due to mismatch of control circuit for time interleaving and sampling switches. This work introduces a very low sampling timing skew circuit and the sampling instant is immune to the mismatch of control circuit for time interleaving. Only matching of the sampling switches dominates the timing skew and it could be minimized by increasing W/L of MOS transistors without degrading performance in the proposed sampling circuit.

ADC Architecture and Building Blocks

Fig. 1 shows the ADC architecture. It consists of 4 pipelined SAR ADCs. TI technique increases the effective sampling rate to 800MS/s. The first stage MDAC of each ADC branch comprises a 5b asynchronous SAR ADC with online offset calibration, a 4b capacitive DAC and a residue amplifier with gain of 8. The second stage 9b SAR is constructed by a 2.8b flash and a 7b asynchronous SAR. Two adjacent ADC channels share the same OP amplifier and reference voltage buffer to save power. The reference is also shared by 5b SAR, MDAC and 9b SAR to avoid gain error. Timing mismatch results in inter-modulation tones and becomes worse at higher input frequency. Proposed input sample-and-hold circuit is shown in Fig. 2. It is merged in MDAC to save power and area, also bottom-plate sampling topology is adopted. Bootstrap switches that follow input signal provide over 12b linearity. At the rising of CK_{PH1}, M_4 is turned on and CK_{S1} is pulled to low to start the input sampling. While CK_{MUX1} is going to high, M_3 and M_4 are turned off. In the meantime, bootstrap switch M_5 is turned on and CK_M is passed to CK_{S1}. Note that M_5 is turned on and becomes a linear resistor with low resistance before the rising edge of CK_M. The resistance mismatch of M_5 between ADC channels is insignificant owing to the large and constant V_{GS} of bootstrap switch. As soon as CK_M goes high, it goes through an equivalent linear RC to turn off sampling switch M_1, M_{2a} and M_{2b}. To keep the sample-and-hold circuit in hold mode during ADC conversion, M_3 is turned on a little later CK_M goes high to keep M_1, M_{2a} and M_{2b} in off state. The major timing mismatch source is reduced to merely M_1 since M_{2a} and M_{2b} are used to set the common mode voltage. Large W/L(several hundreds of um for W) of M_1 is also required for high-frequency input and very good matching is obtained. Because the bottom-plate sampling is adopted in this design, large W/L for sampling switch doesn't produce much signal dependent feedthrough and charge injection compare to the commonly used top-plate sampling.

More details of the proposed MDAC are shown in Fig. 3. Input sampling and online offset calibration for 5b ADC that keeps the OP output around half swing are performed simultaneously within T_0. The DAC value is decided successively in T_1 and residue is amplified in T_2. Ground is used as the lowest reference voltage to have lower reference noise in this ADC. By applying higher reference voltage which is 4/3 of input swing to at most 3/4 of total sampling capacitors C_s, the MDAC transfer curve for full-swing input can be folded to reduce OP output swing while keeping the same loop gain and bandwidth. Higher reference voltage also lowers the on-resistance of PMOS switches and speeds the DAC settling. The two stage OP amplifier schematic and the following 9b SAR ADC block diagram are shown in Fig. 4. The common mode feedback(which is not shown) is performed by using switched capacitor circuit to save power. There are also separate compensation capacitors C_C for I/Q channels. While C_C of I channel compensates the stability, C_C of Q channel is reset. The reset clock CK_{rst} for CDAC of 9b SAR which is triggered at the end of SAR conversion is manually set to about 200ps and maintained by an internal loop. In normal SAR operation and there is no comparator metastability, CK_{rst} resets CDAC at the end of conversion. As soon as metastability occurs, CK_{rst} becomes shorter or not be triggered for worst case. The proposed OP compensation capacitor and output loading reset scheme improves OP speed without sacrificing conversion time and also guarantees $<10^{-15}$ error rate.

Experimental Results

The ADC is implemented in TSMC 28nm technology and occupies active area of 290um×320um. Operating at 800MS/s, this ADC consumes 14.6mW from 1.0V supply including 3.2mW is dissipated in the on-chip references. Measurement results are shown in Fig. 5. The input signal swing is $1.2V_{pp}$ differential. Only offset and gain mismatch calibrations are performed and there is no sampling timing skew calibration for this TI ADC. Measured DNL is +0.88/—0.76LSB and INL is +1.62/—1.45LSB. Measured SFDR stays higher than 75dB up to Nyquist input. It achieves 62.2dB SNDR for low frequency input, 60.8dB SNDR for near Nyquist input, ERBW of greater than 500MHz and Nyquist FOM of 20fJ/conversion-step. Chip micrograph and output spectrum of 409.99MHz input are shown Fig. 6. The estimated timing skew from measurement results is around $60fs_{rms}$ while the simulated timing skew is about $35fs_{rms}$. Table I summarizes the performance of this work and comparison with state-of-art ADCs.

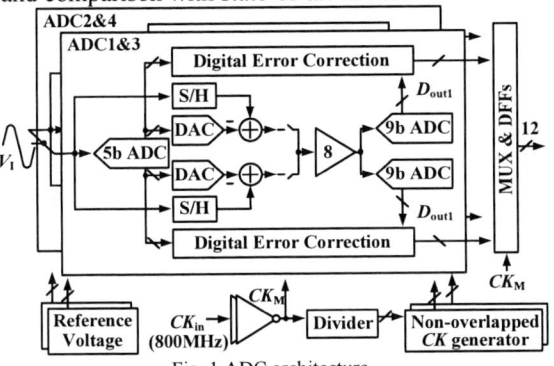

Fig. 1 ADC architecture.

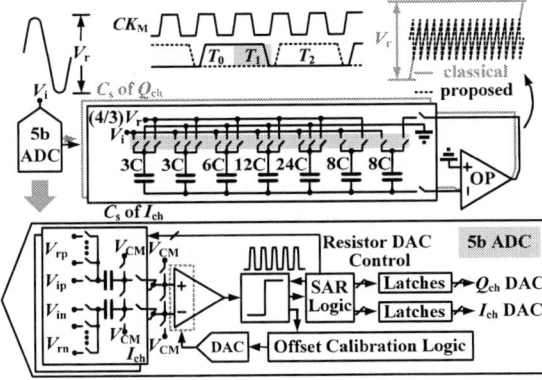

Fig. 2 Proposed very low timing skew sample-and-hold circuit.

Fig. 3 The proposed MDAC.

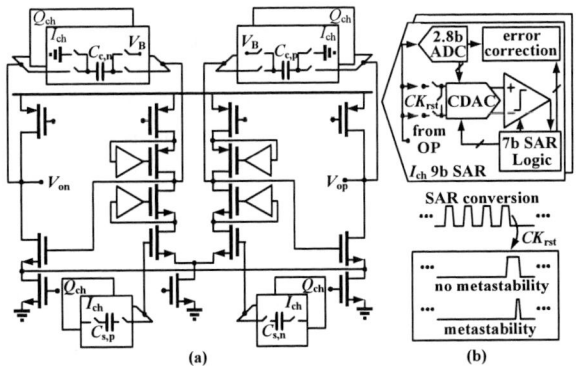

Fig. 4 (a) OP schematic and (b) block diagram of the following 9b SAR ADC.

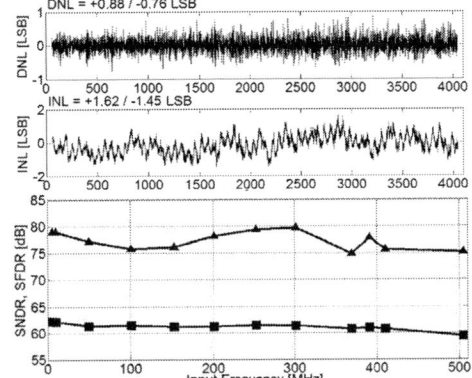

Fig. 5 Measured DNL/INL, SNDR and SFDR at Fs = 800MS/s.

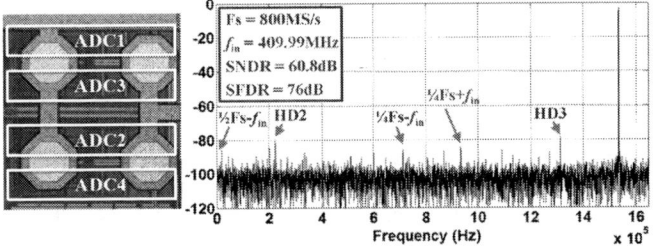

Fig. 6 The chip micrograph and output spectrum of 409.99MHz input. (Output is decimated by 243.)

TABLE I Performance Summary and Comparison

	ISSCC11 J. Mulder[1]	ISSCC14 S. Lee[2]	ISSCC15 B. Sung[3]	This work
Architecture	PIPE, TI	SAR, TI	SAR, TI	PIPE+SAR, TI
Technology	40nm	65nm	45nm	28nm
Fs(MHz)	800	1000	1600	800
Resolution	12b	10b	10b	12b
VDD(V)	1.0/2.5	1.0	1.1	1.0
SNDR@Peak	59dB	53.3dB	57.2dB	62.2dB
SNDR@Nyquist		51.4dB	56.1dB	60.8dB
Power(mW)	105	18.9	17.3	14.6
FOM@Nyquist (fJ/conversion)	180.2	62.3	21	20

References

[1] J. Mulder, et al., "An 800MS/s Dual Residue Pipeline ADC in 40nm CMOS," *ISSCC Dig. Tech. Paper*, pp. 184–185, Feb., 2011.

[2] S. Lee, A. P. Chandrakasan and HS. Lee "A 1GS/s 10b 18.9mW Time-Interleaved SAR ADC with Background Timing-Skew Calibration," *ISSCC Dig. Tech. Paper*, pp. 384–385, Feb., 2014.

[3] Ba-Ro-Saim Sung, et al., "A 21fJ/conv-step 9 ENOB 1.6GS/s 2×Time-Interleaved FATI SAR ADC with Background Offset and Timing-Skew Calibration in 45nm CMOS," *ISSCC Dig. Tech. Paper*, pp. 464–465, Feb., 2015.

An Oscillator Collapse-Based Comparator with Application in a 74.1dB SNDR, 20KS/s 15b SAR ADC

Minseob Shim[1,2], Seokhyeon Jeong[2], Paul Myers[2], Suyoung Bang[2], Chulwoo Kim[1], Dennis Sylvester[2], David Blaauw[2] and Wanyeong Jung[2]

Korea University, Seoul, Korea[1]; University of Michigan, Ann Arbor, MI[2]

sms@kilby.korea.ac.kr

Abstract

This paper presents a new energy-efficient ring oscillator collapse-based comparator, which is demonstrated in a 15-bit SAR ADC. The comparator automatically adjusts comparison energy according to its input difference without any control, eliminating unnecessary energy spent on coarse comparisons. The employed SAR ADC supplements a 10-bit differential main CDAC with a 5-bit common-mode CDAC. This offers an additional 5 bits of resolution with common mode to differential gain tuning that improves linearity by reducing the effect of switch parasitic capacitance. A test chip fabricated in 40nm CMOS shows 74.12 dB SNDR and 173.4 dB FOMs. The comparator consumes 104 nW with the full ADC consuming 1.17 µW.

Introduction

An energy-efficient comparator that automatically scales its conversion energy based on the input signal difference is useful for many applications, including voltage regulator, brown-out detection and analog to digital conversion (ADC). A low-noise comparator is especially important in SAR ADCs. SAR ADCs have enabled energy-efficient data conversion for moderate resolution (<12b). However, at higher resolution (≥12b) their energy efficiency tends to be poor relative to pipeline or sigma-delta ADCs, due to the low noise requirement on the comparator, which wastes comparator energy on those cycles that require only coarse comparison. Hence a comparator that automatically adjusts its energy consumption to the noise requirement would be ideal for SAR ADC and would enable higher resolution SAR ADC with high energy efficiency. To address this need, a dual comparator structure that uses a coarse comparator for MSB bits [1], oversampling and selective majority voting for noise-critical bits [2] and a time-domain comparator with long digital preamplifier stages [3] were proposed. However, these structures impair the simplicity of the SAR structure, worsening design and control complexity. This paper proposes a ring oscillator collapse-based comparator, referred to as an *edge-pursuit comparator* (EPC), which automatically scales comparison energy according to its input difference without external control, tailoring comparison energy to each conversion. This work also proposes a fine-grain 5-bit CDAC with common mode to differential gain tuning to improve linearity.

Proposed Edge-Pursuit Comparator

Fig. 1 shows the structure and operating principle of the EPC, which is composed of two NAND gates and eight inverter delay cells (Fig. 1a). The comparator initiates a comparison when the signal *START* goes high simultaneously at both NAND gates. This injects two propagating edges into the oscillator, which travel around the comparator until one overtakes the other, collapsing the oscillation. Differential input signals (V_{INP}, V_{INM}) are applied to both the top and bottom

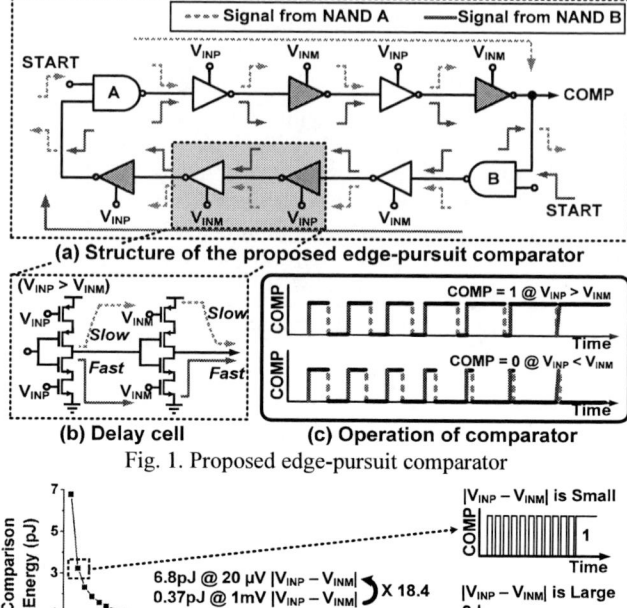

(a) Structure of the proposed edge-pursuit comparator

(b) Delay cell (c) Operation of comparator

Fig. 1. Proposed edge-pursuit comparator

Fig. 2. Simulation results of the edge-pursuit comparator

current-limiting transistors of the delay cells, modulating the pull-up and pull-down edge-propagation delays (Fig. 1b). Propagation delay of these two edges is controlled by mutually exclusive current-limiting transistors such that increasing V_{INP} causes one edge to propagate faster and the other to become slower (and vice versa for V_{INM}). After one propagating edge overtakes the other edge, the oscillation collapses and the stage outputs settle to either V_{DD} or GND, dictated by which edge was slower and overtaken (Fig. 1c). Comparator output *COMP* is sampled from an internal stage that becomes one when $V_{INP} > V_{INM}$ and zero otherwise. When the voltage difference between V_{INM} and V_{INP} is small, the two injected edges have similar propagation delay and the number of cycles required to make a decision automatically becomes longer (Fig. 2). This filters out high frequency noise, as the design performs noise averaging over a long time. On the other hand, if the voltage difference is large, the oscillation inherently collapses quickly, limiting dynamic energy consumption for coarse comparison.

In this manner, the comparator inherently adjusts its energy consumption without any external control (by 18.4× for 20µV to 1mV input voltage difference, Fig. 2) and realizes both high accuracy and low power operation. Also, EPC performance at the same input voltage difference can be easily controlled by changing the supply voltage, MOSFET size in the delay cell, and the number of delay cells.

Proposed CDAC Structure

The proposed EPC was applied to a 15-bit SAR ADC, which

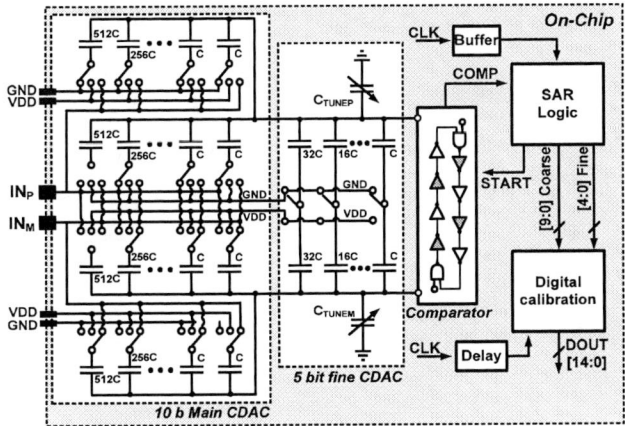

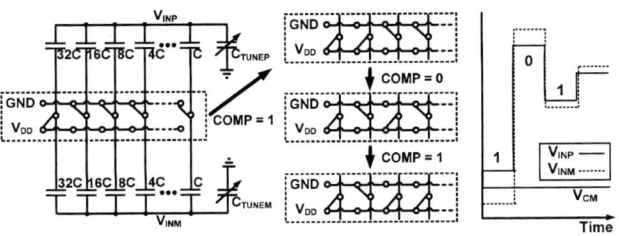

Fig. 3. Overall architecture of the SAR ADC.

Fig. 4. Operation of the proposed 5-bit fine CDAC

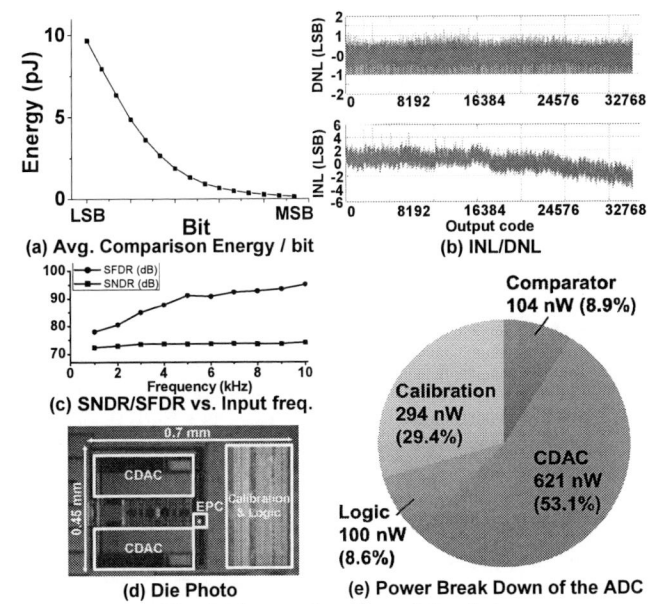

(a) Avg. Comparison Energy / bit

(b) INL/DNL

(c) SNDR/SFDR vs. Input freq.

(d) Die Photo

(e) Power Break Down of the ADC

Fig. 5. Measured results and die photo

Table 1. Comparison table with relevant ADCs.

	This work	Tai, ISSCC 2014 [1]	Harpe, ISSCC 2014 [2]		Lee, JSSC 2011 [3]	Lim, ISSCC 2015 [5]	Bannon, VLSIC 2014 [6]
Technology	40 nm	40 nm	65 nm		180 nm	65 nm	180 nm
Architecture	SAR	SAR	SAR		SAR	Pipelined-SAR	Pipelined-SAR
Resolution (bits)	15	10	12	14	10	13	18
Sampling rate, Fs (kS/s)	20	200	32		100	50000	5000
Input voltage range (Differential)	1.8 V$_{pk-pk}$	Not reported	Not reported		1.2 V$_{pk-pk}$	2.4 V$_{pk-pk}$	10 V$_{pk-pk}$
Area (mm²)	0.315	0.0065*	0.18		0.125	0.0544*	5.74
SFDR	95.1	76.25	78.4	78.5	67	84.6	Not reported
SNDR	74.12	55.63	67.8	69.7	57.7	70.9	98.6
ENOB (SNDR-1.76)/6.02	12.02	8.95	10.97	11.29	9.3	11.5	16.09
INL$_{max}$ (LSB)	5.5	0.45	0.82	3.50	0.8	0.96	0.52
DNL$_{max}$ (LSB)	1.9	0.44	0.58	1.75	0.4	0.4	0.10
Total Power (μW)	1.17	0.084	0.310	0.352	1.3	1000	60520
Critical Power** (μW)	0.104	0.025***	0.124***	0.141***	0.130***	336***	Not reported
FOM$_S$ (dB)	173.4	176.8	174.9	176.3	163.6	174.9	177.7
FOM$_C$ (dB)****	184	181.6	178.9	180.2	173.6	179.7	-

* Active area ** Power for noise critical block (comparator in SAR, amplifier in pipelined SAR)
*** Calculated value from the paper/presentation material **** SNDR+10log(F$_s$/2/(Critical power))

consists of a main and fine CDAC, the EPC, logic, and a digital calibration block (Fig. 3). The 10-bit differential CDAC is implemented using a split capacitor array to reduce switching power. The proposed 5-bit fine CDAC shares top plates with the CDAC (V_{INP}, V_{INM}) and has the same unit capacitor size as the main CDAC. Intentional difference between tuning capacitors C_{TUNEP} and C_{TUNEM} induces a small differential voltage change as shared bottom plates of the fine CDAC change, allowing high resolution without significantly increasing overall CDAC capacitance.

Differing from a conventional bridge technique [4], the proposed 5-bit fine CDAC has shared bottom plates of each pair of capacitors (Fig. 4, left). When a capacitor is switched, it injects the same charge into both CDAC output nodes. Hence, switching a capacitor only shifts the common mode voltage of the two output nodes and does not impact the SAR. However, by creating a small imbalance between the total capacitance to ground of the two CDAC output nodes, this common mode shift will also translate into a small differential voltage difference (Fig. 4, right). This common mode to differential gain is fine-tuned using the two tuning capacitors C_{TUNEP} and C_{TUNEM}. Because the tuning capacitors are connected to the shared top plate nodes, voltage across the tuning capacitors always stays near the input common-mode voltage whenever fine comparison is performed and changes only by a small amount during fine decision. This voltage swing is ~32× smaller than in the bridge capacitor scheme. Hence, the proposed top-plate shared fine CDAC structure shows improved linearity over the bridge technique by reducing non-linearity introduced by the non-linear parasitic capacitance of switches controlling C_{TUNEP} and C_{TUNEM}.

Measurement Results

The ADC with proposed edge-pursuit comparator is fabricated in 40nm CMOS process with a total area of 0.315 mm². Measured SFDR/SNDR at the Nyquist frequency is 95.1/74.12 dB. The measured total power consumption is 1.17μW and max DNL/INL are 1.9/5.5 LSB.

Fig. 5a shows the measured EPC comparison energy. Comparator energy is intrinsically adjusted as each bit comparison progresses. Comparator power is 104 nW, representing only 8.9% of the total ADC power consumption (Fig. 5e). FOM calculated using total power consumption of the ADC is 173.4 dB, however FOM$_C$, calculated based on comparator power only, is 184 dB, which compares favorably to other similar designs (Table 1). This underscores the applicability of the EPC to other low-power SAR ADC topologies.

Acknowledgements

The authors thank TSMC University Shuttle Program for chip fabrication and Analog Devices for partial financial support.

References

[1] H. –Y. Tai et al., *ISSCC Dig. Tech. Papers*, pp. 194–195, 2014.
[2] P. Harpe et al., *ISSCC Dig. Tech. Papers*, pp. 196–197, 2014.
[3] S. Lee et al., *JSSC*, pp. 651-659, Mar. 2011.
[4] A. Agnes et al., *ISSCC Dig. Tech. Papers*, pp. 246–247, 2008.
[5] Y. Lim et al., *ISSCC Dig. Tech. Papers*, pp. 458–459, 2015.
[6] A. Bannon et al., *VLSIC Dig. Tech. Papers*, pp. 33-34, 2014.

A 0.44fJ/conversion-step 11b 600KS/s SAR ADC with Semi-Resting DAC

Sung-En Hsieh, *Chih-Cheng Hsieh

Department of Electrical Engineering, National Tsing Hua University, Hsinchu, Taiwan
*cchsieh@ee.nthu.edu.tw

Abstract

A 0.3V 600KS/s 11b SAR ADC with semi-resting (SR) DAC, cascade-input (CI) comparator, and double rail-to-rail input range is implemented in 90nm CMOS. The SR DAC consumes only 6-13.5% switching energy of the state-of-the-art works. The CI comparator consumes only 49% of power and 66% of decision time with an x3 front-stage gain boost. The prototype achieves a SNDR of 58.7dB, an ENOB of 9.46b, a power of 187nW, and a resulting FoM of 0.44fJ/conv.-step.

Introduction

Internet-of-everything (IoE) applications require high energy-efficient ADC with moderate sampling rate and resolution. Successive-approximation register (SAR) architecture demonstrates a convincing and improving performance in recent years with technical developments of DAC switching [1-3] and low-power comparator [4-5] in a pushing-down supply operation (0.3-0.5V). For DAC switching energy reduction, many works were reported with design tradeoffs including requirements of extra reference voltage [1], sub-ranging operation [2], and reset energy [3]. For comparator power reduction, majority voting [4] and hybrid SAR with TDC [5] were proposed with penalties of complex critical-decision-detection and calibration-supporting circuits. With a lowering-down operational supply, the achieved performance and application of ADC are limited due to the degraded signal swing as well as the decreasing LSB. To overcome the mentioned drawbacks, this work presents a new SAR ADC architecture with semi-resting (SR) DAC, cascade-input (CI) comparator, and double rail-to-rail range. As a result, the achieved FoM performance is 0.44fJ/conv, which is only half of the best-reported result [2].

Architecture and Circuit Implementation

Fig. 1 shows the architecture of the proposed 11b SAR ADC composed of two 10b ADC (AD_0 and AD_1) and one global control unit. Each 10b ADC is implemented with a 9b C-DAC (with redundancy), a comparator, and SAR logic. The AD_0 and AD_1 are designed to handle the conversions of the input range of $0 > V_{ip} - V_{in} > -2V_{dd}$ and $2V_{dd} > V_{ip} - V_{in} > 0$, respectively. By detecting the signal polarity (MSB) after sampling phase, one of the 10b ADC will be disabled (rest) for power reduction, that is, the semi-resting (SR) DAC operation. Moreover, the total input range is effectively doubled to $\pm 2V_{dd}$, which is $\pm 0.6V$ at 0.3V supply in this work with SR DAC switching. Double-boosted sample-and-hold (S/H) and local-boosted switches are implemented for linearity and accuracy requirement with special care to leakage control.

Fig. 2 shows a 4-bit conversion example of the proposed ADC with SR DAC. Since the conversion process is symmetrical, only the case of $V_{ip} - V_{in} > 0$ is illustrated for simplicity. During the sampling phase, the bottom plates of PDAC_1/NDAC_0 and PDAC_0/NDAC_1 are reset to V_{dd} and V_{ss}, respectively. With MSB=1 detected by the comparator of AD_1, AD_0 is disabled and AD_1 takes over the remaining 10b conversion solely. The bottom plates of AD_1 are then merged together to generate a common-mode level of $V_{dd}/2$

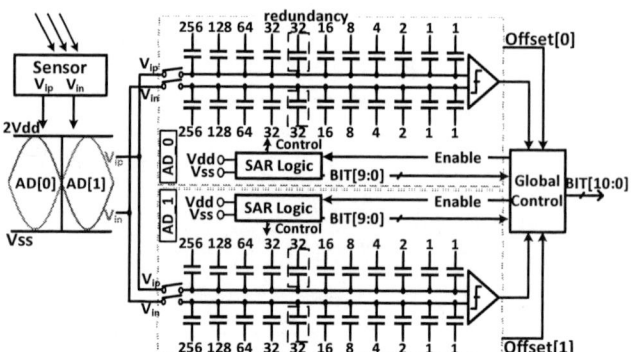

Fig. 1 Proposed SAR ADC architecture.

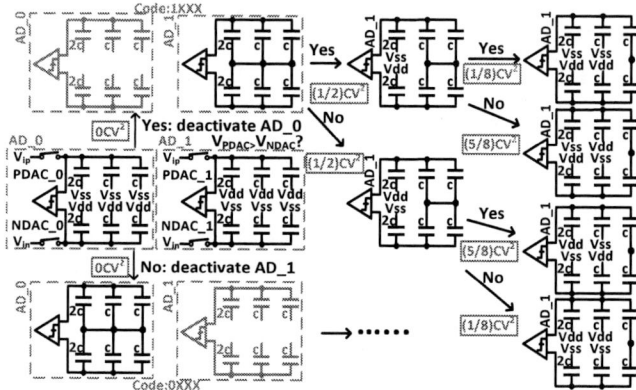

Fig. 2 A 4b conversion example of the semi-resting DAC.

(V_{cm}) by charge averaging operation. In the meantime, the required total voltage shift V_{dd} of top plates in PDAC_1 and NDAC_1 for MSB-1 detection ($V_{ip} - V_{in} > V_{dd}$?) is generated without consuming any switching energy. The remaining conversions of MSB-2 to LSB are accomplished in a MCS-based [1] operation by switching the corresponding bottom plates from V_{cm} to V_{dd} or V_{ss} based on the comparison result. To avoid the missing code error caused by the offset mismatch of comparators in AD_0 and AD_1, a 64LSB redundancy with 32C insertion is implemented to cover an offset distribution of 6-sigma (1-sigma = 1.7 mV). Foreground digital calibration is applied by recording the average of output offset codes of AD_0 and AD_1 with V_{dd} input to cancel it out in digital domain. Fig. 3 shows the switching energy versus output code of 10b conversion example with an identical input range ($\pm V_{ref}$) for comparison purpose. The average switching energy numbers (including conversion and reset operation) of well-known techniques are calculated and listed down based on a differential DAC implementation. Compared to MCS ($170.2CV_{ref}^2$) [1], Subranging ($229.7CV_{ref}^2$) [2], and MS ($233.9CV_{ref}^2$) [3], the proposed SR DAC consumes an average switching energy of $23.1CV_{ref}^2$ (including 32C redundancy), which is only 13.5% of MCS and around 6-to-10% of the others without any extra common-mode reference voltage and reset energy. Considering the number of capacitors for each sub-ADC with SR DAC is only half of the conventional

978-1-5090-0636-6/16 $31.00 © 2016 IEEE   128   2016 Symposium on VLSI Circuits Digest of Technical Papers

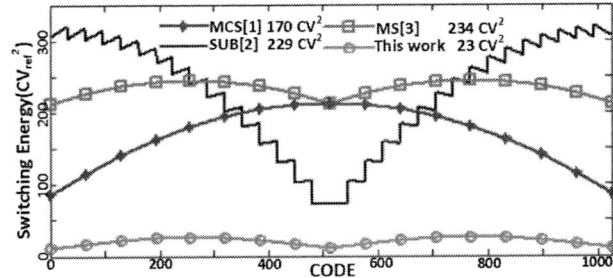

Fig. 3 Switching energy versus output code.

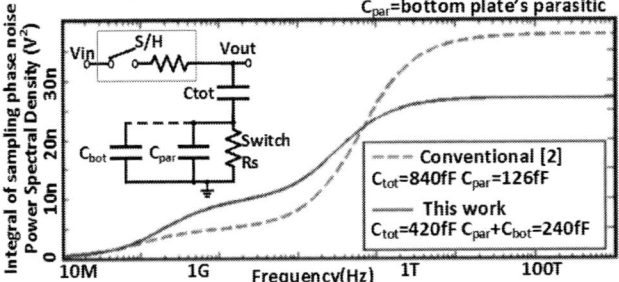

Fig. 4 Sampling cap reduction with an intentional Cbot insertion to get the same kTC noise.

architecture in sampling phase, the unit-C capacitance needs to be doubled for kTC requirement. To avoid the double loading effect of input signal driver, the sampling capacitance (Ctot) is reduced by half in this work by adding an intentional cap (Cbot) at bottom plate to meet the kTC requirement as shown in figure 4. As a result, the power consumption of signal driver for SR DAC in sampling phase keeps the same even with double number of capacitors (from AD_0 and AD_1).

For the higher common-mode level of double rail-to-rail input and noise reduction of comparator, Fig. 5 shows the proposed low-noise low-power cascade-input (CI) comparator with an increasing front-stage gain. At reset phase, all the internal nodes (op1~3, on1~3) of the stacking input pairs (M1p/M1n, M2p/M2n, and M3p/M3n) are reset to V_{dd} for initialization. In the comparison procedure, the input pairs 1(M1p/M1n) to 3(M3p/M3n) are activated sequentially with the corresponding ramping-down output as shown in the transient simulation waveform. The input difference is equivalently amplified in succession from nodes op1/on1 to op3/on3 before latch decision. As a result, the effective front-stage gain is increased by 3-x with a 3-stacking cascaded input pair in this design compared to the conventional architecture with a single input pair (M1p/M1n only). Compared to the conventional noise reduction approach by adding a bypass C-loading at the output of the front stage, the CI comparator consumes only 49% of power and 66% of decision time to achieve the same 1-sigma input-referred noise in simulation.

Experimental Results

A prototype chip is fabricated in 90 nm CMOS with a core area of $0.0354mm^2$ ($295\mu m \times 120\mu m$). Fig. 6 shows the static and dynamic performance of the implemented 11b ADC with a 0.3V supply and 600KS/s. The DNL and INL are +0.37/-0.63 and +0.72/-0.71LSB, respectively. With a Nyquist-rate input, the measured SNDR, SFDR, and ENOB are 58.71dB, 73.35dB, and 9.46-bit, respectively. The measured power consumption is 187nW with a distribution of 4% for S/H, 16% for comparator, 30% for DAC, and 50% for digital control. Table I shows the performance summary and comparison table. With the proposed SR DAC, CI comparator, and corresponding

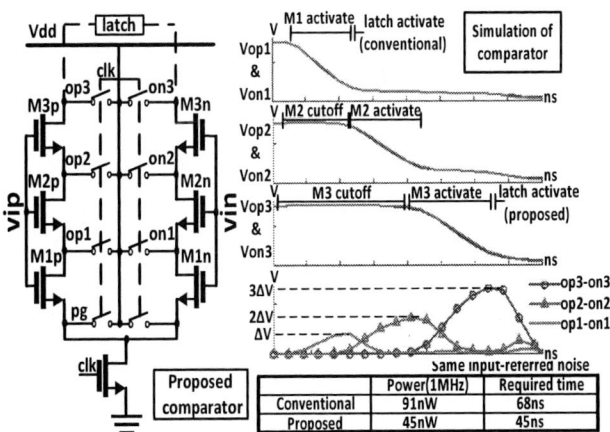

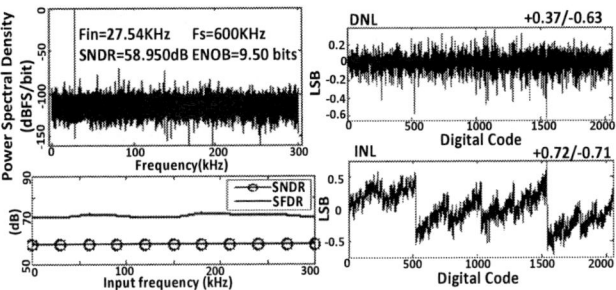

	Power(1MHz)	Required time
Conventional	91nW	68ns
Proposed	45nW	45ns

Fig. 5 Proposed cascade-input comparator.

Fig. 6 Measured static and dynamic performance.

TABLE I Performance summary and comparison

	ISSCC-13[2]	TCAS-I-14[3]	ISSCC-13[4]	VLSI-14[5]	This work
Technology	40nm	90nm	65nm	90nm	90nm
Supply Voltage(V)	0.45	0.3	0.6	0.4	0.3
Ideal input swing	±Vref	±Vref	±Vref	±Vref	±2Vref
Sample rate (kS/s)	200	90	40	250	600
Resolution (bit)	10	10	12	10	11
DNL (LSB)	0.44	0.38	0.97	0.43	0.63
INL (LSB)	0.45	0.66	1.9	0.67	0.72
Power (nW)	84	35	97	200	187
ENOB (bit)	8.95	8.38	10.1	8.63	9.46
FoM (fJ/c.-s.)	0.85	1.17	2.2	2.02	0.44
Active Area (mm²)	0.0065	0.031	0.076	0.04	0.035

design for low-voltage operation, the implemented 0.3V 600KS/s 11b ADC achieves a FoM of 0.44fJ/conv-step.

Acknowledgement

The authors acknowledge the support of National Chip Implementation Center (CIC) Taiwan, SiSAL, EE, NTHU, and. Ministry of Science and Technology, Taiwan under contract MOST 104-2220-E-007-009 and 104-2221-E-007-MY3.

References

[1] V. Hariprasath, et al., "Merged capacitor switching based SAR ADC with highest switching energy-efficiency," Electron. Lett., vol. 46, No. 9, pp. 620-621, Apr. 2010.

[2] J.-Y. Tai, et al., "A 0.85fJ/conversion-step 10b 200kS/s subranging SAR ADC in 40nm CMOS," ISSCC Dig. Tech. Papers, pp. 196–197, Feb. 2014.

[3] J.-Y. Lin and C.-C. Hsieh, "A 0.3 V 10-bit 1.17 f SAR ADC With Merge and Split Switching in 90 nm CMOS," IEEE TCAS-I, vol. 62, no. 1, pp. 70-79, Dec. 2015.

[4] P. Harpe, et al., "A 2.2/2.7fJ/conversion-step 10/12b 40kS/s SAR ADC with Data-Driven Noise Reduction," ISSCC Dig. Tech. Papers, pp. 270-271, Feb. 2013.

[5] Y.-J. Chen and C.-C. Hsieh, "A 0.4V 2.02fJ/Conversion-Step 10-bit Hybrid SAR ADC with Time-Domain Quantizer in 90nm CMOS," IEEE Symp. VLSI Circuits, pp. 1-2, Jun. 2014.

A 35 mW 10 Gb/s ADC-DSP less Direct Digital Sequence Detector and Equalizer in 65nm CMOS

AKM Delwar Hossain[1], Aurangozeb[1], Maruf Mohammad[2], Masum Hossain[1]
[1]University of Alberta, Edmonton, Canada, [2]Qualcomm Atheros
masum@ualberta.ca

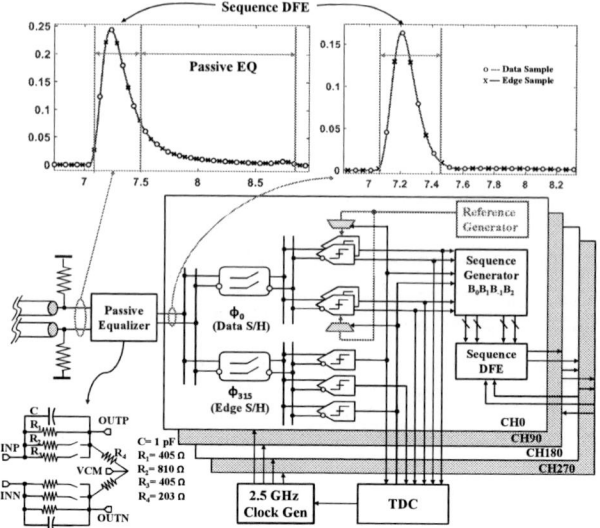

Fig.1 Maximum likelihood sequence detector with passive equalization and timing recovery

Abstract

This paper describes design technique of energy-efficient ADC-DSP less sequence detection and equalization. This scheme takes advantage of the ISI in the channel to reconstruct the time domain bit sequence. This concept is demonstrated with a 4-bit sequence decoder designed and fabricated in 65nm CMOS using only 4-data, 3-edge comparators. Consuming only 35 mW at 10 Gb/s and without any transmit equalization, this receiver is capable of compensating 27 dB channel loss with 90 mV Voltage margin and 25 ps timing margin at BER of 10^{-12}.

Keywords— MLSD, Sequence DFE, Collaborative Sequence and Timing Recovery and SNR Optimized Receiver.

Introduction

In high speed systems frequency dependent channel loss is the main source of ISI. In simple word, ISI is the residue of the current symbol that affects the following symbols (post-cursor) as well as the previous symbols (pre-cursor). Existing techniques such as FFE/DFE mostly rely on reducing these pre and post cursors for error free detection of the main cursor. While reducing the ISI, signal to noise ratio (SNR) also degrades, especially at high channel loss degradation of SNR (both crosstalk and random) limits the performance of symbol-by-symbol detection technique. In such SNR limited cases sequence decoders outperforms symbol-by-symbol detectors. However, existing maximum likelihood sequence detectors (MLSD) requires an analog-to-digital (ADC) converter in the front-end [1]. While ADC alone approaches 10 pJ/bit power consumption, including the DSP significantly exceeds receiver power budget for SerDes solution space. Therefore, an ADC less solution is an attractive low power alternative to existing approach.

Direct Digital Sequence Decoder

The receiver architecture is highly digital– it includes a front-end programmable passive equalizer and a set of data and edge comparators followed by digital back end for sequence detection and equalization (Fig. 1). The receiver is designed considering a voltage mode transmitter where the transmitter differential swing varies from 800mVpp to 1Vpp. Therefore, signal attenuation is needed to scale received signal to match the dynamic range at the comparator bank input. Since high frequency signal is already attenuated by the channel, only low frequency signal is attenuated which translates to passive equalization. For a 27 dB loss channel, only 5 to 7 dB boost (or DC attenuation) is sufficient to contain ISI components within four taps – one pre-cursor, main and two post-cursor taps in the single bit response (SBR) (Fig. 1). This partially equalized signal is then decoded by a 4-bit sequence decoder.

A. Analog to Sequence Conversion

The received signal at any point of time is a combination of pre and pre cursor components of the neighboring bit stream and main cursor component of the current bit. Therefore, from the received signal sample we can reconstruct the corresponding time sequence of previous, current and next bit sequence-$B_{+2}B_{+1}B_0B_{-1}$. Here, $B_{+2}B_{+1}$ are two previous bits, B_0 is current bit and B_{-1} is the next bit. The unique advantage of sequence detection is the constructive use of ISI – we are using the pre-cursor to decode the next bit that is later on used to correctly decode the sequence. To summarize,

rather than reducing the ISI, its information can be used to reconstruct the time sequence of bits and SNR optimized detection.

The simplest approach for sequence detection is to directly calculate the distance from the received sampled value to each sequence constellation. Distance can be calculated by comparing input signal to references generated based on different combination of main (h_0), pre-cursor (h_{-1}) and post-cursors (h_{+1} and h_{+2}). In general, sequence length N depends on the number of un-equalized pre and post cursors in the SBR. In the signal space this N taps can combine in 2^N number of ways creating 2^N signal levels corresponding to 2^N unique sequences. Finding the correct sequence from these 2^N possibilities can be challenging. First, comparing to all reference levels require 2^N-1 comparators costing significant power and area. Second, unlike ADC, these reference levels are not monotonic; therefore, converting comparator output to a sequence requires significant logic complexity. Third, sequence to reference conversion is not unique – meaning two different sequences can converge to same reference levels due to non-binary relationship between tap values creating ambiguity in sequence selection. Several techniques are introduced in this work to overcome these challenges: (1) Sequence DFE and (2) Collaborative Sequence and Timing Recovery.

B. Sequence DFE

Rather than comparing to 2^N number of reference levels, the entire signal space is sub-divided into fewer selected regions. Based on the region where the signal is - a set of possible sequences are generated. These sequences are generated based on two criteria: a) The references corresponding to this set of sequence have the minimum distance from the sampled value and b) These sequences are differentiable based on their previous bits. In this design each region is having four possible sequences, and from these four sequences most likely sequence is selected by comparing B_{+2} and B_{+1} to previously decoded symbols as illustrated in Fig. 2. Therefore, sequence DFE is similar to conventional DFE except it resolves a sequence rather than

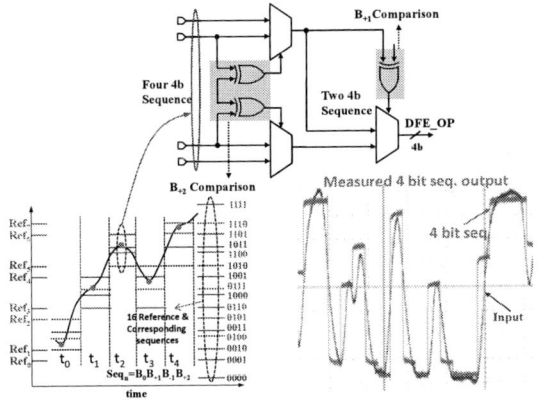

Fig.2 4-bit Sequence DFE implementation. 4-bit sequence generation and corresponding half-rate 4-bit sequence DAC output

detecting only symbol.

In existing receivers, bit error is related to their dependency to a single comparator. Although in M-tap speculative DFE, 2^M comparators are used; only one of them is selected based on previous decisions. Therefore, the bit decision is made by the single selected comparator and its decision error translates to bit error. Sequence DFE proposed in this work is immune to single comparator decision error. Rather than dividing the signal space into minimum 4 regions, adding additional 4 comparators (Ref0 to Ref7 in Fig. 2) we divide the signal space into 8 regions (Fig. 2). However, each comparison still allows 4 possible sequences, and by doing so it provides considerable amount of redundancy. As a result, if the individual comparator decision error causes a different region to be selected, correct sequence can still be part of the possible set of sequences and eventually DFE can make correct symbol decision. For example, in fig. 2 at time t_2 and t_4 sequence1011,1101 are part of likely set although Ref0 to Ref7 comparator outputs are different. Here the comparator outputs are used as a 'soft decision' (as opposed to 'hard decision' in speculative DFE) to determine the most probable location of the received sample to determine the set of likely sequences. Therefore, individual comparator error does not translate to symbol error so long the correct sequence is part of the 'likely' sequence set.

C. Collaborative Sequence and Timing Recovery

For an N-bit sequence detector input signal needs to be compared to 2^N reference levels that results in a power and area penalty similar to a flash ADC. So, for a 4-bit sequence 16 comparators are required. With sequence DFE, as discussed in previous section, comparator requirement reduces to 8. For further power reduction we recycle the comparators based on previous sample. Note that due to ISI, sample to sample signal variation is limited – therefore, it is not necessary to cover the entire signal space. Rather based on previous sample position, covering only 50% of the signal space is sufficient. This observation leads to a simple four comparator front-end implementation where each comparator reference level can be switched between two possible values. Reference muxes are driven directly by two edge comparators. Reference switching behavior in Fig. 3 shows that within 150 ps reference settles within 10% of its final value. Three edge comparators serve dual purposes: (a) to provide the timing error information with higher resolution and (b) to place the data comparators in the vicinity of the next sample. In addition, edge samples with decoded sequence allow us to filter edges with ISI.

Implementation & Measurement

The implemented quad rate receiver shown in Fig. 4 occupies only 0.23 mm². Received signal after 4-bit sequence decoding converts

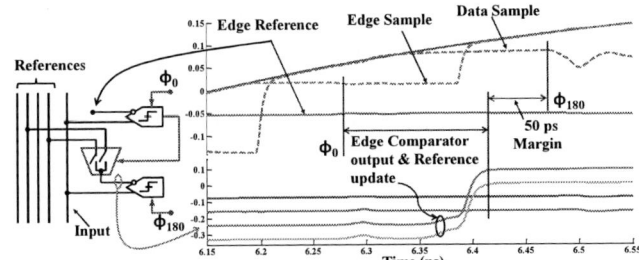

Fig. 3 Reference distribution and settling behavior

to 16 level output from which the current bit is taken out for error detection. Without any transmit equalization, 4-bit sequence decoder operates error free over a 27 dB loss channel with 90 mV voltage and 25 ps timing margin consuming only 35 mW (Fig. 6). Despite the simplicity of the architecture loss compensation capability is comparable to ADC-DSP based solutions in [2,3]. This highly digital architecture is portable and can be scalable to decode longer sequence at higher channel loss. The comparison between this work and existing state of the art ADC based solutions is listed in Table I.

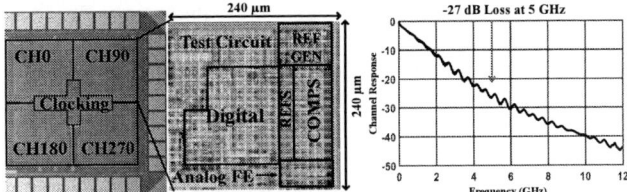

Fig. 4 Implemented prototype in 65nm CMOS and channel response

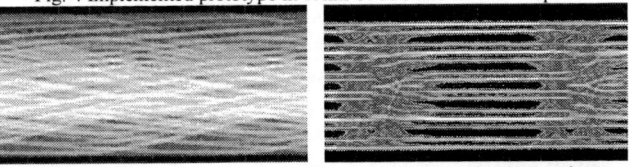

Fig. 5 Measured input eye (left) and 16 levels output eye of sequence detector (right). Due to non-binary tap values, level 3 and 4, and 11 and 12 are merged

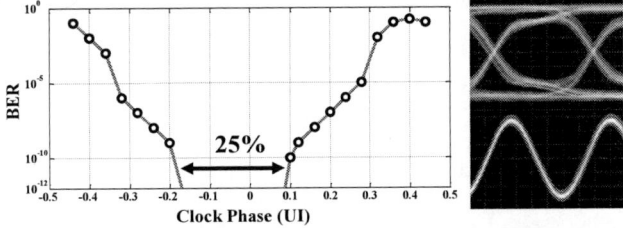

Fig. 6 BER bathtub, eye diagram of recovered data (B0) & clock

TABLE I: Performance Summary

	Chen JSSC'12	Zhang ISSCC'13	Shafik ISSCC'15	This work
Equalizer Architecture	4x Variable Ref. Flash ADC	4x Rectified Flash	32xSAR	4x Sequence DFE
Data Rate	10 Gb/s	10.3125 Gb/s	10 Gb/s	10 Gb/s
Technology	65nm	40nm	65nm	65nm
Compensated Channel loss	-29/-23 dB @10 Gb/s	-34 dB @10.3125 Gb/s	-25.3 dB @10 GS/s	-27 dB @10 GS/s
Timing Recovery	Baud-rate	Baud-rate	None	Data-Edge Sampled
Power Consumption (mW)	Rx– 130	ADC-195 DSP- 500	ADC – 79 Dig. EQ – 8	Sampler – 12 Digital – 14 Clocking – 9
Area (mm²)	0.26	0.82	0.81	0.23
Efficiency (pJ/bit)	13/10.6	67.4	8.7	3.5
FoM (pJ/bit/dB)	0.45/0.46	1.98	0.344	0.13

References

[1] O. Agazzi *et. al.*, JSSC, 2008.
[2] B. Zhang *et. al.*, ISSCC 2013.
[3] E.-H.Chen *et. al.*, JSSC, April 2012.
[4] A. Shafik *et. al.*, ISSCC, 2015.

A 125 mW 8.5-11.5 Gb/s Serial Link Transceiver with a Dual Path 6-bit ADC/5-tap DFE Receiver and a 4-tap FFE Transmitter in 28 nm CMOS

Bharath Raghavan, Aida Varzaghani, Lakshmi Rao, Henry Park, Xiaochen Yang, Zhi Huang, Yu Chen, Rama Kattamuri, Chunhui Wu, Bo Zhang, Jun Cao, Afshin Momtaz, and Namik Kocaman

Broadcom® Corporation, 5300 California Avenue, Irvine, CA 92617, USA, e-mail: bharathr@broadcom.com

Abstract

This paper describes an 8.5-11.5 Gb/s transceiver with a dual path receiver and a voltage-mode transmitter. The RX can operate either in ADC mode for complex loss channels such as optical multimode fiber or in DFE mode for copper-based backplane links. The ADC path implements a 2X interleaved 6-bit rectifying flash ADC using a programmable gain amplifier (PGA) with controlled bandwidth and peaking, comparator pipelining, and super-source follower circuit techniques. The LRM optical sensitivity requirements are met with a > 6 dB margin while achieving an ENOB of 4.59 bits at a 5 GHz input frequency. The TX/RX DFE path achieves copper channel loss compensation of 38 dB with BER $< 10^{-12}$ at 11.5 Gb/s consuming 46mW from a 0.9V supply. The TX/RX ADC path consumes 125 mW at 10.3125 Gb/s. The TX/RX occupies 0.56 mm^2 in a 28nm standard CMOS process.

Introduction

As data center and metro network bandwidth requirements rise, there is a need for higher 10 Gb/s network port density to support traffic over legacy multimode fiber (MMF) and electrical backplanes. Power consumption per lane has to be minimized to expand the available rack power by mitigating cooling requirements. Modal dispersion over MMF fiber creates multiple peaks and a time-varying pulse response. Robust performance thus requires an ADC-based solution with a DSP backend [1-3] while DFE-based RX can efficiently address electrical copper-based links [4-5].

This paper presents a transceiver featuring a dual path AFE receiver, which can be operated in either ADC or DFE mode, a voltage-mode TX driver with 4-tap FFE and an LC-PLL. The transceiver meets requirements of all 10G Ethernet standards for fiber (LRM, SR, LR etc.) and copper-based links (KR, CX1, SFP+ etc.) [6]. This paper mainly focuses on the RX ADC path with the rest described in greater detail in [5].

Circuit Implementation

The top-level block diagram of the AFE RX is shown in Fig. 1. On the RX ADC path, a coarse PGA (PGAc) receives the external input. This drives a 2X interleaved 6-bit rectifying flash ADC with bubble error correction which is followed by a digital 4-tap FFE and 5-tap DFE backend. A baud rate timing recovery circuit based on the Mueller-Muller algorithm is implemented. Separately, the inputs are also connected to the CTLE of the DFE path. The RX DFE path implements a 5-tap DFE and a CDR with 2X-oversampling and baud-rate timing

recovery options. The PLL provides the quadrature differential 5 GHz clocks to the RX phase interpolators which are shared between the ADC and DFE modes and achieve a phase resolution of 2.8°.

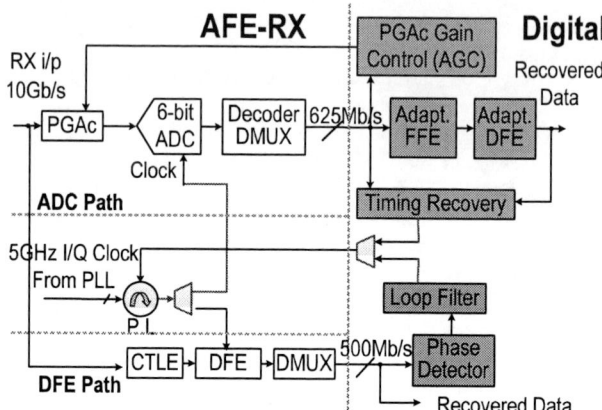

Fig. 1 Block diagram of the AFE receiver

The PGAc shown in Fig. 2(a) is a 3-stage amplifier chain. Programmable gain is realized using resistor degeneration. Digital calibration of the coarse PGA gain ensures constant 400 mV pp-diff signal amplitude at the ADC inputs. To minimize bandwidth variations and to ensure < 3 dB peaking across gain settings, each stage also has a switchable capacitor bank. The last stage shown in Fig. 2(b) further uses shunt peaking to extend its bandwidth. Thus, the PGAc achieves a > 25 dB dynamic range with a 3 dB bandwidth > 5.5 GHz.

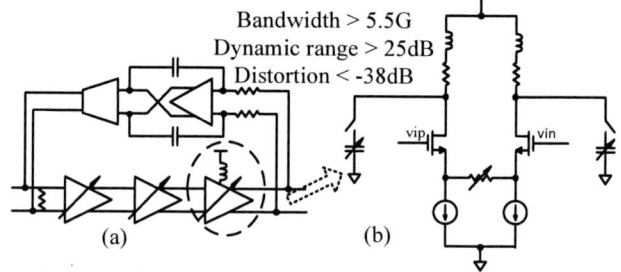

Fig. 2 (a) Block diagram of the PGAc (b) PGAc last stage

A rectifying flash ADC is implemented to halve the number of comparators required. In contrast with [1], a 2X interleaved architecture with an additional sampling stage per channel is implemented to achieve the same throughput but at a doubled channel clock rate. One such channel is shown in Fig. 3. The sampling path at the input of each flash ADC is split into the data path and the MSB path. Based on the sign magnitude of the input signal as detected by the MSB path, the data is rectified using the switching network S4. With only half the regeneration time available at the faster clock rate, efficiently achieving adequate BER using a single MSB comparator is challenging. The techniques used to implement a rectifying flash ADC clocked at up to 5.75 GHz are described below.

Each MSB path uses two of the double-tail comparators [7] shown in Fig. 4(a), which are pipelined as shown in Fig. 3, to obtain additional regeneration gain for good settling. Thus, folding related BER errors are minimized. Correspondingly, sampling switches S1/S1m are inserted onto each channel's input to match the data path delay with the increased MSB path latency. These are driven directly by the PGAc on opposite, non-overlapping clock phases without requiring input buffer stages for channel isolation. Thus, the sampled data path only needs two source follower buffers, which are implemented using the super-source follower circuit topology [8] for fast settling. The fine PGA (PGAf) is a differential pair amplifier with shunt inductive peaking to extend its bandwidth. It has about 1 dB of adjustable gain for interleaving gain mismatch calibration implemented using source degeneration.

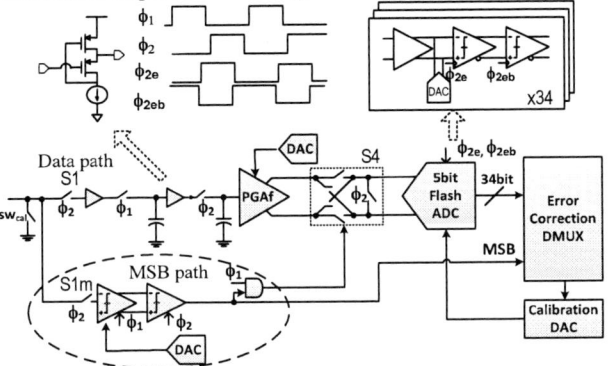

Fig. 3 Block diagram of one ADC receiver interleaving channel

The flash ADC has 34 preamplifiers for comparator kickback isolation, each driving consecutive strong-ARM comparators shown in Fig. 4(b). The comparators, which are pipelined as shown in Fig. 3 to minimize preamplifier loading, achieve a hysteresis of < 1 mV. Their size is minimized to reduce area and power consumption. Their offset is calibrated using a 6-bit current DAC, connected to the corresponding preamplifier output, having an offset resolution of < 1 mV.

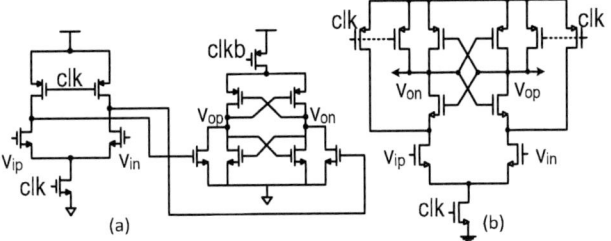

Fig. 4 Single (a) MSB comparator (b) flash ADC comparator

Measurement Results

A dual RX path transceiver is built in a 28 nm standard CMOS process. In ADC mode, the chip achieves error-free operation with a > 6 dB margin for the precursor, post-cursor and symmetric optical stressors, and for a 10 Hz dynamic channel specified by the 10GBASE-LRM standard. Across gain settings, as shown in Fig 5, the PGAc has a measured

bandwidth of 6.3-6.9 GHz and a dynamic range > 26 dB. The ADC achieves an ENOB of 4.8 at 100M and 4.59 for 5 GHz 700 mV pp-diff input at 10.3125 GS/s as shown in Fig. 6.

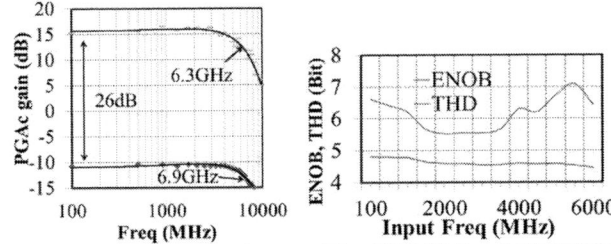

Fig. 5 PGAc measured gain curve   Fig. 6 RX ADC path THD/ENOB

The RX DFE path achieves a > 0.45 UIpp jitter tolerance at 11.5 Gb/s data rates as shown in Fig. 7. The TX/RX DFE path achieves BER < 10^{-12} at 11.5 Gb/s using 2^{31} -1 PRBS data over an electrical backplane link with 38 dB loss at Nyquist while consuming 46mW from a 0.9V supply.

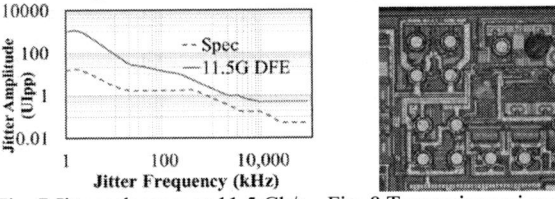

Fig. 7 Jitter tolerance at 11.5 Gb/s   Fig. 8 Transceiver micrograph

A micrograph of the transceiver is shown in Fig. 8. The dual-path TX/RX is 0.56 mm² in area.

MMF applications				Backplane applications			
Design	B. Zhang [1]	S. Verma [2]	This Work	Design	Musah [4]	Shafik [3]	This Work
CMOS Process	40nm	40nm	28nm	CMOS Process	22nm	65nm	28nm
Time-interleaved ADC Architecture	4X rectified flash	4X flash	2X rectified flash	TX EQ		4-tap FFE	No TX	4-tap FFE
ADC (bit)	6	6	6	RX EQ	CTLE+6-tap DFE	32X 6-bit SAR	CTLE+5-tap DFE
Data rate (Gb/s)	8.5 to 11.5	10.3125	8.5 to 11.5	Data rate (Gb/s)	16	10	8.5 to 11.5
LRM Optical Sensitivity at 10.3125Gb/s	6dB margin	6dB margin	6dB margin	Channel Loss at Nyquist	24dB	36dB	38dB
RX Area(mm²)	0.82 (dual path)	0.27	0.47 (dual path)	TX/RX Area(mm²)	0.08	0.81	0.56(dual)
RX Power(mW)	195	242	110	TX/RX efficiency (mW/Gb/s)	~3.8	8.9	4

Fig. 9 Performance comparison with previous work

The performance comparison with previous work is shown in Fig. 9. The RX ADC path consumes 110 mW using 0.9V supply at 10.3125 Gb/s. This is the lowest amongst recent DSP-based solutions for MMF applications [1-2] while achieving similar optical stressor performance. The core ADC has an FoM of 0.33 pJ/conversion step. The TX/RX ADC path consumes 125 mW at 10.3125 Gb/s. The TX/RX DFE path provides superior copper link equalization capability compared to [4], within a transceiver with dual-mode receiver capabilities, at a similar power efficiency of 4 mW/Gb/s.

References

[1] B. Zhang et al., ISSCC, Feb. 2013, pp. 34-35.
[2] S. Verma et al., ISSCC, Feb. 2013, pp. 271-350.
[3] A. Shafik et al., ISSCC, Feb. 2014, pp. 62-63.
[4] T. Musah et al., IEEE JSSC, Dec. 2014, pp. 3079-3090.
[5] T. Ali et al., VLSI Circuits, Jun. 2015, pp. C348-C349.
[6] IEEE Standards: 802.3ae, 802.3aq, and 802.3ap.
[7] D. Schinkel et al., ISSCC, Feb. 2007, pp. 314-315.
[8] R.G. Carvajal et al., IEEE TCAS-I, Jul. 2005, pp. 1276-1291.

A 0.003 mm² 5.2 mW/tap 20 GBd Inductor-less 5-Tap Analog RX-FFE

Ryan Boesch, Kevin Zheng, and Boris Murmann
Stanford University, Stanford, CA, USA

Abstract

A 0.003 mm² 5.2 mW/tap analog receive-side feedforward equalizer (RX-FFE) is demonstrated in 40 nm CMOS for up to 20 GBd ADC-based links. The FFE is constructed entirely with analog-inverter transconductors and capacitors, avoiding the use of area-intensive inductors. The delay element is implemented as a first-order Padé approximant of an ideal delay. The equalization performance is measured to be sufficient to relax the ADC resolution requirement by 1 bit. The total power consumed is less than 26 mW with less than 9.2 nV/√Hz output noise for all configurations.

Introduction

As the data rates for high-speed transceivers continue to increase, inter-symbol interference (ISI) due to channel loss is becoming more pronounced and multiple techniques have been suggested to address this issue [1-4]. One technique that has recently been gaining popularity is the ADC-based receiver [1,2] (see Fig. 1). In ADC-based receivers, a digital feedforward equalizer (FFE) is used in conjunction with a decision feedback equalizer (DFE) to equalize the channel and recover the data. However, in order to recover the data with a high fidelity, a power-hungry ADC is needed to digitize the signal [1]. A continuous-time linear equalizer (CTLE) can be used to perform some equalization prior to digitization, but recent work has shown that an analog receive-side FFE (RX-FFE) prior to the ADC can further reduce the required ADC resolution while achieving the same BER [2]. In order to obtain a net improvement for the system, the RX-FFE must be implemented with low power consumption, low noise, and small chip area.

The objective of the RX-FFE is not to completely equalize the channel and open the eye but instead to reduce the dynamic range of the signal resulting in a relaxation of the ADC resolution requirement. This is equivalent to reducing the peak signal to main cursor ratio (PMR). The peak signal results from the worst case sequence of bits and will be equal to the sum of the magnitude of the main cursor, p_0, and all pre/post-cursor ISI, p_k (i.e. PMR $= \sum_k |p_k|/p_0$). In this paper, we present an RX-FFE that achieves over 2x reduction in PMR (i.e. 1 bit relaxation in ADC resolution) while outperforming state-of-the-art designs in terms of power, noise, and area [3,4].

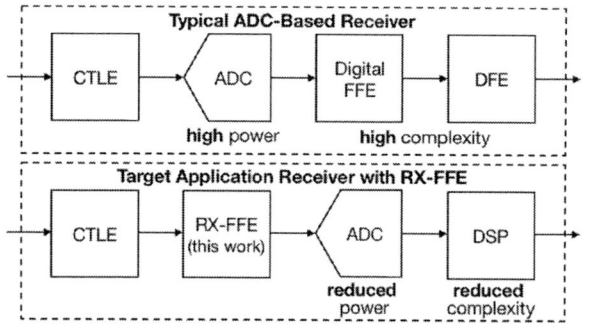

Fig 1. Diagram of a typical ADC-based wireline receiver and of the target system for the RX-FFE.

FFE Architecture and Circuit Design

A. FFE Architecture

Fig. 2 shows the half circuit of our pseudo-differential 5-tap RX-FFE architecture which consists of 4 delays, 5 digitally-programmable coefficients, and a weighted summing circuit. Each of these blocks is constructed from power-efficient class-AB analog-inverter transconductors resulting in an architecture with a small number of nodes and therefore a low-power, low-noise, and high-bandwidth design without the use of inductors [5].

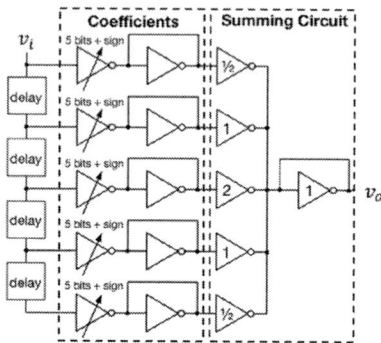

Fig 2. Half circuit block diagram of the inductor-less RX-FFE.

B. Analog Delay Implementation

A common strategy for previous RX-FFE designs was to use inductors to emulate an ideal delay line at the expense of large area [4]. A more recent design implemented the delays with a simple first-order pole/zero Padé approximant of an ideal delay, demonstrating that the coefficients of the FFE can absorb the non-idealities [3,6]. This approach, while an improvement on previous designs, suffers from increased noise and power due to the two signal paths in the delay implementation. The delay in this design achieves this same delay transfer function, but avoids the two-path implementation (see Fig. 2). It is implemented with a transconductor driving a self-biased replica load transconductor to achieve unity low-frequency gain and a feed-through capacitor to create the pole and zero. Although this simple cell implements the desired transfer function, it does not provide a high-impedance input to the previous stage. As a result, a buffer is required in between each delay. As shown in Fig. 3, the buffer is implemented as a replica of the transconductors used in the delay cell. The finite drive strength of this buffer results in an approximately 3x offset in the pole/zero spacing, but it can be shown that this offset can also be absorbed into the FFE coefficients. The load transconductors for both the buffer and the delay stage are degenerated by triode transistors biased to supply and ground to counteract the gain loss due to finite output conductance and to provide some linearization. In the proof of concept design, a 13 fF MOM capacitor was used to achieve an average tap delay of approximately 25 ps which is equivalent to T/2 at 20 GBd.

978-1-5090-0636-6/16 $31.00 © 2016 IEEE   134   2016 Symposium on VLSI Circuits Digest of Technical Papers

C. Coefficients and Summing Circuit

As depicted in Fig. 2, the coefficients of the RX-FFE are composed of 5-bit binary-weighted inverter transconductors driving a self-biased load sized to nominally set the low-frequency gain to a maximum of unity. A replica connected from the negative half circuit allows for negative coefficient values for a total of 5 bits + sign. The summing circuit is composed of a set of weighted transconductors designed to anticipate the relative magnitudes of the coefficient values to appropriately distribute the resolution (see Fig. 2). Additionally, it isolates changes to path gains due to adjacent coefficient value changes.

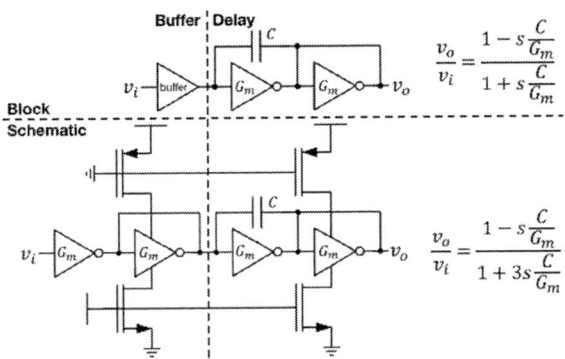

Fig 3. Block diagram and schematic of the unit delay cell half circuit.

Measurement Results

The circuit was fabricated in 40 nm GP CMOS with an FFE area of 0.003 mm² as shown in the die photo in Fig. 5. Table I shows the measured results and comparison with the state of the art [3,4]. The power consumption of the FFE is between 20 mW and 26 mW for all coefficient values. Similarly, the measured noise is between 4.2 to 9.2 nV/√Hz. A BERT scope was used to generate 50 ps (20 Gb/s) pulses through a 20 inch FR4 PCB backplane channel which was probed to the on-chip GSSG pads connected to a 50 Ω differential t-line terminated at the FFE input. After the FFE, the signal was driven off chip by 50 Ω pad drivers through GSSG pads probed out to a high-speed oscilloscope. Fig. 4 shows the measured channel and equalized pulse responses normalized to unity main cursor amplitude. For this channel and FFE coefficients, the attenuation of the main cursor is approximately 3x (not depicted in Fig. 4). The PMR of the channel was measured to be 3.87 which was reduced by the FFE to 1.83 for a 2.07x reduction, which is equivalent to a 1-bit relaxation in the ADC DR requirements. This DR range reduction is illustrated by the inset in Fig. 4 which shows the normalized PRBS responses generated from the associated pulse responses. In summary, the measurement results show that this design has advanced the state of the art in terms of power, noise, and area, while reducing the ADC DR requirement by 1 bit for 20 GBd signals.

Acknowledgments

This project was funded in part by the Broadcom Foundation. Chip fabrication was made possible by the TSMC University Shuttle Program. We thank Hiroshi Takatori, John Duan, and Albert Vareljian from Futurewei for help with chip debugging and for access to test equipment. Finally, we thank Frankie Liu and Vincent Lee from Oracle Corporation for access to test equipment.

References

[1] B. Zhang et al., "A 195mW / 55mW dual-path receiver AFE for multistandard 8.5-to-11.5 Gb/s serial links in 40nm CMOS," ISSCC Dig. Tech. Papers, pp. 34-35, Feb., 2013.

[2] E-Hung Chen et al., "Power optimized ADC-based serial link receiver," IEEE J. Solid-State Circuits, vol. 47, no. 4, pp. 938-951, Apr., 2012.

[3] E. Mammei et al., "A power-scalable 7-tap FIR equalizer with tunable active delay line for 10-to-25Gb/s multi-mode fiber EDC in 28nm LP-CMOS," ISSCC Dig. Tech. Papers, pp.142-143, Feb., 2014.

[4] A. Momtaz and M.M. Green, "An 80 mW 40 Gb/s 7-tap T/2-spaced feed-forward equalizer in 65 nm CMOS," IEEE J. Solid-State Circuits, vol. 45, no. 3, pp.629-639, Mar., 2010.

[5] B. Nauta, "A CMOS transconductance-C filter technique for very high frequencies," IEEE J. Solid-State Circuits, vol. 27, no. 2, pp. 142-153, Feb., 1992.

[6] N. Wiener and Yuk-Wing Lee, "Electrical network system," U.S. Patent 2 124 559, July 26, 1938.

TABLE I
PERFORMANCE SUMMARY AND COMPARISON WITH STATE-OF-THE-ART HIGH-SPEED RX-FFES.

	This Work	**[3]**	**[4]**
Power [mW]	20 to 26	55 to 90	80
Taps	5	7	7
Power/Tap [mW]	4 to 5.2	12.8	9.3
Baud Rate [GBd]	20	10 to 25	40
Process	40 nm	28 nm	65 nm
Supply [V]	1	1	1
Noise [nV/√Hz]	4.2 to 9.2	11 to 26.6	—
Area [mm²]	0.003	0.085	0.75

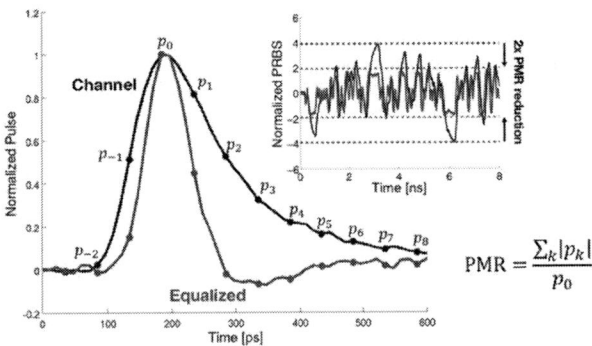

$$PMR = \frac{\sum_k |p_k|}{p_0}$$

Fig 4. Measured channel and equalized pulse responses normalized to unity main cursor amplitude and (inset) PRBS responses generated from the pulse responses illustrating the PMR reduction.

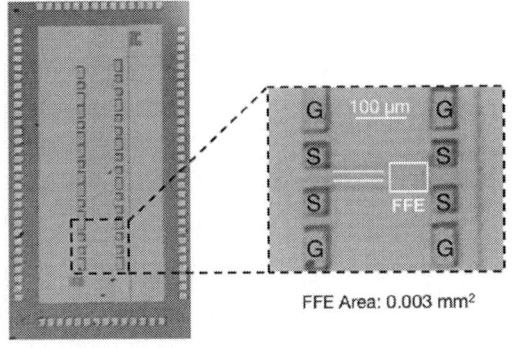

FFE Area: 0.003 mm²

Fig 5. Die photo.

A 16Gb/s 14.7mW Tri-Band Cognitive Serial Link Transmitter with Forwarded Clock to Enable PAM-16 / 256-QAM and Channel Response Detection in 28 nm CMOS

Yuan Du[1], Wei-Han Cho[1], Yilei Li[1], Chien-Heng Wong[1], Jieqiong Du[1], Po-Tsang Huang[1,2], Yanghyo Kim[1], Zuow-Zun Chen[1], Sheau Jiung Lee[1], Mau-Chung Frank Chang[1,2]

[1]University of California, Los Angeles, CA 90095-1594, USA, [2]National Chiao Tung University, Hsinchu, Taiwan
yuandu@ucla.edu

Abstract

A cognitive tri-band transmitter with forwarded clock using multi-band signaling and high-level digital signal modulations is presented for serial link application. The transmitter features learning an arbitrary channel response by sending a sweep of continuous wave, detecting power level, and accordingly adapts modulation scheme, data bandwidth and carrier frequency. The modulation scheme ranges from NRZ/QPSK to PAM-16/256-QAM. The highly re-configurable transmitter is capable of dealing with low-cost serial link cables/connectors or multi-drop buses with deep and narrow notches in frequency domain (e.g. 40dB loss at notches). The adaptive multi-band scheme mitigates equalization requirement and enhances the energy efficiency by avoiding frequency notches and utilizing the maximum available signal-to-noise ratio and channel bandwidth. The implemented transmitter consumes 14.7mW power and occupies 0.016mm^2 in 28nm CMOS. It achieves a maximum data rate of 16Gb/s per differential pair and the most energy-efficient FoM (defined in Fig. 8) of 20.4 μW/Gb/s/dB considering channel condition.

Introduction

The data rate of peripheral serial I/O for PCs and mobile computing platforms continue to scale to meet high-bandwidth applications including high-resolution displays/camera sensors and large-capacity external storage [1]. Recent publications demonstrated a multi-band signaling architecture to meet such stringent requirements in cost and energy efficiency [2-4]. Typical low-cost cables/connectors and multi-drop buses (MDB) impose notches and non-linearity in the frequency domain resulting from the resonance effect. The multi-band signaling takes advantage of such impairments by transferring data via multiple modulated-carriers where there is no such non-ideality. The previous work, however, works only with one specific cable/connector configuration, because not only the carrier frequency is fixed, but also there is no mechanism to gain knowledge on the channel conditions. In order to provide a universal solution capable of handling all different channels, we propose a cognitive tri-band forwarded-clock serial link transmitter (TX) with a frequency response learning algorithm. The TX senses the channel condition by first sending a single tone from 50 MHz to 10 GHz. Then the detector measures the received power on the other side of channel and feeds it back to the TX. With this information, the TX cognitive controller determines the carrier frequencies, modulation scheme, and bandwidth based on the system BER and data rate requirement.

Channel Responses with Notches

The common scenario of low-cost peripheral I/Os and its channel insertion loss is depicted in Fig. 1. When considering a cable-only case, the dielectric and conduction loss would exhibit a simple low-pass characteristic. Whereas in Fig. 1(b), the complete channel including packages, solder balls, wire-bonds, vias, traces and connectors suffers from higher loss at certain frequencies. This leads to the higher dispersion and distortion of signal. The phenomenon is more pronounced in low-cost packaging, PCB, cable and connector technologies. Another example of having such non-idealities is the case of MDB. As shown in Fig. 1(c), there could be multiple notches with more than 40dB loss. The deep and narrow of notches require complicated equalization or sensitive compensation technique, which are not energy and cost efficient solutions.

Transmitter Architecture and Implementation

A. Base-band TX vs Multi-band TX

Conventionally, serial link TX uses base band directly to transmit NRZ signal; however, with ever increasing bandwidth, there is less room to manipulate in time domain. Fig. 2 shows conceptual comparison of base-band TX and multi-band TX. Multi-band signaling offers simultaneous and orthogonal communication channels in the frequency domain, transferring part of the burden from time domain to frequency domain. More attractively, with multi-band signaling, it is easy to avoid notches by smartly choosing carrier frequency.

B. System Architecture

The block diagram of the proposed TX is shown in Fig. 3. The baseband consists of a 16-bit parallel PRBS generator operating up to 1 GHz; a UART serial interface to configure control register and monitor TX operation status; a cognitive controller to determine modulation scheme and carrier frequency allocation based on channel response detected by power detector and ADC; and a modulation mapping block to map PRBS binary code to DAC input. The analog/RF front end is comprised of two I/Q RF band paths -- quadrature dividers ($\div 4$ and $\div 2$ bands) and up-conversion I/Q mixers, one baseband path for forwarded-clock purpose, and a broad band summation block.

C. Circuit Implementation

Fully differential current-mode architecture is utilized to mitigate simultaneous switching noise (SSN) and supply/electromagnetic noise. As shown in Fig. 4(a), 4-bit current-steering DAC and double-balanced mixer are combined to improve energy efficiency. The 4-bit DAC output current swings from 20μA to 937.5μA with around 100mV voltage swing. Fig. 4(b) shows that the summation block consists of five slices, for $\div 4/\div 2$ I/Q bands and baseband clock. A 100 Ohm termination resistor (matching to channel characteristic impedance) and a switch (turning off if termination is not required) are attached in series at the end. The block needs to sum all signals from all bands (provide broadband operation up to 7GHz) and to substrate DC current to avoid desensitizing RX.

Measurement Results

A test chip comprising carrier generation, digital baseband controller, and tri-band front end is fabricated in a 28nm CMOS process and occupies 0.016mm^2 area. As Fig. 5 shows, a commercial power detector LMX2492EVM with 12bit-ADC is used to detect received power through channels from 50MHz to 10GHz during TX frequency sweeping. Detected channel frequency response information is processed by MachX03L FPGA board, based on which the cognitive algorithm will determine carrier frequency allocation, modulation schemes, maximum achievable data rate, and other reconfigurable parameters. Two different channel conditions are tested – 10'' low-cost differential cable by 3M and MDB modeled by open-stub transmission line on PCB. For RX side, down-conversion mixers, low-pass filters, amplifiers and HP 83460A as local oscillator (LO) constitute a high-performance receiver to coherently demodulated TX output signal. Fig. 6(a) is die photo and Fig. 6(b) shows the TX output spectrum of ÷4/÷2 RF bands with 256-QAM running at 1G baud (symbols/second) and forwarded half-rate baseband clock. Fig. 7 shows the demodulated eye diagrams and I/Q constellations with <-30dB EVM by the instrumental receiver. The TX achieves 16Gb/s data rate with 14.7mW power (i.e. 919fJ/b) at 1.2V power supply, which represents the most energy-efficient FoM of 20.4 µW/Gb/s/dB (i.e. power consumption of transmitting per Gb/s data and overcoming per dB worst-case channel loss within Nyquist frequency) compared with other state-of-arts (Fig. 8).

Acknowledgements

This work was funded by Broadcom Foundation. We would like to thank Dr. Afshin Momtaz in Broadcom for valuable advices and TSMC for chip fabrication support.

References

[1] Rajesh Inti, et al., VLSI Circuits, pp.346-347, Jun. 2015
[2] Yanghyo Kim, et al., ISSCC, pp.50-52, Feb. 2012
[3] Wei-Han Cho, et al., CICC, pp.1-4, Sep. 2015
[4] Wei-Han Cho, et al., ISSCC, in press, Feb. 2016
[5] Tamer Ali, et al., VLSI Circuits, pp.348-349, Jun. 2015
[6] Saxena, S., et al., VLSI Circuits, pp.352-353, Jun. 2015

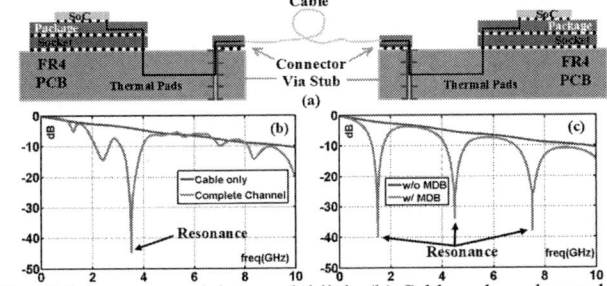

Fig. 1(a) Common periphery serial link, (b) Cable-only and complete channel insertion loss, (c) w/ MDB and w/o MDB insertion loss

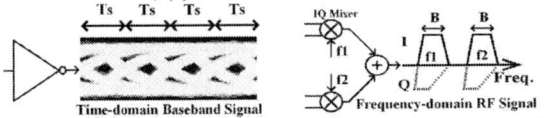

Fig. 2 Conceptual comparison of base-band TX and multi-band TX

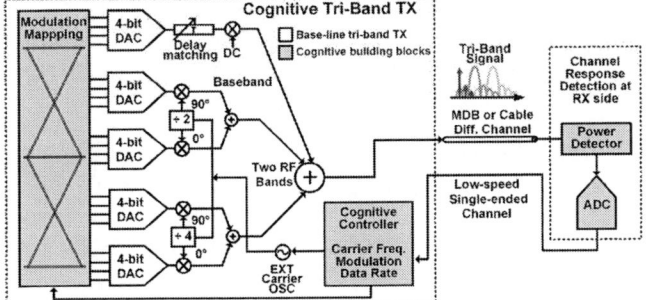

Fig. 3 System architecture of cognitive tri-band serial link TX

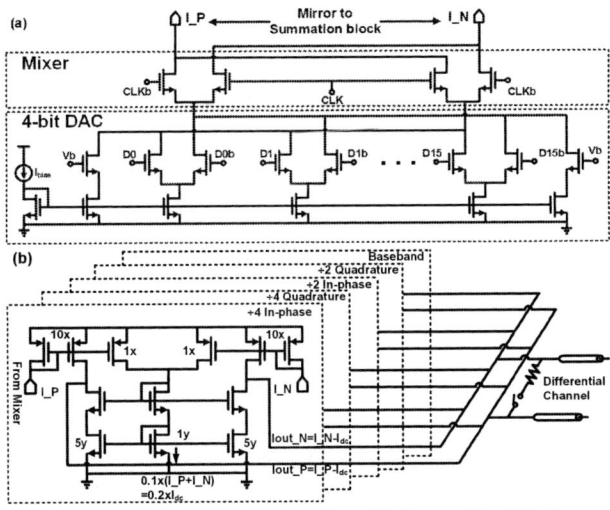

Fig. 4. Schematics of (a) DAC & Mixer (b) Summation Block

Fig. 5 Experiment platform

Fig. 6 (a) Die photo (b) Measured TX output spectrum

Fig. 7 Measured eye diagrams and constellations

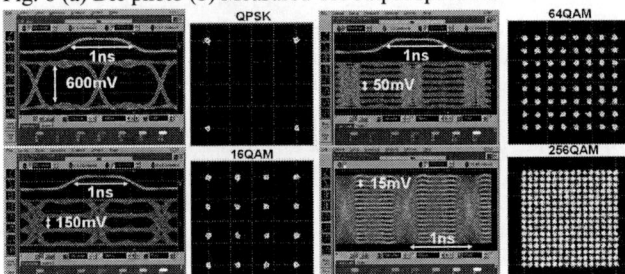

Metric	[1]VLSI 2015	[5] VLSI 2015	[6] VLSI 2015	This work
Technology	22nm CMOS	28nm CMOS	65nm CMOS	28nm CMOS
Data rate/lane	8 Gb/s	13 Gb/s	14 Gb/s	16 Gb/s
Signaling	Base-band NRZ	Base-band NRZ	Base-band NRZ	Multi-band QPSK,16,64,256-QAM
Clocking	Forwarded-clock w/ addition channel	Embedded Clock	Embedded Clock	Forwarded-clock w/o addition channel
TX Area/Lane	--	0.028 mm²	0.061 mm²	0.016 mm²
TX Power	2.56 mW	17.0 mW	12.5 mW	14.7 mW
TX Efficiency	320 fJ/bit	1308 fJ/bit	893 fJ/bit	919 fJ/bit
Worst Channel Loss within Nyquist Freq.	12 dB	35 dB	12 dB	45 dB (Cable) 40 dB (MDB)
FoM (µW/Gb/s/dB)*	26.7	37.4	74.4	20.4 (Cable) 23.0 (MDB)

Fig. 8 Performance summary and comparison table

FOM (µW/Gb/s/dB)* represents power consumption of transmitting per Gb/s data and overcoming per dB worst-case channel loss within Nyquist frequency.

A Low-EMI Four-Bit Four-Wire Single-Ended DRAM Interface by Using a Three-Level Balanced Coding Scheme

Il-Min Yi[1], Seung-Jun Bae[2], Min-Kyun Chae[1], Soo-Min Lee[1], Young-Jae Jang[1], Young-Chul Cho[2], Young-Soo Sohn[2], Jung-Hwan Choi[2], Seong-Jin Jang[2], Byungsub Kim[1], Jae-Yoon Sim[1], and Hong-June Park[1]

[1]Pohang University of Science and Technology (POSTECH), Pohang, Korea, E-mail: hjpark@postech.ac.kr

[2]Samsung Electronics Co., Hwasung, Korea

Abstract

The measured H-field EMI peak was reduced by around 15dB in a 4-wire single-ended DRAM interface by using a 3-level balanced coding scheme with a 100% pin efficiency. Charge-pump circuits are used to generate 3-level channel signals (-100mV, 0, +100mV) at TX. The RX input comparator uses the ground-level (0) as the voltage reference and employs the meta-stability to identify the ground-level input. The energy efficiency was 2.67pJ/b at 6.4Gb/s with a 65nm LP 1.2V CMOS process and 3-inch FR-4.

Introduction

In the mobile DRAM interface, electro-magnetic interference (EMI) is an important issue because mobile devices are packed with many closely-spaced chips including mobile DRAMs and many parallel high-speed signal lanes [1]. For low-EMI, a differential signaling is more efficient than a single-ended signaling due to balanced transitions. However, the differential signaling is not preferred in the DRAM interface because of pin count limitation. A multi-level balanced coding can reduce EMI with a single-ended signaling [2], [3]. In this work, a 3-level balanced coding is proposed to transmit 4-bit data through 4 parallel wires (Fig.1). With a conventional binary coding for 4 parallel wires, the sum of the currents flowing through the 4 wires fluctuates with time; it is constant for the balanced coding. Thus, a lower EMI is expected with the 3-level balanced coding because two currents flowing through the parallel wires in the opposite direction will cancel each other in the EMI field. It is important to maintain a 100% pin efficiency in the DRAM interface. Compared to [2], this work uses 4 parallel wires instead of 8 to reduce routing complexity and the current loop area that is proportional to the EMI field and crosstalk noise. The pin efficiency is 100% in this work; it is 75% in [3].

3-Level 4-Bit 4-Wire Balanced Coding

The 3-level signal values on the wires are -100mV, 0, +100mV in this work. A ground termination (a termination resistor connected between a RX input pin and ground) of this work enables an interface between two chips with different VDD values; this feature is preferred in DRAM interface. The ground-reference is used for the RX input comparator of this work to enhance noise-immunity [4]. Only 19 codes are available as balanced codes among 81 3-level codes on 4-wires. Among the 19 balanced codes, 12 codes of [+1, 0, 0, -1] combinations and 4 codes of [+1, -1, +1, -1] combinations are used in this work to minimize the average power and the temporal total current change of the TX driver circuit. For the '0' code input, the TX driver consumes no static power.

Circuit Implementation

The proposed 3-level 4-bit 4-wire (4b4w) transceiver consists of four TX and RX circuits, four FR-4 wires, encoder and decoder circuits, and two idle-mode charge-pump circuits for the TX circuits (Fig. 2). A TX circuit accepts two 2b half-rate data and a differential clock (CK, CKb) as input, and generates a 3-level output signal (+100mV, 0, -100mV) on a FR-4 wire; this provides a ground-centered 3-level NRZ signaling while [4] provides a ground-centered 2-level RZ signaling. The TX driver is a voltage-mode driver with three supply voltage (VP, VSS, VM); the on-resistance of three NMOS transistors (MP, MG, MM) is adjusted to Z_0. Two capacitive equalizers are used to compensate for ISI; one is activated for single-level transitions and both are activated for two-level transitions. The active-mode charge-pump circuits generate VP(+200mV) and VM(-200mV) for the TX circuits; the idle-mode charge-pump circuits are used to maintain the VP and VM values during the idle time where the clock signals (CK, CKb) are inactive. A RX circuit consists of two half-rate 3-level detectors and a ground termination; it accepts a 3-level signal input and generates two 2b half-rate binary data outputs.

The active-mode charge-pump circuits of TX consist of two parallel half-rate circuits for both VP and VM (Fig. 3(a), (b)). For the odd-mode VP charge-pump circuit (Fig. 3(a)), C_P is pre-charged to VDD while CK is '0', and a $+C_P \times VDD$ charge is pumped into C_{POBC} while both CK and the data input DPO are '1'. By controlling the number of parallel odd slices connected to the VP node, the VP value is set to 200mV in this work. For the odd-mode VM charge pump circuit (Fig. 3(b)), a $-C_M \times VDD$ charge is pumped into C_{MOBC} while both CK and DMO are '1'.

A half-rate RX 3-level detector of Fig. 2 consists of a latch-type comparator, a SR latch, and a D F/F with a preceding logic circuitry; the SR latch detects the '+1' or '-1' signal and the D F/F detects the '0' signal (Fig. 4). Because the RX input comparator uses the VSS level as the reference voltage, it takes a relatively long time for decision due to the meta-stability for the '0' signal input [5]. By appropriately delaying the sampling clock signal (CKRD), the D F/F output becomes '1' for the '+1' or '-1' signal input and '0' for the '0' signal input. By combining the SR latch output and the D F/F output, one can tell easily whether the RX input signal (RXIN) is '-1', '0', or '+1'.

Measurement Results

The proposed transceiver chip was fabricated in a 65nm LP CMOS technology (Fig. 5). The chip was measured with 6.4Gb/s PRBS-31 data, 1.2V supply, and 3-inch FR-4. The measured eye diagrams of the RX input (Fig. 6) reveal that the proposed signaling has lower crosstalk noise and larger voltage and timing margins than the conventional binary signaling with a 200mVpp signal swing (Fig. 6 and Fig.7). The measured power consumption was 54mW and 14.4mW for the entire 4b4w TX and RX chips, respectively. The energy efficiency of the entire transceiver chip is 2.67pJ/b. Because E-field has much undesirable noise, an H-field probe was used to measure the near-field EMI at different heights (Fig. 8). By integrating the square of the measured H-field, the H-field power is calculated. Compared to the binary signaling of Fig. 6, the peak reduction of 15dB at 3.2GHz and the integrated H-field power reduction of 9.5dB was achieved in the proposed transceiver with 0.5mm height. Table I summarizes the performance and compares this work with other 3-level-coded transceivers.

Acknowledgements

This work was supported by the NRF of the MSIP, Korea, under the contract number of 2014-052875, the ITRC support program (IITP-2015-H8501-15-1002), Samsung Electronics Company, and IDEC.

References

[1] Tae-Young Oh, *et al.*, *IEEE ISSCC Dig.*, pp. 430–431, Feb. 2014.
[2] A. Singh, *et al.*, *IEEE ISSCC Dig.*, pp. 442–443, Feb. 2014.
[3] S. Zogopoulos and W. Namgoong, *IEEE JSSC*, pp.549–557, Feb. 2009.
[4] J. Poulton, *et al.*, *IEEE ISSCC Dig.*, pp. 404–405, Feb. 2013.
[5] A. Shikata, *et al.*, *IEEE JSSC*, pp. 1022–1030, Apr. 2012.

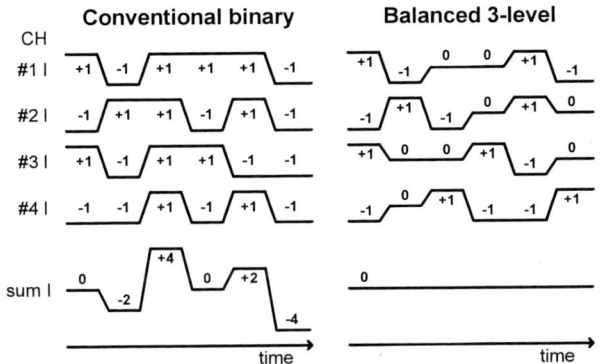

Fig. 1. Comparison of the conventional binary and the 3-level signaling.

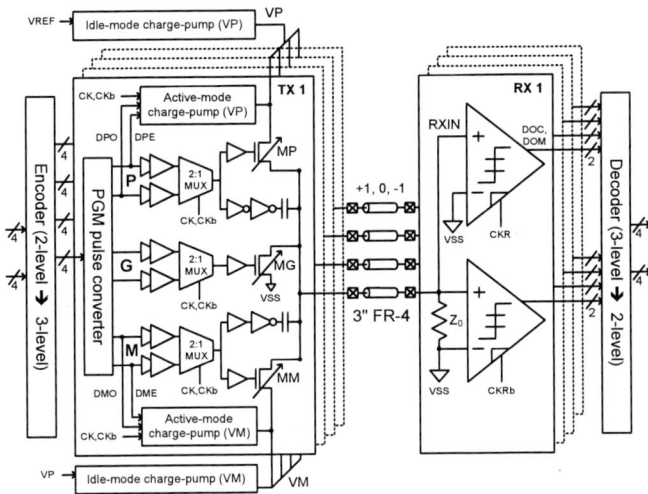

Fig. 2. The proposed 3-level 4b4w transceiver.

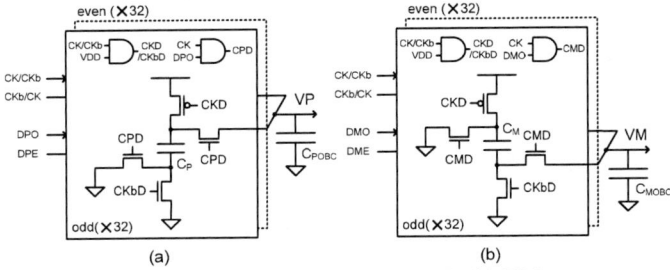

Fig. 3. Active-mode charge-pumps for TX (a) VP, (b) VM.

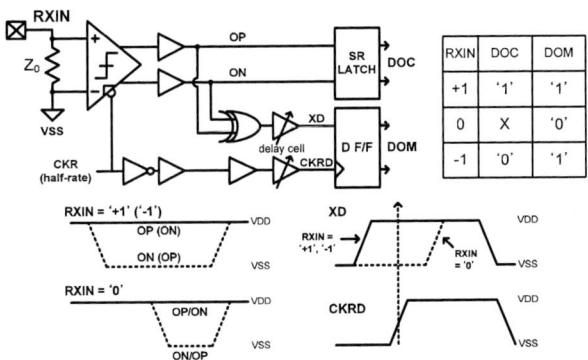

Fig. 4. A half-rate RX 3-level detector.

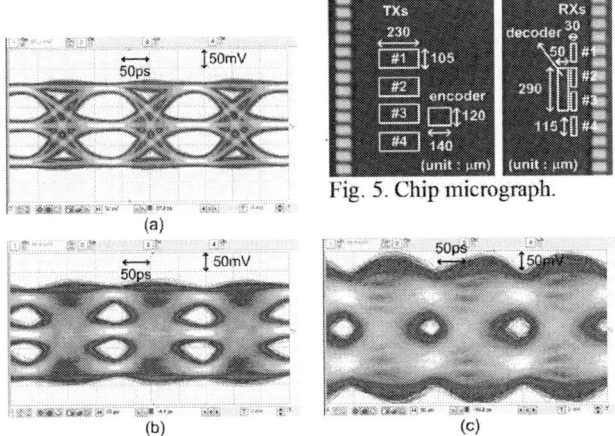

Fig. 5. Chip micrograph.

Fig. 6. Measured 6.4Gb/s eye diagrams (a) the proposed signaling at the TX #2 output with TX only, (b) the proposed signaling at the RX #2 input with TRX, (c) the conventional binary signaling at the RX #2 input with TRX.

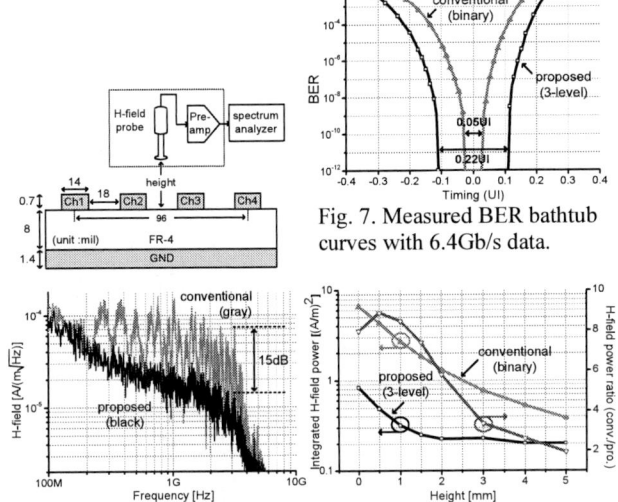

Fig. 7. Measured BER bathtub curves with 6.4Gb/s data.

Fig. 8. Setup for H-field EMI measurement and measured results.

TABLE I. PERFORMANCE COMPARISON

	JSSC'09 [3]	ISSCC'14 [2]	This work
Tech (nm)	180	40 GP	65 LP
Data rate (Gb/s)	4.2	12	6.4
Level	3	3	3
Min. # of wires	4	8	4
Pin efficiency (%)	75	100	100
Energy efficiency (pJ/b)	5.7	*4.29	2.67

* includes the clock generation circuits

A 0.3-2.6 TOPS/W Precision-Scalable Processor for Real-Time Large-Scale ConvNets

Bert Moons, Marian Verhelst

Department of Electrical Engineering, ESAT-MICAS - KU Leuven, Leuven, Belgium
bert.moons@esat.kuleuven.be, marian.verhelst@esat.kuleuven.be

Abstract

A low-power precision-scalable processor for ConvNets or convolutional neural networks (CNN) is implemented in a 40nm technology. Its 256 parallel processing units achieve a peak 102GOPS running at 204MHz. To minimize energy consumption while maintaining throughput, this works is the first to both exploit the sparsity of convolutions and to implement dynamic precision-scalability enabling supply- and energy scaling. The processor is fully C-programmable, consumes 25-288mW at 204 MHz and scales efficiency from 0.3-2.6 real TOPS/W. This system hereby outperforms the state-of-the-art up to 3.9x in energy efficiency.

Introduction

Recently CNN's (Fig. 1) have come up as state-of-the-art classification algorithms, achieving near-human performance in speech-recognition and visual-detection [1-3]. However, they are typically very expensive in terms of energy consumption. In [4], an algorithm-level study, we demonstrated opportunities for drastic energy reductions in CNN's through dynamic word length scaling and sparse guarding. Precision requirements vary across CNN's and even across CNN-layers, as the necessary number of bits can go down from 16 to 5 or even 1 bit for different benchmarks, with less than 1% accuracy loss (Tab. 1). Origami [5], Nvidia Tegra [6], and Eyeriss [7] offer non-optimal embedded solutions, as they keep computational precision constant and do not adapt to varying requirements. This work is the first to exploit these opportunities and to implement them in a state-of-the-art CNN architecture. It optimizes energy consumption for any CNN with any precision requirement up to 16-bit fixed point, without sacrificing flexibility, programmability, accuracy or throughput. We hereby empower low power, high-performance embedded applications of computer vision.

Low Power CNN Processor Design

This CNN-processor achieves scalable low power operation through three key innovations: (**A**) a 2D-SIMD MAC-array with shifted inputs, (**B**) dynamic precision and voltage-scaling and (**C**) guarded data-fetches and -operations. Figure 2 shows the high level processor-overview. It contains a precision-scalable 2D-SIMD array in a voltage-scalable power domain, a total of 148kB on-chip data-, guard- and program memory, max-pool and Rectified Linear Unit vector-arithmetic and a DMA with Huffman compression, all in a fixed power domain. The processor has a custom VLIW and SIMD instruction set and is fully programmable in C using dedicated libraries and a custom generated compiler. The chip is clock-gated and operator guarded where possible to save dynamic power.

A. The 16x16 2D-SIMD MAC array (Fig. 3) generates 256 intermediate outputs per cycle while consuming only 16+16 inputs. These MACs are single cycle and contain a 48-bit accumulation register. In an 11x11 convolution example, the MAC-array takes in 16 subsequent pixels from a single image channel and 16 filter weights from different filters in the first cycle. In the next 10 cycles, 17 words are fetched: 16 filter weights and a single pixel, which is shifted through a shift-register. This sequence is repeated once for every row in the 11x11 kernel. Custom instructions allow parallel

convolution-execution and data-fetching while all intermediate values are stored in the local accumulation registers. This 2D approach requires 16x fewer data fetches than 1D-SIMD.

An on-chip memory optimized for data-locality (Fig. 4) consists of 64x2kB single-port SRAM macros, subdivided into 4 blocks of 16 parallel banks. 3 blocks can be alternately read or written in parallel: 2 by the processor, a 3rd by the DMA.

B. Dynamic precision and voltage scaling can be performed without sacrificing significant accuracy [4]. At lower precision, not only the switching activity drops, but the critical path can become shorter as well if enforced at design-time through a multi-mode optimization (Fig. 3). Positive slack can then be compensated for through a lower supply voltage $V_{scalable}$ while keeping frequency constant. The full 2D-array is placed in a single dynamically scalable power domain. All non-scalable other arithmetic, memories, control and data-transfer are at a fixed supply voltage. E.g. in layer 2 (l2) of the well-established AlexNet benchmark [1], images and filters need 7 bits. This leads to a 1.9x gain compared to full precision and 1.3x additional savings when scaling the supply voltage (Fig. 6).

C. Guarding memory fetches and operations allows further energy reductions in sparse CNN computations. The amount of zero-valued data can go up to 89% (Tab. 1). This sparsity is caused both by ReLu operators and by using low precision words [4]. As zero-valued weights or pixels do not contribute to the CNN output, all computations using these values can be skipped. This is done by preventing memory fetches and MAC-operations (Fig. 3). As all sparsity info is known at the start of a new layer, it is stored in a dedicated guard flag memory (Fig. 2, 4). Only 16+16 1-bit flags are fetched to potentially prevent 256 MACs and 32-SRAMs from switching. In AlexNet l2, 19% and 89% zeroes in filters and images lead to an additional 1.9x energy gain (Fig. 6). The Huffman IO-compression (Fig. 2, Tab. 1) reduces bandwidth up to 5.8x for image data and up to 2x overall.

Measurement Results

This processor was fabricated in a 40nm LP CMOS technology. At room temperature, it runs at a nominal frequency of 204 MHz at 1.1V (Fig.7). It has a total active area of $1.2x2 = 2.4mm^2$. When scaling down from 16- to 8- or 4-bit, the supply voltage $V_{scalable}$ can go down from 1.1V nominally at 204 MHz to 0.9V and 0.8V at 8-bit and 4-bit respectively at the same speed (Fig. 5). All modes run the same non-guarded program, with a typical MAC-efficiency of 77%, achieving up to 2.6 TOPS/W in its most efficient 4-bit mode at 12 MHz.

Three benchmarks: AlexNet [1], LeNet-5 [3] and a general non-scaled/non-guarded 16-bit large scale CNN are run on the processor with measurement results summarized in Table 1. This table shows the precision and guarding opportunities per layer, and the resulting processor memory bandwidth and power reduction. At its nominal frequency the processor dissipates 33mW or 1.6 real TOPS/W for 13400 fps LeNet-5 (28x28 MNIST images) and 76mW or 0.94 real TOPS/W for 47fps AlexNet (227x227 ImageNet images), effectively minimizing energy consumption for both benchmarks. The chip hereby outperforms the non-scalable state-of-the art up to 3.9x in terms of energy efficiency (Fig. 8).

Acknowledgements

This work was partly funded by IWT. Special thanks to S. Redant, L. Folens and E. Wouters (imec IC-link) for back-end support.

References

[1] A. Krizhevsky, et al., "ImageNet classification with deep Convolutional Neural Networks", *Adv. in NIPS*, 2012.

[2] O. Abdel-Hammid, et al., "Applying CNN concepts to hybrid NN-HMM model for speech recognition", *ICASSP*, 2012.

[3] Y. LeCun, et al., "Gradient-based learning applied to document recognition", *Proceedings of the IEEE*, 1998.

[4] B. Moons, et al., "Energy-efficient ConvNets through approximate computing", *WACV*, 2016. *In press.*

[5] L. Cavigelli, et al., "Origami: a convolutional network accelerator", *Great Lakes Symposium on VLSI*, 2015.

[6] Nvidia Tegra K1 GPU.

[7] Yu- Hsin Chen, et al., "Eyeriss: an energy-efficient reconfigurable accelerator for deep Convolutional Neural Networks", *ISSCC*, 2016. *In press, with author approval.*

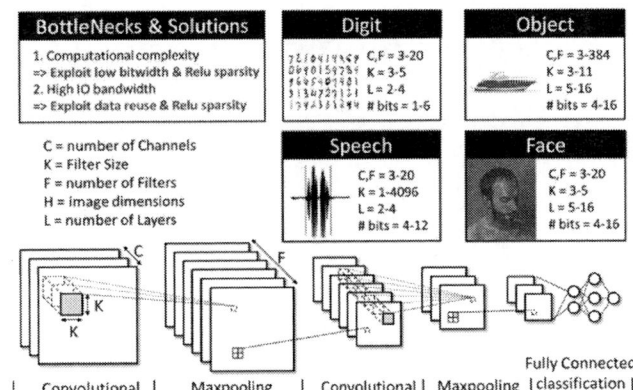

Fig. 1: ***Scalability in deep convolutional neural networks.***

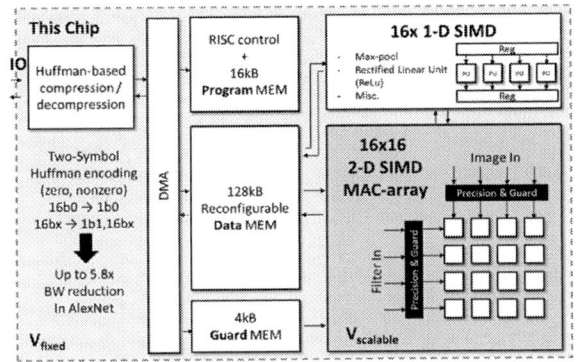

Fig.2: ***Top level architecture***. *All non-scalable logic is in a fixed power domain. The MAC-array is in a scalable power domain.*

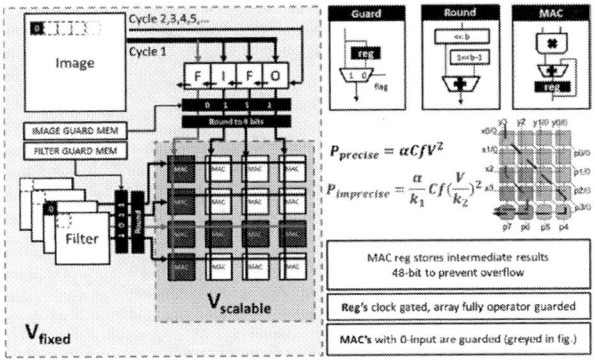

Fig.3: ***2D-SIMD array***. *The switching activity and critical path scale with precision. The latter allows lower supply at constant frequency.*

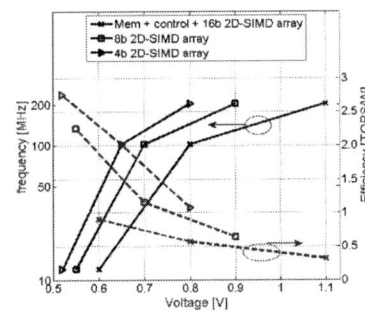

Fig.4: ***Memory architecture***.

Fig.5: ***Processor performance***. *Voltage is for the MAC-array only in the 8b/4b case. Efficiency is for the whole chip.*

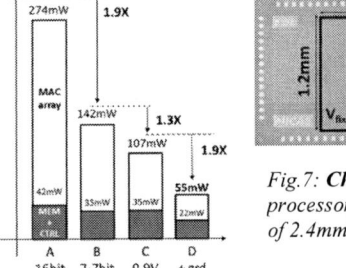

Fig.6: ***Energy saving mechanisms*** *in AlexNet layer 2. (B) 7-bit (filters) to 7-bit (images) are sufficient. (C) Supply voltage scaling. (D) Guarding added.*

Fig.7: ***Chip photograph***. *The processor has an active area of 2.4mm² in 40nm CMOS.*

Technology	40nm LP (1P_8M)
Core Area	1.2mmx2mm
On-Chip MEM Size	144kB
# MAC's	256
# Gates (NAND-2)	1600k
Supply voltage	0.55-1.1 V
Leakage	0.7 mW
Frequency	12-204 MHz
Word bit width	1-16 bit fixed
# Filters	All-programmable
# Channels	All-programmable
Stride	Horizontal: 1-4 Vertical: no limit
Peak performance	102 GOPS
Power (AlexNet)	76 mW
Throughput (AlexNet)	227x227 @ 47fps

Fig.8: ***Chip overview and comparison***. *This work outperforms the state-of-the-art up to 3.9x. Mentioned throughput is for the CNN convolutional layers. [6] is a full Tegra board.[5], [7] and this work are cores only.*

Layer	Filter / Image bits (0%)	Filter / Image BW Reduc.	IO / HuffIO (MB/frame)	Voltage (V)	MMACs/ Frame	Power (mW)	Real (TOPS/W)
General CNN	16 (0%) / 16 (0%)	1.0x	—	1.1	—	288	0.3
AlexNet l1	7 (21%) / 4 (29%)	1.17x / 1.3x	1 / 0.77	0.85	105	85	0.96
AlexNet l2	7 (19%) / 7 (89%)	1.15x / 5.8x	3.2 / 1.1	0.9	224	55	1.4
AlexNet l3	8 (11%) / 9 (82%)	1.05x / 4.1x	6.5 / 2.8	0.92	150	77	0.7
AlexNet l4	9 (04%) / 8 (72%)	1.00x / 2.9x	5.4 / 3.2	0.92	112	95	0.56
AlexNet l5	9 (04%) / 8 (72%)	1.00x / 2.9x	3.7 / 2.1	0.92	75	95	0.56
Total / avg.	—	—	19.8 / 10	—	—	76	0.94
LeNet-5 l1	3 (35%) / 1 (87%)	1.40x / 5.2x	0.003 / 0.001	0.7	0.3	25	1.07
LeNet-5 l2	4 (26%) / 6 (55%)	1.25x / 1.9x	0.050 / 0.042	0.8	1.6	35	1.75
Total / avg.	—	—	0.053 / **0.043**	—	—	33	1.6

Tab. 1: ***Performance overview*** *of 16-bit, AlexNet and LeNet-5 CNN's. In AlexNet and LeNet, the used number of bits in filter- and image weights can reduce drastically down to 1-bit, with less than 1% accuracy loss. This data will have high percentages of zero-values (0%), leading to IO BW reduction and more gains through guarding. Supply voltage can be modulated. Not every layer needs an equal amount of MAC-operations/frame. AlexNet l2 needs most operations, leading to a weighted average power consumption of 76mW.*

A 1.40mm^2 141mW 898GOPS Sparse Neuromorphic Processor in 40nm CMOS

Phil Knag, Chester Liu, Zhengya Zhang

Department of Electrical Engineering and Computer Science, University of Michigan, Ann Arbor

Abstract

Sparsity is a brain-inspired property that enables a significant reduction in workload and power dissipation of deep learning. This work presents a 1.40mm^2 40nm CMOS sparse neuromorphic processor that implements a two-layer convolutional restricted Boltzmann machine (CRBM) for inference and a support vector machine (SVM) classifier. The processor incorporates sparse convolvers to realize sparsity-proportional workload reduction. The architecture is parallelized along a non-sparse dimension to minimize stalling. At 0.9V and 240MHz, the processor achieves an effective 898.2GOPS performance, dissipating 140.9mW. Using sparsity, we reduce the workload, datapath power consumption and area by 3.4×, 3.3× and 1.74×, respectively. The design uses latch-based memory to reduce area and dynamic clock gating to save power.

Introduction

Deep learning is a powerful technique for big data analytics. Popular deep learning algorithms, e.g., convolutional deep belief network [1], rely on filtering an input using multiple layers of specialized kernels for detection and classification. These powerful algorithms demand intense computation for practical applications, as the input is often high in dimensionality, and the number of kernels and the kernel size need to be sufficiently large. Moreover, as we increase the depth of the network, the computational intensity and memory size grow much further. Prior work has demonstrated chip-, package-, board-level integration, and custom ASICs that accelerate deep learning up to 1.93TOPS/W [2], [3]. In this work, we present a sparse neuromorphic processor to exploit sparsity that is inherent in biologically inspired deep learning algorithms to enable the next order of magnitude improvement in performance and efficiency.

A Sparse Neuromorphic Processor

Sparsity is a key advantage of our approach, not only because of the physiological evidence that the brain uses a sparse representation to encode sensory inputs, but also due to its significant reduction of computational complexity. Enforcing sparsity has been shown to learn better features for classification [1], [4]. To introduce sparsity, a term of neuron activity is added to the cost function, and learning is done via sparsity regularization.

We design a sparse neuromorphic processor for object recognition based on convolutional deep belief network trained with sparsity regularization [1]. The processor adopts a representative 3-layer configuration (Fig. 1) that supports common image and video processing tasks. Layer 1 and 2 (L1 and L2) are CRBM layers that perform feature extraction. The outputs of L1 and L2 are max-pooled to create H1 and H2 respectively. Layer 3 (L3) is a SVM that performs classification using the sum of each channel from H2. The majority of the workload, i.e., 76M OP per 100×100 input patch (an OP is defined as an 8b multiply or a 16b add), is done in L2, followed by 18M OP in L1. Trained with sparsity regularization, the outputs of L1 and L2, i.e., H1 and H2, become sparse (Fig. 1). To achieve a high classification accuracy, the sparsity targets of H1 and H2 are tuned to ≥87.5%, i.e., ≥87.5% of H1 and H2 are zeros, enabling power reduction.

Sparse Convolver

The effective use of sparsity to reduce workload is a challenge for parallel architectures, as random sparse inputs result in redundant operations and memory contention. A conventional k×k patch convolver (Fig. 2(a)) that computes the convolution of a k×k kernel with an input works well for dense inputs. However, if the input is sparse, the patch convolver is inefficient, as multiplying by zero is wasteful. We design a sparse convolver (Fig. 2(b)) that utilizes a priority encoder to scan a stream of inputs at a time and forwards only the non-zero entries to a line convolver, skipping redundant operations. A k-pixel line convolver achieves k-way parallelism using k multipliers, and a (2k−1)-entry register bank with a k:(2k−1) selector for data alignment. The line convolver computes the convolution of an input pixel with a row in the kernel (Fig. 2(c)). The row address of the input pixel determines which row of the kernel to use, and the column address of the input pixel sets the selector for data alignment. Thanks to

the predictable data dependency between consecutive line convolutions, the memory contention between consecutive convolutions is resolved by a simple selector. The selector only needs to access a small 1-D line of temporary outputs, as opposed to a large 2-D block. The line convolver outputs a line that is written to a contiguous memory region, which is not accessed again for the current kernel operation to save memory accesses.

With a target sparsity no less than 87.5%, an 8-pixel sparse convolver matches the throughput of an 8×8 patch convolver, but its power and area are 3.3× and 1.74× lower than the patch convolver. Since the input to L1 cannot be assumed to be sparse, L1 is implemented using patch convolvers. The more computationally intensive L2 uses sparse convolvers to save significant power and area.

Parallel Architecture Optimized for Sparse Inputs

A parallel architecture consisting of multiple convolvers could incur inefficiencies in processing a sparse input, as the irregularity of sparse entries causes variable completion times and stalling. A common way to allocate multiple convolvers is along one primary dimension (Fig. 3): pixel dimension (P-parallel), channel dimension (C-parallel), or kernel dimension (K-parallel). If an input is sparse in a dimension (often the P and C dimension), allocating convolvers along the sparse dimension results in uneven workloads among the convolvers. Instead, we choose to parallelize along the non-sparse K dimension. However, aggressively pursuing parallelism along one dimension leads to uneven memory bandwidths (Fig. 3), e.g., K-parallel requires a low input bandwidth, a high kernel and output bandwidth, whereas P-parallel has some opposite characteristics.

Guided by the insights, we design a 3× P-parallel and 2× K-parallel architecture for L1 using 6 8×8 patch convolvers (Fig. 4) to balance the input and weight bandwidths. Since the input to L2 is sparse, we design a 16× K-parallel and 2× P-parallel architecture for L2 using 32 8-pixel sparse convolvers (Fig. 4) to minimize stalling. Note that the 2× P-parallel is used to halve the weight bandwidth at the cost of 11% stalling. The latency of L1 and L2 processing are balanced for interleaving. The fluctuations in sparsity are smoothed out by H1 and H2 buffering. At 160MHz, the architecture meets the 30fps 1920×1080 HD video data rate.

Test Chip Design and Measurement

A 1.40mm^2 sparse neuromorphic processor (Fig. 5) is implemented in 40nm CMOS. The chip uses 40Kb registers to store weights and 12Kb registers to queue L2 outputs. As the weight and L2 output storage are not updated in a pipelined fashion, we replace the registers by latches that are 25% more compact than registers after routing. Since the weight storage, and H1 and H2 interface buffers are infrequently updated, dynamic clock gating is applied to turn off the clock input to save the power by 47%.

With a 0.9V supply and 240MHz frequency, the processor's measured throughput at room temperature is 96.4M pixel/s, consuming 140.9mW (Fig. 6). The processor takes advantage of sparsity to achieve an effective performance of 898.2GOPS. Tested with the Caltech 101 dataset [5] to identify faces (with 50% faces, 50% non-faces, 434 training images and 434 testing images), the processor achieves an 89% classification accuracy. The chip demonstrates a competitive power efficiency of 6.37TOPS/W and area efficiency of 641.6GOPS/mm^2, which are 3.3× and 15.6× higher than the state-of-the-art non-sparse deep learning processor [2] (Table I). With a scaled supply voltage of 0.65V, the efficiency improves to 10.98TOPS/W. The techniques demonstrated in this work are applicable to a class of sparse processing problems.

Acknowledgements

This work was supported in part by Intel, DARPA and SONIC.

References

[1] H. Lee, et al., ACM Commun., Oct. 2011.
[2] S. Park, et al., ISSCC, 2015.
[3] J. Lu, et al., ISSCC, 2014.
[4] J. K. Kim, et al., VLSI Symp., 2015.
[5] L. Fei-Fei, et al., IEEE CVPR, 2004.

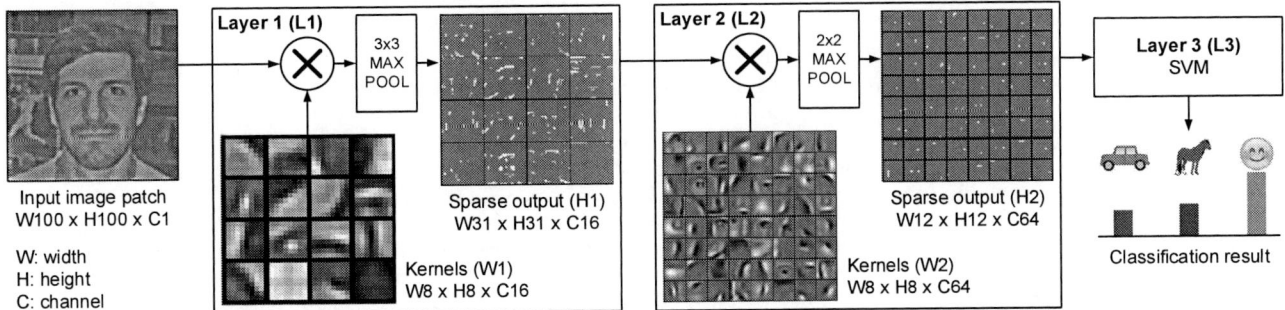

Fig. 1. Convolutional deep belief network composed of two layers of convolutional restricted Boltzmann machine (CRBM) layers and a support vector machine (SVM).

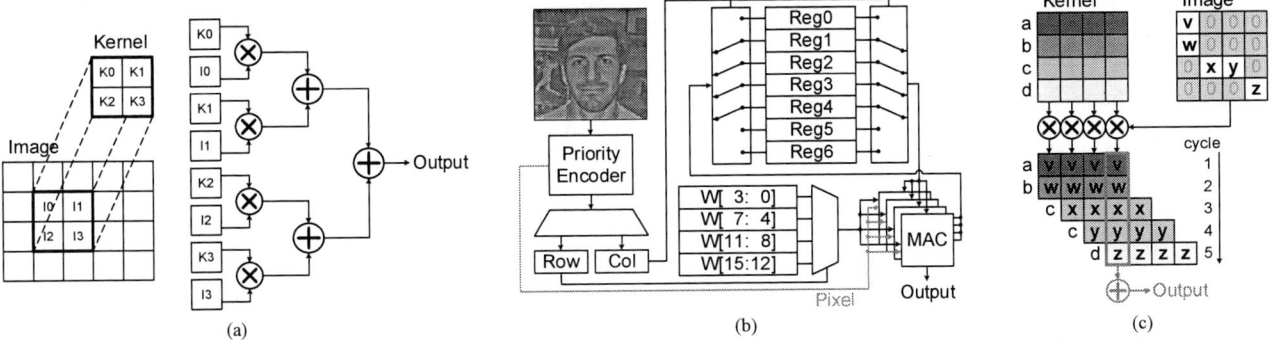

Fig. 2. (a) 2x2 patch convolver, (b) 4-pixel sparse convolver consisting of a priority encoder and a line convolver, (c) illustration of line convolver operation.

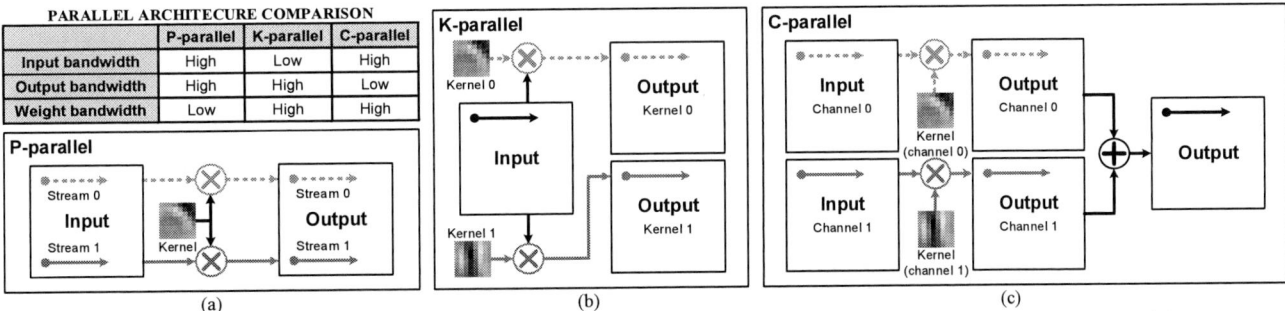

Fig. 3. Comparison of three types of parallel architectures: (a) pixel-parallel (P-parallel), (b) kernel-parallel (K-parallel), and (c) channel-parallel (C-parallel).

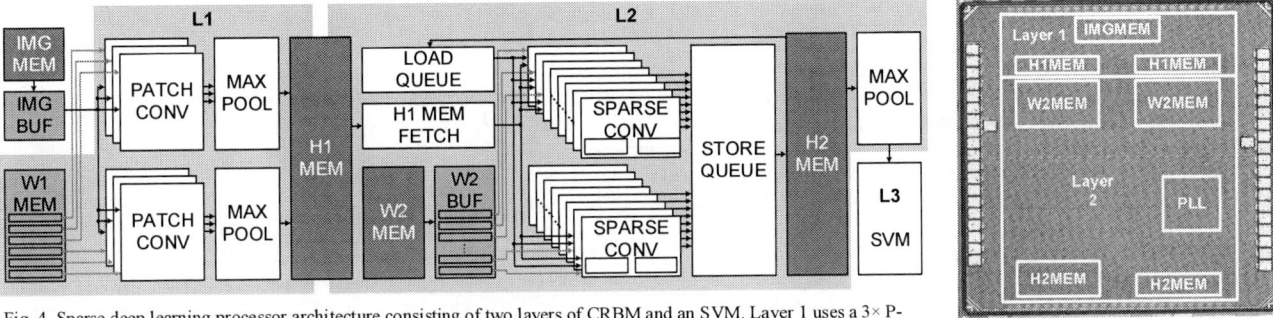

Fig. 4. Sparse deep learning processor architecture consisting of two layers of CRBM and an SVM. Layer 1 uses a 3× P-parallel and 2× K-parallel architecture, and layer 2 uses a 16× K-parallel and 2× P-parallel architecture.

Fig. 5. Chip microphotograph.

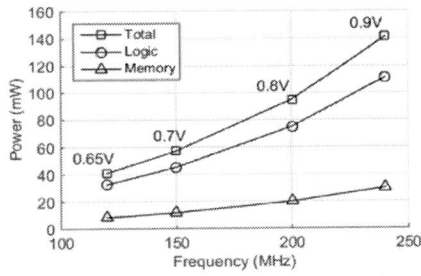

Fig. 6. Measured power and frequency of the test chip at room temperature.

TABLE I: COMPARISON WITH PRIOR WORKS

Reference	This Work		ISSCC'15 Park [2]	ISSCC'14 Lu [3]
Application	Object recognition		Big data analysis	Pattern recognition
Function	Deep neural network		Deep neural Network	Unsupervised online clustering
Technology	40nm		65nm	0.13um
Area	1.4mm²		10.0mm²	0.36mm²
Voltage	0.9V	0.65V	1.2V	3V
Frequency	240MHz	120MHz	200MHz	8.3kHz
Performance	898.2GOPS	449.1GOPS	411.3GOPS	0.012GOPS
Power	140.9mW	40.9mW	213.1mW	11.4uW
Power Efficiency	6.37TOPS/W	10.98TOPS/W	1.93TOPS/W	1.04TOPS/W
Area Efficiency	641.6GOPS/mm²	320.8GOPS/mm²	41.13GOPS/mm²	0.03GOPS/mm²

* Performance, power efficiency, and area efficiency of this work are based on effective number of operations

A 190GFLOPS/W DSP for Energy-Efficient Sparse-BLAS in Embedded IoT

Richard Dorrance, Dejan Marković
University of California, Los Angeles, CA

Abstract

A DSP for sparse-BLAS is realized in 40nm CMOS. Featuring an efficient data stream reordering scheme and an intelligent, CSC-aware memory controller, the DSP achieves a peak energy efficiency of 190 GFLOPS/W at 0.6V, 160MHz, and a peak performance of 4.12 GFLOPS at 1V, 515MHz showing more than 6,600×, 2,700×, 1,100×, and 450× higher energy efficiency than state-of-the-art CPU, GPU, DSP, and FPGA hardware designs, respectively.

Introduction

With the explosive growth of the Internet of Things (IoT), mobile and embedded systems are being asked to support more and more computationally intensive applications such as augmented reality, neural networks, 3D gaming, portable medical imaging, and mobile health monitoring—all of which rely on the manipulation of very sparse data sets [1,2]. The energy efficiency of the sparse basic linear algebra subroutines (sparse-BLAS) has always significantly lagged behind that of their dense counterparts by 2 to 3 orders of magnitude. In existing architectures, sparse matrix-vector and matrix-matrix multiplications (SpMxV and SpMxM) are almost exclusively calculated as series of dot products using the compressed sparse row (CSR) sparse matrix data format. However, due to low computational complexity and the irregular memory accesses of CSR, frequent cache misses and data hazards force the processor to stall for dozens to hundreds of cycles [2] (Fig. 1), further exacerbating the 100× energy gap between computation and memory accesses [3]. In order to improve the energy-efficiency of sparse-BLAS for embedded IoT applications, this paper presents a dedicated sparse linear algebra DSP, based on the compressed sparse column (CSC) data format, in a 40nm CMOS technology.

System Architecture

We designed a scalable sparse-BLAS architecture (Fig. 2) with CSC to roughly halve the number of required memory accesses and eliminate the data hazards present in CSR. It consists of a sparse-BLAS controller (with an integrated memory controller), 4 processing elements (PE), and a 512Kb cache memory. Each PE uses CSC to calculate SpMxV ($y = Ax$) as a series of column-wise vector additions of A weighted by each element of x. Due to the format of the memory references in CSC, the resulting column-major operations allow each element of x to be fetched sequentially and reused (as opposed to multiple random accesses and no reuse for CSR). For example, a sparse matrix with ~10 nonzero elements per column results in a 40% reduction in the memory bandwidth using CSC versus CSR due to reuse. Additionally, every CSR partial product results in a data hazard that stalls the processor for 3-8 clock cycles (fused multiply-add latency) due to the recursive data dependencies of the row-based approach. In principle, CSC avoids these data hazards by accumulating partial products from different rows each clock cycle, but it does not eliminate them. CSC sparse matrices on average see one data hazard for every 15-50 nonzero elements [1,2].

To eliminate the remaining data hazards, a stall-free "Shuffler" has been designed to manage the flow of data in each PE. Fig. 3 shows a block diagram and layout of the PE. In addition to the "Shuffler," each PE occupies 0.055mm², contains a single FPU (32b FP adder and multiplier), and a 16Kb dual-port SRAM (DP-SRAM) to compute and update partial products in the same clock cycle. When enabled, the "Shuffler" first fills a FIFO-like buffer (depth of 4) for each element of the data matrix (A_{ij}), vector (x_j), row address (i), and $Valid$ signal used to

calculate a partial product. The "Shuffler" then monitors the last 4 addresses of i issued to the FPU (the latency of our 32b FP adder) for potential data hazards in the buffer. When the first item in the buffer causes a data hazard, the shuffler substitutes the first available, hazard-free, partial product. If no hazard-free, partial product exists in the buffer, the "Shuffle FSM" stalls the PE until the data hazard is resolved. However, a buffer depth equal to the adder latency guarantees zero data hazards. Using this strategy, data can be continuously streamed into each PE with a small startup overhead equal to the combined latency of the adder, multiplier, and the "Shuffler" buffer. When the "Shuffler" is disabled, the PE must stall for 4 clock cycles to resolve a data hazard.

To provide a scalable, high-speed data interface between the 2GB of DDR2 memory on our FPGA system and each of the 4 PEs (chosen to optimize energy per bandwidth, Fig. 2), an on-chip, 512Kb, DP-SRAM is used as a memory cache. The resulting memory hierarchy is top heavy compared to CPUs, GPUs, and DSPs (Fig. 1). Since each PE can only store a finite number of elements, the PE contains a partial working copy of the vector being computed (the 16Kb DP-SRAM in each PE is chosen to optimize energy per partial product computation, Fig. 2). Blocking is performed along the rows of A by the on-chip, sparse-BLAS controller (i.e. each PE is assigned up to 512 rows of A for computation). The final vector is then assembled by concatenating the output of each PE during memory write-back (requiring no additional latency).

Measurement and Comparisons

SpMxV was performed on 10 unstructured matrices [1,2] using our DSP and the results are compared to existing CPU, GPU, and DSP architectures [2,4-6]. Detailed in Fig. 1, FP resource utilization hovers around 1-2% for CPUs, 0.2-0.5% for GPUs, and 18-20% for DSPs. Our DSP achieves a maximum utilization of 99.92%, with an average of 95.29% (Fig. 4). To measure its effectiveness, the "Shuffler" was disabled and the tests were repeated. The average utilization dropped to 75.82% due to the PEs having to stall for 4 clock cycles once every ~20 nonzero elements due to data hazards.

The power consumption and operating frequency of the sparse-BLAS DSP were measured vs. supply voltage (Fig. 4). The minimum energy point (MEP) is found to be at 0.6V, which corresponds to 6.73mW at 160MHz. At this point, the sparse-BLAS DSP achieves a peak throughput of 1.28 GFLOPS for an energy-efficiency of 190 GFLOPS/W. The maximum operating frequency of the sparse-BLAS kernel is 515MHz at $V_{DD}=1V$, achieving 3.2× higher throughput than the MEP at the cost of 2.9× lower energy efficiency.

Our sparse-BLAS kernel is compared to several CPUs and GPUs, as well as prior FPGA and DSP chip implementations for SpMxV (Fig. 5) [2, 4-6]. Overall, it achieves a 2× higher throughput, while averaging 6,600× better energy efficiency (~100× from reduction in processing time and memory access, 3× from voltage and 20× from frequency scaling), than a CPU running architecture-specific software. Similarly, it achieves 2× higher throughput than the prior DSP chips, with a 1,100× higher energy efficiency. Both the GPU and FPGA SpMxV implementations average ~4× higher throughput, while using ~20× more memory bandwidth than our sparse-BLAS kernel. However, our DSP has 2,700× and 450× higher energy efficiency than the GPU and FPGA designs, respectively. Consuming less than 10mW, the 0.927mm² sparse-BLAS kernel in 40nm CMOS (Fig. 6) can provide an energy savings of 2 to 3 orders of magnitude, enabling a variety of IoT applications.

References

[1] N. Bell, et al., SC'09, pp. 1-11, Nov. 2009.
[2] R. Dorrance, et al., FPGA'14, pp. 161-170, Feb. 2014.
[3] M. Horowitz, ISSCC'14, pp. 10-14, Feb. 2014.

[4] J. Fowers, et al., FCCM'14, pp. 36-43, May 2014.
[5] Y. Gao, et al., ASAP'13, pp. 168-174, Jun. 2013.
[6] Y. Gao, et al., HPEC'14, pp. 1-6, Sep. 2014.

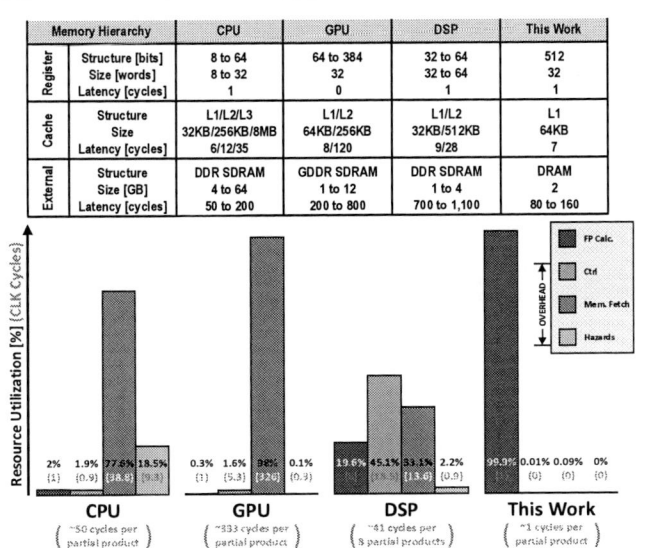

Fig. 1 Resource utilization and memory hierarchy for performing sparse-BLAS.

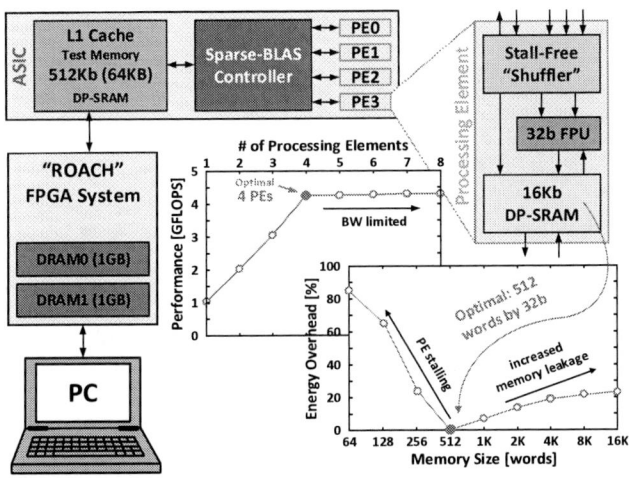

Fig. 2 System architecture and optimization.

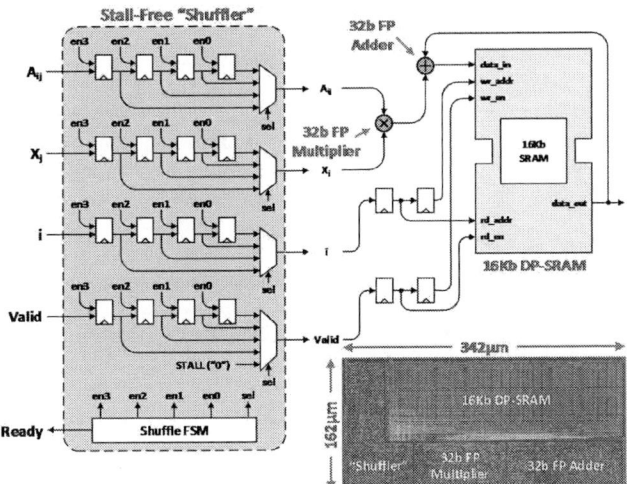

Fig. 3 Block diagram and layout of the sparse-BLAS PE with stall-free, data stream reordering.

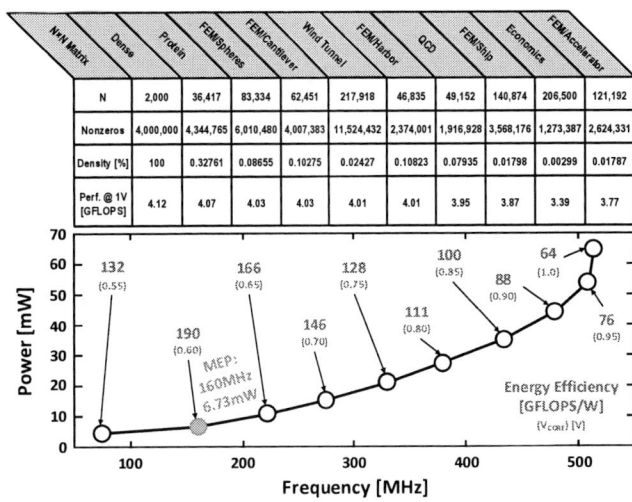

Fig. 4 Measurement results.

Design	CPU [5] i5-650	CPU [2] i7-2600	GPU [4] GTX 580	GPU [2] GTX Titan	DSP [5] C6678	DSP [6] 66AK2H12	FPGA [4] Stratix V D5	FPGA [2] Virtex-5 SX95T	This Work
Tech. [nm]	32	32	40	28	45	28	45	65	40
# of Cores	2/4*	4/8*	512	2,668	8	8	96	64	4
Core Area [mm²]	195	216	520	561	< 576†	< 1600†	~120‡	~150‡	0.927
Freq. [MHz]	3,200	3,400	772	837	1,250	1,200	150	150	≤ 515
Core Perf. [GFLOPS]	1.89	2.01	13.45	14.86	1.67	2.63	9.99	17.64	4.12
Core Power [W]	55.6	77.2	231.9	163.0	9.77	14.9	45.0	5.1	< 0.065
Energy Efficiency [GFLOPS/W]	0.034 (5,597×)	0.026 (7,319×)	0.058 (3,281×)	0.091 (2,091×)	0.171 (1,113×)	0.176 (1,081×)	0.222 (857×)	3.46 (55×)	190 (1×)

* physical/virtual cores   † based on package size   ‡ estimated area based on resource usage

Fig. 5 Comparison with prior CPU, GPU, DSP, and FPGA sparse-BLAS implementations.

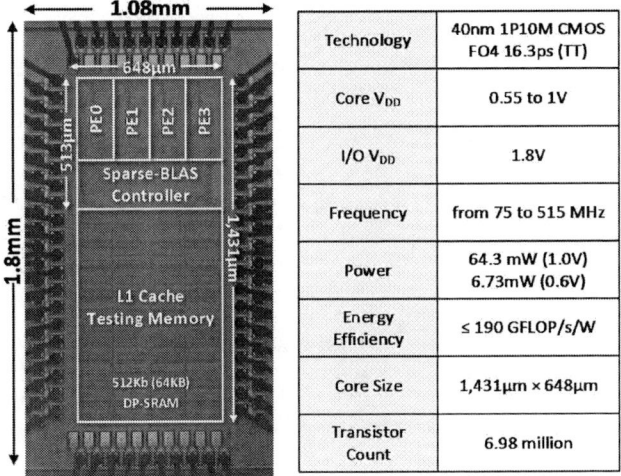

Technology	40nm 1P10M CMOS FO4 16.3ps (TT)
Core V_DD	0.55 to 1V
I/O V_DD	1.8V
Frequency	from 75 to 515 MHz
Power	64.3 mW (1.0V) 6.73mW (0.6V)
Energy Efficiency	≤ 190 GFLOP/s/W
Core Size	1,431µm × 648µm
Transistor Count	6.98 million

Fig. 6 Die photo and chip summary.

978-1-5090-0636-6/16 $31.00 © 2016 IEEE

A 58.6mW Real-Time Programmable Object Detector with Multi-Scale Multi-Object Support Using Deformable Parts Model on 1920x1080 Video at 30fps

Amr Suleiman, Zhengdong Zhang, Vivienne Sze
Massachusetts Institute of Technology, MA, USA

Abstract

This paper presents a programmable, energy-efficient and real-time object detection accelerator using deformable parts models (DPM), with 2x higher accuracy than traditional rigid body models. With 8 deformable parts detection, three methods are used to address the high computational complexity: classification pruning for 33x fewer parts classification, vector quantization for 15x memory size reduction, and feature basis projection for 2x reduction of the cost of each classification. The chip is implemented in 65nm CMOS technology, and can process HD (1920x1080) images at 30fps without any off-chip storage while consuming only 58.6mW (0.94nJ/pixel, 1168 GOPS/W). The chip has two classification engines to simultaneously detect two different classes of objects. With a tested high throughput of 60fps, the classification engines can be time multiplexed to detect even more than two object classes. It is energy scalable by changing the pruning factor or disabling the parts classification.

Keywords: DPM, object detection, basis projection, pruning.

Introduction

Object detection is critical to many embedded applications that require low power and real-time processing. For example, low latency and HD images are important for autonomous control to react quickly to fast approaching objects, while low energy consumption is essential due to battery and heat limitations. Object detection involves not only classification/recognition, but also localization, which is achieved by sliding a window of a pre-trained model over an image. For multi-scale detection, the window slides over an image pyramid (multiple downscaled copies of the image). Multi-scale detection is very challenging as the image pyramid results in a data expansion, which can be more than a 100x in HD images. The high computational complexity of object detection processing necessitates fast hardware implementations [1] to enable real-time processing.

This paper presents a complete object detection accelerator using DPM [2] with a root and 8 parts model as shown in Fig. 1. DPM results in double the detection accuracy compared to rigid template (root only) detection. The 8 parts account for deformation such that a single model can detect objects at different poses (Fig. 6) and increase detection confidence. However, this accuracy comes with a classification overhead of 35x more multiplications (i.e. DPM classification consumes 80% of a single detector power), making multi-object detection a challenge. A software-based DPM object detector is described in [3], which enables detection for 500x500 images at 30fps but requires a powerful fully loaded Xeon 6-core processor and 32GB of memory. In this work, the classification overhead is significantly reduced by two main techniques:

- Classification pruning with vector quantization (VQ) for selective part processing.
- Feature basis projection for sparse multiplications.

Architecture Overview

Fig. 2 shows the block diagram of our detector architecture, including histogram of oriented gradients (HOG) feature pyramid generation unit and support vector machine (SVM) classification engines. A feature pyramid size of 12 scales (4 octaves, 3 scales/octave) is selected as a trade-off between detection accuracy and computation complexity. The pyramid contains 87K

feature vectors, which is 2.7x more features than a typical HD image. To meet the throughput, three parallel histogram and normalize blocks generate the pyramid. Two classification engines share the generated feature to detect two different classes of objects simultaneously. The root and the parts SVM weights can be programmed with a maximum template size of 128x128 pixels. This large size gives the detector the flexibility to detect many objects classes with different aspect ratios. Each SVM engine contains a root classifier for root detection, a pruning block to select candidate roots, and 8 part processing engines for parts detection. Local feature storage in each part engine allows parallelism, reduces the feature storage read bandwidth and enables 7x speedup. Finally, the Deform block uses a coarse-to-fine technique for 2.2x speedup in finding the maximum score in a 5x5 search region for each part after adding the deformation cost.

Classification Pruning and Vector Quantization

With more than 2.6 million features generated per second, on-the-fly processing is used for root classification similar to [3] for minimal on-chip storage, where partial dot products are accumulated in SRAMs. Using the same approach with parts classification would require large accumulation SRAM sizes (more than 800KB for one classification engine). However, it was observed that if the root score is too low, then the likelihood of detecting an object based on parts is also low. Since parts classification requires significant additional computation, we choose to prune the parts classification when the root score is below a programmable threshold. By pruning 97% of the parts classification (i.e. a 33x reduction of the root candidates that are processed), we achieve a 10x reduction in classification power with negligible 0.03% reduction in accuracy.

To avoid re-computation, HOG features are stored in line buffers to be reused by the part processing engines after pruning (Fig. 3). VQ is used to reduce the feature line buffers write bandwidth (from 44.4MB/s to 2.5MB/s), making its size suitable for on-chip SRAM (from 572KB to 32KB) and eliminating any off-chip storage. Three parallel VQ engines are used to meet the throughput. A programmable 256 clusters centers are stored in a shared SRAM to minimize the read bandwidth. The 143-bit HOG feature vector (13-D, 11-bit each) is quantized to 8 bits per vector, giving a 15x reduction in the overall feature storage size. De-quantization is just a memory read from the feature SRAM.

Feature Basis Projection

To further reduce the cost of each classifier, the features are projected into a new space where the classification SVM weights are sparse. Zeros multiplications are skipped and only the non-zero weights are stored on-chip. Fig. 4 shows that the percentage of zero weights is increased from 7% to 56% after projection. The programmable basis vectors are designed such that at least 7 out of the 13 dimensions in the weights are zeros. An overhead of a 13-bit flag is stored to label the zeros positions, resulting in a total reduction of the SVM weight SRAM size and read bandwidth by 34%. The basis projection reduces the number of multiplications by 2x and reduces the overall classification power by 43%.

Evaluation

The accuracy of our accelerator is analyzed on PASCAL VOC 2007 [5], which is a widely used image dataset containing 20 different object classes (aeroplane, bicycle, bird, etc.) in 9,963 images. With 97% pruning, VQ and feature basis projection, 10x fewer classification multiplications and a 3.6x smaller memory size are achieved, leading to 5x reduction in the total power consumption with a drop in the detection accuracy by only 4.8%.

Implementation and Testing

The chip is implemented in 65nm CMOS technology. It is tested to process HD images at 30fps in real-time operating at 62.5MHz and 0.77V while consuming 58.6mW, resulting in a peak performance of 1168 GOPS/W and an energy efficiency of 0.94nJ/pixel. Fig. 5 shows the die photo and the chip specifications, along with a sweep for different throughput and energy consumption. At 1.11V, the chip can process HD images up to 60fps while consuming 216.5mW. The power breakdown for different tested configurations is shown in Fig. 6. The DPM classification power is significantly reduced down to only 15% of the total power of a single detector. With features sharing, detecting an additional object class with DPM increases the total power consumption by only 19%. Fig. 6 also shows the chip output with multi-object detection (cars and pedestrians). Comparing to the detector accelerator in [6], our chip boosts the detection accuracy with multi-scale and detecting 8 deformable parts per object while consuming 30% less energy per pixel.

Acknowledgement

The authors would like to thank TSMC University Shuttle Program for the chip fabrication and DARPA and TI for funding.

References

[1] J. Tanabe, et al., "A 1.9TOPS and 564GOPS/W heterogeneous multicore SoC with color-based object classification accelerator for image-recognition applications," ISSCC 2015.
[2] P.F. Felzenszwalb, et al., "Object Detection with Discriminatively Trained Part-Based Models," PAMI 2010.
[3] M.A. Sadeghi, et al., "30Hz Object Detection with DPM v5," EECV 2014.
[4] A. Suleiman, V. Sze, "Energy-efficient HOG-based object detection at 1080HD 60 fps with multi-scale support," SiPS 2014.
[5] www.pascal-network.org/challenges/VOC/voc2007/workshop/index.html.
[6] K. Takagi, et al., "A Real-time Scalable Object Detection System Using Low-power HOG Accelerator VLSI," JSPS 2014.

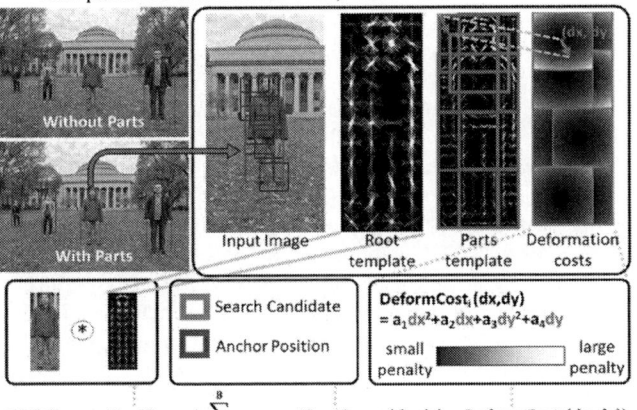

$$DPM\,Score = RootScore + \sum_{i=1}^{B} max_{dx,dy}(PartScore_i(dx,dy) - DeformCost_i(dx,dy))$$

Fig. 1 Detection example with DPM templates and score calculation

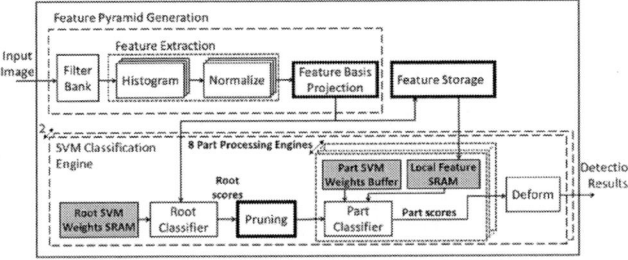

Fig. 2 Block diagram of the DPM object detection accelerator

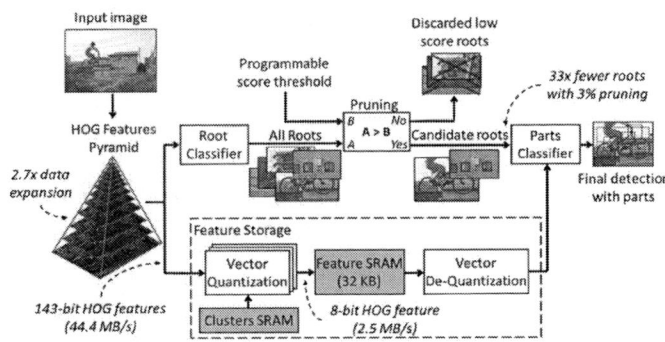

Fig. 3 DPM detection with pruning and vector quantization

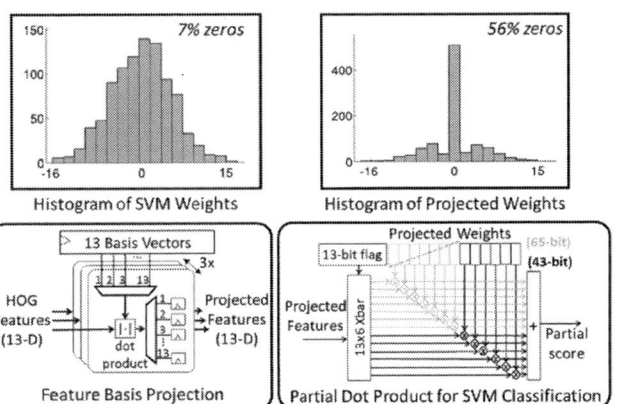

Fig. 4 Feature basis projection for sparse classification

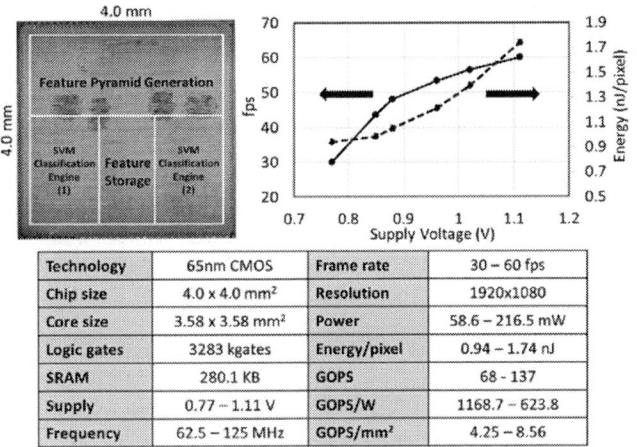

Technology	65nm CMOS	Frame rate	30 – 60 fps
Chip size	4.0 x 4.0 mm²	Resolution	1920x1080
Core size	3.58 x 3.58 mm²	Power	58.6 – 216.5 mW
Logic gates	3283 kgates	Energy/pixel	0.94 – 1.74 nJ
SRAM	280.1 KB	GOPS	68 – 137
Supply	0.77 – 1.11 V	GOPS/W	1168.7 – 623.8
Frequency	62.5 – 125 MHz	GOPS/mm²	4.25 – 8.56

Fig. 5 Die photo and summary of the chip specifications. Numbers are measured with the two detectors running and 97% pruning set.

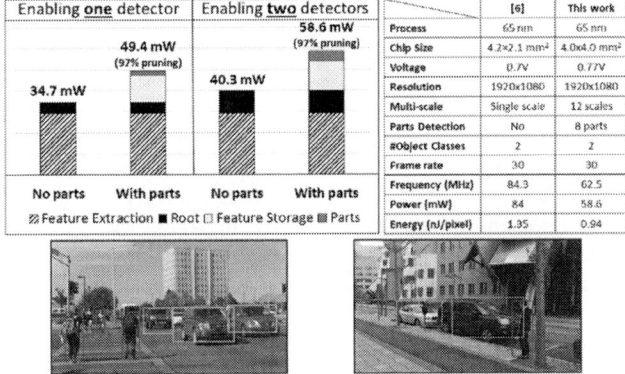

	[6]	This work
Process	65 nm	65 nm
Chip Size	4.2×2.1 mm²	4.0x4.0 mm²
Voltage	0.7V	0.77V
Resolution	1920x1080	1920x1080
Multi-scale	Single scale	12 scales
Parts Detection	No	8 parts
#Object Classes	2	2
Frame rate	30	30
Frequency (MHz)	84.3	62.5
Power (mW)	84	58.6
Energy (nJ/pixel)	1.35	0.94

Fig. 6 Performance comparisons and detection examples.

Adaptive Clocking with Dynamic Power Gating for Mitigating Energy Efficiency & Performance Impacts of Fast Voltage Droop in a 22nm Graphics Execution Core

Minki Cho, Carlos Tokunaga, Stephen Kim, James Tschanz, Muhammad Khellah, Vivek De

Circuit Research Lab, Intel Corporation, Hillsboro, OR, USA

Abstract

Combining adaptive clocking with dynamic power gating in an optimal manner mitigates energy efficiency and performance impacts of fast supply voltage droop in a 22nm graphics execution core more effectively than adaptive clocking alone. Measurements show that there is an optimal V_{MIN} where the combination provides the best improvement -- 14% lower energy at 890MHz vs. 4% with adaptive clocking.

Introduction

Worst-case fast voltage droop in a graphics execution core while running a power virus workload sets the minimum voltage values, V_{MIN}, needed to meet frequency targets across a wide range. This $V_{MIN\ increase}$ degrades energy efficiency as it impacts both dynamic and leakage power components. Several techniques with varying degrees of circuit and architectural complexities have been reported to mitigate impacts of fast voltage droops [1-6]. Adaptive clocking (AC) [4-5] has been shown to be a simple yet effective scheme to recover some of the energy and performance loss due to fast voltage droops. It exploits critical path clock-data delay compensation during a droop event to proactively gate or divide the clock, thus quickly masking the impact of the droop, and enabling lower V_{MIN} as a result. However, the degree of V_{MIN} reduction with AC can be limited, particularly at high frequency targets, and overall performance overhead can be significant in case of frequent droop events. In addition, the worst-case FIFO depth to support clock domain crossing between regions with and without droop may be too large.

In this paper, we combine AC with dynamic power gating (DPG) [6] in an optimal manner to mitigate energy efficiency and performance impacts of fast voltage droops across a wide frequency range for a graphics execution core in 22nm tri-gate CMOS. DPG improves energy efficiency by modulating existing power gates to introduce a dynamic load-line that gives a load-current dependent linear reduction of the voltage swing on the virtual (gated) rail.

Adaptive clocking (AC) and Dynamic power gating (DPG)

AC consists of an adaptive clock distribution (ACD) unit, a tunable replica circuit (TRC), and a clock gating/division logic (Fig.1). ACD is inserted between the clock generator and the global clock distribution. It consists of programmable transistor and interconnect delay components used to *prolong* the natural critical-path clock-data delay compensation at the onset of the droop. TRC1 at the clock distribution root detects the onset of a droop and triggers clock gating or division thus removing or slowing the clock edges to prevent timing failure.

Dynamic power gating (DPG) (Fig. 2) utilizes the existing power gates (PG), used for shutting off leakage during idle periods, to mitigate droop impacts on energy efficiency. In the baseline design, the PG is sized up to minimize IR drop for maximum load current. Then, the IR drop across the PG is smaller than the worst-case for typical load currents. DPG exploits this voltage headroom present for typical loads by dividing the baseline PG into a primary PG (PPG) and a secondary PG (SPG). The PPG is always on during active operation, but since its impedance is higher than the original (bigger) PG, the IR drop through PPG depends strongly on the

load current, thus reducing the virtual rail voltage at typical load current. The SPG is dynamically turned on, in quick response to a sudden higher load current demand, to maintain the same worst-case droop as the baseline design. TRC2, with its cycle-fast reaction, triggers a 3 bit SR that turns on the SPG.

For a target operating frequency, V_{MIN} and energy consumptions of baseline, DPG, AC and AC+DPG schemes (Fig. 3) are compared (Table I). In the baseline design, the minimum voltage needed to meet target cycle time, T_{CYCLE}, under maximum droop event is V_1. DPG operates at the same voltage as baseline, V_1, but lowers the gated rail voltage, V_{GATED}, by appropriately setting PPG to SPG ratio. On the other hand, AC enables a V_{MIN} of $V_2 < V_1$ while meeting the same target frequency. AC incurs some performance and leakage energy overheads since additional idle clock cycles are invoked during worst-case droop events, expected to be relatively infrequent over the workload runtime. Combining AC+DPG can enable both a lower V_{MIN} of $V_3 < V_1$ and a lower virtual rail voltage, $V_{GATED'}$ with the same performance as AC. Note that best energy efficiency is achieved with AC+DPG when $V_3 \times V_{GATED'} < V_2^2$, even as V_3 can be higher than V_2. Alternately, in case of frequent droop events, slightly higher V_3 can be used to reduce AC overheads in terms of clock gated cycle count and clock crossing FIFO depth.

Chip Implementation & Measurements

The 22nm graphics execution core test chip includes the core for performing key floating-point operations for a 3D graphics pipeline, a PLL, a 270KB SRAM array and a test controller for issuing at-speed test vectors, and signature registers for validating correct test results (Figs. 4 & 5). AC and DPG are implemented to operate together with TRC1 programmed to detect a droop and trigger AC, and TRC2 programmed to monitor a droop and activate DPG. For either scheme, optimal TRC programming is critical to achieve maximum energy reduction (Fig. 6). The ACD delay is set to 4 cycles to provide sufficient time to respond to voltage droop. Fig. 7 shows transient capture of baseline at frequency target of 520MHz and V_{MIN} of 0.7V. Corresponding captures for DPG, AC, and AC+DPG are also shown. Note that with AC, V_{MIN} is reduced to 0.66V, while for the combined AC+DPG scheme, a slightly higher V_{MIN} of 0.68V is optimal. While AC alone gives only 4% (3%) reduction in energy (V_{MIN}) compared to the baseline design at 890MHz, AC+DPG increases the energy benefits to 14% despite a slightly higher V_{MIN} (Fig. 8). The combined scheme achieves the best energy efficiency across the entire frequency range.

Acknowledgment

This research was, in part, funded by the U.S. Government (DARPA). The views and conclusions contained in this document are those of the authors and should not be interpreted as representing the official policies, either expressed or implied, of the U.S. Government.

References

[1] T. Fischer et al., JSSC, pp. 218-228, Jan. 2006.
[2] N. Kurd et al., JSSC, pp. 1121-1129, 2009.
[3] A. Grenat et al., ISSCC, pp. 106-107, 2014.
[4] K. Bowman et al., JSSC, pp. 907-916, 2013.
[5] C. Tokunaga et al., ISSCC, pp. 108-109, 2014.
[6] M. Cho et al., ISSCC, 2016.

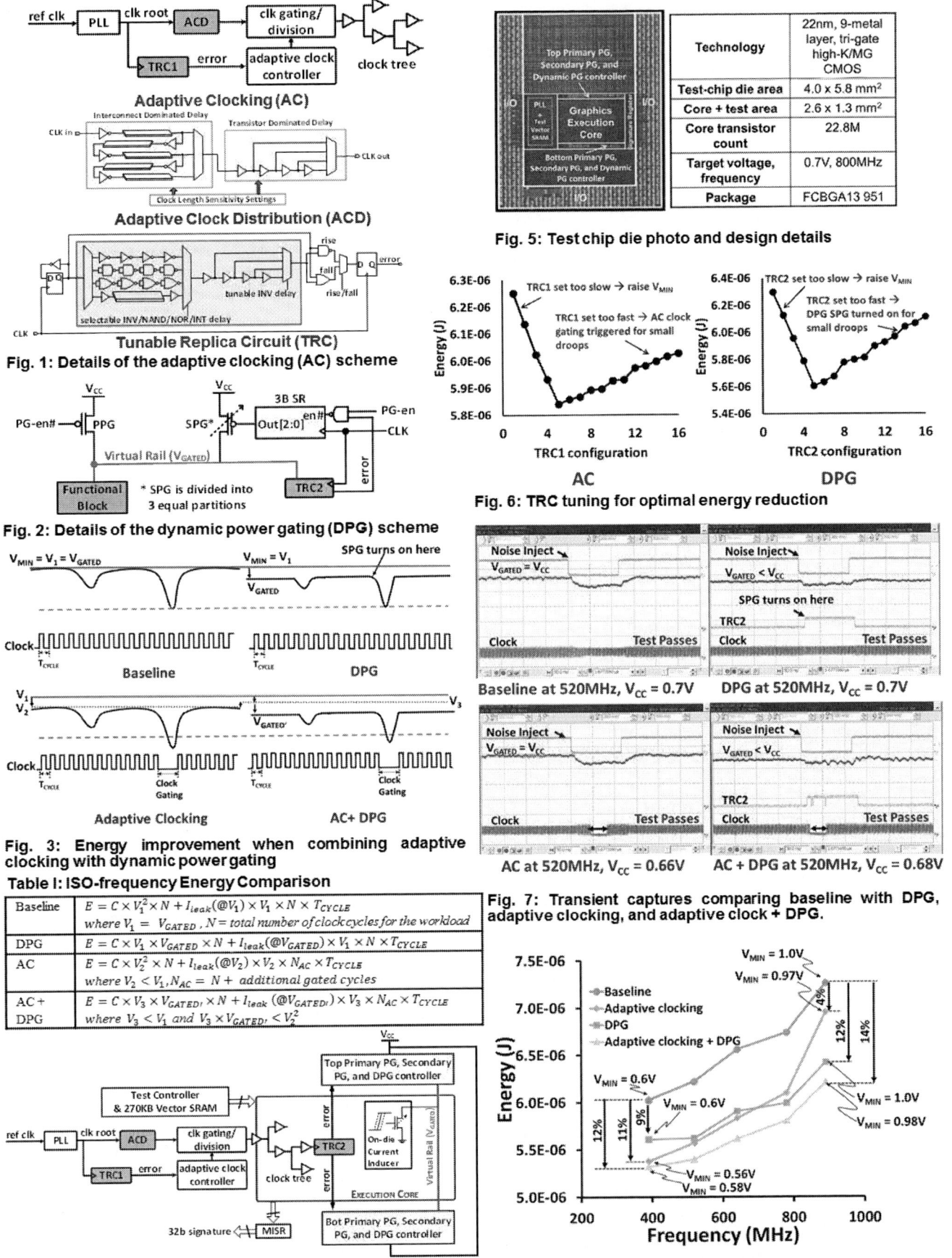

Adaptive Clocking (AC)

Adaptive Clock Distribution (ACD)

Tunable Replica Circuit (TRC)

Fig. 1: Details of the adaptive clocking (AC) scheme

Fig. 2: Details of the dynamic power gating (DPG) scheme

Fig. 3: Energy improvement when combining adaptive clocking with dynamic power gating

Table I: ISO-frequency Energy Comparison

Baseline	$E = C \times V_1^2 \times N + I_{leak}(@V_1) \times V_1 \times N \times T_{CYCLE}$ where $V_1 = V_{GATED}$. $N =$ total number of clock cycles for the workload
DPG	$E = C \times V_1 \times V_{GATED} \times N + I_{leak}(@V_{GATED}) \times V_1 \times N \times T_{CYCLE}$
AC	$E = C \times V_2^2 \times N + I_{leak}(@V_2) \times V_2 \times N_{AC} \times T_{CYCLE}$ where $V_2 < V_1, N_{AC} = N +$ additional gated cycles
AC + DPG	$E = C \times V_3 \times V_{GATED\prime} \times N + I_{leak}(@V_{GATED\prime}) \times V_3 \times N_{AC} \times T_{CYCLE}$ where $V_3 < V_1$ and $V_3 \times V_{GATED\prime}, < V_2^2$

Fig 4.: Overall chip block diagram showing both adaptive clocking along with dynamic power gating

Technology	22nm, 9-metal layer, tri-gate high-K/MG CMOS
Test-chip die area	4.0 x 5.8 mm²
Core + test area	2.6 x 1.3 mm²
Core transistor count	22.8M
Target voltage, frequency	0.7V, 800MHz
Package	FCBGA13 951

Fig. 5: Test chip die photo and design details

AC   **DPG**

Fig. 6: TRC tuning for optimal energy reduction

Baseline at 520MHz, $V_{CC} = 0.7V$ DPG at 520MHz, $V_{CC} = 0.7V$

AC at 520MHz, $V_{CC} = 0.66V$ AC + DPG at 520MHz, $V_{CC} = 0.68V$

Fig. 7: Transient captures comparing baseline with DPG, adaptive clocking, and adaptive clock + DPG.

Fig. 8: Measured energy vs. frequency for baseline, DPG, adaptive clocking, and adaptive clocking with DPG

A 0.23 μg Bias Instability and 1.6 μg/Hz$^{1/2}$ Resolution Silicon Oscillating Accelerometer with Build-in Σ-Δ Frequency-to-Digital Converter

Jian Zhao[1,2], Xi Wang[2], Yang Zhao[1,2], Guo Ming Xia[1], An Ping Qiu[1], Yan Su[1], Yong Ping Xu[2]

[1]Nanjing University of S&T, Nanjing, P.R China, [2]National University of Singapore, Singapore

elfevil007@126.com, yongpingxu@nus.edu.sg

Abstract: This paper presents a silicon oscillating accelerometer (SOA) with CMOS readout circuit. To reduce the bias instability, a PLL is employed to sustain the oscillation instead of the conventional auto-amplitude-control (AAC) circuit. A sigma-delta frequency-to-digital converter (FDC) is built in the PLL to produce the digital output. The MEMS sensor and readout circuit are fabricated in 80 μm SOI and standard 0.35 μm CMOS process, respectively. The SOA achieves 0.23 μg bias instability and 1.6 μg/Hz$^{1/2}$ resolution with ±30 g full-scale, which are equivalent to 4-ppb relative instability and 27-ppb/Hz$^{1/2}$resolution. In addition, it only consumes 2.7 mW under a 1.5 V supply.

Introduction

Inertial navigation sets a very stringent requirement on long-term stability of the accelerometer, which is characterized by bias-instability. For this requirement, SOA has proven to be a promising candidate compared with the capacitive accelerometer. So far, most of readout circuits in SOAs are realized with AAC technique to stabilize the oscillation. However, due to the amplitude-stiffening (A-S) effect, the flicker noise in AAC circuit appears in the drive signal and deteriorates the bias instability. By employing flicker noise attenuation techniques in AAC, a MEMS SOA with 0.4 μg bias instability and 1.2 μg/Hz$^{1/2}$ resolution with ±20 g full scale was reported recently [1]. Since the flicker noises from AAC circuit in two oscillator channels are uncorrelated, though employing attenuation techniques, the residual noises still deteriorate the bias instability, and are difficult to eliminate. An alternative way to sustain oscillation without using AAC topology was reported in [2]. It can eliminate the noise from AAC circuit, however, it suffers from noise aliasing and the performance is poor.

System Description

This paper demonstrates a SOA with a novel architecture, in which the AAC circuit is completely eliminated without noise aliasing. Fig.1 shows the system block diagram of the proposed SOA. Two oscillator channels are formed with MEMS resonators that are connected to the proof mass. When subject to input acceleration, the two oscillation frequencies change in opposite directions and their difference measures the acceleration. In each oscillator channel, a PLL is employed to track the phase of the front-end TIA's output (oscillator output) and provide correct phase shift for the drive signal to sustain the oscillation. Meanwhile, the drive signals for both oscillators are generated by a differential low noise reference whose polarity is modulated by the feedback (phase) signal from PLL through a switch. In other words, the phase and amplitude of drive signal are set by the PLL and external low noise reference, respectively. Thus, the amplitude noise in the drive signal mainly come from the given reference. Since the drive signals for both oscillators are derived from the same reference, the flicker noise in the oscillation amplitudes of the two oscillators are correlated, therefore the A-S effect induced flicker frequency noise can be cancelled at the final output (f_1–f_2). Hence the bias instability can be further improved.

Furthermore, a third-order Σ-Δ FDC is employed with the phase quantizer embedded in the PLL loop, to digitize the frequency output.

To start the oscillation initially, a start-up circuit is required. At start-up phase, S1 is on and S2 is off. The start-up circuit provides feedback signal with correct phase and amplitude to drive the MEMS resonator and start oscillation. Meanwhile, the PLL will lock to the phase of the front-end output. When oscillation is stabilized, the oscillation detector turns S1 off and S2 on, and the system enters operation phase, in which the feedback loop via PLL takes over to sustain the oscillation, and the start-up circuit will be turned off to save power. The oscillation detector has hysteresis characteristics to prevent the start-up circuit from switching back.

PLL Phase tracking and FDC

The PLL in Fig.1 is the key building block, it not only tracks the phase to sustain the oscillation, but also performs frequency to digital conversion.

Generally, the sinusoidal oscillation signal from front-end output needs to be shaped to square wave before feeding to the PLL. As the noise from front-end is relatively wide-band, i.e. at least several times higher than the oscillating frequency, the waveform shaping will introduce noise folding and cause high in-band noise. To avoid problem, a PLL with hybrid PFD is proposed to do away with the nonlinear waveform shaping. This PFD consists of an analog multiplier phase detector (PD) and a 3-states PFD with large deadzone. When sine wave applies to the input, the 3-state deadzone PFD detects the frequency difference, and helps the PLL to converge. Once the frequencies are close to each other, the 3-state PFD will be automatically disabled due to its large dead zone and the analog PD will take over, to detect the phase difference between the input sine wave and the feedback signal continuously. With the hybrid PFD, both frequency and phase can be tracked, while the noise folding due to waveform reshaping is avoided. The switchover behavior works under all corners in simulations, as well as in the experiments.

The schematic of proposed hybrid PFD is shown in Fig.2(A). The multiplier is implemented by a trans conductor (Gm) and an switching multiplier, where the Gm converts the input sinusoidal voltage to current and then multiplied with the quantized feedback signal (*fbq*). Thus, the phase error can be detected. To prevent the flicker noise of Gm cell from contaminating the detected phase error, the V to I conversion is performed in AC domain before the multiplier. And the Gm utilizes an OTA and a capacitive feedback to achieve excellent linearity. A pair of pseudo-resistor is used to set the DC bias of the OTA. The hybrid PFD has a fully differential topology with CMFB, to reject common-mode interference.

To convert the oscillation frequency to digital output, the FDC is implemented by embedding a phase quantizer (Fig.2(B)) in the feedback path of PLL after the divider. In this way, the quantization noise induced by DFF1 will be attenuated by the PLL loop and third-order shaped. Therefore for the same resolution requirement, the clock frequency and hence the power consumption can be significantly reduced. In addition, due to the fully differential topology of SOA, the requirement of clock reference (common to both channels) is also relaxed.

Fig.2(C) shows the front-end TIA, which adopts the first stage of the band-pass front-end in [1], to achieve low power.

Measurement Results

The MEMS sensor is fabricated in SOI process with 80-μm thickness in a wafer-level vacuum package and sealed in a ceramic package. The intrinsic frequency of each resonator is designed to be 18 kHz, while the scale factor is about 200 Hz/g. The quality factor of resonators is around 15,000.

The start-up behavior has been recorded and plotted in Fig.3, which shows that the system can correctly switch over to PLL tracking mode after initial start-up. The measured performance of the SOA are shown in Fig.4. The MEMS sensor can achieve 50 ppm nonlinearity within ±30 g range through centrifugal test. The static performance of the SOA is evaluated at its analog and digital outputs, respectively. The clock frequency is only 750 kHz. The SOA achieves 0.23 μg bias-instability and 1.6 μg/Hz$^{1/2}$ resolution, which is equivalent to 4-ppb relative instability and 27-ppb/Hz$^{1/2}$ resolution if taking full-scale into consideration. At higher frequency, 3rd-order noise shaping can be observed. The SOA, including readout circuit chip, consumes only 2.7 mW under a 1.5 V supply. Fig.5 gives the comparison of this work with state-of-the-art SOA and capacitive accelerometers. The readout circuit is implemented in a standard 0.35 um CMOS process and occupies 10 mm^2 which is smaller than the MEMS chip. Fig.6 is the chip microphotograph.

References

[1] X. Wang, et al., *ISSCC*.pp.476-478,Feb 2015.
[2] Comi, Claudia, et al. *JMEMS*, 19(5), pp. 1140-1152, 2010.
[3] X. Gao, et al., *JSSC*, pp.3253-3263, Dec 2009.
[4] Ullah, P, et al.,*IEEE ISS Symposium*, pp. 1-13, 2015.
[5] H. Xu, et al., *JSSC*, pp.2101-2112, Dec 2015.

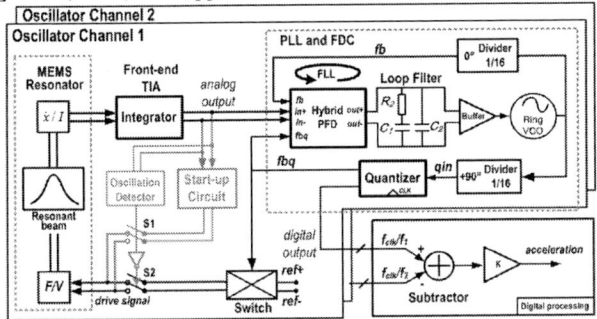

Fig. 1 SOA system block diagram.

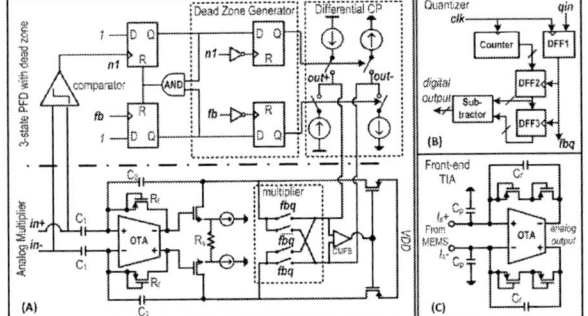

Fig. 2 Schematic of (A) Hybrid PFD, (B) phase quantizer and (C) Front-end TIA,

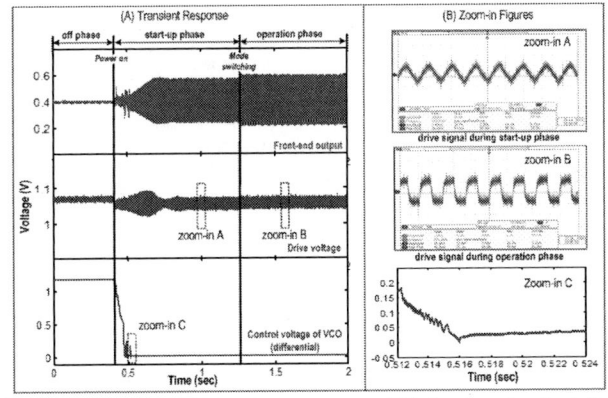

Fig. 3 Measured SOA transient response. (A) front-end output (top trace), drive signal (middle trace), and VCO control voltage (bottom trace), obtained using NI DAQ; (B) Zoom-in waveforms of specified moments in (A).

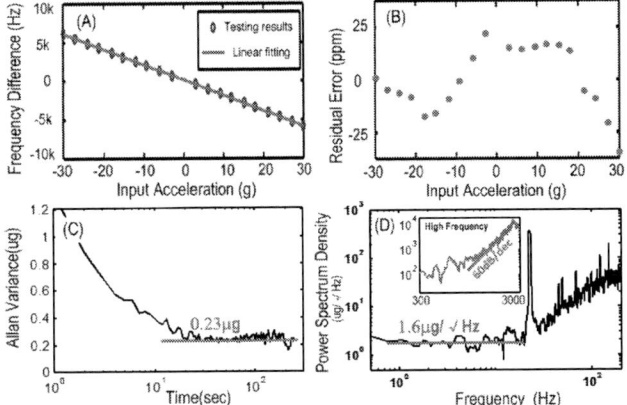

Fig. 4 Measured results: (A) SOA linearity results, (B) relative residual errors of (A), (C) Allan variance from frequency of analog output, (D) Power spectrum density from digital output.

Parameter	[1]	[4]	[5]	This work
Mechanism	SOA	Capacitive	Capacitive	SOA
Process(μm)	0.35	NA	0.5	0.35
Supply (V)	1.5	NA	7	1.5
Full scale (g)	±20	±15	±1.2	±30
Power (mW)	4.4	400	23	2.7
Bias instability (μg)	0.4	0.8	18	0.23
Noise floor (μg/Hz$^{1/2}$)	1.2	1	0.2	1.6
Relative Instability (ppb)	10	27	7500	3.8
Relative resolution (ppb/Hz$^{1/2}$)	30	33	83	27
Readout	CT AAC based oscillator	Σ-Δ ADC	Σ-Δ ADC	CT PLL based oscillator

Fig. 5 Performance comparison.

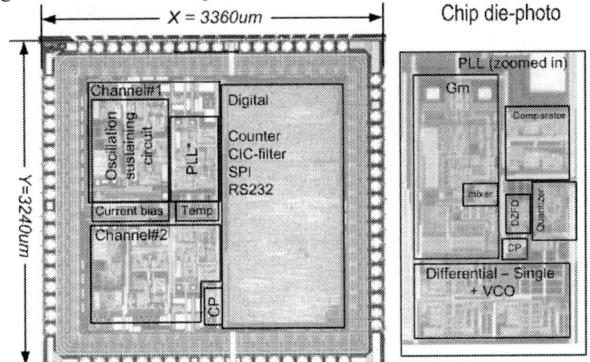

Fig. 6 Chip microphotograph.

A BJT-based Temperature-to-Digital Converter with ±60mK (3σ) Inaccuracy from -70°C to 125°C in 160nm CMOS

Bahman Yousefzadeh, Saleh Heidary Shalmany, Kofi Makinwa
Delft University of Technology, Delft, the Netherlands

Abstract

This paper presents the most accurate BJT-based CMOS temperature-to-digital converter (TDC) ever reported, with an inaccuracy of ±60mK (3σ) from -70°C to 125°C. This is 2× better than the state-of-the-art, despite being implemented in a process (160nm) that only offers low-β_F (<5) PNPs. It is also the most energy-efficient ever reported, with a resolution FOM of 7.3pJ°C^2. This level of performance is achieved by an improved β_F-compensation scheme, the use of dynamic error correction techniques to suppress non-BJT related errors and the use of an energy-efficient zoom-ADC based on current-reuse OTAs. These techniques also result in very low power-supply sensitivity (12mK/V), thus maintaining TDC accuracy for supply voltages ranging from 1.5V to 2V.

Introduction

TDCs are widely used for the temperature compensation of high-performance SoCs, such as frequency [1] or voltage references [2]. In such systems, their inaccuracy is a significant part of the total error budget. TDCs that sense the temperature-dependent base-emitter voltage V_{BE} of substrate PNPs represent the state-of-the-art [3-5], and can achieve inaccuracies as low as ±0.1°C (3σ) over the military range (-55°C to 125°C), after a simple 1-point trim [3].

However, the accuracy of PNP-based TDCs does not seem to benefit from process scaling [3-5]. One reason for this is that scaling decreases the current gain β_F of substrate PNPs. Since PNPs are biased via their emitters, this means that scaling worsens collector-current spread, and hence that of V_{BE}.

The influence of β_F-spread on V_{BE} can be mitigated by β_F-compensation schemes [3,5]. The analog scheme proposed in [3] uses a pair of BJTs to generate a current that biases another pair with β_F-independent collector currents. However, its effectiveness relies on the accuracy of the biasing current and the β_F-matching of the BJTs, both of which are increasingly difficult to achieve as processes scale [5].

This design presents a TDC with a precision biasing circuit that uses DEM to mitigate circuit mismatch, and improved layout to mitigate BJT mismatch. The result is a TDC with state-of-the-art accuracy and power-supply sensitivity.

Proposed Design

The temperature sensing front-end of the proposed TDC consists of a pre-bias (PB) and a bipolar-core (BC) built around four identical substrate PNPs (Fig. 1a). A 1:5 current mirror biases two PNPs (Q_{LB}, Q_{RB}) at an emitter-current ratio $p=5$. An opamp then forces the difference in their base-emitter voltages ΔV_{BE} across a resistor R_b to generate a PTAT bias current $I_b=\Delta V_{BE}/R_b$ (160nA at 25°C). This is used to bias the other two PNPs (Q_L, Q_R) at the same $p=5$, to generate the main sensing voltages V_{BE1} and V_{BE2}. Provided that the PNPs match, the β_F-compensating resistor R_β ensures that $I_c = I_b$, and so V_{BE1} and V_{BE2} are independent of β_F variations [3]. Via switch S_β, β_F-compensation can be turned on and off. β_F can then be extracted by measuring the voltage $V_\beta = I_b \cdot R$ in both modes (Fig. 1a).

PNP mismatch is mitigated by inter-digitating the devices of the PB and BC. The dominant sources of inaccuracy are

then the spread in the PNPs' saturation current I_s and in the nominal value of R_b. Although these can be corrected for by a single PTAT trim [3], this will not correct for non-PTAT errors caused by current-mirror mismatch. To mitigate such errors, the current mirrors in the PB and BC are dynamically matched. Furthermore, to ensure that I_b is accurately copied to the BC, the two mirrors are periodically swapped (Fig. 1a).

As in [4], a 2nd-order zoom-ADC digitizes $X = V_{BE}/\Delta V_{BE}$ in two steps (Fig. 1b). In the first step, a 31-element capacitor DAC and the 1st integrator implement a 6-step SAR algorithm, which finds the integer part of X by successively comparing V_{BE} with $K \cdot \Delta V_{BE}$, where K=1:31. In the second step, a 2nd order ΣΔ modulator balances V_{BE} against reference voltages $(K_{SAR}-1) \cdot \Delta V_{BE}$ or $(K_{SAR}+1) \cdot \Delta V_{BE}$, where K_{SAR}, is the result of the first step. From the resulting bitstream average $\mu_{z\Delta}$, the final result $X=K_{SAR}+2 \cdot \mu_{z\Delta}$ is obtained, from which a PTAT function of temperature, $\mu=\alpha/(\alpha+X)$ can be derived [4]. The ADC can also be configured to digitize $X = V_\beta/\Delta V_{BE}$, thus enabling, for the first time, the direct determination of β. Each of the ADC's integrators (Fig. 1b) is built around an energy-efficient current-reuse OTA (Fig. 2). Low offset (<1μV) and 1/f noise are achieved by the use of CDS (in the 1st integrator) and system-level chopping.

Measurement Results

Realized in 160nm CMOS (Fig. 3), the TDC occupies 0.16μm^2, is ceramic packaged and draws 4.6μA from a 1.8V supply voltage. For flexibility, the sinc^2 decimation filter and the digital backend were realized off-chip. When clocked at 35kHz, the TDC achieves a kT/C–limited resolution of 15mK$_{rms}$ in a conversion time T_{conv} of 5ms (Fig. 4).

Twenty TDCs were characterized from -70°C to 125°C in a climate chamber. A large aluminum block was used to stabilize the temperature of the samples to within 1mK. The measured β_F of the PNPs was found to spread significantly (±10%), and vary from 3 to 9 over temperature (Fig. 5). Without trimming, the TDC achieves an inaccuracy of ±0.4°C (3σ) (Fig. 6a). Offset trimming X (equivalent to PTAT trimming V_{BE}) at 30°C reduces the total inaccuracy to below ±0.1°C (3σ) over the military range (Fig. 6b). As in [4], optimizing α (=15.33) reduces the residual curvature to ±40mK. For greater accuracy, this can be reduced to ±5mK by a fixed 3rd order polynomial obtained by batch-calibration (Fig. 6c). Offset trimming X then results in a total inaccuracy of ±60mK (3σ) (Fig. 6d). This result demonstrates the effectiveness of the proposed β_F-compensation scheme. The TDC's performance is summarized in Fig. 7 and compared to the state-of-the-art. Compared to designs in the same process [4], or in more mature processes (0.7μm, $\beta_F \sim 25$) [3,6], it achieves superior inaccuracy, resolution FOM and power-supply sensitivity.

References

[1] S. Zali Asl et al., ISSCC, 2014.
[2] G. Maderbacher et al., ISSCC, 2015.
[3] M.A.P. Pertijs et al., JSSC, 2005.
[4] K. Souri et al., JSSC 2013.
[5] X. Pu et al., JSSC, 2015.
[6] A. Heidary, et al., ISSCC, 2014.

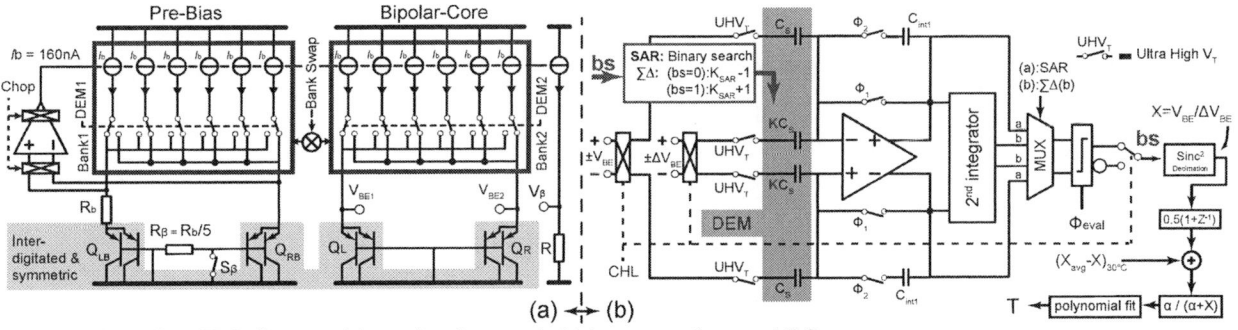

Fig. 1. Complete TDC diagram, (a) sensing front-end, (b) incremental zoom-ADC

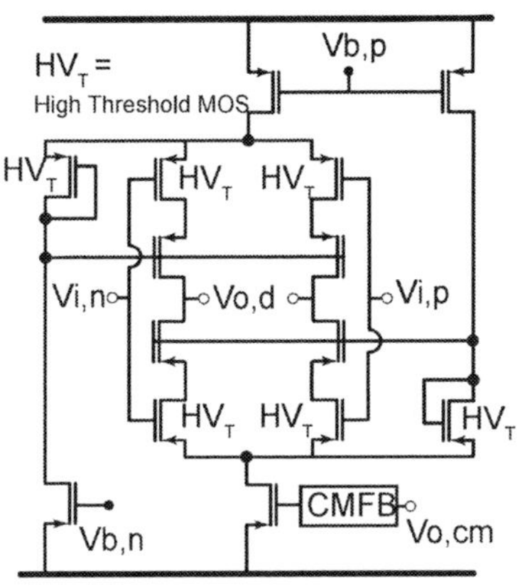

Fig. 2. Current reuse OTA, in 1st and 2nd stage (current scaled) integrators

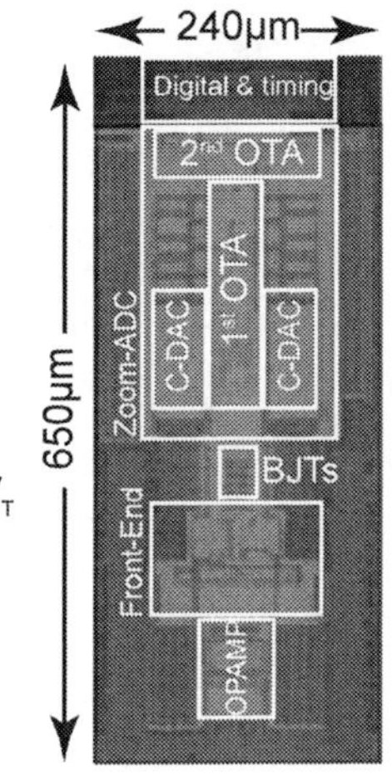

Fig. 3. Chip micrograph

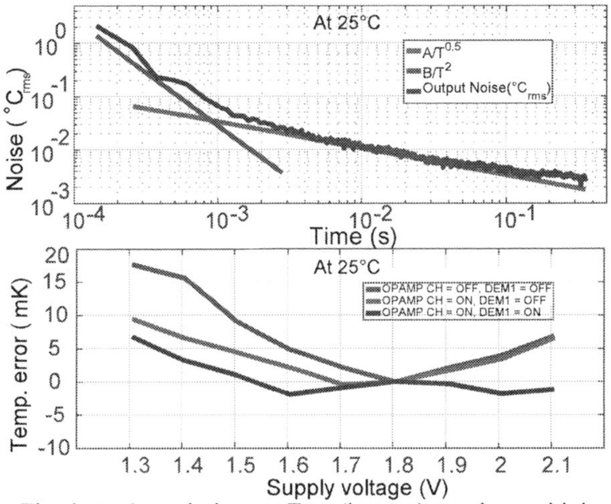

Fig. 4. (top) resolution vs. T_{conv}. (bottom) supply sensitivity

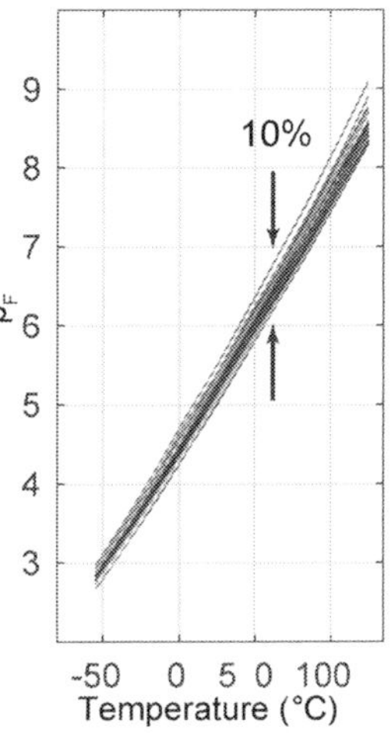

Fig. 5. Measured β of 20 samples over temperature

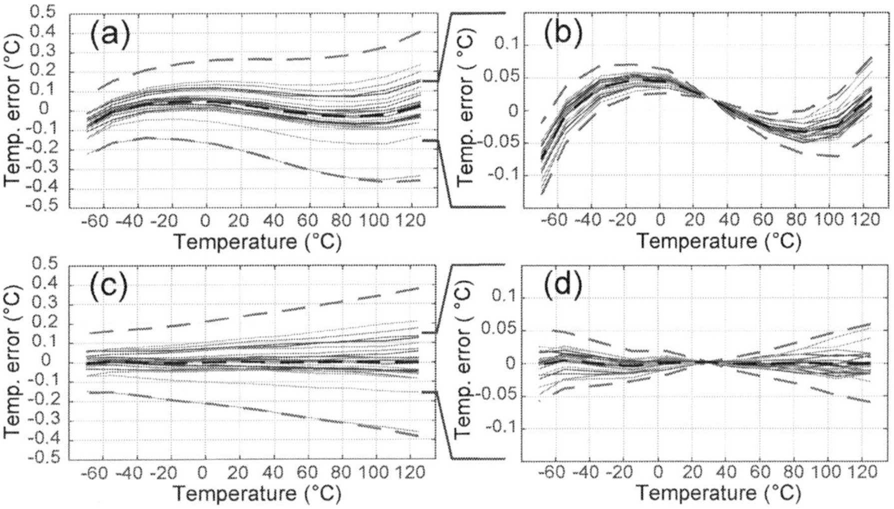

Fig. 6. Temperature error, without curvature compensation. (a) untrimmed, (b) offset trimmed X at 30°C. with 3rd order curvature compensation. (c) untrimmed, (d) offset trimmed X at 30°C. Red dotted lines are the ±3σ limits, black dotted are the average.

Item	Tech (μm)	Area (mm³)	Supply (V)	Current (μA)	T. Range (°C)	Inaccuracy (±3σ error)	PSS (°C/V)	Res. (m°C) T_{conv} (ms)	Res. FOM (pJ°C²)	Relative InAcc (%)	Reference type
This work	0.16	0.16	1.5–2	4.6	-70 – 125	±60mK	0.01	15 (5)	7.3	0.06	Self-referenced
[3]	0.7	4.5	2.5–5.5	75	-55 – 125	±100mK	0.03	10 (100)	1875	0.11	Self-referenced
[4]	0.16	0.08	1.5–2	3.4	-55 – 125	±150mK	0.5	20 (5.3)	11	0.17	Self-referenced
[5]	0.065	0.2	1.5	0.5	-40 – 130	±400mK	N/A	125 (2)	23	0.47	External Voltage
[6]	0.7	0.8	2.9–5.5	55	-45 – 130	±150mK	0.05	3 (2.2)	3.2	0.17	External Frequency

Fig. 7. Comparison table

A 28nm CMOS Ultra-Compact Thermal Sensor in Current-Mode Technique

Matthias Eberlein, Idan Yahav

Intel Deutschland GmbH, Neubiberg 85579, Germany
matthias.eberlein@intel.com

Abstract

This paper describes an innovative architecture for temperature sensors, which achieves excellent linearity at minimum complexity. Fabricated in 28nm, the circuit occupies only 0.0038mm^2 and draws 16µA from 1.8V, capable also for lower supplies. The 8-bit smart sensor operates from -20 to 130°C and utilizes the benefits of parasitic NPN transistor. Current-mode technique is adopted in a new way, which simplifies digital output and makes common error correction dispensable. Excellent PSRR and a raw accuracy of 1.8°C (3σ) is obtained without calibration, due to the inherent robustness against MOS mismatch. Precision can be increased to ~ +/-0.8°C by convenient single point soft-trimming.

(smart thermal sensor, current-mode, NPN bipolar, CMOS)

Introduction

Today, thermal management becomes an increasing challenge not only for CPUs. Most System-on-Chips (SoC) require some kind of temperature tracking to control performance, including mixed-signal functions like power management or RF. Especially mobile platforms need monitors on small grid, due to power optimization at huge computing effort (e.g. LTE). To enable multiple placement and cost reduction [1], such sensors should have small size, digital output and low power to avoid self-heating. For many applications a fair accuracy of +/-1°C to 4°C is sufficient.

Most integrated solutions are based on the temperature characteristic of parasitic BJTs (Fig. 1a), and include some signal processing, ADC plus reference. Since MOS mismatch has the largest impact, conventional sensors typically adopt techniques like dynamic element matching or chopping for error correction. In consequence the circuits achieve high precision after trimming, but are rather complex [1-4], therefore improper for multi-placement. Likewise costly are solutions which need a 2-point trim[1, 5].

A mismatch-insensitive concept can result in much lower complexity and improved robustness. This work presents such innovation within a 28nm SoC, which is respectable in terms of simplicity, size and power. Featuring low voltage and soft-trimming capability, for easy system integration, it combines and exceeds the advantages of previous concepts.

Sensor realization and functionality

A. Principle of operation

Fig. 1b describes the operating principle of the realized sensor: It is based on a current-mode technique which avoids errors introduced by device mismatch. This is in contrast to previous publications, which mostly use voltages, either by direct measurement or through conversion to frequency [2-5]. The impact of resistance and bipolar spread does not degrade linearity and can be easily trimmed out.

Temperature sensing is performed by subtracting two currents with opposite temperature coefficient: A precise PTAT ("proportional-to-absolute temperature") and a CTAT

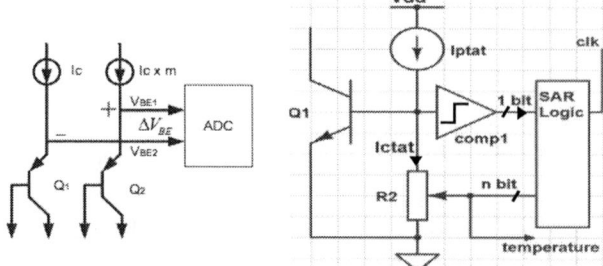

Fig. 1a+b Classic & new sensor concept with SAR ADC and R-DAC

("complementary-to-absolute temperature") current, which is generated by resistor R2 across Vbe-voltage of a NPN bipolar transistor. This parasitic device is generally available in triple-well standard CMOS. At a certain threshold temperature, adjustable by R2, the current difference equals zero and triggers a current-comparator (comp1). This 1-bit output allows simple threshold monitoring, or can be further processed with a successive approximation algorithm yielding n-bit, by means of controlling R2 realized as resistive DAC.

B. Analog main circuit

Schematic implementation of the sensor demonstrates particular simplicity (Fig. 2), since core function includes only 5 active devices plus 2 resistors. Two NPN transistors form a pseudo-differential pair with asymmetry defined by a (size/current) ratio of 1:N. In consequence, a PTAT current (Iptat) is generated such that the base-emitter (Vbe)-difference of Q1 and Q2 appears across R1 within a (positive) feedback loop. In that way the accuracy of Iptat is not degraded by errors of current mirrors (M3), or even amplifier offset: Q1 & Q2 have, unlike MOS pairs, a large gm and negligible mismatch.

Resistor R2 in series to R1 provides a reference current (Ictat) for comparison, without the need for additional current mirrors. The negative temperature coefficient (tc) of Vbe1 allows larger sensitivity than a conventional zero-tc reference. The error due to Q1 base current is effectively compensated by adding resistor R0 in series to Q2 [3]. Further, if Iptat is chosen much larger than the current through Q1, Ib1 gets negligible.

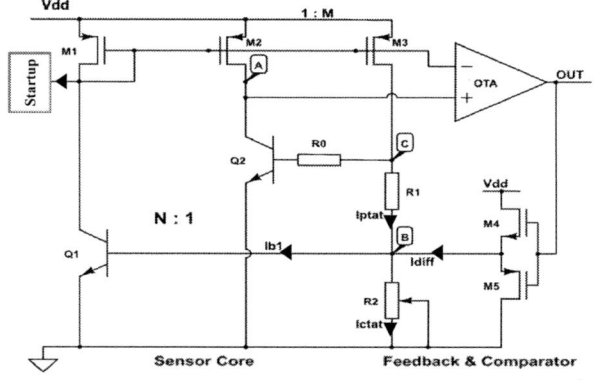

Fig. 2: Simplified analog circuitry including digital output

C. Transfer characteristic

In order to stabilize the circuit, a second (negative) feedback is provided, which delivers current Idiff to the summing node B. By sensing specifically the zero-crossing of Idiff, where Iptat matches Ictat, we determine the temperature threshold. This value depends only on device ratios and Vbe(Q1), calculated as follows:

$$\frac{\ln(N)}{R1} \cdot \frac{kT}{e} - \frac{Vbe1}{R2} = 0; \quad \rightarrow T = \frac{Vgo}{\frac{R2}{R1} \cdot \frac{k \cdot \ln(N)}{e} - tc} \quad (1)$$

Vgo is ~1.2V, related to the silicon bandgap; tc ~ -2mV/K

The function of a current comparator is effectively accomplished by M4/M5, which provide exactly at Idiff=0 a strong nonlinear characteristic inside the loop (like antiparallel diodes): Intermediate node OUT switches sharply when Idiff changes polarity, as explained by I-V plots in Fig. 3.

The module is fully self-biased, including the OTA added to decouple node A from B. Its characteristic is uncritical, but differential input benefits symmetry for good supply rejection.

The striking simplicity of this sensor is due to the fact, that generation of temperature & reference signal, even comparison with A2D-conversion, is executed at once in a single current branch/loop. In that way, and exploiting the BJT advantages, errors through MOS offsets are avoided. Due to its static nature, the sensor works as standalone macro without clock (noise) or control overhead. Compared to previous solutions [1], it achieves similar linearity and speed, with less area and power.

D. Trimming

The mapping of binary code versus temperature is defined by formula (1) and depends on resistive-DAC (R2) structure. A serial DAC is implemented here, yielding a curved mapping (Fig. 4). Though resolution varies with temperature, in fact this simple structure has distinct advantages: Due to the analogy of both effects in the denominator of (1), "tc"-variation and coding of Ictat, calibration procedure is greatly simplified. Process spreads induce that the curve is shifted from the ideal transfer by a ~constant value in y-direction. In consequence effective trimming is possible simply by adding an offset to binary code reading. In that way single-point soft calibration is enabled, without touching hardware, while °C-mapping may be implemented by firmware or lookup-table.

Monte-carlo simulations showed a residual error after (ideal) trim at room temperature of +/-0.8°C (3σ).

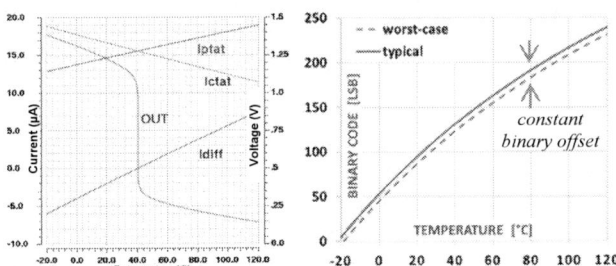

Fig. 3 Simulated signals vs. Temp. Fig. 4 Transfer curve w/ serial

Measurement Results

Several sensor instances were implemented as part of a complex 28nm RF SoC. Fig. 5 provides untrimmed silicon

data, measured from extreme wavers within a process window lot (PWL), and different products. The absolute error of packaged samples (top) is impacted by the uncertainty of thermal resistance and self-heating, therefore the trim-example (DTS1) is referenced to +1.5°C offset.

Statistical data on waver level (bottom) from 630 samples achieved a variation of ≤ 0.59°C (1σ). The small offset vs. simulation, observed at cold, may result from handler or device model, and can be corrected through formula (1).

1.8V supply was chosen according to system constrains, though the circuit works down to ~1.1V without precision impact. Silicon test showed perfect supply rejection (effect < 0.5LSB, simulated =0.1°C/V), proving the inherent robustness of concept.

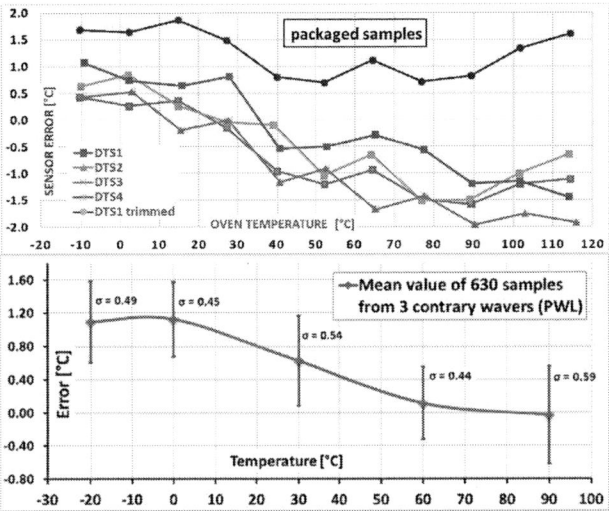

Fig. 5: Measured sensor error vs. simulation w/o calibration

References

[1] Y. W. Li, H. Lakdawala, "Smart integrated temperature sensor – mixed-signal circuits and systems", *Proc. CICC*, pp. 1-8, 2011.

[2] F. Sebastiano et al, "A 1.2V 10mW NPN-based temperature sensor in 65nm CMOS with an inaccuracy of ±0.2°C (3σ) from -70°C to 125°C", *IEEE JSSC*, No. 12, Dec. 2010.

[3] M. A. P. Pertijs, K. Makinwa, J. H. Huijsing, "A CMOS smart temperature sensor with 3σ inaccuracy of ±0.1 °C from -55°C to 125°C", *IEEE JSSC*, No. 12, Dec. 2005.

[4] K. Souri, K. Makinwa, "A 0.12 mm² 7.4mW micropower temperature sensor with an inaccuracy of 0.2°C (3σ) from -30°C to 125°C," *in Proc. ESSCIRC*, pp. 282-285, 2010.

[5] D. Duarte et al, "Temperature sensor design in a high volume manufacturing 65nm CMOS digital process," *Proc. CICC*, pp. 221-224, 2007.

Table1: Performance summary

Range	-20/+130°C
Resolution	8 bit
Supply	1.1 – 2.0V
Power	16µA (typ.)
Raw error @90°C	3σ = 1.8°C
Error[1] with single trim	3σ ~ 0.8°C
PSRR	< 0.5LSB
Speed	< 4µs/bit
Area	0.0038mm²

(1) simulated

Fig. 6: Layout 55µm x 70µm

A 35fJ/Step Differential Successive Approximation Capacitive Sensor Readout Circuit with Quasi-Dynamic Operation

Hesham Omran, Abdulaziz Alhoshany, Hamzah Alahmadi, and Khaled N. Salama

King Abdullah University of Science and Technology (KAUST), Thuwal 23955-6900, Saudi Arabia

Abstract

We propose a successive-approximation capacitive sensor readout circuit that achieves 35fJ/Step energy efficiency FoM, which represents 4× improvement over the state-of-the-art. A fully differential architecture is employed to provide robustness against common mode noise and errors. An inverter-based amplifier with near-threshold biasing provides robust, fast, and energy-efficient operation. Quasi-dynamic operation is used to maintain the energy efficiency for a scalable sample rate. A hybrid coarse-fine capacitive DAC achieves 11.7bit effective resolution in a compact area.

Introduction

Capacitive sensors are widely used in several applications, and are attractive for low-energy microsystems because they do not consume static current [1-6]. However, the capacitive sensor readout circuit, i.e., the capacitance-to-digital converter (CDC), can be the dominant source of energy consumption in the system [1]. Previously reported CDC architectures typically employ capacitance-to-time (C/T) or ΣΔ techniques [1-5]. For C/T CDCs a time-to-digital converter is typically required which increases the power consumption [4]. On the other hand, ΣΔ interfaces typically employ power-hungry op-amps running at a relatively fast oversampling clock [2]. A capacitance-to-voltage (C/V) stage followed by a successive approximation (SAR) ADC was reported in [6], and a direct SAR CDC was reported in [7]; however, both use op-amps that dominate the power consumption and limit the performance.

In this work, we propose a direct differential SAR CDC that uses an energy-efficient inverter-based amplifier with near-threshold biasing to provide robust operation against offset voltages and parasitics. The amplifier is power-gated to achieve quasi-dynamic operation. Wide capacitance range and fine absolute resolution are achieved using a hybrid coarse-fine capacitive DAC (CDAC).

Differential SAR CDC

The schematic of the proposed SAR CDC is shown in Fig. 1, where C_S is the sensor capacitor, C_{DAC} is a CDAC, and C_P is a parasitic capacitance. Fig. 2 shows the operation phases and timing. The circuit consists of two halves that are excited differentially, i.e., $C_{DAC1,ON} = C_{DAC2,OFF}$, where $C_{DAC,ON}$ and $C_{DAC,OFF}$ are the CDAC capacitances connected to V_{DD} and V_{SS}, respectively. In the reset phase, the inverter amplifiers (I1 and I2) are in unity-gain feedback, and their switching thresholds (V_{M1} and V_{M2}, respectively) are stored in the input capacitance (i.e., input offset storage). In the conversion phase, charge is redistributed according to the SAR logic output. The differential input of the dynamic comparator (ΔV) is given by

$$\Delta V = \frac{(C_S - C_{DAC1,ON})}{C_S + C_{DAC} + C_P} \cdot 2A_o V_{DD} + (V_{M1} - V_{M2} + V_{os,cmp}), \quad (1)$$

where A_o is the inverter gain and $V_{os,cmp}$ is the comparator offset. The SAR logic performs binary search till the selected CDAC capacitance $(C_{DAC1,ON} = C_{DAC2,OFF})$ matches C_S within an error equal to CDAC LSB. The effect of the error term $(V_{M1} - V_{M2} + V_{os,cmp})$ is reduced by properly selecting

A_o. Mismatch or variation in A_o, C_P, V_{DD}, or V_M will not affect the output as long as the sign of ΔV is not changed.

A supply voltage (V_{DD}) of 0.8V is used such that each transistor is biased near its threshold voltage (V_T), where $V_M \approx V_{DD}/2 \approx V_T$, which enables best compromise between energy efficiency and speed. The circuit is powered down automatically after conversion completion; however, if a slow clock is used, energy efficiency will deteriorate due to the amplifier static power. In order to maintain constant energy consumption independent of clock frequency the amplifier is power-gated as follows: 1) in the reset phase, the amplifier is enabled for a programmable number of clock cycles (SMPL signal), and 2) in the conversion phase, the comparator decision is detected (CMP_DONE signal), and then used to asynchronously reset EN_AMP signal and power down the amplifier. Hence, quasi-dynamic operation is achieved.

The CDAC is implemented as a hybrid coarse-fine array comprising an 8bit fine CDAC with a 3.75fF MOM unit capacitor and a 4bit coarse CDAC with a 780fF MIM unit capacitor. The three-sigma DNL of the coarse CDAC is 1.5fF (less than 0.5LSB); thus, a single calibration point is sufficient to reconstruct the output from the coarse and fine parts.

Measurement Results

The prototype is fabricated in a 0.18μm CMOS technology and occupies 0.1mm^2 (Fig. 3). The coarse-fine CDAC provides 81% area saving compared to binary-weighted MOM array and has a full-scale capacitance of 12.66pF and LSB of 3.75fF, i.e., 11.7bit effective resolution. Measured DNL/INL of the fine and coarse CDACs are 0.45/0.49fF and 0.98/1.12fF, respectively. Fig. 4 shows the CDC output tested by a dummy capacitive sensor and showing excellent linearity. The power consumption increases with C_S (Fig. 5) and the power consumption at full-scale and 1MHz clock is 7.25μW. The CDC DNL/INL are measured by sweeping the reference voltage of C_S, which emulates sweeping C_S from zero to full-scale [3, 6]. Both DNL and INL are less than 1LSB (Fig. 6). The CDC is tested using a MEMS capacitive pressure sensor (Protron Mikrotechnik), where the CDC output matches the non-linear sensor characteristics (Fig. 7). For a constant C_S the measured output code variation is limited to a single LSB; thus, the resolution is limited by the CDAC quantization noise rather than the thermal noise, i.e., the rms resolution is $LSB/\sqrt{12} = 1.1 fFrms$. The energy efficiency FoM is 35fJ/Step and is maintained at low clock frequencies by virtue of the quasi-dynamic operation (Fig. 8). The design is inherently temperature insensitive because it is independent of analog references. The measured temperature sensitivity is 5.2ppm/°C (Fig. 9), which is 3× better than the calibrated output of [5]. Table I compares this work with the state-of-the-art.

References

[1] M. Ghaed et al., TCAS, 2013. [2] Z. Tan et al., VLSI, 2012.
[3] S. Oh et al., VLSI, 2014.   [4] Y. He et al., ISSCC, 2015.
[5] W. Jung et al., ISSCC, 2015. [6] H. Ha et al., ISSCC, 2014.
[7] H. Omran et al., CICC, 2014.

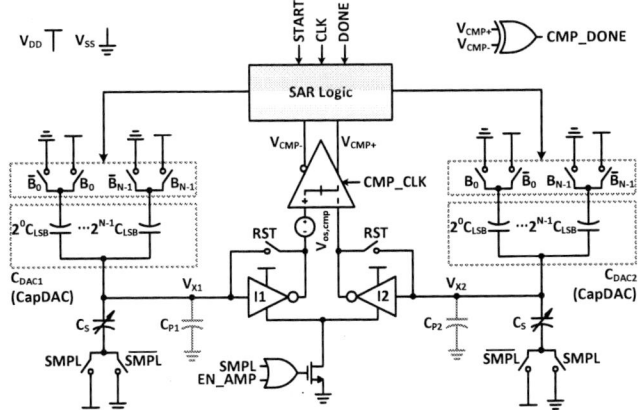

Fig. 1. Schematic of the proposed differential SAR CDC.

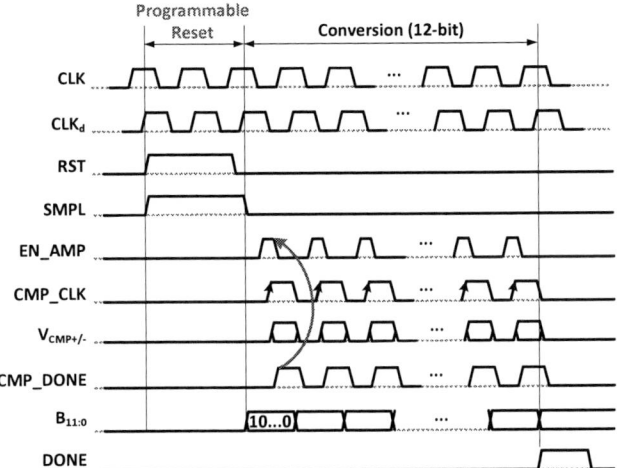

Fig. 2. Timing diagram of the differential SAR CDC.

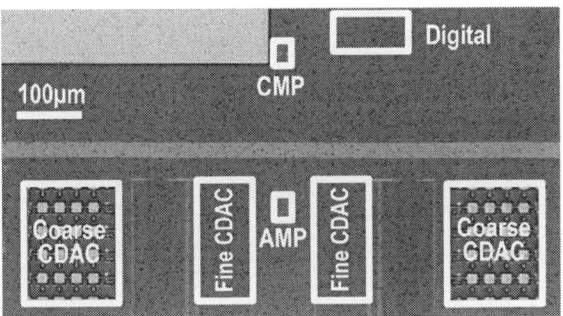

Fig. 3. Die photo.

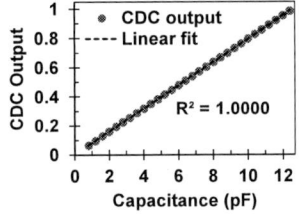

Fig. 4. Measured CDC output (normalized) vs dummy sensor capacitance.

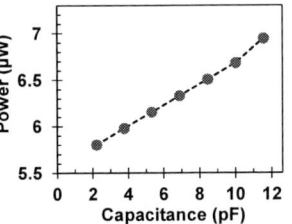

Fig. 5. Measured power consumption vs sensor capacitance.

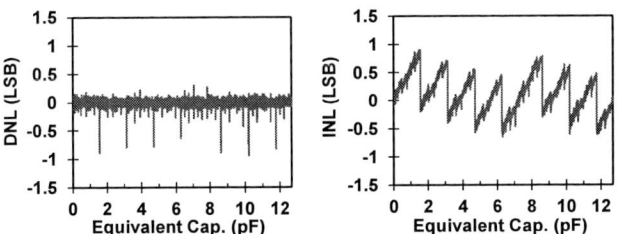

Fig. 6. Measured CDC DNL and INL vs equivalent sensor capacitance. 1LSB = 3.75fF.

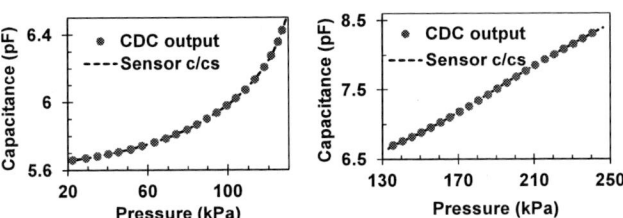

Fig. 7. Measured CDC output vs pressure in barometric and high pressure ranges overlaid on the characteristics of the MEMS pressure sensor (characterized using Agilent E4980A LCR meter).

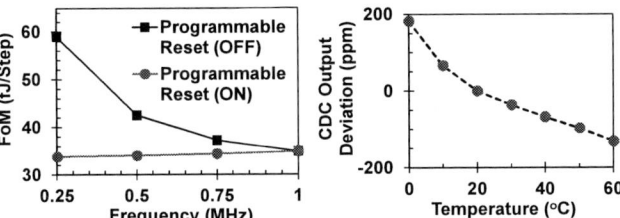

Fig. 8. Measured FoM vs clock frequency with fixed and programmable reset phase.

Fig. 9. Measured CDC output temperature sensitivity.

Table I. Performance summary and comparison.

	VLSI 2012 [2]	VLSI 2014 [3]	ISSCC 2014 [6]	ISSCC 2015 [4]	ISSCC 2015 [5]	This work
Technique	ΣΔ	SAR + ΣΔ	C/V+ SAR	C/T	C/T	**SAR**
Differential	Yes	No [a]	No	No	No	**Yes**
Tech. (nm)	160	180	180	160	40	**180**
Area (mm²)	0.28	0.456	0.49	0.05 [b]	N/A [c]	**0.1**
Power (μW)	10.32	33.7	0.16	14	1.84	**7.25**
Conv. Time (μs)	800	233	4000	210 [d]	19 [d]	**16**
Cap. Range [e] (pF)	0.52	24	10	8	10.6 [f]	**12.66**
Res.(fFrms)	0.07	0.16	6	1.4	12.3	**1.1**
ENOB [g] (bit)	11.1	15.4	8.9	10.6	8	**11.7**
FoM [g] (pJ/Step)	3.76	0.175	1.3	1.87 [d]	0.141 [d]	**0.035**

[a] The C/V stage is not differential.

[b] Off-chip reference capacitor is employed.

[c] Not available because area of on-chip reference capacitor is not reported.

[d] Multiple measurement cycles are required to cancel the effect of references/parasitics. Conv. time and FoM are reported for single cycle only.

[e] Only the cap. range that is covered by the reported conv. time is considered.

[f] Cap. range can be extended up to 10nF but with degraded energy efficiency.

[g] $FoM = Power \times Conv.Time / 2^{ENOB}$, where $ENOB = (SNR - 1.76)/6.02$, and $SNR = 20\log(Cap.Range/2\sqrt{2}/RMS\ Resolution)$ [3-5].

A 9.84–73.2 nJ, 0.048 mm^2 Time-Domain Impedance Sensor that Provides Values of Resistance and Capacitance

Yan Hong[1,2], Yong Wang[1], Wang Ling Goh[1], Yuan Gao[2], Lei Yao[2]

[1]Nanyang Technological University, [2]Institute of Microelectronics, Singapore.

Email: wangyongieee@gmail.com, ywang13@e.ntu.edu.sg

Abstract

A new time-domain impedance sensor readout circuit based on 0.18-μm CMOS technology is presented. A current DAC is used to charge the device under test (DUT) to increase the node voltage of the DUT. Using a time-domain comparator and a counter, a time period between the start of charge till the moment that the node voltage reaches a reference level is recorded and digitally converted. The resistance and capacitance components of the impedance can be quantized by using the time period data. The fabricated prototype consumes only 9.84 to 73.2 nJ of energy and requires merely 3 ms per measurement, where both are >10^3 times' reductions as compared to the state-of-the-arts. Moreover, to the best of the authors' knowledge, this proposed readout chip is the first of its kind that is able to deduce each resistance and capacitance component of the impedance. The chip takes up 0.048-mm^2 of area.

Introduction

Impedance measurement is the key requisite for wireless sensor nodes, body area networks, internet of things, etc. The state-of-the-art impedance sensing circuits [1–5] use I/Q demodulation technique in Fig. 1(a) to decipher both the magnitude and phase information of the impedance. To obtain accurate FFT transformation, long sampling period is required. This leads to considerable energy consumption per measurement. For example, to measure the impedance at 16 kHz, [5] needs to collect 2200 data points for each measurement, which calls for 15 seconds and 87-mJ of energy (i.e. 5.8 mW × 15 s). Also, due to the slow measurement speed, the abovementioned reports are only useful for limited applications (e.g. the impedance in slow motion).

The ability to extract each component value from the measurement results is useful and necessary but it is still beyond the reach of many, e.g. impedance sensing circuits [1–5], capacitance-to-digital converters, etc. The impedance of the DUT is usually a combination of several resistive and capacitive components, as shown in Fig. 1(b). A change in value in either the capacitive or resistive component will always indicate different physical meaning, for example, the change of physical parameters that are relevant to the dielectric leads to the variation of capacitance, while physical parameters relevant to conductivity have influence on resistance.

Proposed Design

The concept of the proposed readout approach is illustrated in Fig. 2. The most complicate type-A RC-combination shown in Fig. 1(b) is use as a DUT impedance example. The current source provides a constant current to charge the DUT. The voltage at the DUT node, V_{IN}, increases with time, and the response of which can be expressed using a mathematic equation. A time period is added up until V_{IN} reached a reference voltage, V_{REF}. Three different time periods, t_1, t_2, and t_3, can be obtained with three different current strengths, I_1, I_2, and I_3. With these, three equations can be constructed as shown in Fig. 2, where the unknown R_S, R_P, and C_S parameters can thereby be unraveled. Obviously, this approach is applicable to other RC combinations on condition that sufficient equations

are created by using many different currents.

The architecture of the proposed readout chip is shown in Fig. 3; and the operation is illustrated using a timing diagram. Prior to measurement, all circuit modules, except for the discharging path, are disabled. When the *TEST* signal arrives, the enable signal, EN_S, turns on to initiate the system. Triggered by EN_S, the switch, SW$_D$, cuts off the discharging path; the 9-bit current DAC (IDAC) generates a constant current that is fed into the DUT; the counter begins recording the time. Once V_{IN} reaches V_{REF}, the output signal of the time-domain comparator, F_{COM}, inverts, and the counter value (i.e. time period t) is latched to the registers. As soon as the data is successfully latched, EN_S turns off the whole circuit to conserve energy. Also SW$_D$ is closed to discharge V_{IN} to zero, in preparation for the subsequent measurement.

The schematic of the 9-bit IDAC is given in Fig. 4. To gain a good accuracy, the output current of the IDAC should maintain constant and irrespective to the load voltage. Therefore, a high output impedance is preferred. In this IDAC, a double cascade structure is adopted and a feedback amplifier is used to further boost the output impedance. From measurements, the output impedance is 66 GΩ at 1 kHz. The current varies <0.1% when V_{IN} rises from 0 to 0.3 V. The output current is programmable from 80 nA to 24 μA, to allow modification to both the resolution and sampling rate to cater for measurements of different impedance ranges.

The schematic of the time-domain comparator is shown in Fig. 5. When *CLK* is at active low, the output nodes of both voltage-controlled delay lines (VCDLs) are discharged to ground. When *CLK* rises to active high, the current-starving transistors provide different currents due to unequal input voltages, V_{in+} and V_{in-}. Therefore, *CLK* propagation delays of these two VCDLs are different. This delay difference is detected by the binary phase detector. The maximum propagation delay is 40 ns, so VCDLs can work up to 25 MHz. The offset will not affect the accuracy and can be calibrated, since V_{REF} is invariant. The counter is implemented using D flip-flops.

The reference voltage V_{REF} is set to a low level of 0.3 V, which presents the following benefits: 1) reducing the power energy drawn by capacitances and speeding up the measurement, 2) relaxing the current variation due to loading voltage, 3) avoiding organism damage in biomedical applications.

Measurement Results and Conclusions

The prototype was fabricated using a 0.18-μm CMOS process, as shown in Fig. 6. The chip was supplied at 1.8 V and clocked at 10 MHz, with sampling rate of 2 kHz. With a DUT sample of R_S = 9.98 kΩ, R_P = 750 kΩ and C_S = 3.192 nF (i.e. typical parameters in neural stimulation application), waveforms of V_{IN}, F_{COM} and EN_S are measured (see Fig. 7). The charging period, t, is ~0.31 ms. The power spectrum density in Fig. 8 is plotted using the digital data output from the prototype, where the SNR is 73.9 dB and the ENOB is 11.98. The chip was calibrated with this DUT. With different typical current settings, a series of time periods were recorded and plotted with an ideal curve as shown in Fig. 9. It can be seen that the results

978-1-5090-0636-6/16 $31.00 © 2016 IEEE   158   2016 Symposium on VLSI Circuits Digest of Technical Papers

fit well with the ideal curve, which implies that the calibration was conducted correctly. Figure 10 measures the type-A DUTs with different R_S, R_P and C_S, ranging from 10 kΩ to 500 kΩ, 10 kΩ to 10 MΩ, and 47 pF to 100 nF, respectively. The equations derived were solved by Matlab. The measured errors of R_S, R_P and C_S are 1.25%, 1.04%, and 0.54%. The measured magnitude and phase errors of impedance are 1.04% and 2.64% respectively. With all circuit turned on, the measured worst-case power consumption ranges from 6.56 μW to 48.8 μW, and corresponding to the current output configuration of IDAC from 80 nA to 24 μA. The energy consumption per measurement ranges from 9.84 nJ to 73.2 nJ.

Table I compares the proposed design with prior works, showing much higher measurement speed and competitive power consumption. Not only is the energy consumption per measurement reduced by $>10^3$ times, the prototype chip occupies a mere 0.048 mm². Most important of all, the proposed sensor is able to obtain all the values of the resistive and capacitive components of the impedance.

References

[1] Giorgio Ferrari, et al., *ISSCC*, Feb. 2014.

[2] H. Jafari, et al., IEEE TBCAS, 2012.

[3] Jing Guo, et al., *VLSIC*, 2013.

[4] Saul Rodriguez, et al., *IEEE TBCAS*, 2015.

[5] Marco Crescentini, et al., IEEE JSSC, 2014.

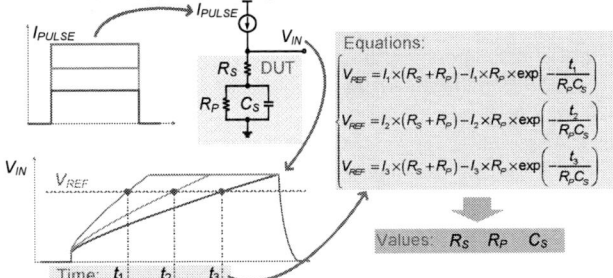

Fig. 1 (a) Overview of conventional impedance sensing circuits using I/Q demodulation, and (b) different impedance types.

Fig. 2 Concept of the proposed impedance measurement approach.

Fig. 3. Architecture of proposed impedance sensing readout and its timing diagram.

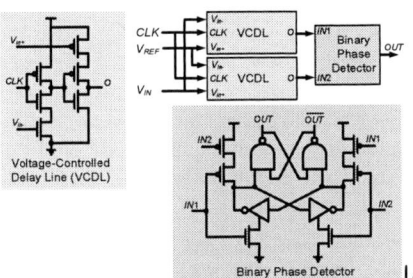

Fig. 5. Schematic of time domain comparator.

Fig. 6. Chip micrograph.

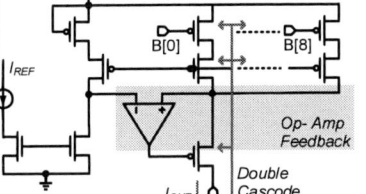

Fig. 7. Measured waveforms of V_{IN}, F_{COM} and EN_S.

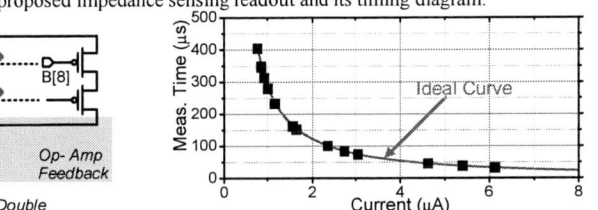

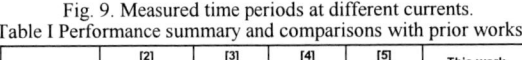

Fig. 4. Schematic of IDAC.

Fig. 8. Measured power spectrum density.

Fig. 9. Measured time periods at different currents.

Table I Performance summary and comparisons with prior works.

	[2] TBCAS 12	[3] VLSI 13	[4] TBCAS 15	[5] JSSC 14	This work
Process	0.13 μm	0.18 μm	0.15 μm	0.35 μm	0.18 μm
Method	IQ	IQ	IQ	IQ	Time domain
Supply (V)	1.2V	1.8V	1.8V	3.3V	1.8V
Power	42μW	144μW	205μW	5.8mw	6.56–48.8μW
Meas. Speed	10s	N.A	15s	1.5ms
Energy/Meas.	420μw	N.A	87mJ	9.84–73.2nJ
Measurement Type	Impedance: Real, Image	Impedance	Impedance	Impedance	Impedance: Capacitance, Resistance
Impedance Error (%)	Magnitude: 8.4 Phase: 7.5	2	1	0.017	Magnitude: 1.04 Phase: 2.64
RC Accuracies of Type-A (%)	N.A				Rs: 1.25 Rp: 1.04 Cs: 0.54
Area (mm²)	1.68	0.48	1.49	9	0.048

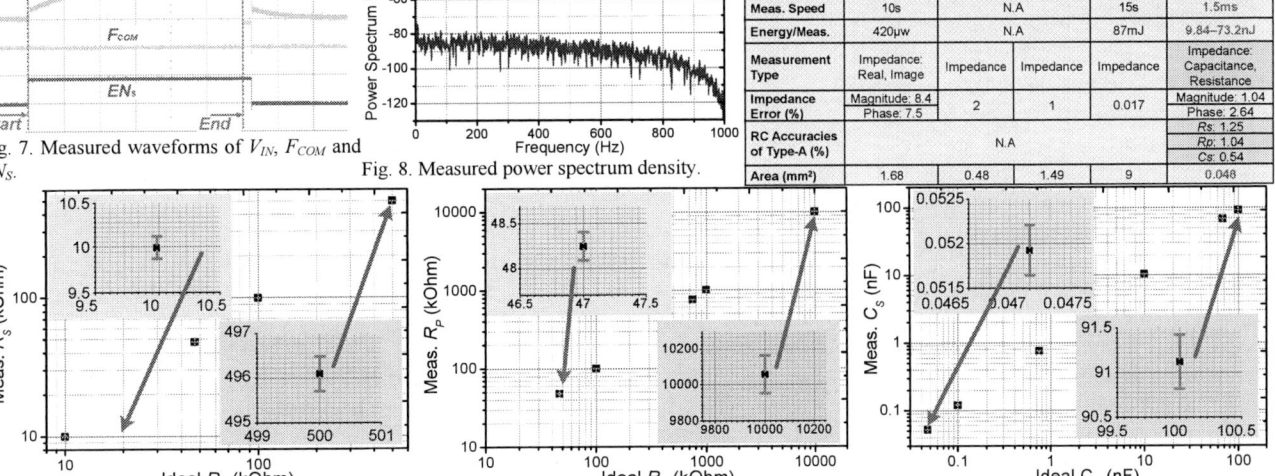

Fig. 10. Measured component values and accuracies of R_S, R_P and C_S in type-A impedance (x-axis is the labeled value of discrete element used for measurement).

978-1-5090-0636-6/16 $31.00 © 2016 IEEE   159   2016 Symposium on VLSI Circuits Digest of Technical Papers

A 23mW 24GS/s 6b Time-Interleaved Hybrid Two-Step ADC in 28nm CMOS

Benwei Xu, Yuan Zhou, and Yun Chiu

University of Texas at Dallas, USA

Abstract

We present a power- and area-efficient 24GS/s, 6b, 16-way time-interleaved (TI) ADC array, featuring a voltage-time (v/t) hybrid two-step structure for high-speed and low-power operation, a crosstalk-free SAR DAC topology and a non-hierarchical sampling frontend obviating reference and input buffers, respectively, for power and area savings. Background timing-skew calibration via dithering a reference ADC is also reported. Fabricated in 28nm CMOS, the prototype ADC array consumes 23mW at 24GS/s and measures an SNDR/SFDR of 35/54dB for a low-frequency input and 29/41dB for a Nyquist input, respectively. The core area of the ADC is 0.03mm^2.

TI-ADC Array Architecture

Time interleaving a large number of SAR ADCs to achieve 10GS/s+ conversion speed is often suboptimal due to the large interleaving factor and the ensuing complexity and high power of the multiphase clock generation and distribution network. Exploiting the resolving time dependence of a comparator on its input voltage, we introduce a v/t hybrid SAR architecture, in which the SAR comparator also functions as a time-domain residue production device, followed by a time-to-digital converter (TDC) as the second quantization stage. Providing a 6b resolution, the v/t hybrid SAR is highly power and area efficient and can be clocked at 1.5GS/s+ (with 80ps allocated for input tracking and 580ps for conversion).

In this work, a 24GS/s aggregate sample rate is achieved by interleaving 16 such hybrid converters, all directly connected to a 50Ω input source without any hierarchical input sampling structure or buffer. Thanks to the hybrid architecture, the input capacitance of each sub-ADC is only 16fF (SE), which greatly simplifies the ADC frontend design.

For timing-skew correction among the 16 sub-ADCs, a reference ADC is added [1]. Fig. 1 shows the TI-ADC array with the reference path. 16 digitally controlled delay lines (DCDL) are employed to fine-tune the clock phases of the 16 TI paths to align with the reference path. The DCDL, featuring a 30ps tuning range and a 100fs step size, is a chain of inverters with variable loading capacitance. The reference ADC is a 6b SAR with a source-follower input buffer to isolate its activity from the main array. The clock generation circuit is similar to [2].

V/T Hybrid Two-Step ADC

The first stage of the two-step sub-ADC is a 4b, asynchronous SAR. After the first three bit cycles, the residue voltage on the summing nodes (SN and SP in Fig. 2) is small and a 4th comparison is performed as a v/t conversion – the resolving time of the comparator is extracted as shown in Fig. 2, which begins by producing a signal EN after the 3rd SAR cycle to enable an edge extractor; the time lapse between the comparator clock Φ_c and the ready signal Rdy represents the resolving time, which is routed to a time amplifier (TA) followed by a standard delay-line-based 4b TDC. Since Rdy always trails Φ_c, the original time-domain residue is unipolar. A time offset is introduced by delaying the Φ_c path by a DCDL, thus converting the residue to bipolar. Finally, the amplified time residue is quantized by the TDC. Two extra arbiters, before the 1st and

after the 15th arbiters, are added in the TDC to detect over-/under-range errors. The DCDL delay and the TA gain are both digitally tunable to overcome PVT variations upon receiving the over-/under-range information.

The nominal residue swing of the SAR is ±15mV, which can be easily saturated if the v/t converter and the SAR comparator exhibit different offsets (as commonly seen in voltage-domain pipelined SAR ADCs). Fortunately, the two are the same device in this work.

Lastly, simulation indicates limited linearity of the v/t conversion as shown in Fig. 2. The 4b+4b two-step architecture provides 2b redundancy to linearize the TDC output using a digital lookup table (LUT).

Timing-Skew Calibration

The timing-skew information is extracted by dithering the reference-path clock by a 1b pseudorandom noise (PN) [3] and correlating it to the differences between the outcomes of the reference ADC and the sub-ADCs. Fig. 3 illustrates three scenarios corresponding to the sub-ADC clock Φ_i leading, trailing, and synchronous to the reference clock Φ_r. Since the skew error is proportional to the skew, the sign of the skew can be derived as sign(Δt_i) = sign[Σ(E-)-Σ(E+)], i.e., only when Δt_i is zero, statistically Σ(E-) = Σ(E+) holds and the skew calibration halts. The PN-directed dither exhibits an added benefit of less likely being correlated to an input pattern [3]. The digital skew calibration logic is shown in Fig. 1 [4].

Crosstalk-Free SAR DAC

Stabilizing the reference voltages of the ADC array (to avoid unwanted crosstalk between the sub-ADCs) dictates power-hungry buffers or large on-chip decoupling capacitors. In this work, a SAR DAC structure that is disconnected from the reference network during bit cycles is reported. Similar to [5], the DAC obviates any buffer or decoupling capacitor; in addition, the crosstalk between the sub-ADCs is also eliminated. Shown in Fig. 4, the SAR sampling capacitor C_S and the DAC capacitors C_{DAC} are separated. They sample V_{in} and V_{ref}, respectively, during the tracking phase; during the conversion phase, the bottom plates of C_S are shorted and those of the DAC capacitors are shorted successively depending on the bit decisions. This process ensures that at the end of the cycles the charge left on C_{DAC} is always nearly constant, resulting in a constant charge drawn from the reference lines during the pre-charging phase of each sample period.

Bottom-plate sampling is used in the SAR due to the benefit that the input (V_{cmi}) and summing-node (V_{cms}) common-mode voltages can be decoupled. A 125mV V_{cmi} is chosen to enable a single NMOS sampling switch without bootstrapping. A V_{cms} of 600mV is selected to speed up the comparator.

The partially floating C_{DAC} during the bit cycles is known to present a time-variant offset problem [5], which can cause errors in SAR. Instead of calibrating the offset, this error is tolerated by the 2b inter-stage redundancy in this work.

Measurement results

The prototype ADC array was fabricated in a 28nm CMOS process. At 24GS/s, the ADC consumes a total of 22.94mW.

Out of which, 12.8mW is from the ADC array with an 850mV supply, 9.5mW is from clock generation with a 950mV supply, and 0.64mW is from the reference ADC. The digital calibration logic is implemented off-chip. Ideally the reference ADC can be used to calibrate the static inter-sub-ADC mismatch and the intra-sub-ADC radix errors and nonlinearity (including the time residue LUT) [1]. But in this work we used it solely for the background timing-skew calibration for simplicity. The static mismatch calibration is performed in foreground using a sine-fit algorithm.

Fig. 5 shows the measured code histogram of the TDC – it should be flat if the v/t conversion is linear, as the SAR residue is a 3b quantization noise. The learned LUT code mapping (Fig. 5) clearly reveals the reversal of the v/t nonlinearity. In experiment, the TA gain and the DCDL delay in the v/t interface were determined and set with a one-shot power-on calibration.

Fig. 6 plots the array output spectra for an 11.9GHz sinusoidal input at 24GS/s (decimated by 125×). The measured SNDR and SFDR before/after calibration are 14.5/28.9dB and 21.6/41.1dB, respectively. Fig. 7 presents the dynamic performance of the TI-ADC array vs. the input frequency, also at 24GS/s. The post-calibration SFDR is limited by HD2 at high input frequencies. A die photo is shown in Fig. 8. The core area of the ADC array is 0.03mm^2 (250μm × 120μm).

Acknowledgement

This work is supported by Semiconductor Research Corporation (SRC) through Texas Analog Center of Excellence at the University of Texas at Dallas (Task ID:1836.148).

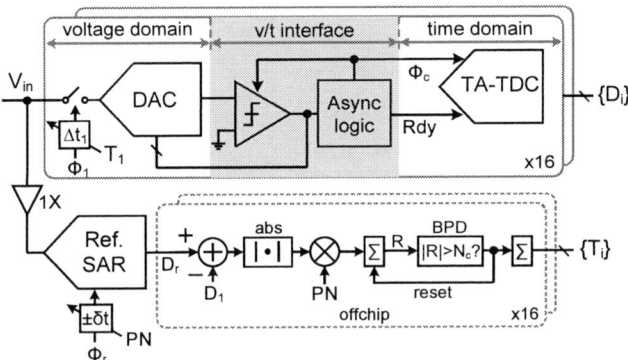

Fig.1 System diagram of the TI-ADC array with skew calibration

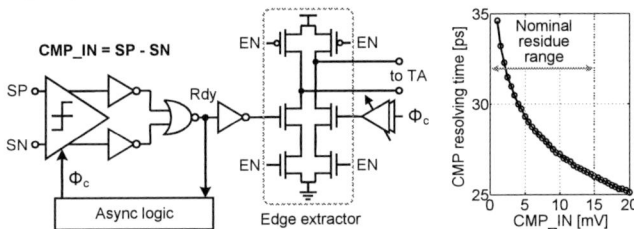

Fig. 2 Voltage-to-time conversion interface and nonlinearity

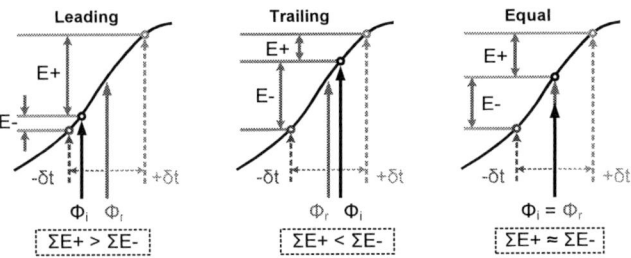

Fig. 3 Illustration of three skew-calibration scenarios

References

[1] W. Liu *et al.*, ISSCC 2009
[2] L. Kull *et al.*, ASSCC 2014
[3] P. Huang *et al.*, CICC 2010
[4] C. Huang *et al.*, VLSIC 2011
[5] J. Craninckx *et al.*, ISSCC 2007

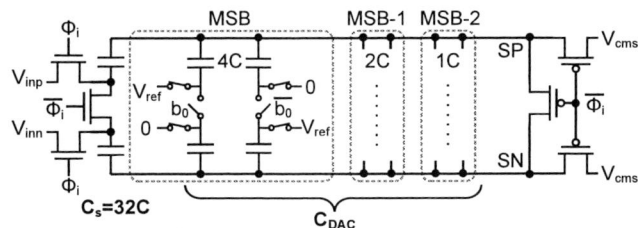

Fig. 4 Circuit diagram of the cross-talk free SAR DAC

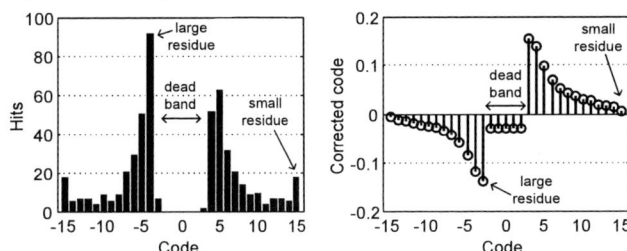

Fig. 5 Measured TDC code histogram (left) and LUT mapping (right)

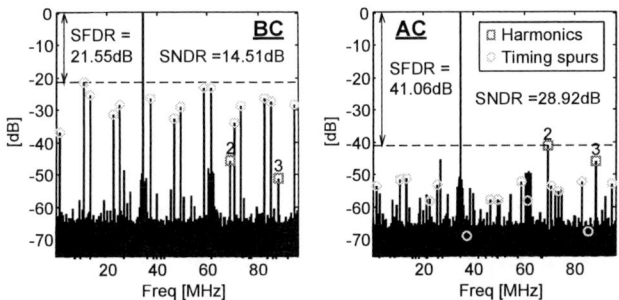

Fig. 6 Measured ADC array spectra for an 11.9GHz input

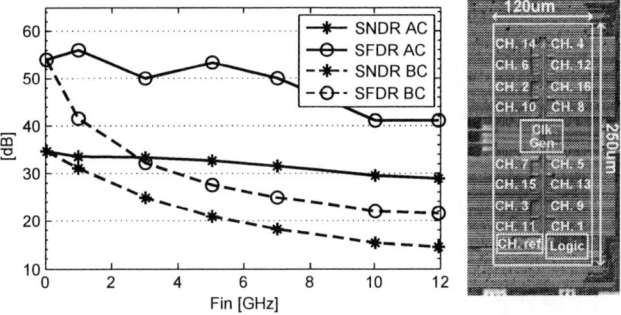

Fig. 7 Measured dynamic performance at 24GS/s   Fig. 8 Die photo

TABLE I Performance Comparison

	This Work	[2]	Cai VLSI'15	Chen ISSCC'15
ADC Architecture	**TI-SAR**	TI-SAR	TI-BS	TI-flash
Technology	**28nm**	32nm SOI	65nm	32nm SOI
Power Supply (V)	**0.85/0.95**	0.9/1	1	0.9
Sample Rate (GS/s)	**24**	36	25	20
Resolution (bits)	**6**	6	6	6
SNDR/SFDR @ LF (dB)	**34.8/53.9**	32.6/47	32/42	34.8/NA
SNDR/SFDR @ HF (dB)	**28.9/41**	31.6/42.8	29.7/40	30.7/NA
Active Area (mm^2)	**0.03**	0.048	0.24	0.25
Power (mW)	**23**	110	88	69.5
FoM @ LF/HF (fJ/c-s)	**21/42**	88/98	108/143	76/124

A 8.2-mW 10-b 1.6-GS/s 4× TI SAR ADC with Fast Reference Charge Neutralization and Background Timing-Skew Calibration in 16-nm CMOS

Ying-Zu Lin, Chih-Hou Tsai, Shan-Chih Tsou, Chao-Hsin Lu

MediaTek Inc., Taiwan

Abstract

This paper presents a 4-way 1.6-GS/s time-interleaved (TI) SAR ADC with fast reference charge neutralization (CN) and background timing-skew calibration. The SAR sub-ADC uses a flip-flop-less digital control unit to achieve 400MS/s operation. The prototype in 16-nm CMOS occupies an active area of 0.023 mm². From a 0.95-V supply, the power consumption is 8.2 mW at 1.6 GS/s. The peak SNDR is 55 dB and HF FOM is 19 fJ/conversion-step.

Introduction

With the popularity of portable devices, such as cell phones and tablets, Wi-Fi plays an important role since it provides wide transmission bandwidth at a range of several hundred meters. In the upcoming 802.11ax, the ADC sampling rate will likely increases to > 1 GS/s but will still require low power to extend battery life time. To achieve high-speed and good power-efficiency, a TI SAR ADC is the preferred architecture [1-4]. High-speed SAR ADCs require reference voltage sources with rapid settling speed because the capacitor array discharging during bit cycling disturbs the reference voltage. However, an external reference voltage (pin count) or large decoupling capacitor (cost) is impractical for highly integrated commercial chips. Reference sharing in a TI ADC also introduces inter-channel interference. With any increase in channel count, a TI ADC requires more reference circuits, each of which usually consumes large dc current to reduce the RC time constant. For high-speed SAR ADCs, the period of a bit cycle can be as short as 100 ps. The proposed TI ADC utilizes a reference charge neutralization technique for very high-speed SAR operation.

TI SAR ADC and Circuit Techniques

In [2], a coarse flash ADC calibrates timing skew of fine SAR ADCs but limits the overall speed. In [3], an additional clock phase is required. Figure 1 depicts the simplified block diagram of the TI ADC which consists of 4 sub-ADCs, a reference ADC, reference circuitry and a clock generator. The reference ADC can be used to calibrate non-linearity errors, as in [5], but this work employs it to calibrate timing skew error. The reference ADC is identical to the sub-ADC, but operates at $f_s/25$, versus $f_s/4$ for each sub-ADC. The sampling time of the reference ADC aligns with that of each sub-ADC in turn. The input signals of the reference ADC and the aligned sub-ADC cancel each other. The remaining code difference is due to mismatches. After removing the offset and gain error, the accumulated unsigned code difference is the skew error information. Based on the code difference, a variable delay in the clock path is adjusted to shift the sampling point of the sub-ADC. An LMS-based calibration is run in the background to track PVT. The cost of the reference ADC is only (40 μm × 45 μm) in 16 nm CMOS.

The drop of the reference voltage during MSB switching possibly causes incorrect subsequent comparisons. This work introduces a data-driven charge compensation technique to neutralize the discharging current during bit cycling. Figure 2 depicts the simplified block diagram of the SAR sub-ADC

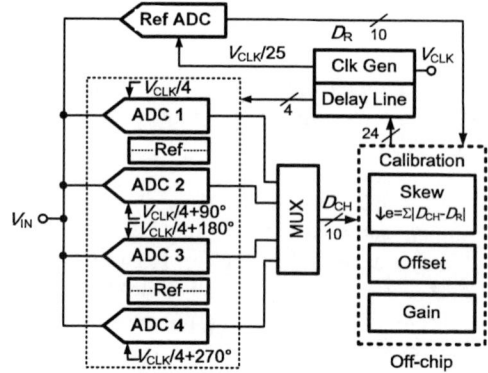

Fig. 1. Block diagram of proposed TI SAR ADC.

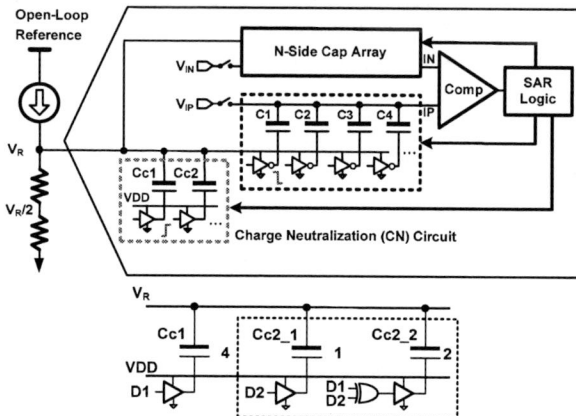

Fig. 2. Block diagram of channel ADC and compensation circuit.

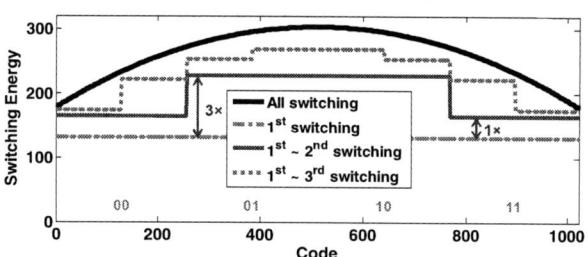

Fig. 3. Switching energy of MSB switching vs. code.

and charge neutralization circuit. The control signals of the compensation circuit are the opposite polarity of the capacitor switching signals. With similar propagation paths and proper signal gating, the charge provided by the compensation circuit neutralizes the charge needed by the capacitor switching. As depicted in Fig. 3, switching charges are code dependent. The switching energy of the 2nd MSB for codes 11/00 is 3× smaller than that for 01/10. Figure 4 compares the reference voltage drop of the 01 and 11 code cases with and without neutralization. As Fig. 4 demonstrates, the drop in the reference voltage is greatly reduced with the neutralization circuit enabled. For different codes, the neutralization circuit supplies a different amount of charge.

978-1-5090-0636-6/16 $31.00 © 2016 IEEE   162   2016 Symposium on VLSI Circuits Digest of Technical Papers

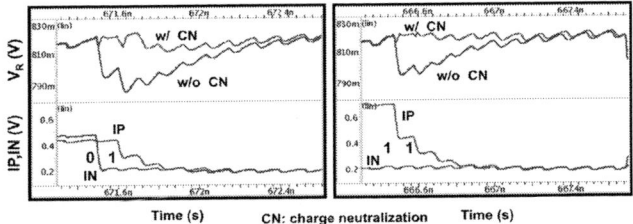

Fig. 4. Timing waveforms with and without charge neutralization.

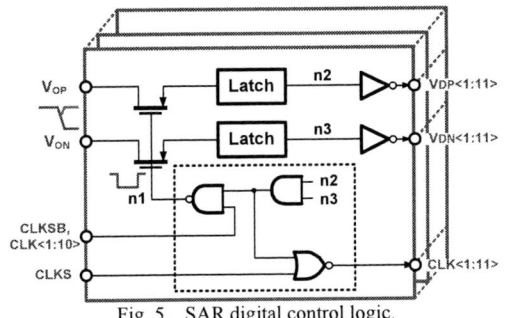

Fig. 5. SAR digital control logic.

Fig. 6. Chip micrograph.

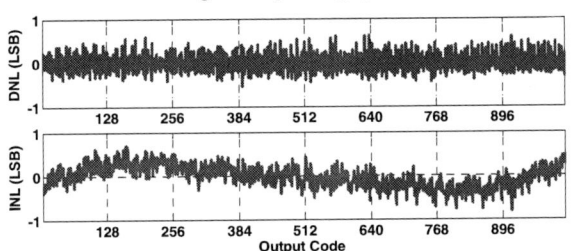

Fig. 7. Measured DNL and INL.

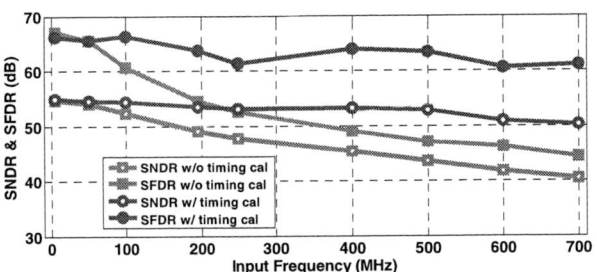

Fig. 8. SNDR and SFDR vs. f_{in} at 1.6 GS/s.

Table 1 Performance summary and comparison to other works

	ISSCC 2014	ISSCC 2014	ISSCC 2015	ISSCC 2015	This Work
Process (nm)	40	65	45	45	16
Architecture	12X TI SAR	8X FATI SAR	12X FATI SAR	4X TI SAR	4X TI SAR
Resolution (b)	9	10	10	10	10
Sampling Rate (GHz)	1.6	1	1.6	1.7	1.6
# of SAR Comparison	9	1 flash + 7	1 flash + 7	8	11
Supply (V)	1.1	1.1	1.1	1.2	0.95
Power (mW)	21.5*/27.5**	18.9*	17.3*	15.4*	8.2* / 9.8**
Input Swing (Vpp)	1	2	N/A	N/A	1
SNDR (dB)@LF	50.7	53.3	58.8	55.3	55
SNDR (dB)@HF (GHz)	48 @ 0.75	51.4 @ 0.5	56.1@ 0.8	51.2 @ 0.9	50.3 @ 0.7
FOM$_{LF}$ (fJ/conv.-step)	51.3/65.6	50	15.5	19	11.1 / 13.3
FOM$_{HF}$ (fJ/conv.-step)	64.7/82.8	62.3	21	30.5	19.2 / 23
Active Area (mm²)	0.83	0.78	0.36	0.057	0.023

*Total power w/o reference circuit   **Total power w/ reference circuit

Experimental Results and Conclusion

Figure 6 displays the micrograph of the TI SAR ADC. The active area is 460 μm × 50 μm and the supply voltage is 0.95 V. The four sub-ADCs consume 7 mW in total; clock distribution and output multiplexing consumes 1.2 mW. The reference ADC is only enabled during timing calibration. With the neutralization circuit enabled, the reference circuits of the ADC consume 1.6 mA. Figures 7 and 8 show the measurement results of the 10-b ADC. The DNL is within −0.54 and 0.62 LSB; INL is within −0.81 and 0.7 LSB. Figure 8 plots SNDR and SFDR at 1.6 GS/s versus input frequency with and without timing calibration. The low-frequency SNDR is 55 dB and SNDR at 700 MHz input is 50.3 dB. For input > 800 MHz, we cannot obtain full-swing output codes even with the maximum power of the signal generator due to the parasitic RC of the input routing. With timing calibration, there is a 9.8-dB improvement in SNDR at 700-MHz input. The LF and HF FOMs without reference power are 11.1 and 19.2 fJ/conversion-step, respectively. Including reference, the HF FOM is 23 fJ/conversion-step.

Table 1 shows the performance summary and comparison to state-of-the-art high-speed TI SAR ADCs. With the help of the charge neutralization technique, the reference voltage remains stable during the 100-ps period bit cycling. With the reference ADC, the skew errors between the sub-ADCs are reduced. Due to the low-latency and power-efficient digital control logic, the sub-ADC achieves 400-MS/s operation even though 11 SAR conversion cycles are necessary. Other ADCs need 8 or 9 cycles. This TI SAR ADC achieves the lowest power consumption and best FOMs reported to date.

As depicted in Figs. 2 and 3, the compensation for the first 2 bits achieves a good balance between efficiency and logic complexity. Timing imbalance between normal and neutralization paths results in poor compensation. The amount of neutralization capacitance depends on the ratio between the supply and reference voltage, and the amount of input attenuation capacitance. The ratio of each neutralization capacitor to its corresponding capacitor is 1/3 in this design. The CN capacitors and logic occupy 5% area of the sub-ADC.

Figure 5 depicts the SAR control unit. Rather than using a flip-flop, the digital control circuit uses 3 gates to generate a pulse at node n1. The previous stage i sends a CLK<i> signal to enable the current stage by pulling n1 low. The enable signal sends the comparator outputs to the latches. Once the latches store the comparator outputs, the latch outputs will pull n1 back high to block the comparator outputs of the other cycles. The pulse window opens briefly after the previous comparison finishes and closes after the current comparison completes without requiring a global valid signal. Compared to a flip-flop-based pulse generator, there is a 30% power reduction for the digital control circuit in the same process node and only 3 gate delays between the comparator outputs and the capacitor drivers [6]. Thanks to the power efficiency and low latency digital control, the 400MS/s sub-ADC is the fastest reported 1-comparison/step 10b SAR ADC despite requiring 11comparison cycles [6].

References

[1] N. Le Dortz, et al., ISSCC Dig. Tech. Papers, pp. 386-387, Feb. 2014.
[2] S. Lee, et al., ISSCC Dig. Tech. Papers, pp. 384-385, Feb. 2014.
[3] B.-R.-S. Sung, et al., ISSCC Dig. Tech. Papers, pp. 464-465, Feb. 2015.
[4] H. Hong, et al., ISSCC Dig. Tech. Papers, pp. 470-471, Feb. 2015.
[5] W. Liu, et al., ISSCC Dig. Tech. Papers, pp. 82-83, Feb. 2009.
[6] C. C. Liu, et al., JSSC, Nov. 2015, pp. 2645-2654.

A 14-bit 2.5GS/s and 5GS/s RF Sampling ADC with Background Calibration and Dither

Ahmed M.A. Ali, Huseyin Dinc, Paritosh Bhoraskar, Scott Puckett, Andy Morgan, Ning Zhu, Qicheng Yu, Chris Dillon, Bryce Gray, Jon Lanford, Matt McShea, Ushma Mehta, Scott Bardsley, Peter Derounian, Ryan Bunch, Ralph Moore and Gerry Taylor

Analog Devices, Greensboro, NC, USA
ahmed.ali@analog.com

Abstract

We describe a 14-bit 2.5GS/s non-interleaved pipelined ADC that relies on correlation-based background calibrations to correct the inter-stage gain, settling (dynamic), kick-back and memory errors. A new technique is employed to inject a large dither signal on the input to dither the non-linear kick-back on the ADC driver, and another large dither signal is injected to dither any residual non-linearity in the pipeline. In order to correct the effect of aging on the comparators, a new background calibration technique is employed to correct the comparator offsets. The ADC is fabricated as a dual in a 28nm CMOS process. An optional interleaved mode is provided, where the two ADCs on chip are time-interleaved to obtain a single 14-bit 5GS/s ADC. Background calibration of offset and gain mismatch and fixed calibration of timing mismatch between the two channels are implemented on chip.

Introduction

Wireless communication and instrumentation applications require ADCs with high sample rates, RF sampling, and high performance. This work describes an RF sampling ADC, with an input bandwidth of 5GHz, which can be configured to operate in two modes. In the non-interleaved mode, the chip is configured as two independent non-interleaved pipelined ADCs clocked at 2.5GS/s. In the interleaved mode, the two ADCs are ping-ponged by shifting the clock phase of one of the cores by 180°, while combining the analog inputs and digital outputs to give a single 5GS/s ADC.

Architecture and Design

A block diagram of the ADC is shown in Fig. 1. Each ADC has a SHA-less pipelined architecture with five 3-bit stages and a 4-bit backend flash. The input buffer uses a 2.5V supply, the reference and MDAC amplifiers use a 1.8V supply, while the flashes, the clocks, and the digital circuitry use a 0.9V supply. The input buffer is a source follower with an active cascode current source and two-level bootstrapping between the input and the drain. This new buffer design supports RF sampling with low distortion and an input bandwidth larger than 5GHz.

At 2.5GS/s, the gain/hold phase of the MDAC is 200ps. The comparators use about 50ps. The DAC and slewing take up about 50ps. This leaves only 100ps to the residue amplifier (RA) for small signal settling, which calls for a closed loop amplifier bandwidth of about 15GHz. The MDAC's RA is a two-stage Miller-compensated amplifier with active cascodes

in the first stage, and a level-shifted push-pull structure in the second stage. This helps reduce power consumption and improves the speed compared to a differential pair.

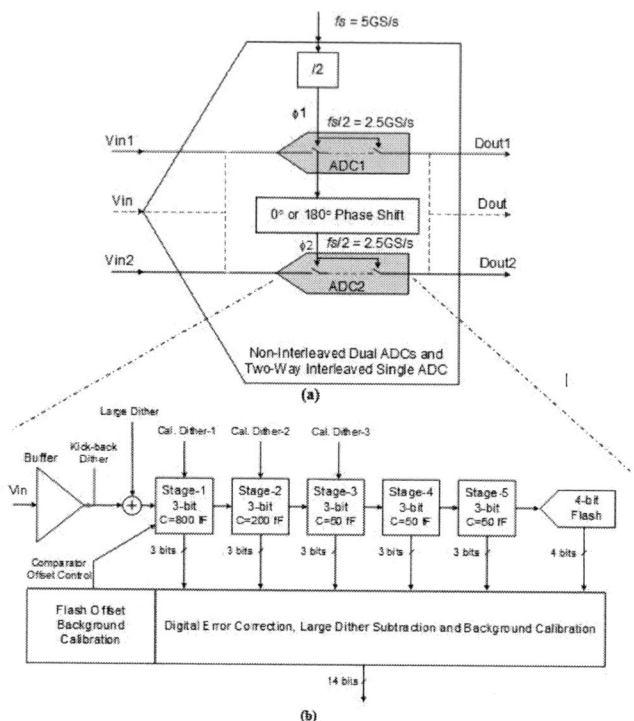

Fig. 1. (a) A block diagram showing the two non-interleaved pipelined ADCs sampling independently at 2.5GS/s (solid path) or interleaved to give a single ADC operating at 5GS/s (dotted path). (b) A block diagram of each of the two pipelined ADCs.

The MDAC amplifier's gain (65dB) and bandwidth (10GHz) are not adequate to achieve the target performance. Therefore, background calibrations of inter-stage gain, settling, memory and kick-back errors are implemented [1]. In addition, an 8-level dither signal is injected on the input in order to dither the non-linear kick-back. A 16-level large dither signal is also injected in the first stage MDAC and flash in order to correct for any residual non-linearity in any of the front-end stages, and to improve the accuracy of the background calibration. The total peak-to-peak amplitude of the large dither signals is equal to a full sub-range of the first stage. As shown in Fig. 2, they occupy a portion of the dynamic range that is beyond the ADC's full scale and is usually not utilized, since the first stage flash is implemented with 8 comparators. The large dither signals are background calibrated and subtracted digitally using the LMS algorithm to track changes with supply, sample

978-1-5090-0636-6/16 $31.00 © 2016 IEEE   164   2016 Symposium on VLSI Circuits Digest of Technical Papers

rate, aging and temperature.

One of the limitations of the 28nm process is the sensitivity of its devices to aging. The NBTI and PBTI (Negative- and Positive-Bias Temperature Instability) aging mechanisms change the device thresholds, which leads to shifts in the comparators' offsets with time. To maximize the RF sampling bandwidth in SHA-less architectures, the correction range usage due to offsets needs to be minimized. In this work, we employ a new background calibration algorithm that uses the residue signal to correct the comparator offsets. It does not require analog changes and it averages out errors due to bandwidth and timing mismatches. This separation of the static offsets from the phase mismatch errors is key to achieving accurate offset correction. It is also independent of the input signal characteristics and distribution. The technique operates on the first stage flash code and the residue of the first stage, as shown in Fig. 2 to generate an error function given by:

$$\varepsilon = E\{[maxL(V_R)|_{code=x}] + [minL(V_R)|_{code=(x+1)}]\} \quad (1)$$

where V_R is the first stage digital residue, ε is the error function used to control the threshold, $E\{z\}$ is the expectation (or average) of all elements of z. The functions maxL() and minL() are leaky peak and trough detectors respectively. The offset correction is fed back to the comparator to control its offset.

In the interleaved mode, the gain and offset background calibrations perform the mismatch correction in the digital backend. The timing mismatch is corrected by adjusting the sampling clock edge.

Results

The ADC is implemented with the entire digital processing needed on chip. The MDAC and dither capacitors are factory calibrated and the background calibrations and dithering are verified across process, sample rate, aging, temperature and supply variations. The flash offset background calibration proved to effectively correct the offsets with an accuracy that is better than 4mV. The non-interleaved pipelined ADC was evaluated up to 2.5GS/s and the interleaved mode of the two ADCs was evaluated up to 5GS/s. The measured results are summarized in Fig. 3 and an example FFT is shown in Fig. 4 at 5GS/s. The power consumption is 1.15W at 2.5GS/s and 2.3W at 5GS/s. The die size is 3.8mmx3.8mm for the whole chip. A comparison with the state-of-the-art is shown in Table I. At 2.5GS/s, this work represents the fastest, highest performance and lowest power non-interleaved 14-bit ADC, which is 2.5x faster than the state-of-the-art. At 5GS/s, this work represents the fastest and highest performance 14-bit ADC in the literature, which is 2x faster than the state-of-the-art. Moreover, the Figure-of-Merit is in line with, or better than, the state of the art.

References

[1] A.M.A. Ali, et al., "A 14b 1GS/s RF sampling pipelined ADC with background calibration", ISSCC Dig. Papers, pp. 482-483, Feb. 2014.
[2] P. Huang et al., "SHA-Less Pipelined ADC with In Situ Background Clock-Skew Calibration", IEEE JSSC, 46(8), Aug. 2011.
[3] M. Brandolini et al., "A 5GS/s 150mW 10b SHA-Less Pipelined/SAR Hybrid ADC in 28nm CMOS", ISSCC Dig. Tech. Papers, pp. 468-469, Feb. 2015.
[4] B. Setterberg, et al., "A 14b 2.5GS/s 8-Way-Interleaved Pipelined ADC with Background Calibration and Digital Dynamic Linearity Correction", ISSCC Dig. Papers, pp. 466-468, 2013.

[5] J. Wu, et al., "A 5.4GS/s 12b 500mW Pipeline ADC in 28nm CMOS", IEEE Symp. VLSI Circuits, pp. C92-C95, 2013.
[6] C.-Y. Chen, et al., "A 12-Bit 3 GS/s Pipeline ADC With 0.4 mm and 500 mW in 40 nm Digital CMOS", IEEE JSSC, 47(4), April, 2012.

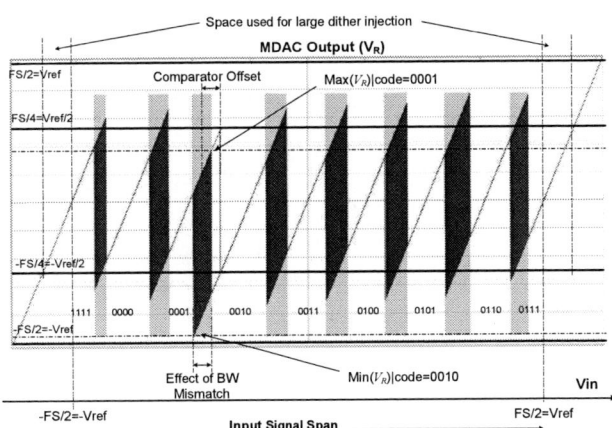

Fig. 2. The output residue of stage-1 showing offsets, BW/timing mismatch and the space where the large dither signals are injected.

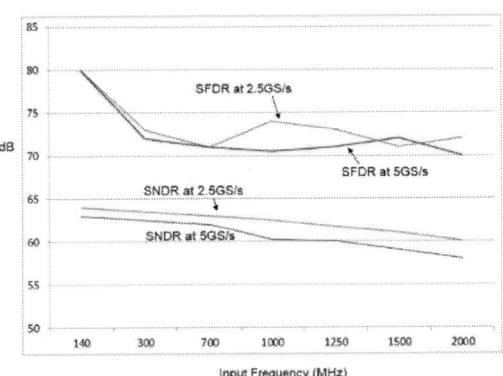

Fig. 3. Measured performance versus input frequency.

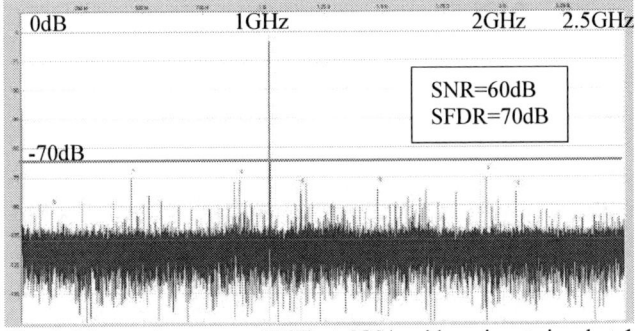

Fig. 4. Measured spectrum (FFT) at 5GS/s with an input signal at 1 GHz and 1.3Vpp.

TABLE I
Comparison of this work to the state-of-the-art

	This Work Non-Interleaved	This Work Interleaved	[3]	[4]	[5]	[6]
Resolution	14b	14b	10b	14b	12b	12b
SNDR	64 dB	63 dB	52.2 dB	61 dB	57 dB	58 dB
SFDR	80 dB	80 dB	65 dB	78 dB	60 dB	62 dB
Noise Spectral Density	155 dB/Hz	157 dB/Hz	146 dB/Hz	152 dB/Hz	155 dB/Hz	150 dB/Hz
Power	1.15 W	2.3 W	0.15 W	24 W	0.5 W	0.5 W
Sample Rate	2.5 GS/s	5 GS/s	5 GS/s	2.5 GS/s	5.4 GS/s	3 GS/s
FOM=SNDR+10log(BW/P)	154.4 dB	153.4 dB	148/154.2 dB	138 dB	154 dB	152.8 dB
Interleaved	No	Yes	Yes	Yes	Yes	Yes

A 14-bit 8.9GS/s RF DAC in 40nm CMOS achieving >71dBc LTE ACPR at 2.9GHz

Vishnu Ravinuthula, William Bright, Mark Weaver, Ken Maclean, Scott Kaylor,
Sidharth Balasubramanian, Jesse Coulon, Robert Keller, Bao Nguyen, Ebenezer Dwobeng

Texas Instruments Incorporated, Dallas, TX, USA

Abstract

We show for the first time an 8.9 GS/s RF current-steering DAC, with an on-chip 1:1 Balun, and an 8-lane 12.5 Gbps JESD204B compliant SerDes, with a measured LTE ACPR >71 dBc in the adjacent 20 MHz band for a 2.9 GHz channel. The DAC has IM3 <−65 dBc for output frequencies up to Nyquist. This performance is accomplished using a novel DAC switch driver and data/dummy-data scheme to minimize the pattern dependent sourcing/sinking of current on the DAC driver supply and ground. The DAC is fabricated in a 40nm dual-oxide CMOS process and dissipates 1.2W, with the contribution of the synthesized digital block and SerDes excluded.

Introduction

Communication systems for wireless multi-band transmitters that use a direct RF approach, require conversion rates ranging from 5 GS/s to 9 GS/s, and output frequencies greater than 1 GHz. These systems have stringent requirements on noise density, phase noise and linearity. Adjacent Channel Power Ratio (ACPR), also known as Adjacent Channel Leakage Ratio (ACLR), is a good FOM measurement for such RF transmitters. The 8.9 GS/s DAC(s) presented have measured LTE ACPR >71 dBc with channel centered at 2.9 GHz.

Recent publications of high conversion rate range RF DACs have demonstrated excellent linearity performance with external Balun at high output frequencies. Methods that reduce data dependency in the switch driver update to improve linearity have been published with good results using Quad Switch [1], Data/Dummy Data [2], and a single output current switch [3]. The RF DAC presented in this work builds on the data/dummy-data scheme in [2], and uses a novel DAC switch driver to achieves IM3 <−65 dBc for output frequencies up to Nyquist with an on-chip 1:1 Balun.

DAC Design

The DAC receives digital input data via an 8-lane 12.5 Gbps SerDes. The JESD data received from the SerDes is further processed by the Digital Up Converter (DUC) in the synthesized digital block to achieve data samples at the DAC full conversion rate. The DUC is capable of processing 2 complex IQ data streams per DAC at frequencies ranging from Baseband to RF Frequency through interpolation rates from 1X to 24X. The processed data from the DUC can be sent to either DAC core at quarter rate where it is encoded and interleaved to the full-rate of the segmented DAC.

The 14-bit DAC's unit current source shown in Fig. 1 is based on a triple power supply. A double cascoded current source stack M5, M6, M7 is powered between ±1.8V using 150nm NMOS devices [1]. The NMOS differential switch pair M1, M2 is cascoded with keep alive currents for increased output impedance [4]. The switch driver circuits use 40 nm core devices and a 1 V power supply to switch the differential pair between 1 V and ground. The DAC consists of a 5-bit thermometer MSB and 9-bit binary LSB segmentation of current sources. The DC linearity of the DAC can be trimmed using fuses and analog currents, however, ±1.5 lsb of INL is achieved by device matching alone and all measurements taken are from untrimmed parts.

The switch driver circuitry and implementation is key to maintaining linearity at high output frequencies. The novel switch driver used in this DAC uses dummy data paths to switch a replica driver when the real data state does not change in order to minimize the pattern dependent sourcing/sinking of current on the DAC driver supply and ground. The switch driver of the DAC shown in Fig. 2 consists of a differentially clocked CMOS inverter pair with cross coupled inverters providing positive feedback to hold the latched output state. This switch driver produces faster rise/fall times than an NMOS only alternative [4]. The addition of devices MPA, MPB, MNA, and MNB controlled by the next state data helps discharge the source of the off device for the next clocked output state. These devices also aid in reducing the data dependency of switch driver output. The switch drivers are paired with replicas operating on dummy data as shown in Fig. 1 to help improve IM3. With this scheme, the IM3 improve by >15 dB as the two tone spacing approaches the supply/ground resonant frequency of about 60 MHz.

The DAC transforms the differential output current to a single-ended voltage output using an integrated 1:1 balun. The integrated 1:1 balun uses a stacked metal topology, with 6 turns and 10 nH each for the primary and secondary coils and is terminated on the primary (DAC) side with a 67Ω differential resistor to increase the power delivered to the load. The balun has an insertion loss of 2.5 dB at its 1.2 GHz center frequency and a −3 dB bandwidth from 500 MHz to 3.3 GHz.

Measurement Results

Full-scale output current of each DAC set to 40 mA for all measurements listed. Fig. 3 shows the measured LTE ACPR to be >71 dBc in the adjacent 20 MHz band with channel centered at 2.9 GHz. Fig. 4 shows that DAC has IM3 <−65 dBc for output frequencies up to Nyquist. SFDR of the DAC is limited by HD2, however the DACs clocking and wide $1X$-$24X$

TABLE I
Performance comparison with state-of-the-art RF DACs

Parameter	Process	Power	ACPR	IM3 @500MHz
Engel [1]	65nm	0.8W @10GS/s	65dBc	−85dBc
McMahill [3]	180nm	2.3W @4.6GS/s	55dBc	−79dBc
This work	40nm	1.2W @8.9GS/s	71.4dBc	−87dBc

interpolation options provide the flexibility to frequency plan such that HD2 does not fall in-band or near the wireless bands of interest. The full-scale output power of DAC with 40 mA output into a 50Ω load is 0.75 dBm at 1.2 GHz.

Conclusion

The presented DAC utilizes a novel DAC switch driver and current source stack replicas operating on dummy data to achieve an LTE ACPR >71 dBc in the adjacent 20 MHz band for a channel centered at 2.9 GHz, and IM3 <−65 dBc for output frequencies up to Nyquist. This performance and the DUC's $1X$-$24X$ interpolation filters make this DAC suitable for a wide variety of wireless infrastructure applications. Table 1 shows a comparison with recently published state-of-the-art RF-DACs. A die photo of the chip is shown in Fig. 5.

References

[1] G. Engel *et al.*, "A 16-bit 10Gsps current steering RF DAC in 65nm CMOS achieving 65dBc ACLR multi-carrier performance at 4.5GHz Fout," *Symposium on VLSI Circuits*, Jun. 2015.

[2] S. Spiridon *et al.*, "A linearity improvement technique for overcoming signal-dependent induced switching time mismatch in DAC-Based transmitters," *European Solid-State Circuits Conference*, Sep. 2015.

[3] D. R. McMahill *et al.*, "A 160 Channel QAM Modulator With 4.6 Gsps 14 Bit DAC," *IEEE Journal of Solid-State Circuits*, vol. 49, no. 12, pp. 2878-2890, Dec. 2014.

[4] C. H. Lin *et al.*, "A 12b 2.9GS/s DAC with IM3 <60dBc beyond 1GHz in 65nm CMOS", *IEEE Solid-State Circuits Conference*, pp. 74-75,75a, Feb. 2009.

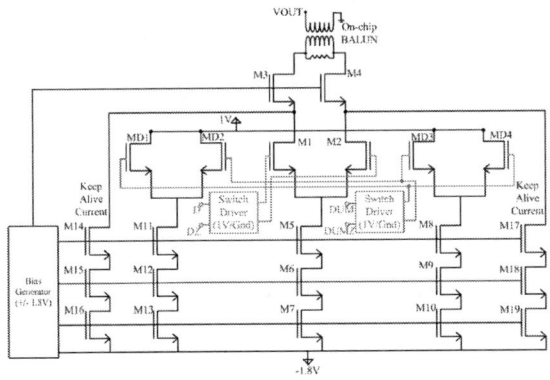

Fig. 1. DAC output stack, with on-chip 1:1 Balun

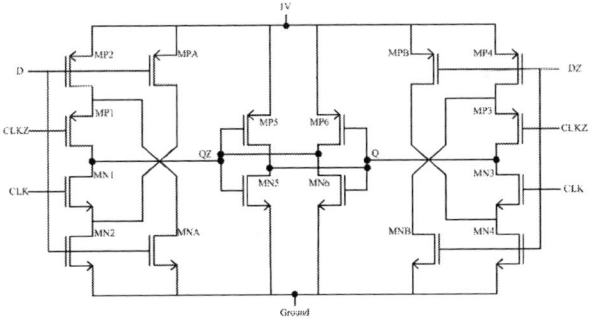

Fig. 2. DAC Switch Driver

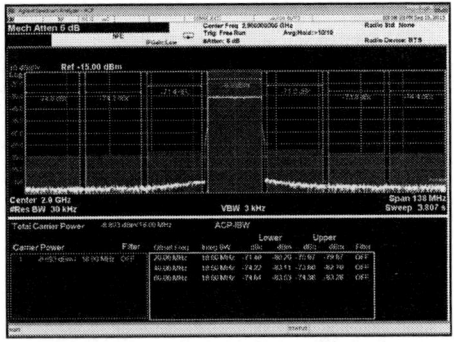

Fig. 3. LTE ACPR >71 dBc in the adjacent 20 MHz band with channel centered at 2.9 GHz.

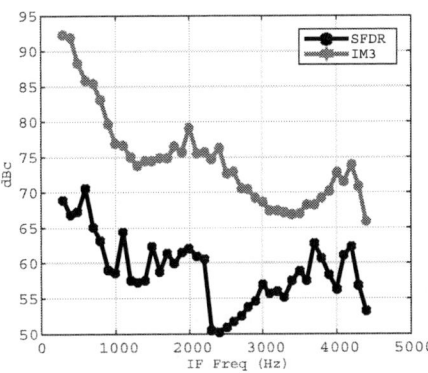

Fig. 4. IM3 with 10 MHz spacing between input tones and SFDR of the DAC at 8.9GS/s.

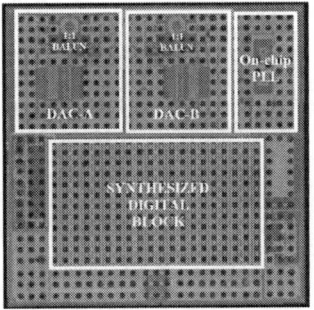

Fig. 5. Chip photograph of dual DAC showing integrated balun and 8-lane SerDes

A 7-to-18.3GHz Compact Transformer based VCO in 16nm FinFET

Mayank Raj, Parag Upadhyaya, Yohan Frans and Ken Chang
Xilinx Inc., San Jose, California, USA
mayankr@xilinx.com

Abstract

A dual-mode wide-band transformer based VCO is proposed. The two port impedance of the transformer based resonator is analyzed to derive the optimum primary to secondary capacitor load ratio, for robust mode selectivity and minimum power consumption. Fabricated in a 16nm FinFET technology, the design achieves 2.6X continuous tuning range spanning 7-to-18.3 GHz using a coil area of $120 \times 150~\mu m^2$. The absence of lossy switches helps in maintaining phase noise of -112 to -100 dBc/Hz at 1 MHz offset, across the entire tuning range. The VCO consumes 3-4.4 mW and realizes power frequency tuning normalized figure of merit of 12.8 and 2.4 dB at 7 and 18.3 GHz respectively.

Introduction

LC VCOs have been widely used for multi-band, multi-standard, and broadband applications like FPGAs due to their superior phase noise. However, they are plagued by limited oscillation range. Typically two or more inductors are used to achieve more than 2X operation range, resulting in large silicon area. Several approaches have been introduced to break this area vs. tuning range tradeoff. [1-2] use a switching inductor based approach, while [3] uses a switching mutual inductance. Both these techniques have significant tank Q degradation due to a series resistance, either directly in the inductor path or through the secondary, owing to impedance transformation. Resonant mode switching techniques [4-5] obviates the need of a switch. However, [4] requires two Gm cells to be active at the same time hence lead to higher power consumption. Moreover, careful design is needed to make them immune to unwanted mode ambiguity [5] and multimode oscillations [6]. We propose a low-power transformer based VCO which uses the two modes of the transformer to achieve 2.6X tuning range. Based on impedance analysis of the transformer tank circuit, optimum capacitor load ratio between secondary and primary is used for robust mode selectivity.

Proposed Wide-band VCO

Fig. 1 shows the magnitude and phase of the two port input impedances Z_{11} and Z_{00} of a transformer based tank circuit. Unlike an inductor based resonator, a transformer has two modes of resonance. A simple oscillator can be constructed by loading either sides of transformer based resonator by a negative conductance (-Gm). For such a design to work as a wide-band VCO, mode 1 should be activated when loading Z_{11} with -Gm and mode 0 when loading Z_{00} (Fig. 1). When -Gm is applied to Z_{11} then the oscillation startup requires $|Z_{11}||Gm|>1$ and $\angle Z_{11}=0$. Fig. 1(a) shows that if the |Gm| is large enough the oscillation startup criteria can be satisfied for both modes. A previous work [5] suggests that mode selection is based only on the fact $|Z_{11}|$ amplitude is higher for mode 1 than mode 0. But [6] shows that multimode oscillations and mode ambiguity may occur in a mutually coupled oscillator in presence of higher order non-idealities.

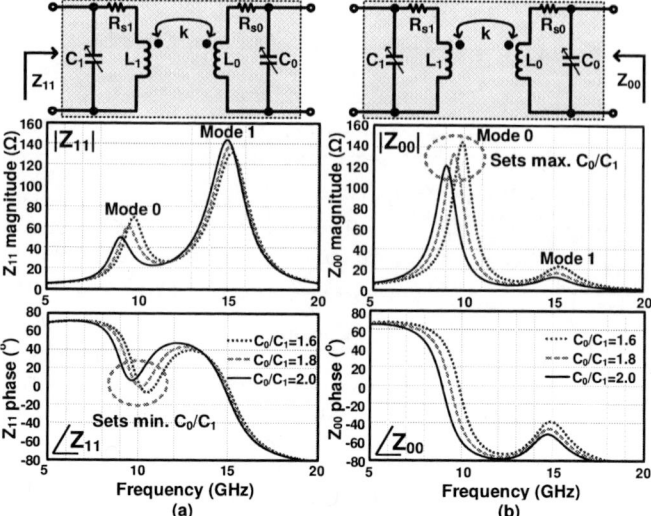

Fig. 1 (a) Z_{11} and (b) Z_{00} of a lossy transformer for different C_0/C_1.

To ensure reliable mode selection and avoid multimode oscillations, we use the fact that in a lossy transformer with $L_1=L_0$, the phase of Z_{11} crosses 0° only once if C_0/C_1 ratio is sufficiently high. This value of C_0/C_1 is dependent on the coupling coefficient (k) as well as the Q of the transformer. For our design this value was 1.8 (Fig. 1 (a)). So, as long as the capacitor ratio is greater than 1.8 the startup condition is satisfied for only one mode leading to robust mode selection. However, if C_0/C_1 is too high then Z_{00} reduces (Fig. 1(b)). Thus a higher |Gm| would be required to sustain oscillation in mode 0. This power constraint sets the upper limit for C_0/C_1.

The simplified circuit architecture of the wide-band VCO is shown in Fig. 2. The two coupled inductors L_0 and L_1 form the transformer core. Capacitive coarse tuning is achieved by switching a 5-bit metal-to-metal finger capacitor bank and fine control by NMOS varactors. The size of the capacitor banks and varactors on L_1 side (C_1) are half of that on the L_0 side (C_0). To tune the frequency, the coarse and fine controls to both L_0 and L_1 tanks are changed together to maintain C_0/C_1 ratio across the entire tuning range. In order to shift between modes, NMOS switches are used to divert current from the negative Gm cell on the L_0 side to the L_1 side. Only one Gm cell is active at once, thereby saving power. Additionally, since the switch is not in the inductor path it does not lead to Q degradation. The output of the VCO is ac coupled to an inverter based self-biased buffer and the appropriate VCO output is selected using the mode signal. The unused buffer is switched off to save power. The transformer comprises of a $120 \times 150~\mu m^2$ sized double-turn outer inductor L_1 and inner inductor L_0. L_0 consists of a single turn top metal layer in series with double-turn lower metal layers. This structure ensures that both L_0 and L_1 have similar inductances. The legs of L_1 and L_0 are symmetrically designed so as to make the layout more amenable to conventional LC VCO design (Fig. 3).

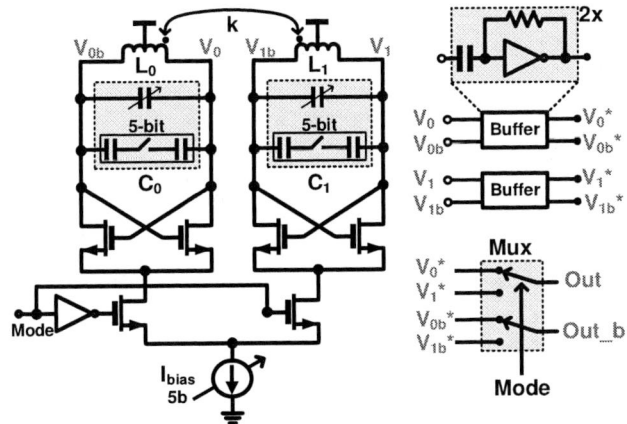

Fig. 2 Circuit architecture of the proposed wide-band VCO.

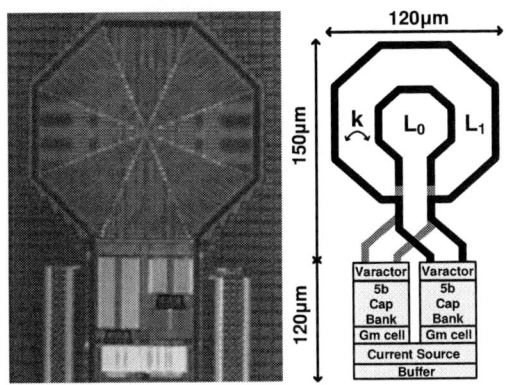

Fig. 3 Chip micrograph and layout details.

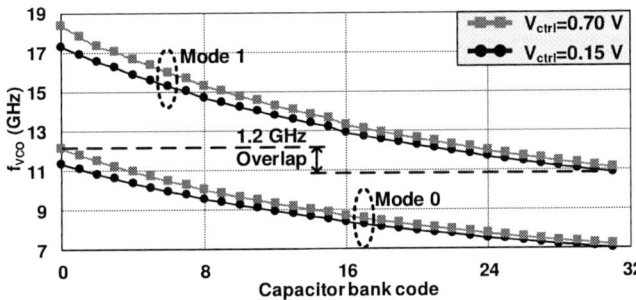

Fig. 4 Measured VCO frequency range.

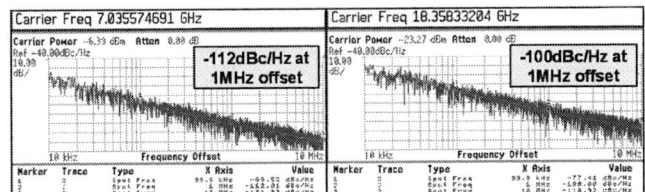

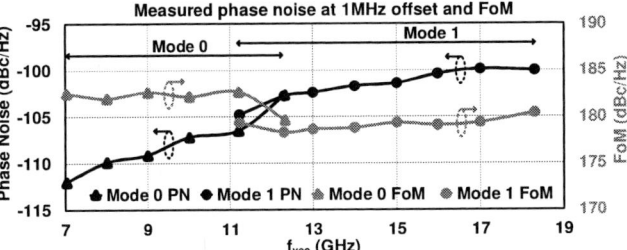

Fig. 5 Measured phase noise across f_{vco}.

TABLE I. Performance Summary and Comparison

	[1]		[2]		[3]	**This Work**	
Number of Modes	4		5		3	**2**	
Process	90nm CMOS		90nm CMOS		45nm SOI	**16nm FinFET**	
Supply (V)	1.2		1.2		1	**0.8**	
Power (mW)	7.67		5.2-7.1		7-14	**3-4.4**	
f_{vco} (GHz)	8.1-15.4		9.9-20.3		7.3-17.5	**7-18.3**	
TR (f_{max}/f_{min})	190 %		205 %		240 %	**260 %**	
PN 1MHz (dBc/Hz)	f_{vco}(GHz)	8.1	15.4	9.9	20.3	12.25	7	18.3
	PN	-112	-104	-103	-84	-93	**-112**	**-100**
	FoM	181	179	176	162	163	**182**	**180**
PFTN-FoM (dB)	6.45/-1.7		0.6/-18.3		-12	**12.8/2.4**	
Coil Size (μm^2)	220×220		65×45		175×175	**120×150**	

Table I compares this work with state-of-the-art wide-band VCOs of similar operating frequencies. This work achieves the best tuning ratio (TR) and phase noise FoM at lowest power consumption. The measured power frequency tuning normalized (PFTN) FoM [7] ranges from 12.8 to 2.4 dB, which is better than [1-3]. [2] realizes a smaller coil size but at an expense of significant phase noise degradation. In conclusion, the proposed design demonstrates the possibility of achieving 2.6X tuning ratio with an area of a single inductor and excellent phase noise and low power consumption at the same time in highly scaled 16nm FinFET process.

Measurement Results

The VCO is implemented in a 16nm FinFET technology. Fig. 3 shows the chip micrograph. The total active area is 120×270 μm^2. Fig. 4 shows the measured VCO frequency across the 5-bit control word of the capacitor bank and varactor control voltage (V_{ctrl}). The measured tuning range in mode 0 is 7-12.1 GHz and mode 1 is 10.9-18.3 GHz. A healthy band overlap of 1.2 GHz is measured between modes 0 and 1 ensuring continuous tuning across PVT variations. The VCO consumes 3-4.4 mW power from a 0.8 V supply. Its bias current is optimized for each setting of capacitor bank code. Fig. 5 depicts the measured phase noise curves at 7 and 18.3 GHz. The measured phase noise at 1 MHz offset for 7 GHz and 18.3 GHz are -112 dBc/Hz and -100 dBc/Hz respectively. The phase noise measured across the entire tuning range at 1 MHz offset and corresponding phase noise figure-of-merit (FoM), is also shown in Fig. 5. The VCO maintains FoM of 182.6 to 178.3 dBc/Hz at 1 MHz offset, across the tuning range.

References

[1] M. Demirkan, S. P. Bruss, and R. R. Spencer, "Design of wide tuning range CMOS VCOs using switched coupled-inductors," *IEEE JSSC*, vol. 43, no. 5, pp. 1156–1163, May 2008.

[2] A.Tanabe *et al.*, "A Novel Variable Inductor Using a Bridge Circuit and Its Application to a 5-20 GHz Tunable LC-VCO," *IEEE JSSC*, vol. 46, no. 5, pp. 883-893, Apr. 2011.

[3] M. Kossel *et al.*, "LC PLL With 1.2-Octave Locking Range Based on Mutual-Inductance Switching in 45-nm SOI CMOS," *IEEE JSSC*, vol. 44, no. 2, pp. 436-449, Feb. 2009.

[4] G. Li *et al.*, "A Low-Phase-Noise Wide-Tuning-Range Oscillator Based on Resonant Mode Switching," *IEEE JSSC*, vol.47, no.6, pp.1295-1308, Jun. 2012.

[5] A. Bevilacqua *et al.*, "A 3.4-7 GHz Transformer-Based Dual-mode Wideband VCO," *ESSCIR*, pp.440-443, Sep. 2006.

[6] S. Datardina and D. A. Linkens, "Multimode oscillations in mutually coupled van der Pol type oscillators with fifth-power nonlinear characteristics," *IEEE TCAS*, pp. 308-315, May 1978.

[7] D. Ham and A. Hajimiri, "Concepts and methods in optimization of integrated LC VCOs," *IEEE JSSC*, pp. 896–909, Jun. 2001.

-197dBc/Hz FOM 4.3-GHz VCO Using an Addressable Array of Minimum-Sized NMOS Cross-Coupled Transistor Pairs in 65-nm CMOS

A.Jha[1], A. Ahmadi[1], S. Kshattry[1], T. Cao[2], K. Liao[3], G. Yeap[3], Y. Makris[1] and K. K. O[1]

[1]Texas Analog Center of Excellence and Dept. of Elec. Eng., The University of Texas at Dallas, Richardson, TX , USA,
[2]Cornell University, Ithaca, NY, USA, [3]Qualcomm Technologies, San Diego, CA, USA
E-mail: amit.jha@utdallas.edu

Abstract

A 4.3-GHz voltage controlled oscillator (VCO) using an addressable array of cross-coupled minimum size NMOS transistor pairs for post fabrication selection is demonstrated in 65-nm CMOS. An algorithm based on Hamming distance using the phase noise measurements of ~1,500 array combinations was used to identify combinations that have record phase noise of -130dBc/Hz at 1-MHz offset from a 4.3-GHz carrier, while consuming 5.2 mW from a 1-V supply. The operating frequency of circuits using post fabrication selection in its high frequency path is increased to 5 GHz.

Keywords: CMOS, VCO, phase noise, array, cross-coupled transistor pair and process variations.

Introduction

The number of intended dopants and un-intended defects in a minimum sized device is reduced with technology scaling. One missing dopant or having an additional defect can have dramatic impact on device characteristics including threshold voltage, current and noise. The defect density related to 1/f noise in nano-scale CMOS is ~1x10^{12}cm^{-2} or higher [1], which translates to on the average, one trap in ~100 nm^2. In the 65-nm or more scaled CMOS node, the minimum transistor gate oxide area could be on the order of 100 nm^2 or less. In this paper, a VCO topology that embraces the variability of nano-scale transistors to reduce the phase noise of VCO's by taking advantage of minimum sized transistors with a fewer defects or traps through post-fabrication selection [2] is presented.

VCO Design

Fig. 1 shows the VCO topology that uses an addressable array of NMOS cross-coupled transistor pairs for post

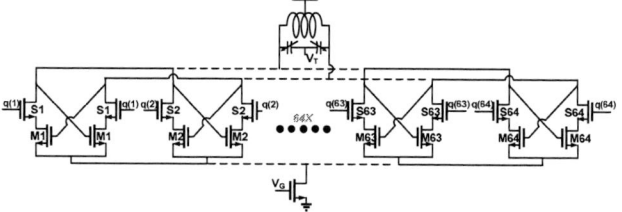

Fig. 1. Array of cross-coupled minimum NMOS transistors.

fabrication selection of pairs with lower noise. To develop this array based design, a conventional VCO using a cross-coupled NMOS pair (8-μm width) with an NMOS tail current source is first designed and then the 8-μm width core transistors are divided into arrays of cross-coupled pairs of minimum sized transistors in a 65-nm process. The top plate of varactors is connected to V$_{DD}$ through the inductors. This minimizes the modulation of output common-mode level by the bias noise current and hence, reducing the phase noise contribution from the AM-PM modulation by the varactors [3]. The inductor is a 5-turn 2.1-nH circular symmetric center tapped structure. The inductor has Q of ~10 at 5GHz. The varactors are of an

accumulation mode type implemented as an NMOS structure in an n-well. The tail current transistor channel length of 0.5μm is chosen to reduce the 1/f noise impact of tail current transistor [4].

The transistor forming a cross-coupled pair is divided into 32 transistors of width and length of 250nm and 60nm, respectively. To provide redundancy, 32 additional pairs are added. Each unit also includes 2.5-μm wide switches for selection of a given pair. The switch can be placed at the gate drain or source of the transistor as shown in Fig. 2 [5]. The

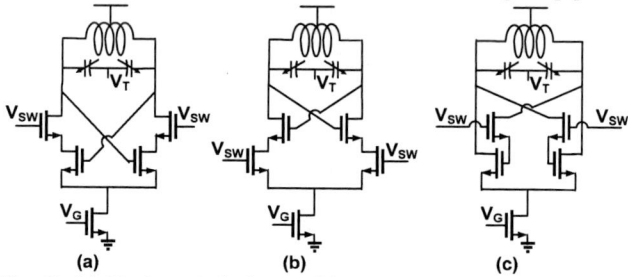

Fig. 2. (a) Drain switched core (b) source switched core (c) gate switched core for VCO.

phase noise with the switch at the gate of the transistor was at least 20 dB worse than that for the VCO with the switch at the drain of the transistor for the same switch size. The switch at the gate can be sized larger to reduce the series resistance but this however unacceptably increases parasitic capacitances of the switch as well as cell area. The switch at the source side of transistor was ruled out as the series resistance of the switch will degenerate the cross-coupled transistor and degrade phase noise by 2.5dB from that with the switch on the drain side. The switches are controlled by a 64-bit serial-in parallel-out falling edge triggered D-flip-flop chain.

Measurement Results

A setup for measuring the phase noise of VCO with different combinations of cross-coupled pair units is shown in Fig. 3. A set of 24/32/40/48 "1" bits is randomly sent through Labview to the DUT using an SPI/I^{2}C interface and the phase noise measured with a Keysight E4440A spectrum analyzer is collected using Labview. Around 500 combinations were measured for each count of "1" bits. Phase noise of VCO's using two NMOS cross-coupled transistors with 8-μm width were also measured. The minimum phase noise at 1-MHz offset is ~ -118 dBc/Hz. From the histograms (Figs. 4 & 5) of phase noise at 600-kHz and 1-MHz offset from the carrier two observations can be made: (i) variation of the phase noise is close to ~20 dB, and (ii) combinations of cross-coupled minimum sized transistors with lower phase noise than that of the conventional VCO do exist and they can be used to reduce phase noise.

A smart search using a Hamming-based algorithm on the data

This work is supported by Qualcomm Technologies.

978-1-5090-0636-6/16 $31.00 © 2016 IEEE

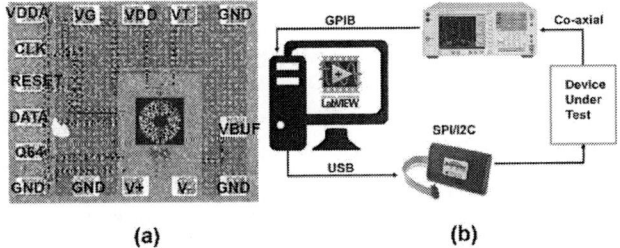

Fig. 3. (a) Die Photograph of chip. (b) Measurement setup.

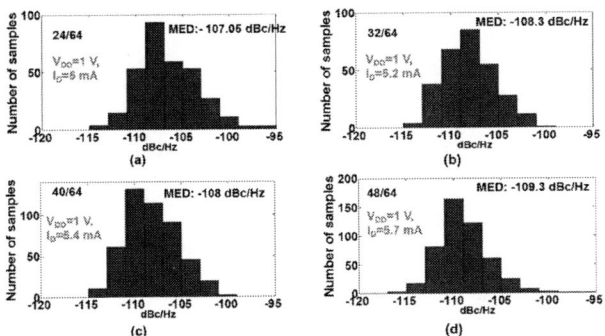

Fig. 4. Histograms of phase noise at 600-kHz offset from the carrier for (a) 24 (b) 32 (c) 40 (d) 48 randomly chosen set of cross-coupled transistor pairs.

is used to identify combinations with reduced phase noise. Using the commonly employed Hamming distance metric, four selection vectors with a minimum distance to the one which produces the lowest offset noise are identified, and then majority vote and absolute majority vote are employed to decide whether the i-th bit in the new selection vector should

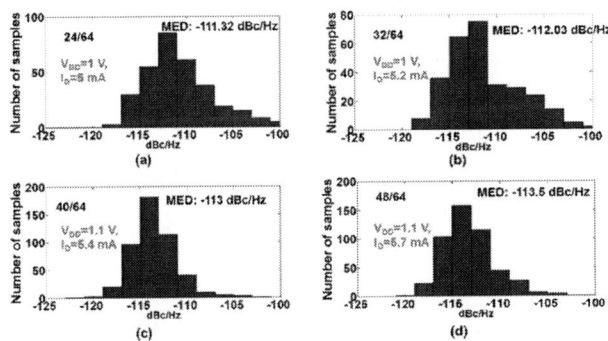

Fig. 5. Histograms of phase noise at 1-MHz offset from the carrier for (a) 24 (b) 32 (c) 40 (d) 48 randomly chosen set of cross-coupled transistor pairs.

be "0" or "1". This procedure is then repeated for the top five selection vectors according to their phase noise from the measurements of randomly generated combinations to create a new set of selection vectors that should result lower noise. Figs. 6(a) and (b) show the histograms of phase noise at 600-kHz and 1-MHz offset for the vectors generated by the proposed algorithm measured using a Keysight E5052B (Signal Source Analyzer). The median phase noise in Fig. 6 is ~7-8 dB lower than the lowest measured using a spectrum analyzer in Figs. 4 & 5. Around 4-5 dB of this difference is due to the use of Keysight E5052B, which better accounts for the drift of VCO output during measurements. The phase noise range is 17.5dB lower. If 32 arrays are randomly chosen from 64 possible arrays, then there are 1.8×10^{18} combinations. The 500 measured combinations is a tiny portion. The result in Fig. 6

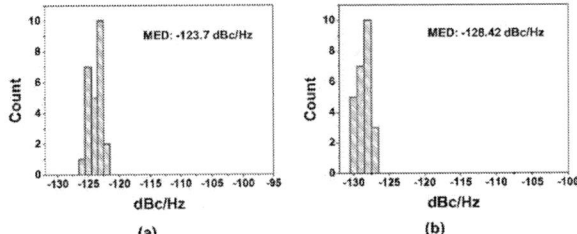

Fig. 6. Histograms of phase noise of combinations found using the hamming metric (a) @ 600 kHz (b) @ 1 MHz offset from the carrier

suggests that the number of measurements needed to select the combination with low phase noise is not impractically high.

The phase noise of VCO with one of the combinations selected by the Hamming distance-based algorithm is measured using an E5052B and shown in Fig. 7(a). The phase noise of VCO is -130dBc/Hz at 1-MHz offset from the carrier frequency of 4.3GHz. The commonly used Figure of Merit (FOM) [11] is plotted in Fig. 7(b). The performance of VCO in this work is compared to that of VCO's from the literature in Table I. The VCO has the lowest phase noise at 1-MHz offset among all VCO's operating around 5 GHz. The VCO has the state-of-art FOM across its tuning range and achieves the lowest FOM of -197dB at 4.3GHz. This work also demonstrated that the operating frequency of circuits using post fabrication selection in its high frequency signal path can be increased to 5GHz. Lastly, an on-chip phase noise measurement technique is needed.

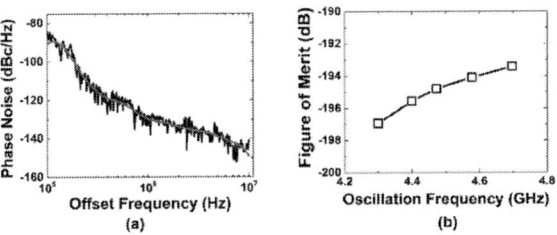

Fig. 7. (a) Phase noise plot of VCO. (b) Figure of Merit of VCO across the tuning range

TABLE I: VCO Measured Performance

	Freq. (GHz)	Tech (nm)	Power (mW)	Lowest Phase Noise at 1-MHz offset (dBc/Hz)	Minimum FOM (dB)
[6]	5-5.42	180	4.2	-127	-195
[7]	9.2-10.4	180	3.6	-115.7	-190.5
[8]	4.84-5.32	180	3.6	-125.8	-194
[9]	2-4-5.3	65	6	-119	-189
[10]	1.97-2.04	350	1	-122.4	-189
This work	**4.3-4.7**	**65**	**5.2**	**-130**	**-197**

References

[1] G. Guisi *et al.*, *TED 2006*, vol. 53, no. 4, pp. 823-828.
[2] L. Pileggi *et al.*, *CICC 2008*, pp. 9-12.
[3] B. Razavi, *RF Microelectronics*, Prentice Hall, 2011.
[4] A. L. McWorther, PhD thesis, MIT, Cambridge, MA, 1955.
[5] D. Hauspie *et al.*, *JSSC 2007*, vol. 42, no. 7, pp. 1472-1480.
[6] Z.-B. Li *et al.*, *JSSC 2005*, vol. 40, no. 6, pp. 1296-1302.
[7] I.-S. Shen *et al.*, *MTT 2012*, vol. 60, no. 2, pp. 318-328.
[8] K.-I. Wu *et al.*, *VLSI 2015*, pp. 236-237.
[9] L. Fanori *et al.*, *ISSCC 2014*, pp. 370-371.
[10] S.-J. Yun *et al.*, *ISSCC 2005*, pp. 540-616.
[11] P. Kinget *et al.*, pp. 353-381, 1999, Kluwer

A 10Gb/s, 342fJ/bit Micro-Ring Modulator Transmitter with Switched-Capacitor Pre-Emphasis and Monolithic Temperature Sensor in 65nm CMOS

Saman Saeedi[1,2], and Azita Emami[2]

[1]Oracle Labs, Redwood Shores, CA, Email: saman.saeedi@oracle.com and [2]California Institute of Technology, Pasadena, CA

Abstract

In this work, a CMOS-SiPh optical transmitter based on carrier-injection ring modulators is presented. It features a novel low-power switched-capacitor-based pre-emphasis that effectively compensates the modulator bandwidth limitation. A wavelength stabilization technique via direct measurement of ring temperature using a monolithic PTAT sensor is also presented. The optical transmitter achieves energy efficiency of 342fJ/bit at 10Gb/s and the wavelength stabilization circuit consumes 0.29mW.

Keywords: CMOS, transmitter, silicon photonics, micro-ring, resonator, pre-emphasis, thermal tuning

Introduction

Electro-optic modulators (EOM) that are CMOS-compatible, compact and low power are essential elements in realization of chip-to-chip optical signaling. Carrier injection micro-ring modulators (MRM) are one of the promising candidates [1], [2]. Compared with carrier depletion MRM, they can operate with higher extinction ratio and with CMOS-compatible drive voltages. However, the speed of carrier-injection rings is limited to slow carrier dynamics and necessitates pre-emphasis to compensate their nonlinear transient behavior. Both types of MRMs are also susceptible to temperature variations and need wavelength stabilization loops. In this paper we present a hybrid-integrated CMOS-SiPh transmitter that tackles these challenges. A novel low-power switched-capacitor-based (SC) pre-emphasis technique that effectively compensates the modulator bandwidth limitation is proposed. A feed-forward bias-based wavelength stabilization technique via direct measurement of ring temperature using a monolithic PTAT temperature sensor is also presented.

Principles of Operation

Carrier-injection MRMs are inherently slow and limited by

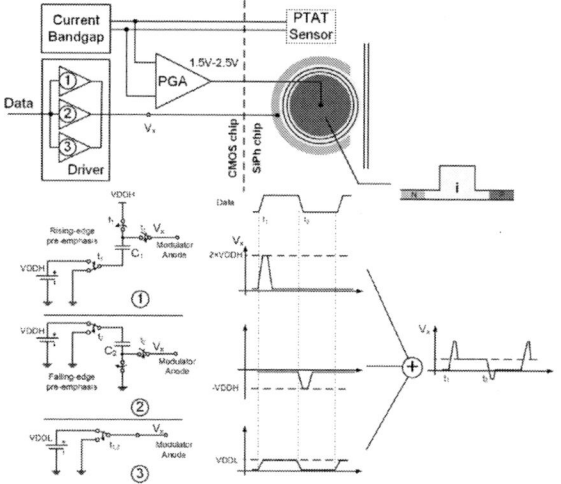

Fig. 1 Top-level block diagram of the transmitter.

recombination lifetime of carriers in the intrinsic region of the p-i-n junction (Fig. 1). Nonlinear pre-emphasis has proven to

be an effective way of reducing carrier dynamic rise-time and fall-time [1], [2]. By increasing the pre-emphasis voltage, rise-time/fall-time is shortened and higher data-rate is achievable. Prior pre-emphasis techniques relied on stacked output drivers that are highly power-inefficient and have a maximum pre-emphasis voltage drive of 2×VDDL, where VDDL is the thin-oxide transistors' voltage. In this work, we use a low-power SC-based pre-emphasis technique that can boost the output voltage to 4×VDDL. A top-level block diagram of the transmitter with proposed SC-based pre-emphasis technique is shown in Fig. 1. The driver consists of three main elements, a conventional voltage driver and two pre-emphasis blocks for rising and falling data edges. There are two voltage levels required for operation of this scheme, VDDL=1V, set by the standard thin-oxide transistors' voltage and VDDH=2V. The conventional voltage driver provides a steady state voltage to keep the junction in forward bias when needed. The two pre-emphasis blocks work by first accumulating charge on C_1 and C_2 up to VDDH. Subsequently, these capacitors are switched so that for the rising-edge pre-emphasis the output is at 2×VDDH and for falling-edge pre-emphasis the output is at -VDDH. The charge on these capacitors is used as pre-emphasis to inject and extract charge from the intrinsic region of the junction.

Implementation and Measurements

Fig. 2 shows the schematic circuit details of the SC-based pre-emphasis technique. A 2V pulsed-cascode stage, similar to

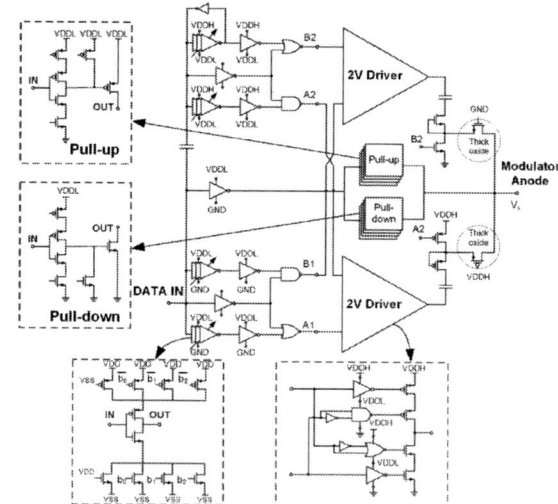

Fig. 2 Schematic circuit details of the proposed MRM driver with switched-capacitor-based pre-emphasis.

[1], is used to charge capacitors C_1 and C_2. Tunable delays are used to adjust the charge time and therefore strength of the pre-emphasis. A voltage driver with digitally adjustable pull-up and pull-down strengths is incorporated to maintain the junction in forward bias region and in off region according to the data. Fig. 3 shows optical measurement results of the transmitter. The static transmission of the MRM shows a

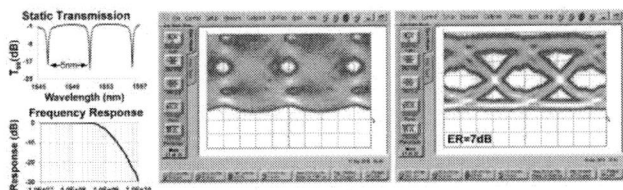

Fig. 3 Measured characteristics of the MRM and output optical eye diagram of the optical transmitter with and without pre-emphasis.

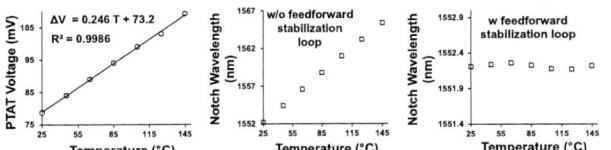

Fig. 5 Characteristics of the PTAT sensor and MRM's resonance wavelength w and w/o wavelength stabilization.

quality factor (Q) of ~6000 and free spectral range (FSR) of 5nm. The measured optical frequency response of the MRM in forward-bias shows a -3dB bandwidth of about 900MHz. When a 10Gb/s, PRBS 2^7-1 data stream is transmitted the output optical eye is completely closed without pre-emphasis. Enabling pre-emphasis opens the eye to have 7dB extinction ratio. The transmitter consumes 3.42mW resulting in per-bit energy of 342fJ/b.

Another challenge for robust operation of MRMs is their sensitivity to temperature fluctuations. Previously reported wavelength stabilization techniques, such as output optical power feedback using bias voltage [1] or heater [3] require extra optical power on the silicon-photonic chips and have excessive power overhead. In this work, we propose wavelength stabilization by direct measurement of temperature through a monolithic distributed PTAT sensor. Fig. 4 shows the schematic block diagram of the feed-forward bias-based wavelength stabilization technique and the SiPh MRM with on-chip PTAT temperature sensor. The monolithic PTAT temperature sensor, used for directly measuring the temperature of the ring, is described in [4]. In [4], the PTAT temperature sensor operation was demonstrated in a carrier-depletion MRM with heater-based wavelength stabilization and without using a CMOS chip. In this work, the PTAT sensor works by measuring the voltage difference between two diodes with different current densities. The PTAT voltage is then applied to an on-chip programmable gain amplifier (PGA) implemented in the CMOS chip. This PGA sets the bias voltage of the MRM. As calculated in Fig. 4, a gain of 5-10 (depending on β_{Temp} and variation of currents) cancels the temperature dependency of MRM's notch wavelength. The PTAT sensor currents, I_1 and I_2, are provided by the CMOS chip using a current bandgap circuit. Measurements verify that these currents vary less than 5% in a range of 25-150°C. Note

that process and voltage variation can be compensated by adjusting gain of the programmable gain amplifier (PGA). Fig. 5 shows measured operation of the feed-forward bias-based wavelength stabilization technique. First, the temperature dependency of the MRM's resonance wavelength is measured to be about 0.11nm/K. The linear operation of the PTAT sensor is independently verified from 25°C to 150°C. Next, the optimal PGA gain is found to be 8.2 to make the notch wavelength temperature-independent. The maximum tuning power is 290µW for a resonance wavelength range of 0.4nm. In order to cover the complete FSR, this technique can be used as a fine-tuning in combination with heater-based thermal control as coarse-tuning [5].

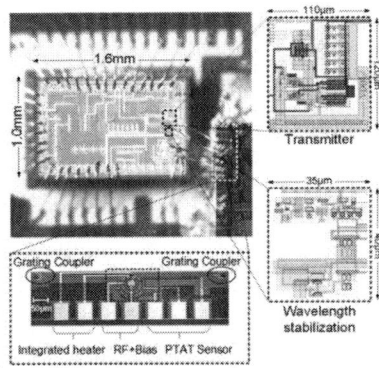

Fig. 6 CMOS and SiPh chip micrographs.

TABLE I
TRANSMITTER PERFORMANCE SUMMARY

Electronics technology	65nm CMOS
TX data-rate	10Gb/s
Extinction ratio	7dB
Tuning power consumption	0.29mW
Active area	0.15mm^2
TX energy/bit	342fJ/b

Conclusion

The optical transmitter CMOS chip is fabricated in a 65nm bulk process and the silicon photonic device is fabricated in OpSIS IME-5 process. The silicon photonic MRM with integrated PTAT sensor is connected to the CMOS chip through wirebonds as shown in Fig. 6. The optical transmitter achieves energy efficiency of 342fJ/bit at 10Gb/s. The feed-forward bias-based wavelength stabilization circuit consumes 0.29mW. Table 1 summarizes the system performance and compares it to prior art.

Acknowledgement

The authors would like to acknowledge ST Microelectronics for CMOS chip fabrication and OpSIS for SiP chip fabrication.

References

[1] C. Li, et al, *ISSCC Dig. Tech. Papers*, pp. 124-125, Feb. 2013.
[2] Y. Chen, et al, *ISSCC Dig. Tech. Papers*, pp. 402-404, Feb. 2015.
[3] H. Li, et al, *ISSCC Dig. Tech. Papers*, pp. 410-411, Feb. 2015.
[4] S. Saeedi, et al, *Optics Express*, Aug 2015, pp. 21875-21883.
[5] C. Chen, et al, *OI Dig. Tech. Papers*, pp. 121-122, May. 2014.

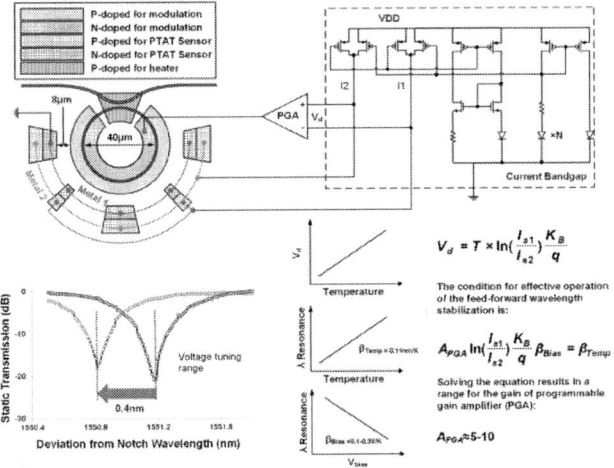

Fig. 4 Schematic of the feed-forward bias-based wavelength stabilization technique. Measured MRM resonance versus bias.

A 50.6-Gb/s 7.8-mW/Gb/s –7.4-dBm Sensitivity Optical Receiver based on 0.18-μm SiGe BiCMOS Technology

Takashi Takemoto[1], Yasunobu Matsuoka[1], Hiroki Yamashita[1], Yong Lee[1], Kenichi Akita[1], Hideo Arimoto[1], Masaru Kokubo[1], and Tatemi Ido[2]

1. Research & Development Group, Center for Technology Innovation – Electronics, Hitachi Ltd., Tokyo, Japan
2. Global MONOZUKURI Division, Information and Telecommunication Systems Company, Hitachi, Ltd., Kanagawa, Japan
E-mail: takashi.takemoto.tj@hitachi.com

Abstract

A 50.6-Gb/s optical receiver (RX) based on the 0.18-μm SiGe BiCMOS technology was fabricated and evaluated. To improve phase margin and sensitivity of the RX without degrading its power efficiency, it was configured with a two-stage pre-amplifier (preamp) with high gain (56 dBΩ) and power-supply-variation (PSV) canceller for improving sensitivity. The RX achieves sensitivity of –7.4 dBm and phase margin of 0.51 UI at 50.6 Gb/s, while its power consumption is 7.8 mW/Gb/s.

Introduction

Continuing rapid growth in data traffic requires low-power 50-Gb/s optical interconnects for intra- and inter-rack transmissions inside data centers. In addition to enhancing throughput by applying PAM4, which has recently been adopted by the IEEE802.3bs 400GbE task force, developing NRZ transmission beyond 50 Gb/s is also important (because it is achieved by utilizing a simple circuit structure) [1, 2]. However, to handle the severe jitter and loss budgets of 50-Gb/s NRZ optical transmission, two problems concerning optical receiver (RX) must be solved. The first problem concerns decrease in phase margin due to signal reflections, which are mainly caused by the inductive component of wire bonding (W/B) at the output electrical transmission line between the optical RX and subsequent LSI. The second problem is improving sensitivity of the RX by compensating PSV and offset voltage without sacrificing power consumption. In this study, a 50.6-Gb/s optical RX, consisting of a photodiode (PD) operating at 1310-nm wavelength [3] and a TIA, which are flip-chip mounted on a ceramic package (PKG), is proposed. To solve the above-described problems, two key approaches are proposed: (1) a two-stage preamp, which achieves a high-gain (56-dBΩ) characteristic by alleviating the effects of input capacitances, and (2) a low-power high-sensitivity scheme utilizing automatic decision-threshold control (ATC) with PSV and offset cancellers. The TIA was implemented on the basis of the 0.18-μm SiGe BiCMOS process. The developed optical RX attained large phase margin of 0.51 UI and high sensitivity of –7.4 dBm at 50.6 Gb/s with power efficiency of 7.8 mW/Gb/s.

Architecture and Circuit Design

To avoid the effects of signal reflection due to W/B, the developed optical RX adopts a flip-chip assembly, in which a buck-illuminated PD is directly connected to the TIA through transmission lines inside a PKG. Because a flip-chip structure is adopted instead of W/B, the output electrical-transmission loss is reduced from –13.5 to –9.5 dB at 25 GHz. As a result, the loss can be compensated by only a simple linear equalizer inside the optical RX. A block diagram of the developed 50-Gb/s TIA for flip-chip assembly, which consists of a PD regulator (PD-REG), a preamp, an ATC, a post-amplifier

(post-amp) stage, and an output driver (driver), is shown in Fig. 1. The PD-REG generates a stable bias voltage for the PD (VPD) from power-supply voltage VDD_{PD}. The post-amp stage is composed of two differential amplifiers with negative feedback, which provides variable gain amplifier (VGA), and a limiting amplifier (LA). The driver provides both equalization (EQ), which provides maximum peaking gain of 11 dB at 25 GHz and compensates for electrical-transmission loss, and 50-Ω impedance matching.

A. Two-stage pre-amplifier

A well-known conventional preamp is a feedback-loop (FBL) preamp, which consists of a pseudo-differential amplifier (PDA) and feedback resistance R_f. However, if its bandwidth cannot be enhanced (due to the inductive component of W/B), it cannot easily achieve high-speed operation (over 50 Gb/s) because gain R_f and bandwidth $G(s)/[(s \cdot C_{IN}R_f)]$, where $G(s)$ and C_{IN} are the transfer function of the PDA and total input capacitances (including PKG), have a direct trade-off relationship. As a result, C_{IN} significantly reduces bandwidth, leading to large jitter. To alleviate this bandwidth degradation due to C_{IN}, , as shown in Fig. 2, a two-stage preamp, which consists of an input stage (common-base amplifier) and an amplifier stage (FBL preamp), is proposed. The transimpedance of the two-stage preamp is approximately given by R_f when $G(s)$ is sufficiently large, namely, $G(s) \cdot (R_f//Z_L(s)) << R_f$, where $Z_L(s)$ is output impedance of the input stage. On the other hand, bandwidth is mainly determined by input parasitic capacitance of the amplifier stage, C_L ($<<C_{IN}$) (i.e., not by C_{IN}). The two-stage structure can increase both gain and bandwidth. Moreover, to increase bandwidth, the PDA has a peaking characteristic, which is generated by a zero point due to a capacitance C_p [4]. The simulated frequency responses, which include that of a PD equivalent circuit and PKG, of the FBL and two-stage preamps are plotted in Fig. 3(a). According to this figure, the two-stage preamp achieves transimpedance of 56 dBΩ and bandwidth of 28 GHz while bandwidth of the conventional FBL preamp is 19.1 GHz.

B. Low-power high-sensitivity scheme

The sensitivity of the optical RX is mainly determined by not only thermal noise but also PSV at VDD_{PRE} (PSV_{PRE}) and offset voltage of the post-amp stage. Especially, the sensitivity of the input stage is degraded by PSV_{PRE}. Cancelling scheme, which consists of on-chip voltage regulator and noise canceller, effectively compensates the PSV_{PRE} [5]. However, unlike the CMOS preamp [5], the voltage regulator consumes large power because the SiGe preamp requires high power-supply voltage (3.3 V). To address this power-consumption issue, a feed-forward ATC consisting of a PSV and offset cancellers, as shown in Fig. 4, is proposed. The PSV canceller (consisting of a differential amplifier) generates a PSV dummy signal

(PSV$_d$) with the same phase as that of PSV$_{PRE}$. Consequently, PSV$_{PRE}$ is canceled by propagating PSV$_d$ at the input of the first post-amp through capacitance C$_{high}$. The PSV canceller does not need high-speed operation over several gigahertz because an on-chip bypass capacitance C$_{PS\_PRE}$ (>300 pF) suppresses PSV$_{PRE}$ in the high-frequency range (down to 1 GHz). As a result, the PSV canceller suppresses PSV$_{PRE}$ without increasing power consumption (only 6.6 mW). As for the operating frequency range of the 50-Gb/s TIA (down to 200 kHz), PSRR of less than −20 dB at the input of the post-amp is confirmed by incorporating a PSV canceller and C$_{PS\_PRE}$ of 300 pF (Fig. 3(b)). The offset canceller, which consists of a LPF and two voltage followers utilizing NMOS transistors, is based on our previous work [6]. Unlike a previous CMOS TIA, the second voltage follower is added to suppress the deterioration of offset compensation due to the base current of the first post-amp. Post-layout simulation of the RX confirmed offset voltage of ±24.5 mV can be reduced to less than ±0.77 mV.

Experimental Results

The optical RX was fabricated on the basis of the 0.18-μm SiGe BiCMOS process. The TIA and PD, which has 0.8-A/W responsivity, 60-fF PD capacitance, and 30-GHz bandwidth, were directly mounted on a PKG (Fig. 5(a)). The measured differential 50.6- and 56-Gb/s eye diagrams for 2^9−1 PRBS are shown in Fig. 5(b). To evaluate the optical characteristics of the RX, a 40-Gb/s LN modulator with a DFB-LD was used as a light source. The measured sensitivities and bathtub curves are respectively shown in Figs. 6(a) and (b). The sensitivities of the RX at BER of less than 10^{-12} for 50.6 and 56 Gb/s were respectively −7.4-dBm and −5.6-dBm OMA, respectively. The phase margins at BER of less than 10^{-12} for 50.6 and 56 Gb/s were respectively 0.51 and 0.37 UI. The preamp and ATC, post-amp stage, and driver respectively consume power of 95, 224, and 76 mW at 50.6 Gb/s. The total power consumptions at 50.6 and 56 Gb/s are respectively 395 and 419 mW (excluding power consumption of the PD-REG, i.e., 28 mW), and power efficiencies at 50.6 and 56 Gb/s are respectively 7.8 and 7.5 mW/Gb/s. The performance of the proposed optical RX is summarized and compared with that of previous works [1, 2] beyond 50 Gb/s in Table 1. Though its bit rate is inferior to that of previous works, its bit rate per product of power consumption and minimum input current (R_b/(P_{DC}· $I_{in,min}$)) is increased by reducing power consumption by almost half.

References

[1] D. M. Kuchta, et al., Proc. OFC 2014, Th3C.2, 2014.
[2] D. M. Kuchta, et al., IEEE PTL, vol. 27(6), pp. 577-580, 2015.
[3] Y. Lee et al., IEICE Trans. Electron., pp. 116-119, E94-C, 2010.
[4] Sedighi, B. and Scheytt, J. C., IEEE TCAS II, vol. 59(8), pp. 461-465, 2012.
[5] T. Takemoto et al., IEEE JSSC, vol. 49(2), pp. 471-485, 2014.
[6] T. Takemoto et al., IEEE JSSC, vol. 49(10), pp. 2259-2276, 2014.

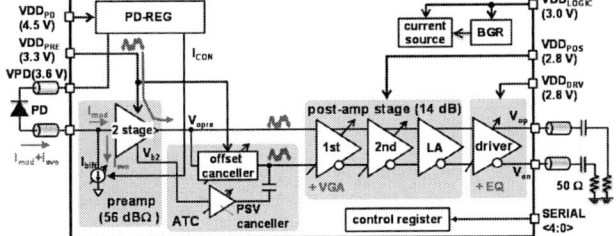

Fig. 1: Block diagram of 50-Gb/s TIA

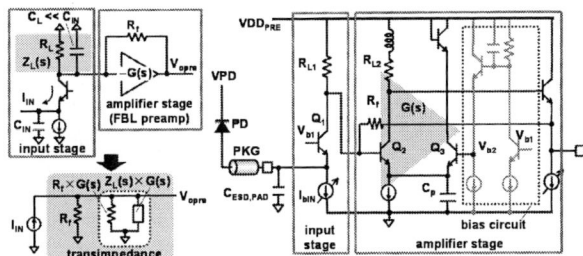

Fig. 2: Circuit diagram of proposed two-stage preamp

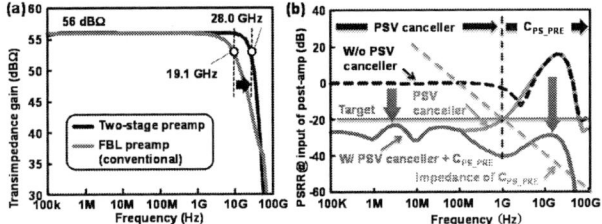

Fig. 3: Simulated frequency responses of (a) two-stage preamp and (b) PSV canceller

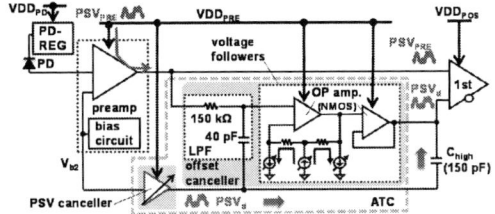

Fig. 4: Block diagram of proposed feedforward ATC with PSV and offset cancellers

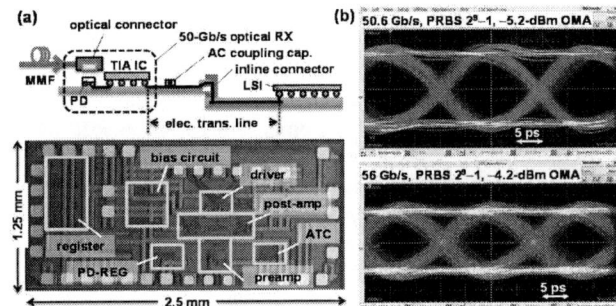

Fig. 5 (a) Chip micrograph and (b) measured 50.6 and 56-Gb/s eye diagrams

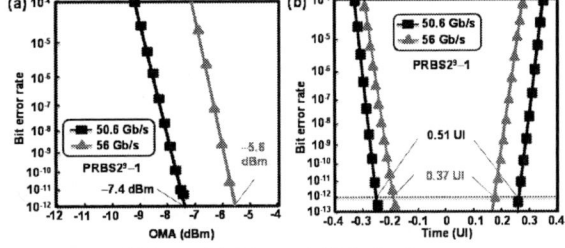

Fig. 6: Measured (a) sensitivities and (b) bathtub curves

Table I: Performance comparison with previous works

	[1]	[2]	This work
SiGe technology	0.13 μm	0.13 μm	0.18 μm
bit rate (Gb/s)	64	71	50.6
connection between PD and TIA	W/B	W/B	Flip chip
power consumption (mW)	780	840	395
power efficiency (mW/Gb/s)	12.2	11.8	7.8
phase margin (UI)	N/A	0.08	0.51
sensitivity (dBm)	-3.8	-4.8	-7.4
resposivity (A/W)	0.55	0.48	0.8
minimum input current*) (μApp)	229	159	146
R_b/(P_{DC}·$I_{in,min}$) [(Gb/s) / (mW·mApp)]	0.36	0.53	0.68

*) Values converted by using PD responsivities

An 8.3M-pixel 480fps Global-Shutter CMOS Image Sensor with Gain-Adaptive Column ADCs and 2-on-1 Stacked Device Structure

Yusuke Oike[1], Kentaro Akiyama[1], Luong D. Hung[1], Wataru Niitsuma[1], Akihiko Kato[1],
Mamoru Sato[1], Yuri Kato[1], Wataru Nakamura[1], Hiroshi Shiroshita[1], Yorito Sakano[1],
Yoshiaki Kitano[2], Takuya Nakamura[2], Takayuki Toyama[1], Hayato Iwamoto[1], Takayuki Ezaki[1]

[1]Sony, Atsugi, Japan. [2]Sony Semiconductor, Kumamoto, Japan. E-mail: Yusuke.Oike@jp.sony.com

Abstract

A 4K2K 480 fps global-shutter CMOS image sensor has been developed with super 35 mm format. This sensor employs newly developed gain-adaptive column ADCs to attain a dark random noise of 140 μV_{rms} for the full-scale readout of 923 mV. An on-chip online correction of the error between two switchable gains maintains the nonlinearity of output image within 0.18 %. The 16-channel output interfaces with 4.752 Gbps/ch are implemented in 2 diced logic chips stacked on a sensor chip with 38K micro bumps.

Introduction

The demand for highly realistic video system for 4K digital imaging has been increasing. Low noise and high bit-resolution ADCs are required to capture both highlights and detailed shadows. In addition to the image quality, higher frame rate is also desired, which can offer slow-motion movies for instant replays in sports broadcasting. High-speed image sensors with a large optical format [1-5] suppress the readout noise with a high analog gain, but they cannot maintain the low noise while receiving the full scale of pixel output. To meet the demands, we propose gain-adaptive column ADCs based on a single-slope ADC (SS-ADC). The previous gain-adaptive functions [6,7] have employed column-parallel automatic gain control (AGC) pre-amplifiers to change the gain according to each pixel output. While the column AGCs make the error correction of column gains complicated, the proposed gain-adaptive column ADCs can suppress the sensor nonlinearity by on-chip online gain correction and realize small vertical fixed pattern noise (FPN) without any other calibration. This paper presents an 8.3M-pixel 480 fps global-shutter CMOS image sensor with super 35 mm optical format.

Circuit Configuration

Fig. 1 shows a block diagram of the sensor. The pixel array has 4624 × 2296 5.86 μm pitch pixels. Each pixel has a charge storage node of MEM under a transfer gate of TRX with a light shield layer. The sensor has 16 channels of 4.752 Gbps/ch scalable low-voltage scaling interface with an embedded clock.

Fig. 2 shows a circuit diagram of gain-adaptive column ADCs. A slope generator provides a high-gain (HG) slope and a low-gain (LG) slope. The two slopes are selected by D_{flag} in each column block. One is provided to a comparator, and the other is connected to a capacitance of C_{dm}. C_{dm} is designed to be of the same value as the comparator input capacitance to maintain the total load capacitance of each slope stable and independent of D_{flag}. The hybrid column counter presented in [5] is extended for digital CDS of gain-adaptive ADC. The data of column buffers can be restored to the counters during a horizontal scanning period as per D_{flag} from a comparator.

Fig. 3 shows a timing diagram of a horizontal scanning period. The pixel reset level, settled in (1), is converted to digital codes as V_{RST_H} with the HG slope selected by resetting D_{flag}. The counter codes of V_{RST_H} are temporarily stored into the column buffers in (3), and the pixel reset level is converted again to V_{RST_L} with the LG slope by setting D_{flag} in (4). Here the gain of LG slope is 12 dB lower than that of HG slope. Further, the operation procedure branches off into (a) and (b) while an analog output of pixel signal level is settling in (5). Along the way of settling, each comparator determines whether the pixel level is reaching a higher or a lower level than a threshold level of gain-switching point provided by the HG slope. D_{flag} is updated by each comparator output of D_{co}. As a consequence, the HG and LG slopes are adaptively selected for darker and brighter pixels, respectively. For the CDS operation in (6), the digital codes of V_{RST_H} are restored into the column counters in a case of (a). Conversely, the digital codes of V_{RST_L} are still stored in a case of (b). Finally each column provides digital outputs of $V_{SIG\_H} - V_{RST\_H}$ or $V_{SIG\_L} - V_{RST\_L}$ with D_{flag}. The counter operates with a 1.336-GHz clock.

This sensor has additional horizontal pixel lines with a preset output on the gain-switching point. At the beginning of each frame scan, the preset output is converted by both HG and LG slopes to calculate the gain ratio. The on-chip online digital calibration can follow a temperature drift, supply voltage conditions and device variations. It helps to make the connecting point invisible on a captured image where the switching points are somewhat scattered by analog temporal noise and variations of pixel transistors.

Experimental Results

Fig. 4 shows the device structure composed of a sensor chip and 2 diced logic chips with 38K micro bumps. The column ADCs are split into an analog part on the sensor chip and a digital part on the logic chips. A sample image was captured by a 14-bit 480-fps global-shutter mode. Table I summarizes the chip characteristics. The sensor achieves 140 μV_{rms} random noise at the gain offset of 0 dB. The vertical FPN is 0.11 e_{rms} at 480 fps without additional correction circuits.

Fig. 5 shows the sensor linearity. The gain-adaptive column ADCs count 12-bit codes for the pixel signal level. The output codes obtained by the LG slope are multiplied by a gain ratio of the HG slope to the LG slope at the on-chip digital block, where the linear 14-bit outputs are reconstructed in the nonlinearity within 0.18 % of the full scale without special trimming or tuning calibration parameters.

Conclusion

The adaptive-gain column ADCs attain 140 μV_{rms} dark noise at 0 dB gain for the full-scale readout of 923 mV. The 2-on-1 stacked device structure achieves the combination of a global-shutter pixel with -99.6 dB parasitic light sensitivity and high-speed digital circuits for 4.752 Gbps/ch interface. Table II shows a performance comparison among [1-5].

978-1-5090-0636-6/16 $31.00 © 2016 IEEE

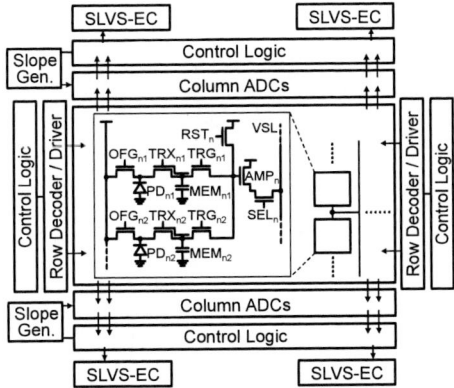

Fig. 1 Block diagram with global-shutter pixels.

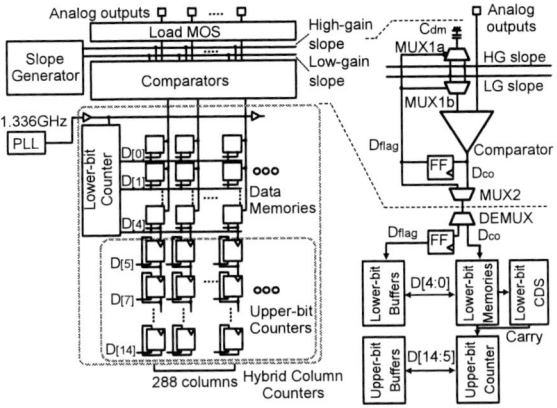

Fig. 2 Circuit diagram of pixel-wise gain-adaptive ADCs.

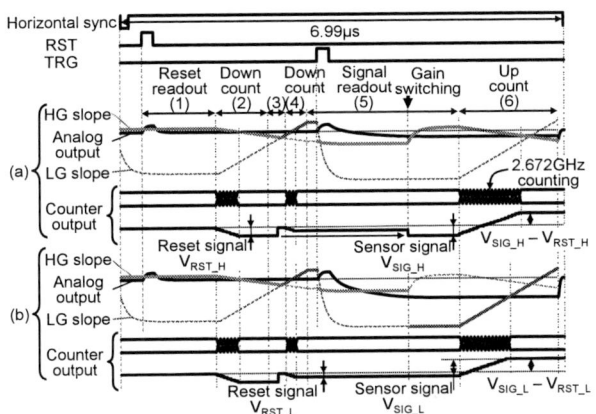

Fig. 3 Timing diagram. The operation is switched between (a) and (b) during the signal readout (5) according to each pixel output.

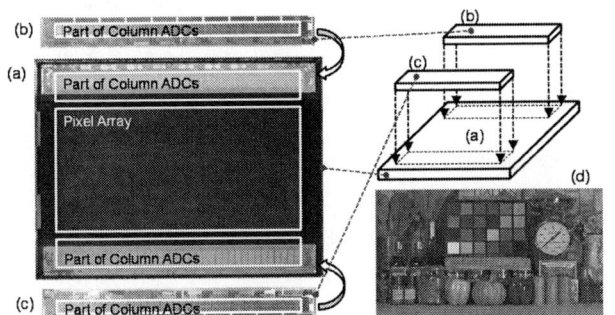

Fig. 4 Chip microphotograph and 2-on-1 stacked structure with micro bumps: (a) a sensor chip, (b-c) logic chips. (d) Sample image.

TABLE I   CHIP CHARACTERISTICS

Fabrication process	90nm 1P5M + Light Shield; 65nm 1P9M Logic
Supply voltage	3.3V / 3.0V / 1.25V / 1.2V
Num. of effective pixels	3840 (H) x 2160 (V)
Num. of total pixels	4624 (H) x 2296 (V)
Pixel size	5.86μm (H) x 5.86μm (V)
Output interface	16ch x 4.752Gbps/ch SLVS-EC
Max. frame rate	480fps
Power consumption	5.23W (480fps); 2.15W (60fps)
Saturation signal	30,450e-
Sensitivity	17,500e-/lx-s (green pixel, 3200K light with IR cut filter)
Parasitic light sensitivity	-99.6dB
Conversion gain	30.3μV/e-
Analog gain	Adaptive gain: 0dB / 12dB Gain offset: +0dB ~ +12dB (for each color independently)
RMS random noise	140μVrms (Gain offset 0dB); 100μVrms (Gain offset 12dB)
RMS vertical FPN	0.11e-rms at 480fps w/o additional correction circuit
Dynamic range	76.3dB

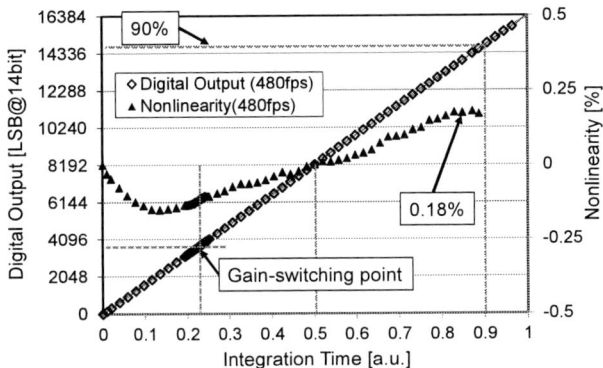

Fig. 5 Sensor linearity with on-chip gain calibration.

TABLE II   PERFORMANCE COMPARISON

		This Work	[1] 2015	[2] 2015	[3] 2013	[4] 2013	[5] 2011
Optical format		Super35	Super35	Full35	2/3-inch	1-inch	Super35
Pixel pitch	mm	5.86	3.2	2.45	5	2.86	4.2
Global/Rolling shutter		GS	RS	RS	GS	GS	RS
Num of pixels (H)	pixels	3840	7680	15360	1920	4620	8192
Num of pixels (V)	pixels	2160	4320	8640	1080	3084	2160
Frame rate	fps	480	120	60	240	80	120
Power consumption	mW	5,230	3,200	11,050	1,100	1,100	3,000
Conversion gain	mV/e-	30.3	61.0	80.0	95.0	45.0	57.0
Analog gain (Low)	times	1	3.5	1	2	1	1
Noise	mVrms	140	317	614	380	315	250
	e-rms	4.6	5.2	7.7	4.0	7.0	4.4
Saturation	mV	923	933	800	1,425	720	1,197
	e-	30,450	15,300	10,005	15,000	16,000	21,000
Dynamic range	dB	76.3	69.7	62.3	71.5	67.2	74.4
FoM1	e-·pJ/DRU1	0.92	4.97	8.18	4.72	2.96	1.29
FoM2	e-·pJ/DRU2	0.92	1.42	8.18	2.36	2.96	1.29

FoM = (power x noise) / (num.pixels x fps x DRU);   DRU1 = {(saturation / gain) / noise}
DUR2 = saturation / noise

References

[1] T. Yasue *et al.*, "A 14-bit, 33-Mpixel, 120-fps Image Sensor with DMOS Capacitors in 90-nm/65-nm CMOS," *in Proc. of IISW*, Jun. 2015.

[2] R. Funatsu *et al.*, "133Mpixel 60fps CMOS Image Sensor with 32-Column Shared High-Speed Column-Parallel SAR ADCs," *ISSCC Dig. Tech. Papers*, Feb. 2015.

[3] P.Centen *et al.*, "A 4e-noise 2/3-inch Global Shutter 1920x1080P 120 CMOS-Imager," *in Proc. of IISW*, Jun. 2013.

[4] H. Honda *et al.*, "A 1-inch Optical Format, 14.2M-pixel, 80fps CMOS Image Sensor with a Pipelined Pixel Reset and Readout Operation," *in Symp. VLSI Circuits Dig. Tech. Papers*, Jun. 2013.

[5] T. Toyama *et al.*, "A 17.7Mpixel 120fps CMOS Image Sensor with 34.8Gb/s Readout," *ISSCC Dig. Tech. Papers*, Feb. 2011.

[6] S. Kawahito *et al.*, "A Column-Based Pixel-Gain-Adaptive CMOS Image Sensor for Low-Light-Level Imaging," *ISSCC Dig. Tech. Papers*, Feb. 2003.

[7] J. Solhusvik *et al.*, "A 1.2MP 1/3" Global Shutter CMOS Image Sensor with Pixel-Wise Automatic Gain Selection," *in Proc. of IISW*, Jun. 2011.

A Dead-time Free Global Shutter CMOS Image Sensor with in-pixel LOFIC and ADC using Pixel-wise Connections

Hidetake Sugo[1], Shunichi Wakashima[1], Rihito Kuroda[1], Yuichiro Yamashita[2], Hirofumi Sumi[2], Tzu-Jui Wang[2], Po-Sheng Chou[2], Ming-Chieh Hsu[2] and Shigetoshi Sugawa[1]

[1]Graduate School of Engineering, Tohoku University, Sendai, Japan, Email: rihito.kuroda.e3@tohoku.ac.jp
[2]Taiwan Semiconductor Manufacturing Company, Hsinchu, Taiwan

Abstract

An almost 100% temporal aperture (dead-time free) global shutter (GS) stacked CMOS image sensor (CIS) with in-pixel lateral overflow integration capacitor (LOFIC), ADC and DRAM is developed using pixel-wise connections. The prototype chip with 6.6μm-pitch VGA LOFIC pixel dead-time free GS mode and 1.65μm-pitch 4.9M sub-pixel high resolution rolling shutter (RS) mode was fabricated with a 45nm 1P4M CIS technology for PD substrate and a 65nm 1P5M CMOS technology for ASIC substrate.

Keywords: CIS, global shutter, pixel-wise connection, LOFIC.

Introduction

For image sensors used in automobile and machine vision applications, the GS function is strongly demanded in order to avoid the image distortion. To implement the GS function into CIS, pixel image signal must be temporally stored until sequential readout. In general, it is stored either in the charge domain or the voltage domain. For the former type, a CIS with in-pixel photo-charge storage node was reported[1]. For this type, the storage must be formed in the same substrate as the photodiode (PD) to achieve a full charge transfer for correlated-double-sampling (CDS) at the floating diffusion (FD). Therefore the trade-off relationships arise between full well capacity (FWC), aperture ratio, light shielding performance and dark current. Recently, a GS operation was demonstrated using a voltage control of organic photoconductive film in 3-transistor type pixel[2]. For the latter type, a backside illumination (BSI) CIS and a BSI stacked CISs with in-pixel voltage sample/hold capacitors for reset and signal levels have been reported[3-4]. For this type, the reset noise at the sample/hold capacitor remains. Also, since the dynamic range is limited by the signal voltage range, miniaturized logic transistor with a low supply voltage (V_{DD}) is difficult to be applied even though a stacked technology is employed. In addition, for all of the above mentioned GS CISs, the photo-carrier integration during pixel signal readout was not carried out, which results in a small temporal aperture. In order to solve the reset noise issue, the in-pixel ADC and digital signal storage is attractive[5], although achieving a high aperture ratio was a challenge. A recent work presented a BSI stacked CIS with in-pixel pulse frequency modulation ADC[6].

This paper presents a BSI stacked CIS with in-pixel lateral overflow integration capacitor (LOFIC), single slope (SS) ADC and DRAM, enabled by pixel-wise connections[7]. To develop a scalable architecture and achieve a low power consumption and a wide dynamic range, the charge domain dynamic range extension with LOFIC[8], and 1.2V V_{DD} 65nm technology-node transistors for in-pixel ADC and DRAM were simultaneously introduced.

Device Structure, Operation and Measurement Results

Fig.1 shows the structure of the developed stacked CIS. It consists of a BSI PD substrate with an array of PDs and pixel transistors including LOFIC and an ASIC substrate with an array of in-pixel ADCs, DRAM and sensing amplifier/repeater

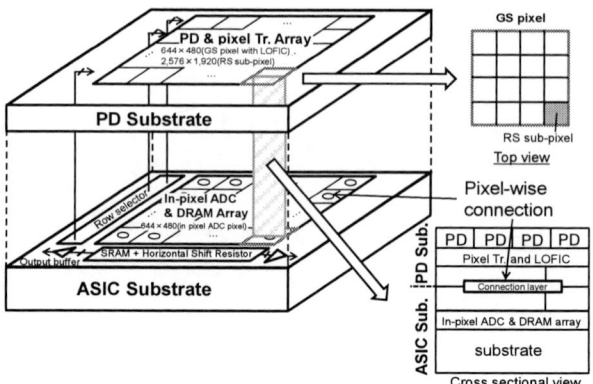

Fig.1 Structure of developed CIS.

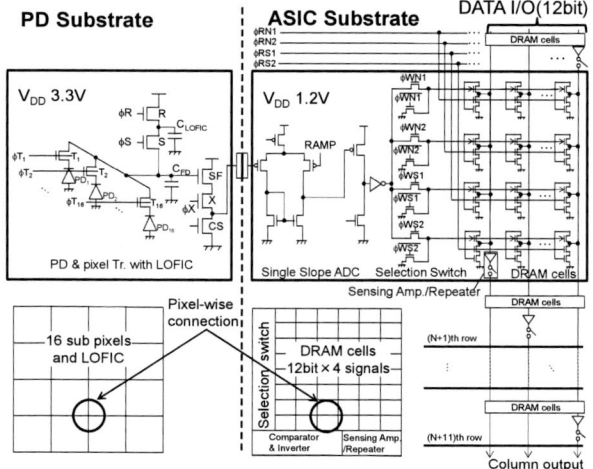

Fig.2 Pixel circuit schematic and layout diagrams.

circuit, row selector, pipeline SRAM, and horizontal shift resistor (HSR) and output buffers. The two substrates are stacked with pixel-wise connections. The PD substrate formation process was optimized for PD and pixel nMOSFET. The stacked image sensor operates in two modes; 6.6μm-pitch VGA pixel GS mode with almost 100% temporal aperture (dead-time free) and wide dynamic range enabled by LOFIC, and 1.65μm-pitch 4.9M sub-pixel high resolution rolling shutter (RS) mode. Using a more miniaturized technology-node for ASIC substrate, the pixel pitch is to be scaled down furthermore. Fig.2 shows the schematic and layout diagrams of pixel circuit. The PD substrate side consists of FD shared sixteen sub-pixels, LOFIC, reset switch (R), switch between FD and LOFIC (S), source follower (SF), current source (CS) and CS control switch (X). The LOFIC is employed to resolve the trade-off between conversion gain and FWC by generating high conversion gain signal with C_{FD} and high FWC signal with $C_{FD}+C_{LOFIC}$ using overflow

978-1-5090-0636-6/16 $31.00 © 2016 IEEE
178
2016 Symposium on VLSI Circuits Digest of Technical Papers

photoelectrons under a single exposure[8]. The ASIC substrate side consists of a 12bit SS ADC and four 12bit 3T-DRAM cells and signal selection switches to select high conversion gain signals (N1 and S1) and high FWC signals (N2 and S2), respectively. The SF output and ADC input were connected by pixel-wise connections. The LOFIC structure captures a wide dynamic range signals by the charge domain, thus, the small voltage swing is sufficient. The power supply voltages for PD and ASIC substrates are 3.3V and 1.2V, respectively. A column 12bit data I/O lines are used for both writing and reading operations of DRAM. In each bit line, a sensing amplifier/repeater circuit composed of a set of a CMOS buffer and a switch is placed in every twelve rows. During the writing operation, all of the switches are ON, and they work as repeaters. During the reading operation, switches placed in-between the selected row and column output are ON, the CMOS buffer nearest the selected row works as a sensing amplifier and others work as repeaters. The high speed writing and reading operations are achieved by this architecture. Fig.3 shows the pulse timing diagram for GS and RS operation modes. In GS mode, during the exposure period of N-th frame, photoelectrons overflow from PD and FD are accumulated at LOFIC. Before the end of the exposure period, X and WN1 are turned ON, and AD conversion of N1 is carried out. Then T1~T16 are turned ON and OFF, all photoelectrons are transferred from PD to FD. X and WS1 are turned ON, and AD conversion of S1 is carried out. All of the charges are mixed at FD and LOFIC by turning ON S, and AD conversion of S2 is carried out. R and T1~T16 are turned ON to reset PD, FD and LOFIC. After turning off T1~T16 and R, exposure period of (N+1)-th frame starts. Then, X and WN2 are turned ON and AD conversion of N2 is carried out for the (N+1)-th frame. Each digitized signal is stored in 12bit DRAMs until readout. During the exposure period of (N+1)-th frame, digital signals are readout as follows. After one row is selected by row selector, N1, S1, S2 and N2 signals stored in DRAM are readout sequentially to column pipeline SRAM (two 12bit/column). SRAM signals are output in serial by HSR. By this operation, exposure and readout are carried out simultaneously, thus the dead-time free integration is achieved. In the RS mode, during each row selecting period, N1 and S1 of sub-pixels 1 are digitized, stored and readout. Then, sub-pixels 2 through 16 sequentially operate in the same manner.

Fig.4 shows the micrograph of the fabricated sensor chip and the digital camera system. Fig.5 shows the captured sample images obtained by GS mode. A distortion-free, dead-time free image by high CG and high FWC signals is captured. Table 1 shows the performance summary of the chip. The FWC was 220ke⁻ in GS mode with linear response. In GS mode the temporal aperture ratio becomes 99% at 120fps.

Conclusion

A stacked CIS with in-pixel LOFIC, 12bit SS ADC and DRAM with dead-time free wide dynamic range GS mode and high resolution RS mode was developed using pixel-wise connections. A sample image capturing by the prototype chip fabricated with a 45nm 1P4M CIS technology for PD substrate and a 65nm 1P5M CMOS technology for ASIC substrate was successfully carried out.

Acknowledgments

The authors would like to thank TSMC CIS teams for their technical assistance.

References

[1] K. Yasutomi et al., ISSCC, pp. 398-399, 2010.
[2] M. Takase et al., IEDM, pp.775-778, 2015.
[3] G. Meynants et al., Int. Image Sensor Workshop, pp. 305-308, 2011.
[4] T. Kondo et al., Symp. on VLSI Circuits, pp.C90-91, 2015.
[5] S. Kleinfelder et al., IEEE JSSC, Vol.36, pp. 2049-2059, 2001.
[6] M. Goto et al., IEDM, pp.84-87, 2014.
[7] C. C.-M. Liu et al., Symp. on VLSI Circuits, 4.4, 2014.
[8] S. Sugawa et al., ISSCC, pp.352-353, 2005.

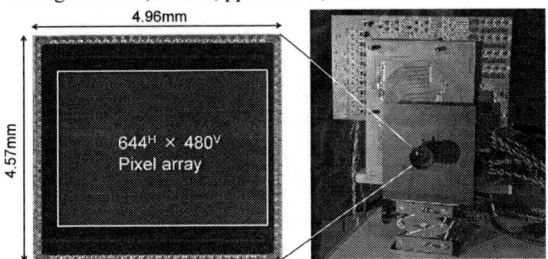

Fig.4 Chip micrograph (left) and digital camera system (right).

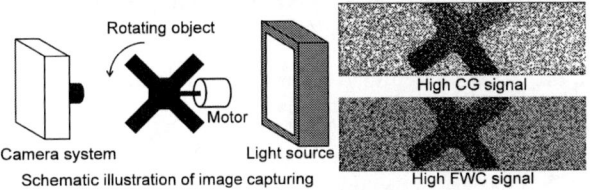

Fig.5 Captured sample images by GS mode (high CG and high FWC signals). A part ($320^H \times 94^V$) of the full image is shown.

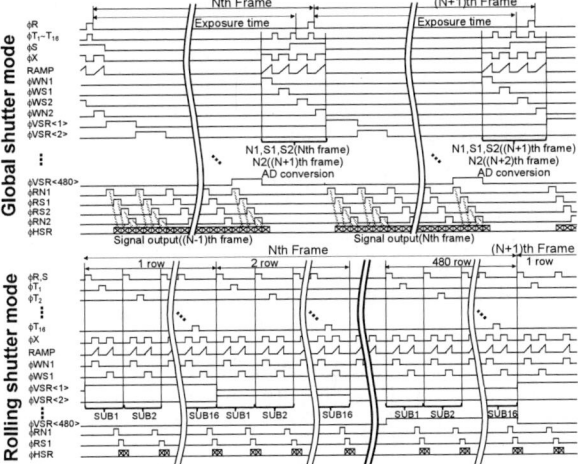

Fig.3 Pulse timing diagram.

Table 1 Performance summary.

Fabrication. Process (PD die / ASIC die)		45nm 1P4M CMOS / 65nm 1P5M CMOS
Chip size		$4,960 \mu m^H \times 4,570 \mu m^V$
# of pixels (GS / RS)	Total	$644^H \times 480^V$ / $2,576^H \times 1920^V$
	Effective	$640^H \times 476^V$ / $2,560^H \times 1904^V$
Pixel pitch (GS / RS)		$6.6\mu m$ / $1.65\mu m$
# of connection per pixel		1 connection / pixel
Shutter mode		GS / RS
ADC type		12bit, Single Slope in pixel ADC
Pixel Memory		12bit × 4 channel
VDD (PD sub. / ASIC sub.)		3.3V / 1.2V
Full Well Capacity (GS mode)		220ke⁻
Integration dead-time per frame		80μs (time aperture ratio: 99% @ 120fps)

A 220pJ/Pixel/Frame CMOS Image Sensor with Partial Settling Readout Architecture

Suyao Ji, Jing Pu, Byong Chan Lim, and Mark Horowitz

Stanford University, Stanford, CA, United States
suyao@stanford.edu, jingpu@stanford.edu, bclim@stanford.edu, horowitz@stanford.edu

Abstract

To reduce power consumption in a CMOS imager readout path, we use partial settling of the column values into a SAR-ADC, creating a 320Hx240V prototype sensor with two column-shared 10-bit ADCs, which consumes 2.2mW at 130 fps. The measured INL and DNL with a third order correction of partial settling behavior is +1.855LSB/-1.855LSB and +0.337LSB/-0.179LSB, respectively. The input referred readout noise is 5e$^-$ with a conversion gain of 90uV/e$^-$.

Introduction

In conventional designs, in order to achieve fast frame rate and high dynamic range, the bitline is often buffered by an energy expensive, high-speed gain stage and settled to 10-14 bits before being read out [1-7]. This static power stage in an ADC readout path consumes a significant amount of power compared to other blocks [3]. Many imagers also use a per-column, single-slope ADC, which can fit into a pixel pitch and achieve high resolution [1,5]. However, based on the survey of A/D performance [8], this converter architecture is not very energy efficient.

We remove the gain stage, use a SAR A/D, pulse the current used to settle the imager bitlines, and minimize this current by only partially settling the SAR input capacitance to the bitline voltage. The residual systematic error is calibrated in the digital domain. Gain adjustment can be accomplished by using in-pixel C_{FD} variation [7]. A SAR-ADC architecture is chosen because of its competitive energy efficiency and high speed allowing it to read multiple columns sequentially [8].

Image Sensor Architecture and Operation

Fig. 1 shows a block diagram of the developed image sensor. The pixel array consists of 320H x 240V pixels. To calibrate the partial settling, a dummy row is added to the array which allows us to directly drive a voltage on the floating diffusion nodes, and bitlines are loaded with extra 2pF to reflect a 10Mpixel imager settling performance. To achieve low power consumption, each bitline driven by the source follower in a selected pixel is directly muxed into a 10-bit SAR-ADC which samples the partially settled level of the bitline. This partial settling readout technique allows us to use one ADC for each 120 bitlines. The converted data of two inputs to an ADC, i.e. the reset signal and pixel value, are post-processed for both ADC calibration and partial settling calibration, and then subtracted to eliminate kT/C noise to produce the correlated double sampling (CDS) output.

Fig. 2 shows a simplified timing diagram for the signal chain of Fig. 1. The source follower current is dynamically turned on by Φ_{ISF} and its pulse is long enough so the bitlines fully settle before the mux switch turns on by Φ_{MUX}. The column-shared voltage nodes (V_L) and ADC are reset before the mux and sampling switch (controlled by Φ_{SMP}) turn on respectively so that the partially settled value is independent of previous readout. A first partial settling occurs when the mux switch turns on, followed by a second partial settling during the sampling phase of the ADC. The voltage loss caused by the charge sharing and then partial exponential settling is determined by the cap ratio of the mux and ADC cap to the bitline, and the total capacitive load along readout path and the g_m of the source follower. This deterministic error can be calibrated out digitally. Finally, the source follower current is turned off to save power, and the bitlines float after readout.

Measurement Results

Fig. 3 and Fig. 4 shows the die photograph of the prototype and its captured image when operating at 130fps with a 10.4MHz sampling clock and off-chip calibration. The imager core excluding test logic, clock divider, and I/O consumes 2.2mW with a FoM of 220pJ/pixel/frame, which is 4 times more energy efficient than previous designs as shown in TABLE I. This readout speed is equivalent to a 10M-pixel sensor operating at 30fps illustrated in the last column of the table.

With the conversion gain of 90uV/e$^-$, the input referred readout noise is 5e$^-$ when sampling at 40ns with 2pF bitline load compared to 3.5e$^-$ when fully settled consistent with our 70% partial setting gain. Fig. 5 shows Fixed-Pattern Noise (FPN) result, and partial settling introduces additional 60% FPN contributed from source follower variation compared to full settling configuration because of larger current when sampling. Clock jitter is not critical in this design as 20 psec clock jitter in the measurement causes a V_{RMS} =0.07LSB which is negligible compared with other components of readout noise.

To verify we can correct partial settling effects in a larger imager size, the prototype chip can be configured to add more capacitive loads to the bitlines. Fig. 6 shows the calibration results configured at 40ns sampling time with a 2pF bitline capacitive load.

The noise performance can be improved by reducing column mux and SAR input capacitance so that the voltage loss due to charge sharing is smaller; a breaking the 100-to-1 mux into two 10-to-1 stages as well as adding one more segmentation in SAR input will improve the settling gain from 70% to 90%. In this work, the noise performance of the 10bit SAR-ADC with 42fJ/conv-step is overdesigned to an 11bit envelope; to match dynamic range and readout noise in peer designs, a 12bit SAR-ADC design with 2 segmentation stages in the DAC can be used without significantly changing the DAC energy [8].

Conclusion

With the partial settling technique, fast frame rate is achieved without a high bandwidth buffer, for given energy constraint. The ADC sharing and dynamic bitline current source driver further reduces power. Deterministic error, introduced by partial settling readout, is calibrated in digital domain.

Acknowledgement

The authors would like to thank members of Bosch Corporation for their support of this work, especially Christoph Lang, Pedram Lajevardi, and Xinyu Xing. We would also like to show our gratitude to Google for their suggestions and help during this pixel characteristics measurement.

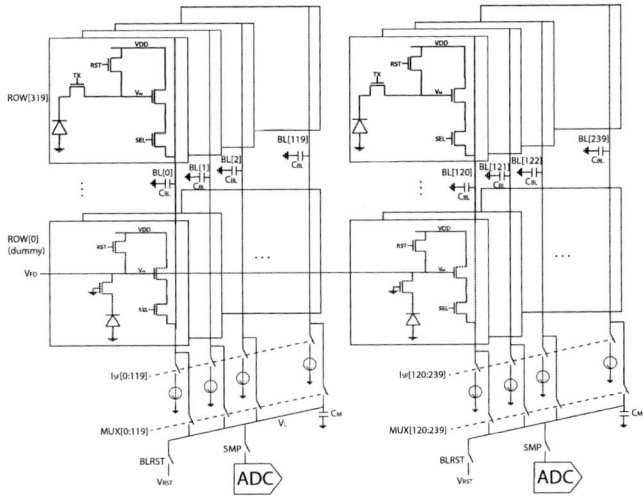

Fig. 1 Circuit block diagram of the developed image sensor.

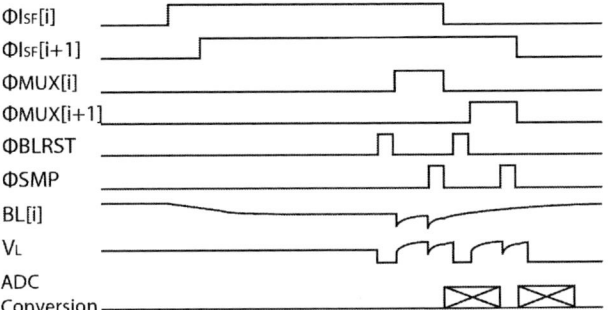

Fig. 2 Timing diagram of the partial settling readout circuitry. $\Phi I_{SF}[i]$ and $\Phi MUX[i]$ represent source follower current clock and mux clock at column i; $\Phi BLRST$ is the bitline reset control clock and ΦSMP is the ADC sampling clock; BL[i] is the i'th column bitline voltage; V_L is the column-shared voltage.

Fig. 3 Chip microphotograph. Fig. 4 Sample image.

Fig. 5 FPN performance under full settling and partial settling configuration.

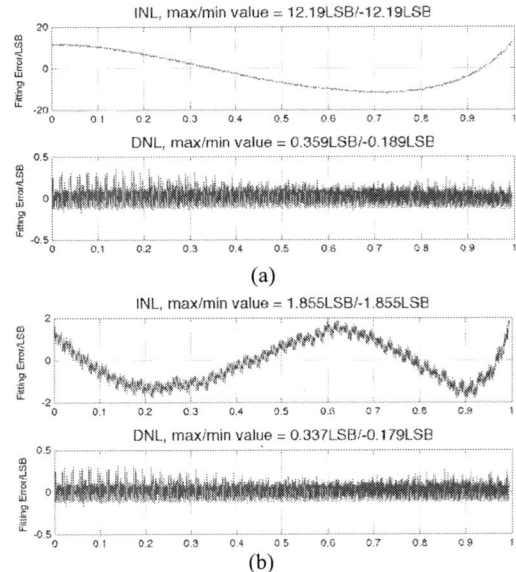

Fig. 6 Measured INL/DNL after (a) 1st order calibration and (b) 3rd order calibration. Higher order fitting gives better INL result.

TABLE I
COMPARISON WITH PREVIOUS DESIGNS

Parameter	[1]	[2]	[3]	[4]	This Work	This work*
Year	2015	2015	2013	2013	2015	2015
Process	90/65nm	180nm	180nm	180nm	180nm	180nm
Supply Voltage	2.9V/1.8V/1.1V	3.3V/1.8V	3.3V/1.8V	3.8V/1.8V	3.3V/1.8V	3.3V/1.8V
ADC type	SS	SAR	SAR	SAR	SAR	SAR
ADC resolution	12	12	11	10	10	10
Frame Rate	30	60	35	17	130	30
Pixel Number	20.6M	133M	960K	1.4M	76.8k	10M
Power	532mW	11W	40mW	51mW	2.2mW	74.7mW
FOM(pJ/pixel/frame)	860	1378	1190	2142	220	249

* By scaling the current design up, the power of a 30fps 10Mpixel image sensor including LVDS is estimated.

References

[1] S. Atsushi et al., "A 1/1.7-inch 20Mpixel Back-Illuminated Stacked CMOS Image Sensor for New Imaging Applications," ISSCC Dig. Tech. Papers, pp. 110-111, Feb. 2015.

[2] F. Ryohei et al., "133Mpixel 60fps CMOS Image Sensor with 32-Column Shared High-Speed Column-Parallel SAR ADCs," ISSCC Dig. Tech. Papers, pp. 112-113, Feb. 2015.

[3] T. Fang, C. Denis, W. Bo, and B. Amine, "Low-Power CMOS Image Sensor Based on Column-Parallel Single-Slope/SAR Quantization Scheme", IEEE Trans. Electron Devices, vol. 60, no.8, pp. 2561-2566, August 2013.

[4] D. Jun et al., "A 187.5μVrms-Read-Noise 51mW 1.4Mpixel CMOS Image Sensor with PMOSCAP Column CDS and 10b Self-Differential Offset-Cancelled Pipeline SAR-ADC," ISSCC Dig. Tech. Papers, pp. 494-495, Feb. 2013.

[5] Y. Shang, C. Kuo, T. Hon, C. Calvin, and H. Fu, "A 0.66e-rms Temporal-Readout-Noise 3D-Stacked CMOS Image Sensor with Conditional Correlated Multiple Sampling (CCMS) Technique" Symp. VLSI Circuit, Tech. Dig., pp.184-185, 2015.

[6] S. Sukegawa et al., "A 1/4-inch 8Mpixel Back-illuminated Stacked CMOS Image Sensor," ISSCC Dig. Tech. Papers, pp. 84-85, Feb. 2013.

[7] H.Hidenari et al., "A 1-inch Optical Format, 14.2M-pixel, 80fps CMOS Image Sensor with a Pipelined Pixel Reset and Readout Operation," Symp. VLSI Circuit, Tech. Dig., pp.c4-c5, 2013.

[8] B. Murmann, "ADC Performance Survey 1997-2015," [Online]. Available: http://web.stanford.edu/~murmann/adcsurvey.html.

A 260µW Infrared Gesture Recognition System-on-Chip for Smart Devices

Sechang Oh[1], Ngoc Le Ba[2], Suyoung Bang[1], Junwon Jeong[3], David Blaauw[1], Tony T. Kim[2], Dennis Sylvester[1]

[1]University of Michigan, MI; [2]Nanyang Technological University, Singapore; [3]Korea University, Korea

Abstract

This paper presents a low-power infrared motion detection system suitable for smart devices such as wearables. The SoC incorporates instrumentation chopper amplifiers (ICA), LPFs, ADCs, and a DSP. The low-noise ICAs amplify very low frequency µV-level thermopile outputs with 2.0 NEF and provide programmable gain modes. To reduce standby power the ICA uses lower current when the system is in idle mode. Wakeup can be triggered by detection of a simple gesture. For the LPF, source degeneration by pseudo-resistors and g_m division techniques are used for both improved linearity and 30Hz bandwidth. The DSP employs a motion history image technique to achieve low-power detection. The system consumes 260µW in active mode and 46µW in idle mode while processing 16×4 infrared data at 30fps. A complete system demonstration is shown.

Introduction

Recent demand for natural human-computer interfaces such as gesture recognition has increased, particularly for compact wearable devices. Cameras are currently the most common platform for gesture sensing, but they are highly sensitive to environmental light conditions. Extended range capacitive sensing [1] and ultrasonic techniques [2] have been explored but they consume significant energy due to their excitation source. In contrast, an infrared sensing system, in which a thermopile array directly converts incoming infrared radiation energy into electrical energy, is an appealing low-power choice since the sensor array itself is passive. It can be fabricated in CMOS technology and generates voltages linearly proportional to the temperature difference between an object and the background environment [3, 4]. However, array sensitivity is just a few µV/°C and its time constant is several ms. To achieve ultra-low power gesture recognition, we propose an SoC including a low-noise instrumentation chopper amplifier for low frequency signals, a low-power LPF for filtering out-band noise including the chopper frequency and its harmonics, an ADC, and a motion history image based [5] low-power DSP.

Proposed Gesture Recognition SoC

This paper targets a gesture sensing system using a thermopile array (Figs. 1 and 2). A hand emits infrared radiation with wavelength representing its temperature; this forms an image incident upon a 16×4 thermopile array. Each thermopile signal connects to an AFE path that consists of an ICA and LPF. The four row ADCs digitize the amplified/filtered signals using time-division multiplexing and the DSP then analyzes the waveform to detect gestures.

Fig. 3 shows the proposed ICA. Since the gesture signals are significantly impacted by 1/f noise, they are chopped to remove this 1/f noise and then sent through two amplifiers. Overall gain needs to be up to 80dB for a power-efficient high dynamic range system. C_1/C_2 (15pF/150fF) and C_3/C_4 (C_4=20fF) set the gains for the Low-noise Amplifier (LNA) and Programmable-gain Amplifier (PGA), respectively. C_3 is programmable (200fF–3pF) for system flexibility. OTA_1 and OTA_2 are implemented with inverter-based cascode amplifiers to maximize g_m and gain at a given current. The common-mode feedback (CMFB) amplifiers consume a fraction of the power using ratioed transistor sizes. As in typical noise-limited designs, the first amplifier stage consumes the majority of the total power (up to 2.5µA current) to achieve sub-µV noise while the PGA consumes just 90nA, constrained by the chopper bandwidth. Transistor sizes are chosen for optimal noise efficiency factor (NEF) and chopper frequency is 1kHz. The ICA high-pass corner is set by $(R_3C_5)^{-1}$ in the DC servo loop. R_1 and R_2 paths set the input common mode voltages and cancels the offsets. Fast-settling switches (FS_{1-3}) are selectively turned on to reduce settling time when ICA settings are changed, decreasing the corresponding resistance by 100× in simulation.

The ICA outputs show ripple at the 1kHz chopping frequency and its harmonics. These are removed with the proposed Gm-C LPF in Fig. 4. The two biquads are connected in series to form a 4th order filter. Since the gesture information resides in a low frequency range, the LPF bandwidth is set to 30Hz to achieve high SNR. C_{LPF} is a capacitor array and is set to 8.9pF to approximately match AFE and thermopile pixel size. Considering $f_{LPF3dB}=g_m/(2\pi C_{LPF})$, g_m in the nS range is required. To achieve this bias current must be extremely low, leading to potentially poor linearity. Thus, source degeneration and g_m division techniques [6] are used in the LPF. The Gm-stage input current is divided by the series-parallel current mirror to effectively obtain $g_m/32$. To enhance linearity, input pair sources are degenerated by pseudo-resistors whose gates are controlled by inputs. Simulation results show the resulting g_m is linear within ±100mV input range (defined by full width at half-maximum). The CMFB amplifier replicates voltages in the main Gm stage and sets the common mode output voltage. LPF outputs in each row are time-multiplexed via a 16:1 analog multiplexer and connected to a differential 8b SAR ADC (Fig. 5). The ADC sampling rate is 1kS/s.

Fig. 6 describes the overall structure of the proposed motion recognition DSP. There are three separate memories to store frame data. The first memory contains the motion history image (MHI), which is the difference between the current and previous frames (Fig. 7). The second and third memories are used to store two continuous frames once motion is detected. Detection modules use data from the three memories to analyze the gesture. Fig. 8 shows the top level design for the proposed gesture detection algorithm. Motion is detected by counting the number of pixels having significant change in value (i.e., ADC output code) between the current and previous frame. If there is no motion for a period of time the processor goes into an idle mode with only a simpler motion detecting circuit enabled to save power.

When motion is detected, a sweeping algorithm uses two motion history image frames to analyze the motion. In this process each row and column of the MHI frames are first summed. The type of movement (diagonal, up-down, or left-right) is then discerned based on the number of peaks found in the row- and column-wise sums. In a diagonal sweep both row and column sums will exhibit clear peaks (i.e., four total peaks detected) whereas in up-down or left-right sweeps only two peaks are observed due to constant behavior in either the horizontal or vertical direction. This is shown in Fig. 7, which illustrates the principle of detection for sweeping gestures. Up-down or left-right direction can be determined based on the relative positions of negative to positive peaks, as seen in Fig. 7. This approach allows the DSP to accurately identify specific gestures.

Measured Results

The proposed gesture recognition SoC is implemented in 65nm CMOS. The ICA input noise density is 31nV/$\sqrt{Hz}$ in active mode and 130nV/$\sqrt{Hz}$ in idle mode (Fig. 9), and chopping successfully suppresses 1/f noise. LPF bandwidth is adjustable between 10–150Hz by C_{LPF} changes (Fig. 10). Fig. 11 shows the LPF noise spectrum and HD3 is 48.9dB at In_{LPF}=0.1Vpp. Fig. 12 shows ADC performance. The system is demonstrated with an external 16×4 thermopile and lens, and Fig. 13 shows detection of a hand sweeping across the field of view. Fig. 15 summarizes measured results and compares with recent works. This work represents the first SoC for gesture sensing applications using a thermopile array. Its size (8.1mm², Fig. 14) and power (260µW and 46µW) are suitable for emerging smart devices.

References

[1] Yingzhe Hu, ISSCC 2014.
[2] R. Przybyla, ISSCC 2011.
[3] M. Hirota, SPIE 2003.
[4] H. Kawanishi, ISSCC 2008.
[5] C. Hsieh, ICSPS 2010.
[6] A. Arnaud, JSSC 2006.
[7] Qinwen Fan, JSSC 2011.
[8] Yen-Po Chen, VLSIC 2014.
[9] P. Bruschi, JSSC 2007.
[10] S.-Y. Lee, TBCAS 2009.

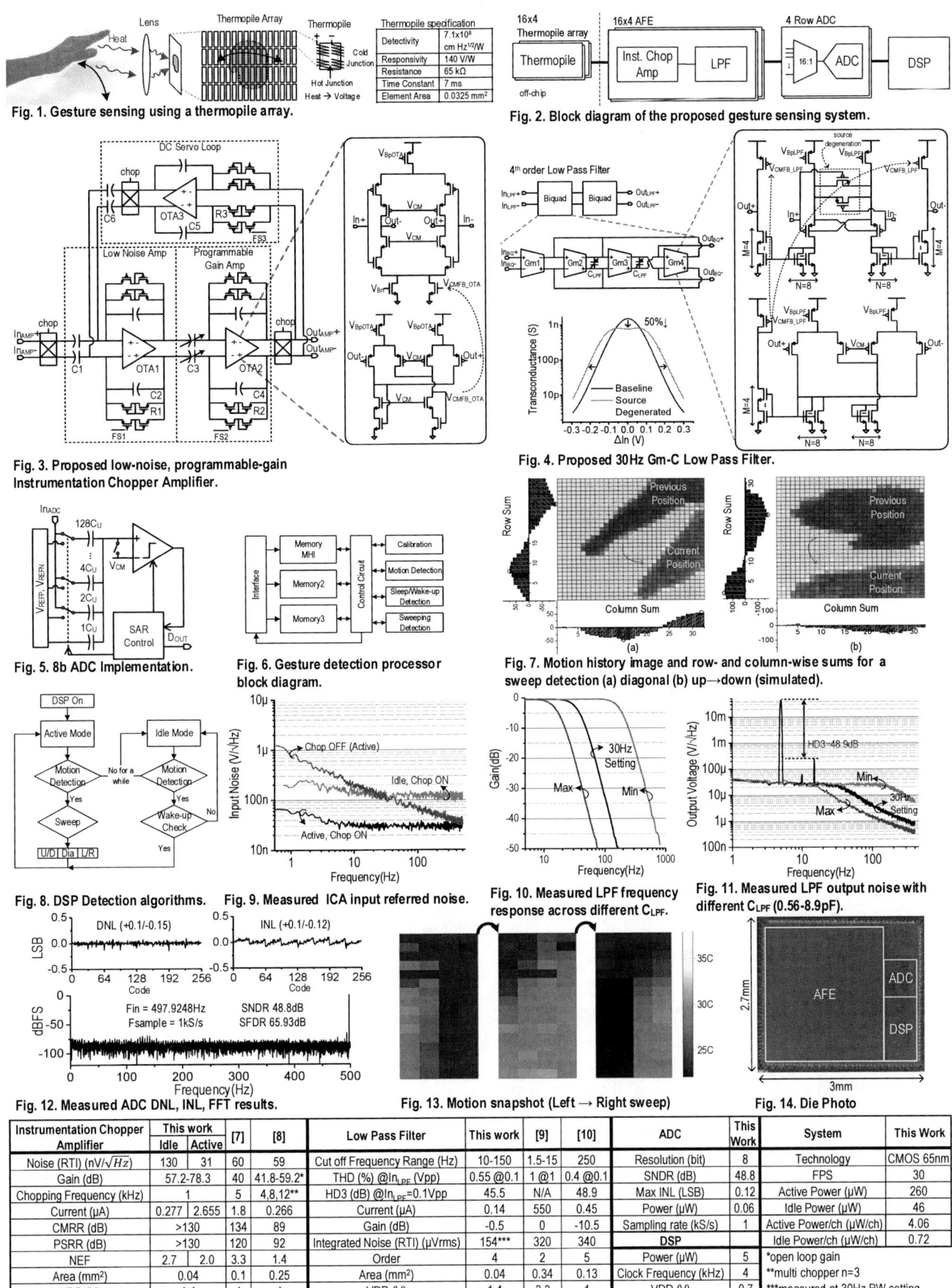

Fig. 1. Gesture sensing using a thermopile array.

Fig. 2. Block diagram of the proposed gesture sensing system.

Fig. 3. Proposed low-noise, programmable-gain Instrumentation Chopper Amplifier.

Fig. 4. Proposed 30Hz Gm-C Low Pass Filter.

Fig. 5. 8b ADC Implementation.

Fig. 6. Gesture detection processor block diagram.

Fig. 7. Motion history image and row- and column-wise sums for a sweep detection (a) diagonal (b) up→down (simulated).

Fig. 8. DSP Detection algorithms.

Fig. 9. Measured ICA input referred noise.

Fig. 10. Measured LPF frequency response across different CLPF.

Fig. 11. Measured LPF output noise with different CLPF (0.56-8.9pF).

Fig. 12. Measured ADC DNL, INL, FFT results.

Fig. 13. Motion snapshot (Left → Right sweep)

Fig. 14. Die Photo

Thermopile specification

Detectivity	7.1×10^8 cm Hz$^{1/2}$/W
Responsivity	140 V/W
Resistance	65 kΩ
Time Constant	7 ms
Element Area	0.0325 mm²

Fig. 15. Performance summary and comparison with prior works.

Instrumentation Chopper Amplifier	This work		[7]	[8]
	Idle	Active		
Noise (RTI) (nV/√Hz)	130	31	60	59
Gain (dB)	57.2-78.3		40	59
Chopping Frequency (kHz)	1		5	4,8,12**
Current (μA)	0.277	2.655	1.8	0.266
CMRR (dB)	>130		134	89
PSRR (dB)	>130		120	92
NEF	2.7	2.0	3.3	1.4
Area (mm²)	0.04		0.1	0.25
VDD (V)	1.4		1	1

Low Pass Filter	This work	[9]	[10]
Cut off Frequency Range (Hz)	10-150	1.5-15	250
THD (%) @InLPF (Vpp)	0.55 @0.1	1 @1	0.4 @0.1
HD3 (dB) @InLPF=0.1Vpp	45.5	N/A	48.9
Current (μA)	0.14	550	0.45
Gain (dB)	-0.5	0	-10.5
Integrated Noise (RTI) (μVrms)	154***	320	340
Order	4	2	5
Area (mm²)	0.04	0.34	0.13
VDD (V)	1.4	3.3	1

ADC	This Work
Resolution (bit)	8
SNDR (dB)	48.8
Max INL (LSB)	0.12
Power (μW)	0.06
Sampling rate (kS/s)	1
DSP	
Power (μW)	5
Clock Frequency (kHz)	4
VDD (V)	0.7

System	This Work
Technology	CMOS 65nm
FPS	30
SNDR (dB)	48.8
Active Power (μW)	260
Idle Power (μW)	46
Active Power/ch (μW/ch)	4.06
Idle Power/ch (μW/ch)	0.72

*open loop gain
**multi chopper n=3
***measured at 30Hz BW setting

An Inductor-less Fractional-N Injection-Locked PLL with a Spur-and-Phase-Noise Filtering Technique

Alvin Li, Yue Chao, Xuan Chen, Liang Wu, Howard Luong

The Hong Kong University of Science and Technology, Clear Water Bay, Hong Kong

Abstract

Utilizing a novel phase-averaging filtering technique capable of wide-band spur-and-phase-noise suppression of up to 20dB, a 1.2-GHz inductor-less fractional-N injection-locked PLL achieves phase noise as low as -146dBc/Hz at 30MHz offset with 2MHz resolution allowing for inductor-less alternatives to LC-based PLLs in wireless applications. The 65nm CMOS prototype improves 10-MHz phase noise from -115 to -135dBc/Hz, injection spurs from -40.5dB to -57dB, and integrated jitter from 3.57ps to 1.48ps while occupying an area of $0.6mm^2$ and consuming 19.8mW from a 0.85V supply resulting in FoM and FoM_{Jitter} of -163dB and -223.6dB respectively.

I. Introduction

Ring-oscillator-based inductor-less frequency synthesizers offer much wider tuning range, smaller area, multiple output phases, and better process scalability compared to LC-based PLLs. However, their typically poor phase noise and integrated jitter performance still presents significant design challenges. Recent work on injection-locked PLLs (IL-PLLs) has demonstrated -20dB/decade suppression of close-in phase noise below the injection bandwidth (IL-BW) of up to $0.4f_{REF}$ through periodic edge retiming to reset the ring oscillator's accumulated phase error [1]. Despite this, ring-based IL-PLLs have still been limited to high-speed SoC clocking [2-3] due to their poor out-band spot phase noise. While edge retiming improves the close-in phase noise, it also degrades the RO's already poor high-frequency phase noise due to noise peaking >3dB around the IL-BW [1] and large spurs caused by both sub-harmonic injection and injection timing errors [4]. This work presents a phase-noise filtering (PNF) technique based on continuous phase-domain averaging to reduce both the high-frequency phase noise and spurs generated by injection-locked PLLs. The prototype PNF is demonstrated using a fractional-N IL-PLL based on a capacitive-coupled ring oscillator to achieve an overall phase noise and spur performance comparable to LC- PLLs.

II. Proposed Phase-Noise Filtering Technique

Figure 1 shows the proposed M-stage phase-noise filter composed of M cascaded phase averagers $(\phi_1+\phi_2)/2$, followed by a delay line of delay $\Delta\tau$ placed in a feedback loop. Intuitively, similar to digital domain IIR filters, input phase variations are continuously averaged such that zero-mean phase errors are filtered out. As derived and plotted in Fig. 1, the noise transfer function (NTF) of the input, $H_{PNF}(s)$, exhibits a sinusoidal response with maximum attenuation notches and no attenuation peaks occurring at odd and even multiples of $\omega=1/(2\Delta\tau)$, respectively. Therefore, the attenuation band can be adjusted by controlling the delay time $\Delta\tau$. By using more phase averagers, the input phase noise can be attenuated more, but the close-in phase noise of the delay line, $H_{DL(s)}$, would also be degraded further resulting in a trade-off between the input

and delay line noise contribution. Moreover, to increase the attenuation at peak offset frequencies and reduce the delay line noise contribution, an FIR response is created by combining time-shifted replicas of the PNF's output signal from multiple taps of the delay line.

III. Proposed Inductor-less Fractional-N IL-PLL

The detailed block diagram of the proposed inductor-less IL-PLL with the proposed phase-noise filter is shown in Figure 2. The fractional-n IL-PLL is based on [5] where the FSM creates a sequence of windowing signals determined by the FCW to direct the injection edges to a specific ring oscillator delay cell each reference cycle. To reduce the injection spurs and increase frequency resolution, the VCO utilizes a passive large-geometry MIM capacitor ring to couple three identical 8-phase ROs together as shown in Fig. 3. Phase matching between delay cells is improved since any mismatch in delay cells will propagate through the capacitive ring and be distributed among all output phases. Coarse frequency alignment is achieved through the fractional-N PLL while a variable delay line controls fine injection timing.

The phase noise filter uses two differential phase domain averagers, a single-ended delay line, and a single-to-differential converter (SDC). Each averager (Fig. 4) is composed of a limiter to eliminate amplitude information followed by a slewing stage and a phase interpolator for linear phase averaging of input signals less than ±90 degrees. A single-ended delay line is utilized to save power in the otherwise differential PNF. Furthermore, the delay cells are designed with long channel length and body biasing to enable a long delay time of up to 50ns under low V_{DD} of 0.85V while also minimizing flicker noise contributions.

To recover a differential signal for the PNF and to reject duty cycle errors, a balanced SDC is required. Conventional SDCs employing pseudo-differential pairs work well for small input signals, but they incur large DC operating point errors between the two output nodes when driven by large rail-to-rail signals from the delay line's output. Therefore, a large signal balanced SDC shown in Fig. 5 is proposed. Taking advantage of the highly uniform delay $\Delta\phi$ between adjacent inverters, three single-ended signals from the delay line can be interpolated together to create an accurate differential output. This is achieved by interpolating v_1 and v_3 together to create v_{o+} while interpolating v_2 with itself creates v_{o-} as illustrated in the phasor diagram.

The overall filtering response applied to a ring-based 1.2-GHz IL-PLL is shown in Fig 6. With M=2 and $\Delta\tau$=50ns, the PNF provides up to 20dB attenuation starting from 1MHz offset without significantly affecting the close-in phase noise.

IV. Measurement Results

The proposed IL-PLL is fabricated in a 65nm CMOS process and occupies a core area of 0.80mm × 0.75mm. The die micrograph is shown in Fig. 7. Using a 48MHz reference,

978-1-5090-0636-6/16 $31.00 © 2016 IEEE

the ILPLL achieves a fractional resolution of 2MHz. Figure 8 shows the measured output spectrum and phase noise performance at 1.152GHz with and without the PNF. The 2-stage PNF's attenuation can be observed starting at 1MHz offset and increasing up to 20dB at 10MHz offset, which closely matches the theoretical attenuation. Without the PNF, the IL-PLL achieves a phase noise of -98.4dBc/Hz @ 100 kHz and -115dBc/Hz @10MHz offset with an injection spur of -40.5dBc at 48MHz. With the PNF, the close-in 100-kHz phase noise remains at -98.4dBc/Hz while the 10-MHz phase noise and spur are reduced to -134.8dBc/Hz and -57dBc respectively. The integrated jitter from 100kHz to 100MHz, including reference spur, is reduced from 3.57ps to 1.48ps while consuming 19.8mW resulting in an FoM and FoM$_{Jitter}$ of -163dB and -223.6dB respectively. The degraded in-band phase noise compared to Fig. 6 is due to only partial rather than complete edge retiming caused by an insufficiently large injection device size and shared common current tail similar to [1] resulting in a narrower IL-BW and degrading the FoM$_J$. Table 1 summarizes the performance compared to other recent inductor-less injection-locked PLLs. While maintaining the close-in noise suppression from injection locking, the proposed inductor-less synthesizer significantly attenuates the far-out spur and phase noise to levels as low as -145dBc @ 30MHz offset which is comparable to LC-based synthesizers.

Acknowledgement

This work is supported by the Hong Kong General Research Funding (16206614).

References

[1] S. Ye, L. Jansson, and I. Galton, "A multiple-crystal interface PLL with VCO realignment to reduce phase noise," IEEE J. Solid-State Circuits, vol. 37, no. 12, pp. 1795–1803, Dec. 2002.

[2] J. Chien, et. al., "A Pulse-Position-Modulation Phase-Noise-Reduction Technique for a 2-to-16GHz Injection-Locked Ring Oscillator in 20nm CMOS," ISSCC Dig. Tech. Papers, pp. 52-54, Feb. 2014.

[3] W. Deng, et. al., "A 0.048mm² 3mW Synthesizable Fractional-N PLL with a Soft Injection-Locking Technique," ISSCC Dig. Tech. Papers, pp 252-254, Feb. 2015.

[4] J. Lee, et. al. "Study of subharmonically injection-locked PLLs," IEEE J. Solid-State Circuits, vol. 44, no. 5, pp. 1539–1553, May 2002

[5] P. Park, et. al., "An All-Digital Clock Generator Using Fractionally Injection-Locked Oscillator in 65nm CMOS," ISSCC Dig. Tech. Papers, pp 336-337, Feb. 2012.

[6] G. Marucci, et. al., "A 1.7GHz MDLL-Based Fractional-N Frequency Synthesizer with 1.4ps RMS Integrated Jitter and 3mW Power Using a 1b TDC", ISSCC Dig. Tech. Papers, pp. 360-362, Feb. 2014.

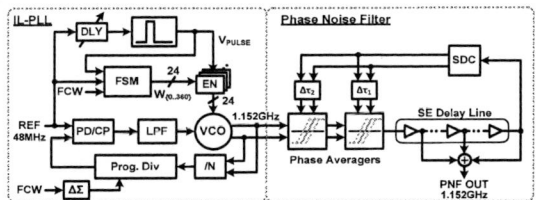

Fig. 2: System block diagram of IL-PLL with phase noise filter

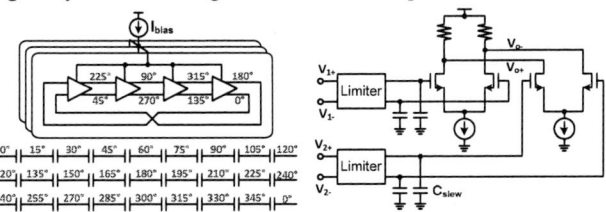

Fig. 3: VCO schematic composed of three capacitive-coupled ROs

Fig. 4: Schematic of phase averager

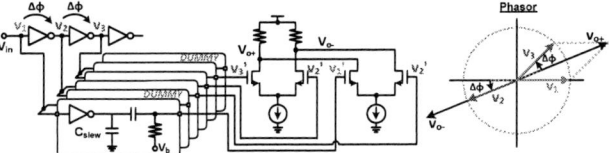

Fig. 5: Proposed large-signal single-to-differential converter

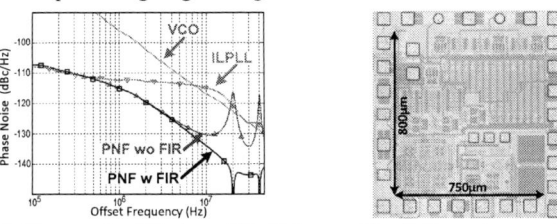

Fig. 6: Phase noise filtering of IL-PLL   Fig. 7: Die micrograph

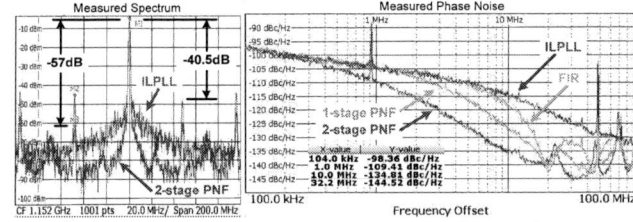

Fig. 8: Measurements spectrum and phase noise performance of ILPLL with and without PNF.

Table 1: Performance summary and comparison of inductor-less PLLs

	This Work	Liang ISSCC 2011	Park ISSCC 2012	Deng ISSCC 2015	Marucci ISSCC 2014
CMOS Process	65nm	55nm	65nm	65nm	65nm
Vdd (V)	0.85	1.2 / 3.3	n/a	0.8	n/a
Freq. (GHz)	1.152	0.216	0.580	1.522	1.651
f$_{REF}$ (MHz)	48	27	32	380	50
f$_{RES}$ (MHz)	2	27	1	∞	0.19 kHz
N	24	8	18.125	4	32
Power (mW)	19.8	6.9	10.5	3	3
Jitter (ps)	1.48 100k-100M	2.4 1k-40M	4.23 100-40M	3.6 100k-100M	1.395 30k-30M
$\mathcal{L}\{\Delta\omega\}$ (dBc) — 1MHz	-109.4	-122.78	-113.5	-115	-116.69
$\mathcal{L}\{\Delta\omega\}$ (dBc) — 10MHz	-134.81	-128.03	-133	-128	-114.96
$\mathcal{L}\{\Delta\omega\}$ (dBc) — 30MHz	-144.52	n/a	n/a	-133	-117.65
FoM$_{Jitter}$	-223.62	-224	-217.26	-224.2	-232
FoM @10MHz	-163.06	-146.33	-158.057	-166.88	-154.54
Area (mm²)	0.6	0.03	0.158	0.048	0.4

FoM = $\mathcal{L}\{\Delta\omega\}$ - $20\log(\omega_0/\omega_m)$ + $10\log(P_{DC[mw]})$
FoM$_{Jitter}$ = $20\log(\sigma)$ + $10\log(P_{DC[mw]})$

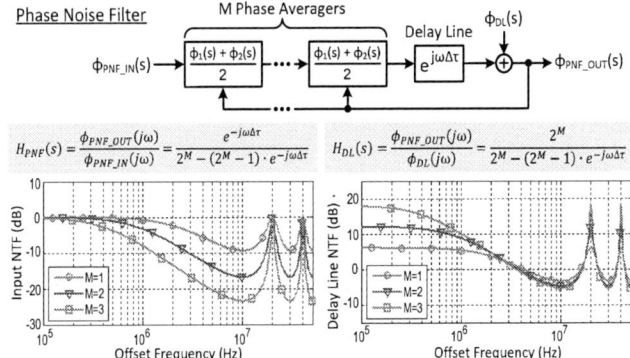

$$H_{PNF}(s) = \frac{\phi_{PNF\_OUT}(j\omega)}{\phi_{PNF\_IN}(j\omega)} = \frac{e^{-j\omega\Delta\tau}}{2^M - (2^M-1)\cdot e^{-j\omega\Delta\tau}}$$

$$H_{DL}(s) = \frac{\phi_{PNF\_OUT}(j\omega)}{\phi_{DL}(j\omega)} = \frac{2^M}{2^M - (2^M-1)\cdot e^{-j\omega\Delta\tau}}$$

Fig. 1: Proposed M-stage PNF and derived noise transfer functions.

An 8.865-GHz -244dB-FOM High-Frequency Piezoelectric Resonator-Based Cascaded Fractional-N PLL with Sub-ppb-Order Channel Adjusting Technique

Sho Ikeda[1], Hiroyuki Ito[1], Akifumi Kasamatsu[2], Yosuke Ishikawa[1], Takayoshi Obara[1],
Naoki Noguchi[1], Koji Kamisuki[1], Yao Jiyang[1], Shinsuke Hara[2], Dong Ruibing[2],
Shiro Dosho[1], Noboru Ishihara[1], and Kazuya Masu[1]

[1]Tokyo Institute of Technology, Yokohama, Japan,
[2]National Institute of Information and Communications Technology, Tokyo, Japan

Abstract

This paper proposes a high-frequency piezoelectric resonator (PZR)-based cascaded fractional-N PLL featuring channel adjusting technique with sub-ppb-order frequency resolution, which can overcome the difficulty using the narrow range GHz PZR. Moreover, undesirable oscillation induced by parasitic inductance of interconnects is suppressed by negative inductance technique. A power-efficient divider contributes to save power of the 2nd-PLL that suppresses output phase noise by the 1 GHz reference. The prototype PLL was fabricated in a 65nm CMOS and achieved 8.484GHz to 8.912GHz output, 180 fs rms-jitter, and -244 dB FOM while consuming 12.7mW.

Introduction

One of the significant bottlenecks for the modern wireless systems in reducing cost and footprint is existence of two bulky crystal devices with extremely high frequency-accuracy; a 32 kHz clock for a timer, and a MHz frequency reference for RF PLLs. The high-Q PZR in GHz-band is a potent alternative of these devices and can concurrently realize a very-low phase-noise synthesizer for RF application in addition to its small-factor. However, there are still some fundamental issues which were not solved in previous works [1-3]; (i) narrow tuning range and large process variation (P-variation) of the PZR-based oscillator, and (ii) undesirable harmonic oscillation caused by parasitic inductance of the bonding and on-board wires between CMOS and the PZR. (iii) Large power of high-frequency loop components in the RF PLL is also the unavoidable challenge. This paper proposes the 1-GHz high-Q PZR-based cascaded PLL as a GHz-order frequency reference, featuring system and circuit level approaches. (i) The channel adjusting technique (CAT) allows the PZR-oscillator to have narrow tuning range and large P-variation, and provides fine output-frequency resolution. (ii) Harmonic oscillation suppression technique (HOST) cancels parasitic inductance and enables to connect CMOS-chip and the PZR through mm-long wiring. (iii) A power-efficient divider realizes both low power and stable operation of the low phase noise 2nd PLL referring high frequency signal output from the 1st PLL.

Proposed Architecture

Fig. 1 shows concepts of the conventional and proposed timing devices. Proposed structure employs two schemes for achieving both sub-ppb-order frequency resolution and low power operation; (i) a high-Q GHz oscillator in a 1st PLL locking on to 32 kHz, and (ii) the CAT, i.e. digital frequency channel calibration, of cascaded fractional-N PLLs to compensate the P-variation of the high-Q oscillator, which allows the high-Q oscillator to have narrow tuning range. The proposed approach has two advantages. The first is that only the 32 kHz clock requires high voltage and temperature stability, while the conventional one in Fig. 1 (a) needs both 32 kHz and MHz references to have high stability. The second is that the bit-width of the DSM clocked by a GHz signal in the 2nd PLL can be reduced, which can significantly reduce power consumption, while fine frequency resolution can concurrently be achieved. Moreover, to achieve low output phase noise, bandwidth of the 1st PLL is narrowed to prevent 32 kHz signal noise from outputting. In this scheme, the 32 kHz signal is used as a *frequency* reference, and a GHz signal of the piezoelectric-based oscillator is used as a *phase* reference.

Fig. 2 details a block diagram of the proposed PLL. In this work, a 915MHz SAW resonator is used for high-Q PZR. Thanks to high reference frequency of the 2nd PLL, i.e. output of the 1st PLL, there is no need to use a trendy high performance ADPLL; a classical analog PLL can be applied to achieve low phase noise. Occupied die area of a loop filter is not critical in our architecture because bandwidth of the 2nd PLL can be wide due to high reference frequency. The DSM in the 1st PLL has 20-bit width, therefore, frequency resolution of the 1st PLL is 32 kHz * 2^{-20}, which is under 1-ppb resolution in 915MHz output. Note that if a non-cascaded PLL with the same resolution and 915-MHz-reference were used, the DSM requires 35-bit widths, which causes significantly large power consumption.

Fig. 3 shows a timing-chart of the proposed CAT. At first, all registers are initialized, and frequency of the PZR-DCO is set to the center of its tuning range (f_{dco_center}). Next, the channel of the 1st PLL is temporarily decided by the channel adjuster as $N_{1st}=N_{1stCAL}=round\ (f_{dco\_center}/f_{32k})$. Then, requirement of the DCO tuning range can be reduced. In the 3rd step, the channel of the 2nd PLL is decided by using N_{1stCAL}. In the 4th step, N_{1st} is re-calculated by using the result of the 3rd step so that the rounding-off error in the 2nd and 3rd step can be compensated in the 5th step. This is matched to the fact that frequency resolution is determined by the 1st PLL.

Fig. 4 (a) shows detailed schematic of the differential pierce PZR-DCO. Since the proposed PLL allows narrow tuning range of the DCO, only fine-tuning bank is equipped, which can reduce parasitic capacitance. The HOST exploits negative-inductance composed by inductors L_S and negative-g_m of the cross-couple transistors [4]. Fig. 4 (b) shows a proposed latch used in a pre-scaler for low power and high speed division. Added NMOS M1 and M2 adaptively weaken the PMOS latch strength, and accelerate the transition from high (low) to low (high) of the outputs. The size of M1 and M2 is smaller than other transistors so that latch function of the PMOS is not disturbed. The proposed latch can achieve 34 % power reduction at 9 GHz input in post-layout simulation.

Measurement Results

The prototype is fabricated in 65nm CMOS and occupies 1.77mm^2 including IO circuits and pads as shown in Fig. 5 (a). Although the DCO and the resonator are connected through bonding wires, on-board wires and vias, no harmonic oscillation is occurred thanks to the HOST. Fig. 5 (b) shows measured spectrum at 8.865GHz. The reference spur were -59 dBc and -60 dBc. Fig. 6 shows measured phase noise, and the RMS-jitter of the 2nd PLL from 10 kHz to 40 MHz is 180 fs. Total power consumption is 12.7 mW. Summary and comparison is given in Fig. 7 and table I. The proposed PLL achieved FOM of -243.9 dB which is one of the best FOMs among the fractional-N PLLs.

Acknowledgements

This work was partly supported by STARC, KAKENHI, and VDEC in collaboration with Agilent Technologies Japan, Ltd., Cadence Design Systems, Inc., and Mentor Graphics, Inc.

References

[1] M. H. Perrott, et al., "A temperature-to-digital converter for a MEMS-based programmable oscillator with < ±0.5-ppm frequency stability and < 1-ps integrated jitter", *IEEE J. Solid-State Circuits*, vol. 48, no. 1, pp. 276-291, Jan. 2013.

[2] S. Zaliasl, et al., "3 ppm 1.5 × 0.8 mm2 1.0 µA 32.768 kHz

MEMS-Based Oscillator", *IEEE J. Solid-State Circuits*, vol. 50, no. 1, pp. 276-291, Jan. 2015.

[3] K. Sankaragomathi, J. Kool, R. Ruby, B. Otis, "A ±3 ppm, 1.1mW FBAR frequency reference with 750 MHz output and 750 mV supply," *ISSCC Dig. Tech Papers*, pp. 454-455, 2015.

[4] H. Ito, et al., "A 1.7-GHz 1.5-mW digitally-controlled FBAR oscillator with 0.03-ppb resolution", *ESSCIRC*, pp. 98-101, 2008.

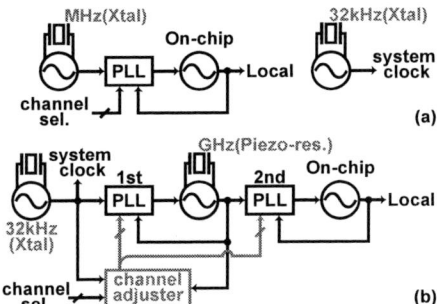

(a)

(b)

Fig. 1: (a) Conventional and (b) proposed configuration.

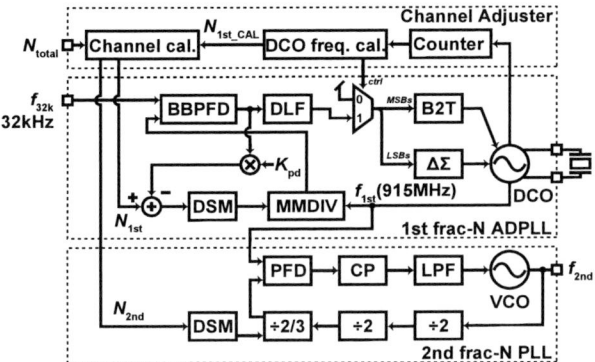

Fig. 2: A block diagram of the proposed PLL.

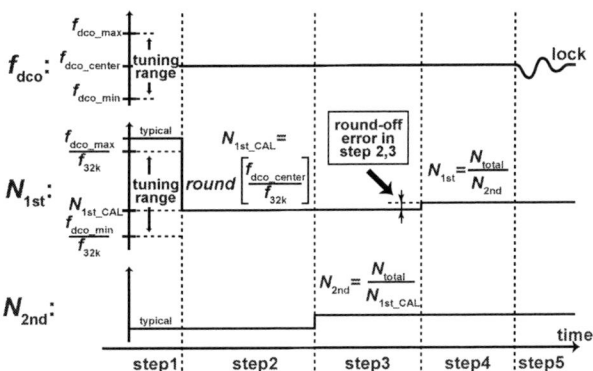

Fig. 3: Timing-chart of the channel adjusting technique (CAT).

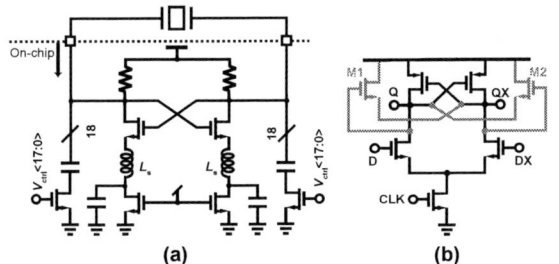

Fig. 4: Schematic of (a) a PZR-DCO, and (b) a latch in a prescaler.

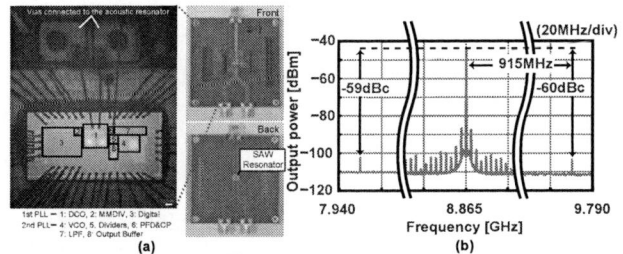

(a) **(b)**

Fig. 5: (a) A chip micrograph and a testing board. (b) Measured spectrum at 8.865GHz.

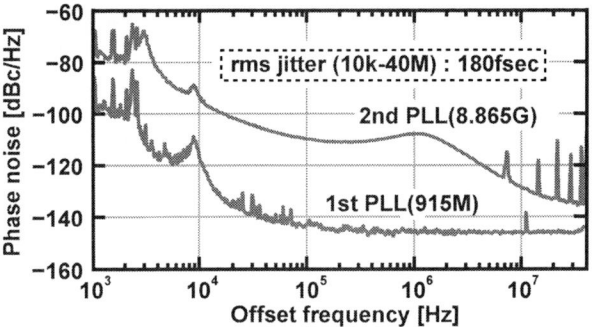

Fig. 6: Measured phase noise at 8.865GHz.

TABLE I: PERFORMANCE SUMMARY AND COMPARISON

	This work	X. Gao ISSCC'15	M. Lee JSSC'09	C.-M. Hsu JSSC'08	D. Tasca ISSCC'11	J. Borremans ISSCC'10
Tech [nm]	65	28	90	130	65	40
Ref. [MHz]	0.032	40	25	50	40	N/A
Freq. [GHz]	**8.865**	5.825	1.68	3.67	4	7
Ref. spur [dBc]	**-59**	-70	-54	-65	-72	-56
Area [mm2]	**0.83**	0.35	2.25	0.95	0.22	0.28
Bandwidth [MHz]	**1.83**	0.7	0.4	0.5	0.312	0.5
RMS jitter [fsec]	**180**	174	495	300	560	560
Power [mW]	**12.7**	9.5	110	46.7	4.5	30
FoM [dB]	**-243.9**	-245.5	-225.7	-233.8	-238.3	-230.3

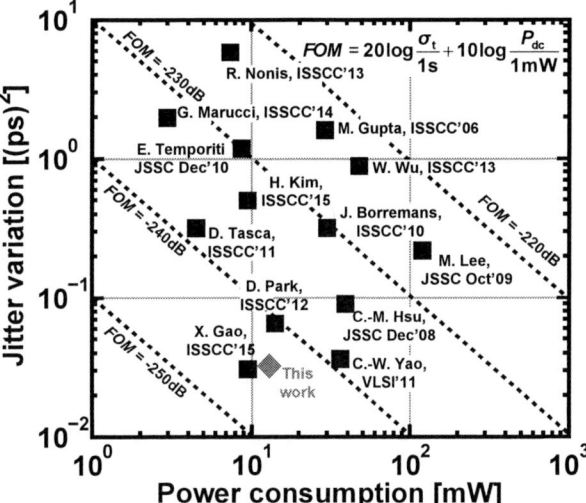

Fig. 7: Performance comparison of fractional-N PLLs.

978-1-5090-0636-6/16 $31.00 © 2016 IEEE

A 2.4-GHz 6.4-mW Fractional-N Inductorless RF Synthesizer

Long Kong and Behzad Razavi

University of California, Los Angeles, CA 90095, USA
longkong@ucla.edu, razavi@ee.ucla.edu

Abstract — A ring-oscillator-based cascaded synthesizer architecture incorporates a digital synchronous delay line and an analog noise trap to suppress the quantization noise of the $\Sigma\Delta$ modulator. Realized in 45-nm digital CMOS technology, the synthesizer exhibits an in-band phase noise of -109 dBc/Hz and an integrated jitter of 1.68 ps$_{rms}$.

It has been demonstrated that a wideband type-I integer-N phase-locked loop (PLL) architecture can achieve a bandwidth close to $f_{REF}/2$, thereby suppressing the phase noise of ring oscillators to levels commensurate with 2.4-GHz wireless standards while drawing moderate power [1]. The useful attributes of ring oscillators, such as a wide tuning range and more compact design, motivate us to extend this concept to fractional-N operation as well. However, we face the basic trade-off between the loop bandwidth and the $\Sigma\Delta$ modulator quantization noise contribution, an issue that has severely limited the former even in the presence of various noise cancellation techniques. For example, the design in [2] provides a bandwidth of about $f_{REF}/13$, which does not adequately reduce the oscillator phase noise if f_{REF} is around 20 MHz.

Synthesizer Design Rather than deal with fractional-N issues such as charge pump nonlinearity, DAC gain error and nonlinearity, etc., one can contemplate inserting a noise filter immediately after the feedback divider [Fig. 1(a)]. If the filter sufficiently attenuates the phase noise peaking below and above f_{REF}, then the loop bandwidth can be widened. The design therefore becomes nearly as simple as that of an integer-N synthesizer.

The noise filter must (a) exhibit a band-pass response precisely centered at f_{REF}, with a steep roll-off before reaching the peaks of phase noise, (b) contribute negligible noise and, (c) be linear enough to avoid folding the quantization noise peaks. These issues discourage the use of an analog implementation. More fundamentally, we must also recognize that a narrowband noise filter can degrade the loop stability, thus posing its own limitations on the synthesizer bandwidth.

We propose a digital solution that resolves all of the above issues. Illustrated in Fig. 1(b), the idea is to delay the divider output phase, ϕ_1, to obtain ϕ_2 and then add ϕ_2 to ϕ_1. If the delay is long enough to invert the phase noise components of interest, then $\phi_1 + \phi_2$ contains less noise. More accurately, the filter transfer function is equal to $1 + \exp(-T_D s)$, where T_D denotes the delay, exhibiting notches at $f = (2n+1)/(2T_D)$ for $n = 0, 1, \cdots$. Since these operations occur in the phase domain, the voltage-domain transfer function is centered at f_{REF}, as required.

Two aspects of the proposed solution merit remarks. First, while asynchronous delay lines suffer from trade-offs between

the delay value, phase noise, and power consumption, their synchronous counterparts do to a much lesser extent. Our delay line employs 24 static flipflops that are clocked by the VCO output, generating negligible phase noise and drawing 300 μW at 2.4 GHz. Second, unlike the simple noise filter shown in the top part of Fig. 1, the proposed technique does not affect the loop stability. This can be seen by assuming a phase step at the VCO output and noting that this step directly reaches the end of the delay line by clocking the last flipflop. In other words, only the quantization noise—and not the desired signal—experiences the notch.

The delay-line-based filter entails two issues. First, with only 24 flipflops, the first notch appears at 50 MHz, failing to suppress the quantization noise if $f_{REF} \approx 20$ MHz. Second, the filter frequency response periodically rises to a peak value of 2 between the notches, causing noise peaking at the synthesizer output. Shown in Fig. 2, the overall synthesizer architecture resolves both issues. An integer-N PLL based on the work in [1] multiplies f_{REF} by about a factor of 50, delivering a 1-GHz signal to the fractional-N loop. Thus, the latter can be designed for a wide loop bandwidth (≈ 12 MHz), with its $\Sigma\Delta$ modulator running at 1 GHz and producing phase noise peaks at 500-MHz offset. The notches created by the delay-line-based filter therefore suppress the noise as it begins to rise with frequency. (We should point out that the design in [3] also employs cascaded PLLs but with an LC-VCO and a fractional-N loop bandwidth of about 1/800 times its input frequency.)

In order to suppress the quantization noise peaking between the first two notches (at 50 MHz and 150 MHz), we introduce on the VCO control line in Fig. 2 a "noise trap" circuit. The trap employs a G_m-based integrator that is loaded by a gyrator. This combination presents an imaginary zero—and hence a notch—at 100 MHz, and a real pole at 120 MHz, ensuring that $|Z_T|$ remains low beyond this frequency. Since V_{cont} assumes a wide range, the integrator consists of complementary differential pairs, G_{m1} and G_{m2}. The bias voltage, V_b, is generated using a scaled, heavily-filtered replica of the main path.

Both PLLs in Fig. 2 employ three-stage ring oscillators with varactor tuning for continuous control and capacitor banks for discrete control [1]. The 1-GHz and 2.4-GHz VCOs respectively consume 2.7 mW and 2.25 mW and have a phase noise of -130 dBc/Hz and -121 dBc/Hz at 10-MHz offset.

Experimental Results The cascaded synthesizer of Fig. 2 has been fabricated in TSMC's 45-nm digital CMOS technology. Shown in Fig. 3(a) is the die photograph, with an active area of about 300 μm $\times$ 100 μm. Plotted in Figure 3(b) are the measured output spectra before and after the delay-line-based filter and the noise trap are turned on. No injection-pulling has been observed between the two PLLs.

Figure 4 plots the measured output phase noise profile. The in-band plateau is at −109 dBc/Hz and the integrated jitter from 10 kHz to 50 MHz is equal to 1.68 ps$_{rms}$. Plotted in Fig. 5 is the magnitude of the fractional spurs as a function of the fractional frequency offset. Note that the spur levels satisfy both IEEE 802.11 a/g and Bluetooth blocking requirements. Table I summarizes the performance of our prototype and compares it to state-of-the-art synthesizers in the range of 1.9 GHz to 2.4 GHz. Compared to the ring-based fractional-N design in [5], the proposed architecture achieves 11 dB lower phase noise with 30% less power consumption.

Acknowledgments The authors thank the TSMC University Shuttle Program for chip fabrication.

References

[1] L. Kong, B. Razavi, *ISSCC Dig.*, pp. 450-451, Feb. 2015.
[2] T. K. Kao et al., *ISSCC Dig.*, pp. 416-417, Feb. 2013.
[3] D. Park, S. Cho, *IEEE JSSC*, pp. 2989-2998, Dec. 2012.
[4] P. C. Huang et al., *ISSCC Dig.*, pp. 362-363, Feb. 2014.
[5] C. F. Liang, P. Y. Wang, *ISSCC Dig.*, pp. 190-191, Feb. 2015.

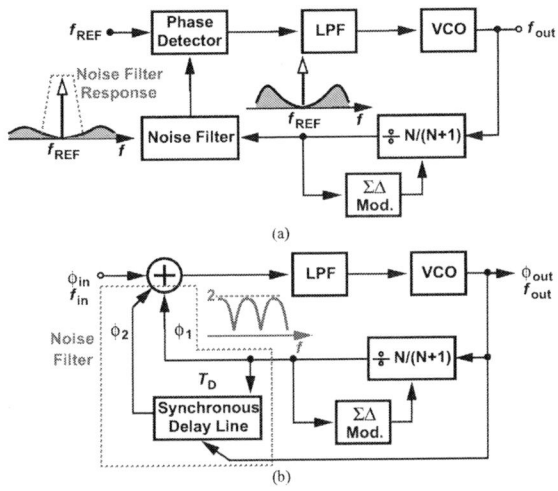

Fig. 1. (a) Conceptual synthesizer with noise filter, and (b) filter implementation using a delay line.

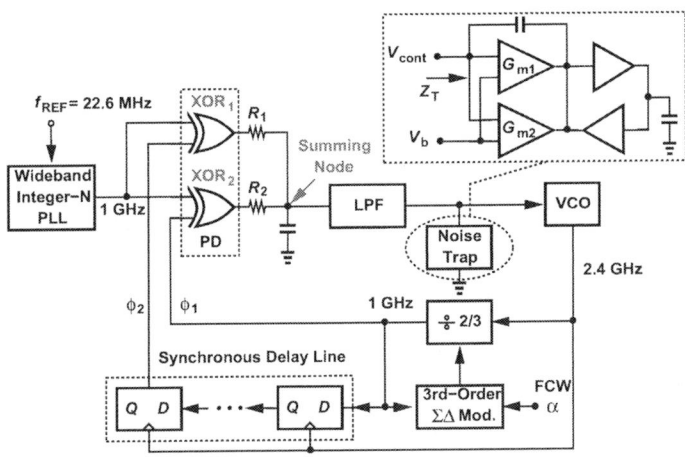

Fig. 2. Proposed synthesizer architecture.

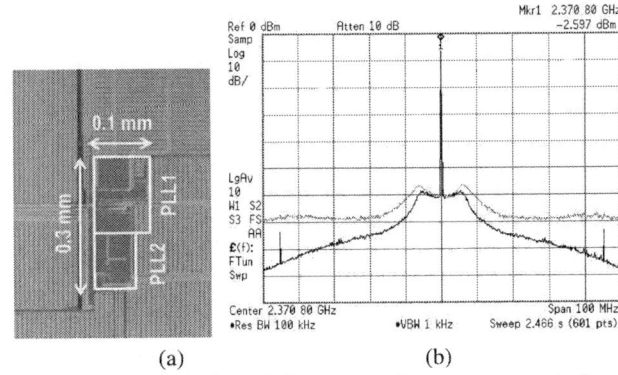

(a)   (b)

Fig. 3. (a) Die photograph, and (b) measured output spectra before (grey) and after (black) delay-line-based filter and noise trap are on.

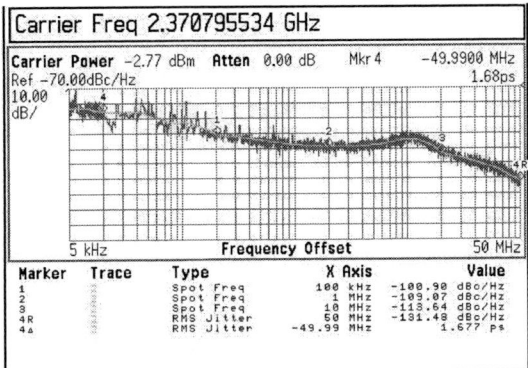

Fig. 4. Measured phase noise.

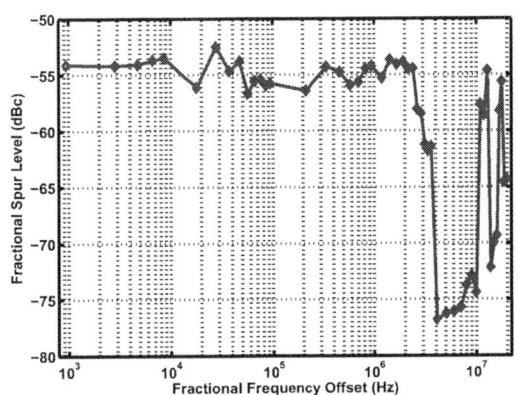

Fig. 5. Measured fractional spurs.

TABLE I. Performance summary.

	ISSCC'13 [2]	ISSCC'14 [4]	ISSCC'15 [5]	This Work
Oscillator Topology	Ring	LC	Ring	Ring
Reference Freq. (MHz)	26	48	26	22.6
Frequency Range (GHz)	1.87 ~ 1.98	2.2 ~ 2.4	2	2.3 ~ 2.6
Phase Noise @ 1MHz offset (dBc/Hz)	−98	−117	−98	−109
RMS Jitter (ps) Integ. range (MHz)	3.4 (0.004~2)	0.3 (0.01~30)	2.36 (0.001~40)	1.68 (0.01~50)
In-band Frac. Spur (dBc)	−50	−53	−70	−52.5
Ref. Spur (dBc)	−67	−55	−87	−70
Power (mW)	10	17.3	9.1	6.4
Area (mm²)	0.047	0.75	0.046	0.03
Tech. (nm)	40	180	40	45
FoM₁ (dB)	−219.4	−238	−223	−227.4
FoM₂ (dB)	153.9	171.9	154.4	168.4

$$FoM_1 = 10\log_{10}[(\frac{Jitter}{1\ s})^2(\frac{Power}{1\ mW})] \quad FoM_2 = 10\log_{10}[(\frac{f_{osc}}{\Delta f})^2(\frac{1\ mW}{Power})] - Phase\ Noise\ (dBc/Hz)$$

A PVT-Robust −59-dBc Reference Spur and 450-fs$_{RMS}$ Jitter Injection-Locked Clock Multiplier Using a Voltage-Domain Period-Calibrating Loop

Yongsun Lee, Heein Yoon, Mina Kim, and Jaehyouk Choi

Ulsan National Institute of Science and Technology (UNIST), Ulsan, Korea

yongsun@unist.ac.kr

Abstract

This paper presents a low-reference-spur and low-jitter injection-locked clock multiplier (ILCM). To secure these performances over PVT-variations, we propose the use of a voltage-domain period-calibrating loop (VDPCL) in the ILCM that monitors the intrinsic period of the VCO and stores this information as the charges in a capacitor. By evaluating the voltage of the capacitor, it is possible to correct the free-running frequency of the VCO. By iteratively accumulating charges, the precision of the calibration can be increased. The measured reference spur and RMS jitter were −59 dBc and 450 fs, respectively, and their degradations over the PVT were less than 1.5 dB and 50 fs, respectively.

Introduction

An ILCM is a promising solution to the need for the low-cost generation of a high-frequency clock signal with a low-jitter profile. In an ILCM, the phase of the VCO output can be periodically realigned by injected reference-pulses. Thus, an ILCM can reduce phase noise (or nondeterministic jitter) dramatically if the free-running frequency of the VCO, f_{VCO}, can be regulated to be close to the target frequency, $N\cdot f_{REF}$, where N is the target harmonic index and f_{REF} is the frequency of the reference clock. Nevertheless, ILCMs still suffer from a large reference spur, caused by the periodic phase-shift of the VCO when it is injection-locked, as shown in Fig. 1(a). This is because the increase in the reference spur is extremely sharp with respect to the deviation of f_{VCO} from $N\cdot f_{REF}$, f_D, i.e., $\text{Spur}_{dBc} \approx 20\log(N\cdot|\alpha|)$ where $\alpha = f_D/(N\cdot f_{REF})$ [1]. As shown in Fig. 1(b), when N is 8, as α increases to just 1%, the spur-level soars by 40 dB and reaches −22 dBc. Therefore, if an ILCM is intended to minimize the reference spur as well as phase noise, it must be equipped with a very precise frequency calibrator. To detect the frequency deviation, the calibrators in [2] and [3] used a replica-VCO and a delay-locked loop having replica cells, respectively. However, even slight mismatches between the delay cells can reduce the accuracy of the calibration, and result in large reference spurs. The calibrators used in [1] and [4] measured instantaneous phase shifts due to the f_D in the time domain. In [4], a pulse-gating technique and a bang-bang PD (BBPD) were used to detect phase shifts, but the minimum spur-level was limited by the resolution of the BBPD. To enhance the calibration resolution, [1] used a GRO-based TDC, but this consumed a large power and in-band noise was high.

This paper presents a voltage-domain period-calibrating loop (VDPCL) that can calibrate the VCO's frequency with high precision. The proposed VDPCL continuously converts the difference between the intrinsic period of the VCO, T_{VCO}, and the target period, T_{REF}/N, to corresponding amounts of charges. Using an iterative process, the voltage of the capacitor deviates significantly from the initial value, even if f_D is very small. By evaluating the voltage of the capacitor using a simple comparator, f_{VCO} can be corrected to near the target frequency. Thus, with high calibration precision, the ILCM can maintain very low levels of reference spur and jitter over PVT.

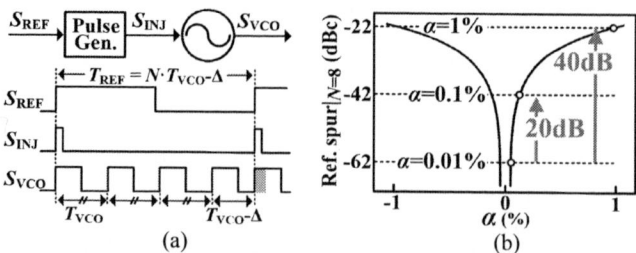

Fig. 1. (a) Phase realignment by injection (b) reference spur over α's

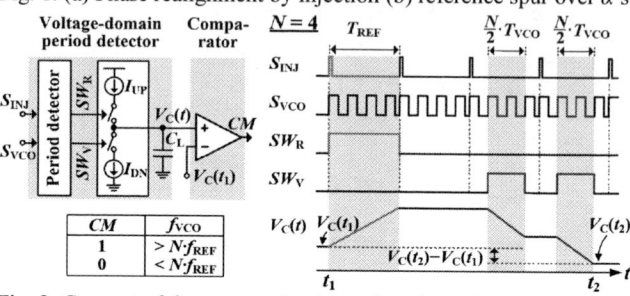

Fig. 2. Concept of the proposed voltage-domain period calibrator

Concept of the Voltage-Domain Period-Calibration

As shown in Fig. 1(a), in the interval where a reference pulse is not injected, the rising edges of the VCO signal preserve the information of the VCO's intrinsic period, T_{VCO}. Thus, by comparing N-times T_{VCO} to the period of the reference clock, T_{REF}, the frequency deviation of the VCO can be detected. Fig. 2 shows the conceptual diagram and operation of the proposed VDPCL, when N is 4. First, the period detector provides the CP with a stream of pulses, SW_V, where the total pulse-width equals $N\cdot T_{VCO}$, since it is not affected by the injection. Concurrently, it provides the pulse of SW_R, where the pulse-width is T_{REF}. According to SW_R or SW_V, the CP charges or discharges the loop capacitor, C_L, in which its voltage, $V_C(t)$, represents $T_{REF} - N\cdot T_{VCO}$. By iterating this process, the voltage difference, $V_C(t_2) - V_C(t_1)$, corresponding to a particular f_D is extended. Finally, using the extended voltage difference, the comparator can accurately determine the relationship between f_{VCO} and $N\cdot f_{REF}$, even if the difference between them is small.

Implementation of the ILCM with the Voltage-Domain Period-Calibrating Loop

An essential precondition for the conceptual operation of the VDPCL in Fig. 2 is that the I_{UP} and I_{DN} of the CP be identical. However, depending on $V_C(t)$, the channel-length modulation may cause a mismatch between I_{UP} and I_{DN}, which reduces the accuracy of the decision. For instance, when f_{VCO} is higher than $N\cdot f_{REF}$, $V_C(t_2)$ must be higher than $V_C(t_1)$ as well. However, in the presence of a current mismatch, $V_C(t_2)$ could be lower than $V_C(t_1)$, which would then lead to a wrong decision. In this work, we resolve this issue by implementing the VDPCL as a differential architecture, as shown in Fig. 3. As shown in Fig. 4(a), C_1 is charged by SW_{R1} (I_{UP}) during T_{REF}, and discharged by SW_V (I_{DN}) during $N\cdot T_{VCO}$. On the other hand, C_2 is charged by SW_V (I_{UP}) during $N\cdot T_{VCO}$, and discharged by SW_{R2} (I_{DN}) during

978-1-5090-0636-6/16 $31.00 © 2016 IEEE

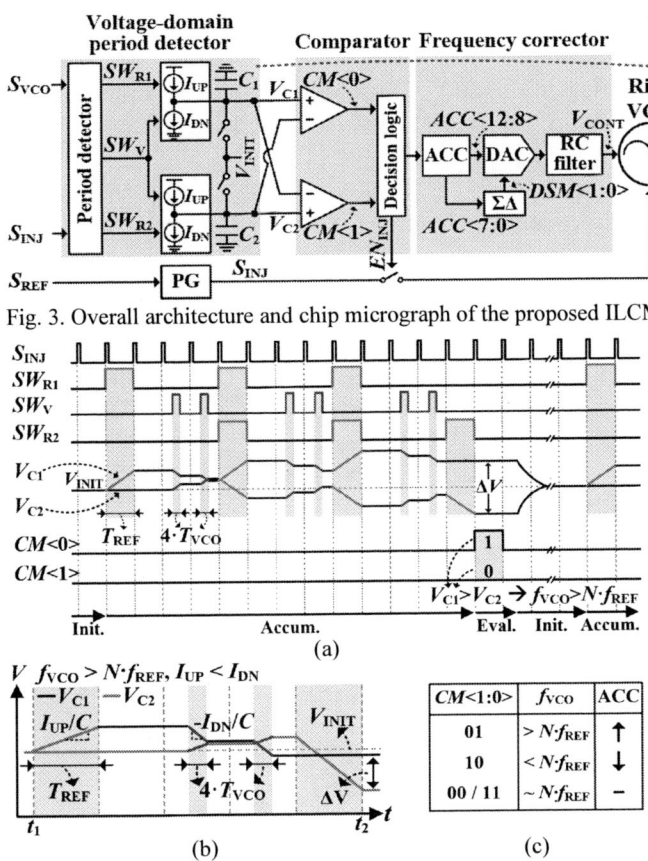

Fig. 3. Overall architecture and chip micrograph of the proposed ILCM

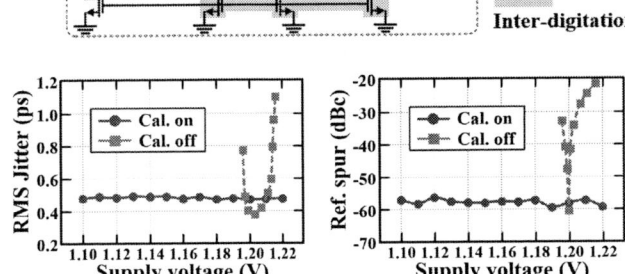

Fig. 4. (a) Timing diagram (b) case with a mismatch in CP's current (c) table for the decision logic

CM<1:0>	f_{VCO}	ACC
01	$> N \cdot f_{REF}$	↑
10	$< N \cdot f_{REF}$	↓
00 / 11	$\sim N \cdot f_{REF}$	–

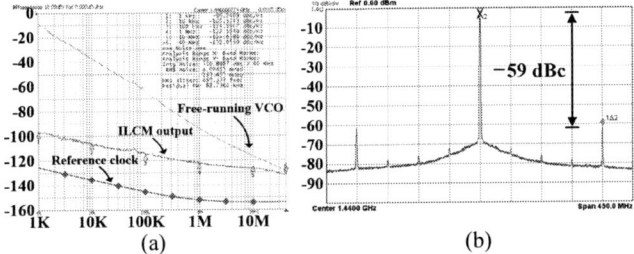

Fig. 5. (a) Measured phase noise (b) measured spectrum

T_{REF}. Finally, by differentially comparing the voltages of C_1 and C_2, V_{C1} and V_{C2}, respectively, the comparator can make a decision. When V_{C1} and V_{C2} are compared, the effect of the mismatch between I_{UP} and I_{DN} is canceled out in $|V_{C1}(t_2) - V_{C2}(t_2)|$; thus, the decision is accurate regardless of the mismatch in the CP's current. Fig. 4(b) shows a case wherein I_{DN} is larger than I_{UP}. In this case, both V_{C1} and V_{C2} finally become less than V_{INIT} at t_2, but the subsequent comparator makes a correct decision since V_{C1} is still much higher than V_{C2}. The CPs and the loop capacitors, C_1 and C_2, are designed symmetrically, as shown in Fig. 3. To minimize potential local mismatches, we used the inter-digitation technique in the layouts of the CPs and the capacitors. As shown in Fig. 4(c), when the comparator outputs $CM<1:0>$ of '01', the code of the frequency corrector, ACC, increases to lower f_{VCO}. In contrast, when $CM<1:0>$ is '10' ACC decreases. When the comparator fails in the decision due to limitations of resolution, $CM<1:0>$ becomes '00' or '11', and ACC is not changed. The DSM is used to enhance the resolution of the DAC, and its noise is suppressed by the subsequent RC-filter.

Fig. 6. Variations over supply voltages: (a) jitter$_{RMS}$ (b) reference spur

Measurement Results

The proposed ILCM was fabricated in a 40-nm CMOS technology, and the active area was 0.061 mm^2 (Fig. 3). The total power consumption was 2.8 mW, while the VDPCL only consumed 1.7 mW. Fig. 5(a) shows the measured phase noise of a 1.44-GHz signal. When injection-locked, the VCO achieved a phase noise of -122.5 dBc/Hz at a 1-MHz offset and an integrated jitter of 450 fs. According to the noise analysis, noise contribution of the building blocks of the VDPCL including the DSM was less than 1%. Fig. 5(b) shows the spectrum of a 1.44-GHz output signal, where the level of the reference spur was -59 dBc. Low-level spurs at the multiples of $f_{REF}/4$ were by the substrate coupling of SW_{R1} and SW_{R2}. Fig. 6(a) shows the variations of the RMS jitter, when the supply voltage of the VCO changed from 1.10 to 1.22 V. When the VDPCL was off, the jitter level was degraded significantly, but when it was on, a low jitter level was maintained regardless of the supply voltages. In Fig. 6(b), the calibrator was also able to regulate the level of the reference spur, which was sharply degraded when it was off. Table I compares the performance of this work with state-of-the-art ILCMs with a real-time calibrator. The proposed ILCM achieved the excellent jitter performance and the lowest reference spur.

TABLE I. Performance Summary and Comparison

	This Work	ISSCC'13[2]	VLSI'15[3]	ISSCC'15[4]
Process	40 nm	65 nm	65 nm	65 nm
VCO Type	Ring	Ring	Ring	LC
Cal. Method	VDPCL	Replica-VCO	Replica-DLL	Pulse gating
f_{REF}	180 MHz	300 MHz	400 MHz	106.25 MHz
f_{OUT}	1.44 GHz	1.2 GHz	1.36 GHz	6.8 GHz
RMS Jitter (σ_t)	450 fs (1k–40 MHz)	700 fs (10k–40 MHz)	448 fs (1k–40 MHz)	190 fs (10k–100MHz)
Reference Spur	-59 dBc	-57 dBc	-39 dBc	-40 dBc
Power (P_{DC})	2.8 mW	0.97 mW	3.6 mW	2.25 mW
Active Area	0.061 mm^2	0.022 mm^2	0.041 mm^2	0.25 mm^2
FOM	-242.5 dB	-243.2 dB	-241.4 dB	-250.9 dB

*FOM: $10 \cdot \log_{10}((\sigma_t/1s)^2 \cdot (P_{DC}/1mW))$ (dB)

References

[1] B. M. Helal *et al.*, *JSSC*, vol. 44, no.5, pp. 1391–1400, May. 2009.
[2] W. Deng *et al.*, *ISSCC*, pp. 248–249, Feb. 2013.
[3] M. Kim *et al.*, *Symp. on VLSI Circ.*, pp. C142–C143, Jun. 2015.
[4] A. Elkholy *et al.*, *ISSCC*, pp. 188–189, Feb. 2015.

A 0.034mm², 725fs RMS Jitter, 1.8%/V Frequency-Pushing, 10.8-19.3GHz Transformer-Based Fractional-N All-Digital PLL in 10nm FinFET CMOS

Chao-Chieh Li[1], Tsung-Hsien Tsai[1], Min-Shueh Yuan[1], Chia-Chun Liao[1], Chih-Hsien Chang[1], Tien-Chien Huang[1], Hsien-Yuan Liao[1], Chung-Ting Lu[1], Hung-Yi Kuo[1], Kenny Hsieh[1], Mark Chen[1], Augusto Ximenes[2], Robert Bogdan Staszewski[2,3]

[1]TSMC, Hsinchu, Taiwan.   [2]Delft University of Technology, The Netherlands.   [3]University College Dublin, Ireland.
Email: jjliv@tsmc.com, R.B.Staszewski@tudelft.nl

Abstract

A tiny LC-tank-based ADPLL in 10nm FinFET CMOS achieves an area comparable to that of inverter-based ring-oscillator PLLs. A DCO occupying 0.016mm² uses a controllable multi-turn magnetic coupling transformer to extend its tuning range to 10.8–19.3GHz (56.5%). A diversity of fine-tune capacitor banks limits the max/min step-size ratio to 2.3x. A new metastability-resolution scheme allows to use the frequency reference (FREF) clock directly instead of a retimed FREF (CKR) of conventional ADPLLs. A low-complexity estimator calculates inverse of the TDC. The fractional phase jitter (725fs) reaches sub-ps for the first time among PLLs of <0.1mm². Frequency pushing is 1.8%/V, which is at least 50x better than in traditional ring-type PLLs.

Introduction

All-digital PLLs (ADPLL) are widely used in advanced CMOS, where they exploit the naturally fine conversion resolution of time-to-digital converters (TDC) and digitally controlled oscillators (DCO), thus further reducing area and power dissipation vis-à-vis analog PLLs [1-5]. Inverter-based ring oscillators (RO) have been the norm for general-purpose wide-tuning-range non-RF oscillator designs but suffer from poor phase noise (PN) and high supply frequency pushing [1,2], especially in advanced CMOS. An LC-based DCO can achieve better PN with much lower supply pushing, but it suffers from narrow tuning range (TR) and occupies large chip area due to the inductor. We propose an ADPLL with an LC-DCO based on a controllable-magnetic-coupling transformer that retains all the above advantages.

ADPLL Architecture and Circuit Design

Fig.1 shows the proposed DCO. The tank is composed of two independent pseudo-differential single-ended transformers, $T_{1,2}$, each with 3 windings. The transformer's *main* turns ratio (N=2) and coupling coefficient (k_m=0.75) provide a passive voltage gain from the drains to the gates of the transconductor pair $M_{1,2}$, thus boosting the loop gain by 50% and improving its start-up at a lower Q-factor [6]. To enhance the DCO tuning range towards higher frequencies, the tertiary winding has 2 turns for the strongest inversed magnetic field cancellation to lower the overall inductance and boost the frequency from 16.0GHz to 19.3GHz, a 20.6% enhancement. By using two different MOS switches with different turn-on resistances, the Q-factor can stay higher than the worst Q-factor at the lowest frequency.

An NMOS-only dc-coupled buffer (M_{3-6}) converts the dc level of the DCO core output (V_D, V_G) of V_{DD}=800mV to ~V_{DD}/2. The W/L ratio of M_{5-6} is 4x that of M_{3-4} to support a max swing of 400mV (BUF). Due to the full swing requirement of the following TSPC divider, a high-speed differential-to-single-ended buffer (M_{7-14}) is adopted. M_{11-14} regenerate BUF- to a square-like clock signal and set its dc level closer to V_{DD}/2. It is then combined with its differential counterpart BUF+ through M_{7-10} to produce a single-ended rail-to-rail output clock (FOUT). The point-symmetric transformer layout allows the switched-capacitor banks to efficiently fill the remaining

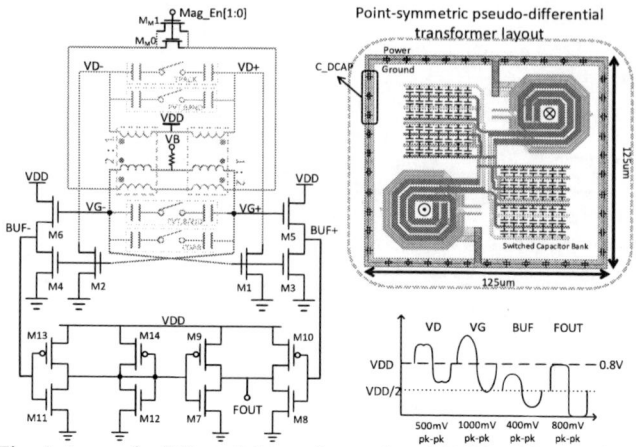

Fig. 1 A pseudo-differential transformer-based DCO with a tertiary winding to extend the tuning range.

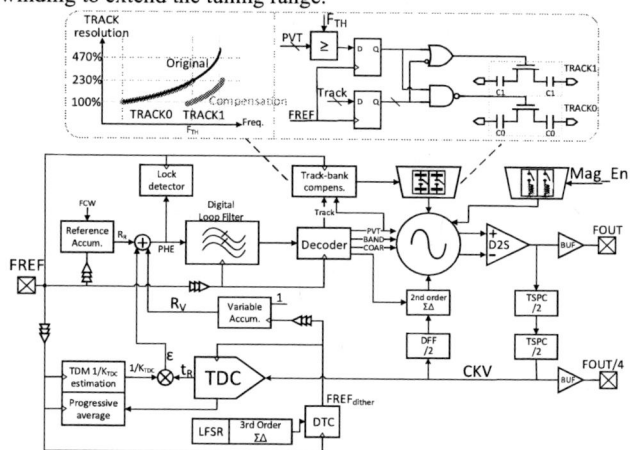

Fig. 2 Proposed ADPLL topology with compensation of tracking band resolution over tuning range.

50% area. Power ring with a distributed high-density MOM/MOS decoupling capacitor C_{DCAP} provides a good AC ground (<1Ω impedance) for both transformers.

The coarse PVT bank is a binary-weighted switched capacitor array split into the transformer's primary and secondary to achieve the maximum Q enhancement [6]. To improve the fine-tune resolution without degrading the total tank Q-factor, TRACK bank is connected to the primary coil to benefit from the capacitance transformation of $1/N^2$. The PVT bank provides large steps of 81MHz/LSB and dominates the DCO tuning range. The COAR bank and TRACK banks have a resolution of 31MHz/LSB and 2.5MHz/LSB, respectively. A time-averaged resolution of 78.1kHz is achieved by 5 fractional tuning bits undergoing a 2nd order $\Sigma\Delta$ dithering, feeding a 3-bit unit-weighted capacitor bank at the transformer's primary.

The LC-DCO fine step-size K_{TRACK} is a strongly non-linear function of the resonating frequency, f_{osc}, exacerbated by the attempt for a high tuning range (TR). K_{TRACK} increases *cubically* with f_{osc} and results in, e.g., 13.4dB quantization noise variation

978-1-5090-0636-6/16 $31.00 © 2016 IEEE   192   2016 Symposium on VLSI Circuits Digest of Technical Papers

corresponding to 470% K_{TRACK} change within the TR of 56.5%. To maintain the quantization noise at least 10dB below the DCO PN, the K_{TRACK} increase within the TR should be <230%. We employ 2 different fine-tune capacitor banks (TRACK0, TRACK1) to reduce the fine-tune max/min step-size ratio to 2.3x. The PVT tuning word is monitored as a proxy for the resonating frequency to select between these banks. Fig. 2 shows the proposed ADPLL. The DCO frequency is divided by 4 by two TSPC÷2 dividers to provide feedback variable clock CKV. In this technology, the TSPC divider can handle >20GHz clocks with only ~100uW. A dithered FREF$_{dither}$ clock is used for TDC to reduce reference spurs and minimize PN at near-integer FCW. A lock detector is constructed by monitoring the integer part of phase error.

Fig. 3 shows the proposed metastability resolution scheme that allows to use the frequency reference (FREF) clock directly instead of a retimed FREF (CKR) of conventional ADPLLs [5]. The FREF is resampled by two paths: A and B. Path A (B) uses DFFs clocked by rising (falling) CKV edges to provide clock when FREF is close to the falling (rising) CKV edge. An edge select mechanism judges the relationship between CKV and FREF. This way, the variable accumulator cross-domain clock issue is solved without any metastability.

Fig. 4 proposes a low-complexity estimator that calculates inverse of the TDC gain ($K_{TDC} = Tv / t_{inv}$) to be used as a multiplier of the TDC output of Fig. 2. A progressive-average calculator smoothens the TDC output from the quantization noise and is preferred over the moving-average method. The Newton–Raphson method is proposed here to solve the near integer FCW convergence problem while maintaining the finest resolution. This method guarantees absolute convergence taking max 3-4 iterations even in face of a large step input. The original method would also need to use 3 multipliers to calculate the reciprocal with high input word-length, thus making it area/power expensive. Instead, applying a time-division multiplexing multiplier reduces the hardware complexity from 3 multipliers to 1.

Measurement Results and Conclusion

The proposed LC-DCO based ADPLL is implemented in TSMC 10nm FinFET process. The free-running DCO PN at 15GHz is -120dBc/Hz @10MHz offset while consuming 9mW. The DCO TR is 10.8-19.3 GHz. The TDC and digital core consume 1.6mW and 1.3mW at 150MHz FREF, respectively. Figure 5 shows the integrated jitter is 725fs at fractional FCW= 2^{-1} (it is 669fs at integer FCW) but it increases when closer to integer-N FCW due to the fractional spurs of -35dBc entering the ADPLL's in-band. Table 1 summarizes the ADPLL performance and compares it state-of-the-art of small-area fractional-N PLLs. The FOM is -232 dBc/Hz with a tiny core size of 0.034mm². The frequency pushing is 1.8%/V, which is at least 50x better than (un-calibrated – all, except for [5]) traditional ring-type PLLs. This topology combines the best advantages of RO and LC oscillator types: the tiny area comparable to RO and jitter performance comparable to an LC-tank, without the need for a large LDO.

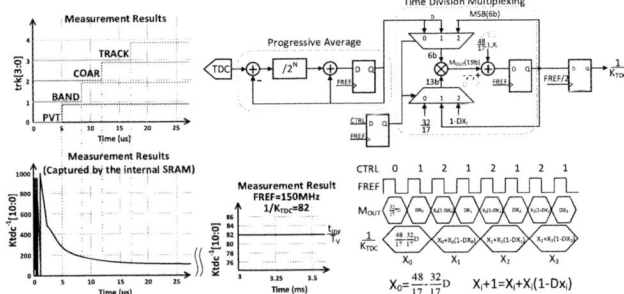

Fig. 4 Low complexity implementation of the $1/K_{TDC}$ estimator.

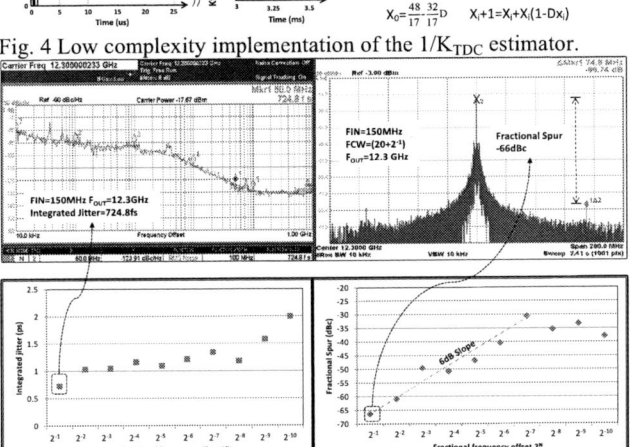

Fig. 5 Measured integrated jitter and fractional-N spurs.

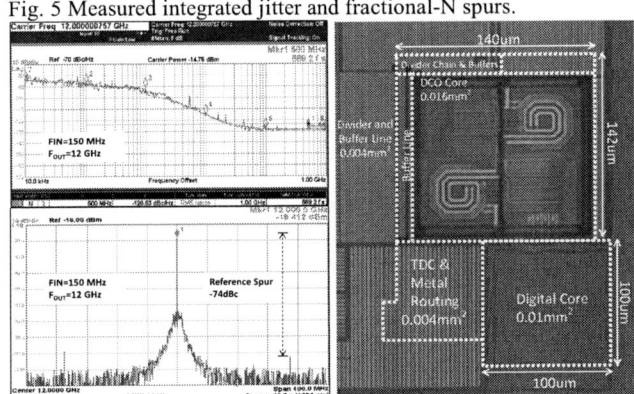

Fig. 6 Measured integer-N integrated jitter and reference spurs. Die photo of 10nm FinFET testchip. ADPLL core size is 0.034mm²

	This	Song [1]	Tsai [2]	Huang[3]	Lee[4]
		ISSCC'15	ISSCC'15	ISSCC'14	VLSI'14
Process	10nm	14nm	16nm	40nm	28nm
PLL Type	ADPLL TDC-based	ADPLL Bang-bang	ADPLL TDC-based	ADPLL Bang-bang	Analog PLL Charge-pump
Osc. Type	LC DCO	Ring DCO	Ring DCO	Ring DCO	LC VCO
Frequency (GHz)	10.8-19.3	0.032-2	0.1-4	2.4	2.7-7
Power Consumption(mW)	11.9	2.1	9.3	6.4	14
Supply Voltage (V)	0.8	1	0.75	1.1	1.8
Core Size (mm²)	0.034	0.009	0.029	0.013	0.07
Phase Jitter(ps)	0.725	18.8	1.4	3.3	1.1
Frequency Pushing (%/V)	1.8	--	233.3	0.9	--
FOM*(dB)	-232	-211	-227	-221	-228

Table.1 Performance comparison with to state-of-the-art of small-area (<0.1mm²) fractional-N PLLs.

$$*FOM = 10\log_{10}\left(\frac{\sigma_t^2}{1s}\right) + 10\log_{10}\left(\frac{P_{DC}}{1mW}\right)$$

References

[1] M. Song, et al., ISSCC, pp. 266-267, 2015
[2] T.H. Tsai, et al., ISSCC, pp. 260-261, 2015
[3] Y.C. Huang, et al., ISSCC, pp. 270-271, 2014
[4] C.H. Lee, et al., "VLSI Symp. Circ., pp.1–2, 2014
[5] R.B. Staszewski, et al., ISSCC, pp. 272-273, 2004
[6] M. Babaie and Staszewski, JSSC, pp. 679-692, Mar. 2015

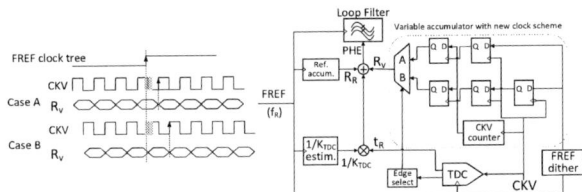

Fig. 3 New metastability resolution scheme.

250mV-950mV 1.1Tbps/W Double-Affine Mapped Sbox based Composite-Field SMS4 Encrypt/Decrypt Accelerator in 14nm Tri-gate CMOS

Sudhir Satpathy, Sanu Mathew, Vikram Suresh, Mark Anders, Himanshu Kaul, Amit Agarwal, Steven Hsu, Gregory Chen, Ram Krishnamurthy
Circuits Research Lab, Intel Corporation, Hillsboro, OR USA
sudhir.k.satpathy@intel.com

Abstract

A 10K-gate 4Gbps unified encrypt/decrypt SMS4 Chinese cryptographic accelerator is fabricated in 14nm tri-gate CMOS, operating at 1GHz, 750mV, 25^0C with total power consumption of 12mW. Double-affine mapped Sbox circuits enable inverse computation using $GF(2^4)^2$ data-path, resulting in 33% reduction in accelerator area by elimination of look-up tables (LUT). Optimal composite-field reduction polynomials, counter-assisted round constant generation circuits, and a hybrid data-path with in-line key-expansion provide additional 14% area saving over traditional designs resulting in a compact layout occupying $2445\mu m^2$. Low voltage optimizations enable robust sub-threshold operation down to 250mV, with peak energy-efficiency of 1.1Tbps/W measured at 330mV.

Introduction

SMS4 is a symmetric key block cipher that is mandated for use by the Chinese National Standard for Wireless Local Area Network Authentication and Privacy Infrastructure [1] with security characteristics equivalent to AES-128. The critical 8 Sboxes in the SMS4 round data-path constitute 60% of accelerator area and 55% critical path delay, and are typically implemented using LUT-based memory-elements [2-3]. However, the large area and power overhead of such sequential macros (Fig. 1) render these implementations unsuitable for area/energy-constrained platforms, limiting ultra-low-voltage operation and portability to highly scaled processes. This paper presents a fully-synthesizable combinational composite-field Sbox circuit using optimal $GF(2^4)^2$ arithmetic, enabling a unified encrypt/decrypt 8 Sbox SMS4 implementation with 5× higher throughput and 2.3× lower area over prior designs [2-4].

SMS4 Accelerator Organization

The SMS4 engine encrypts 128b plain-text with a 128b key using 32 rounds of addition, substitution (Sbox) and shift operations in a unified encrypt/decrypt data-path for concurrent round computation and key-expansion accomplished using 4 Sboxes each (Fig. 2). The 3 words that constitute Sbox input are shifted left by 32b and bypassed to the output during encryption. The inverse computations for decryption can therefore be accomplished using the same Sbox followed by a 32b right-shift. Unlike AES data-path, this eliminates the need for separate inverse Sbox circuits allowing encryption and decryption operations to be mapped onto the same hardware using bi-directional shift circuits. Mode-switch multiplexers interleave data and key registers to decongest interconnect routing, reducing reconfiguration area and latency overhead by 24%.

Double-Affine Mapped $GF(2^4)^2$ Sbox

SMS4 Sbox operation involves two back-to-back affine transformations interspersed with $GF(2^8)$ inverse computation using reduction polynomial of $x^8+x^7+x^6+x^5+x^4+x^2+1$. Mapping of Sbox inputs to a composite-field $GF(2^4)^2$ representation enables inverse computation using compact combinational logic circuits resulting in 33% reduction in accelerator area compared to a LUT-based implementation (Fig. 3). Arithmetic operations in $GF(2^4)^2$ composite-field are governed by a pair of reduction polynomials ($x^4+a_3.x^3+a_2.x^2+a_1x1+a_0$, $x^2+\alpha x+\beta$, $a_i \in GF(2)$, $\alpha,\beta \in GF(2^4)$) [5]. These polynomials determine data-path logic complexity and mapping overheads for transforming elements to and from the composite-field. In contrast to the single affine transform used

in AES datapath, the presence of two affine transforms in the SMS4 Sbox provides the opportunity to merge mapping (M) and inverse-mapping (M^{-1}) matrices into existing affine matrices (A_1, A_2). This reduces mapping area overheads by 26% and improves critical-path delay by 29% compared to a composite-field design with explicit mapping transforms.

Design Space Exploration for Optimal GF Polynomials

The 120 pairs of irreducible extension-field polynomials generate 2880 valid mappings of $GF(2^8)$ to $GF(2^4)^2$, wherein each mapping corresponds to a unique SMS4 implementation (Fig. 4). A set of 21 RTL parameters capture all polynomial dependencies of affine, inverse, scaling and multiplier circuits, enabling an exhaustive design space exploration through gate-level synthesis/APR flow using 14nm tri-gate CMOS standard-cell libraries [6]. The minimum area design was obtained for the polynomial pair of x^4+x^3+1(0x9), x^2+2x+1($\alpha=2$, $\beta=1$) and mapped affine constant 0xE1, with total encrypt/decrypt design occupying $2445\mu m^2$. The use of optimal polynomials reduces multiplier critical path by 40%, and reduces squaring ($x^2\times\beta$) and scaling ($\times\alpha$) circuit gate counts to 1 and 3 respectively, providing 21% additional area savings (Fig. 5).

Counter-Assisted Round Constant (CK) Generation

Each SMS4 round requires a 128b key derived from the prior key using 32b round constant CK. An accumulating register increments/decrements by 28 every cycle to generate round constants during encrypt/decrypt respectively, while the two LSBs of each byte remain fixed. The next two LSBs exhibit periodic repetitions during encrypt and decrypt modes enabling a 2b counter-assisted round constant generation circuit that downsizes adder data-path width to 4b, eliminating 14 flip-flops, and resulting in 44% area savings over a conventional 8b adder-based design (Fig. 6). The compact CK generation circuit along with the mapped Sbox enables concurrent key-expansion in a parallel data-path, with 2× higher SMS4 throughput over prior shared data-path implementations [3].

14nm CMOS Measurement Results and Summary

In a 750mV, 14nm tri-gate CMOS process [6], the SMS4 accelerator operates at F_{max} of 1GHz with encrypt/decrypt throughput of 4.04Gbps, latency of 32 cycles and total power consumption of 12mW. Leakage component of total power is $145\mu W$ (Fig. 7). Ultra-low voltage optimizations [7] and elimination of LUT sequentials enable robust operation over a wide supply voltage range of 250-950mV, with scalable performance of 13Mbps-5.64Gbps and total power consumption of $22\mu W$-29mW. Peak energy-efficiency of 1.1Tbps/W is obtained at near-threshold voltage operation of 330mV, 25MHz, with $89/29\mu W$ total/leakage power consumption (Fig. 8). The use of optimal $GF(2^4)^2$ arithmetic combinational circuits offers 5× higher throughput at 2.3× lower area over previously-published implementations (Fig. 9).

Acknowledgments

The authors thank M. Haycock, M. Mayberry, V. De, J. Tschanz, G. Taylor, W. Feghali, K. Yap, and V. Gopal for encouragement and discussions.

References

[1] Office of State Commercial Cipher Administration of China SMS4 cipher. http://www.oscca.gov.cn/UpFile/200621016423197990.pdf
[2] W. Yan et al., *Int. ASIC Conference*, pp 135-138, 2009.
[3] M. Shang et al., *Int. Conf. on Advanced App. Informatics*, pp 86-90, 2014.
[4] X. Bai et al., *Int. Conf. On Networks Security*, pp. 345-348, 2009.
[5] S. Mathew et al., *VLSI Circuits Symp.*, pp. 166-167, 2014.
[6] C-H. Jan et al., *VLSI Tech. Symp.*, pp. 12-13, 2015.
[7] S. Hsu et al., *ISSCC Dig of Tech. Papers*, pp. 216-219, 2012.

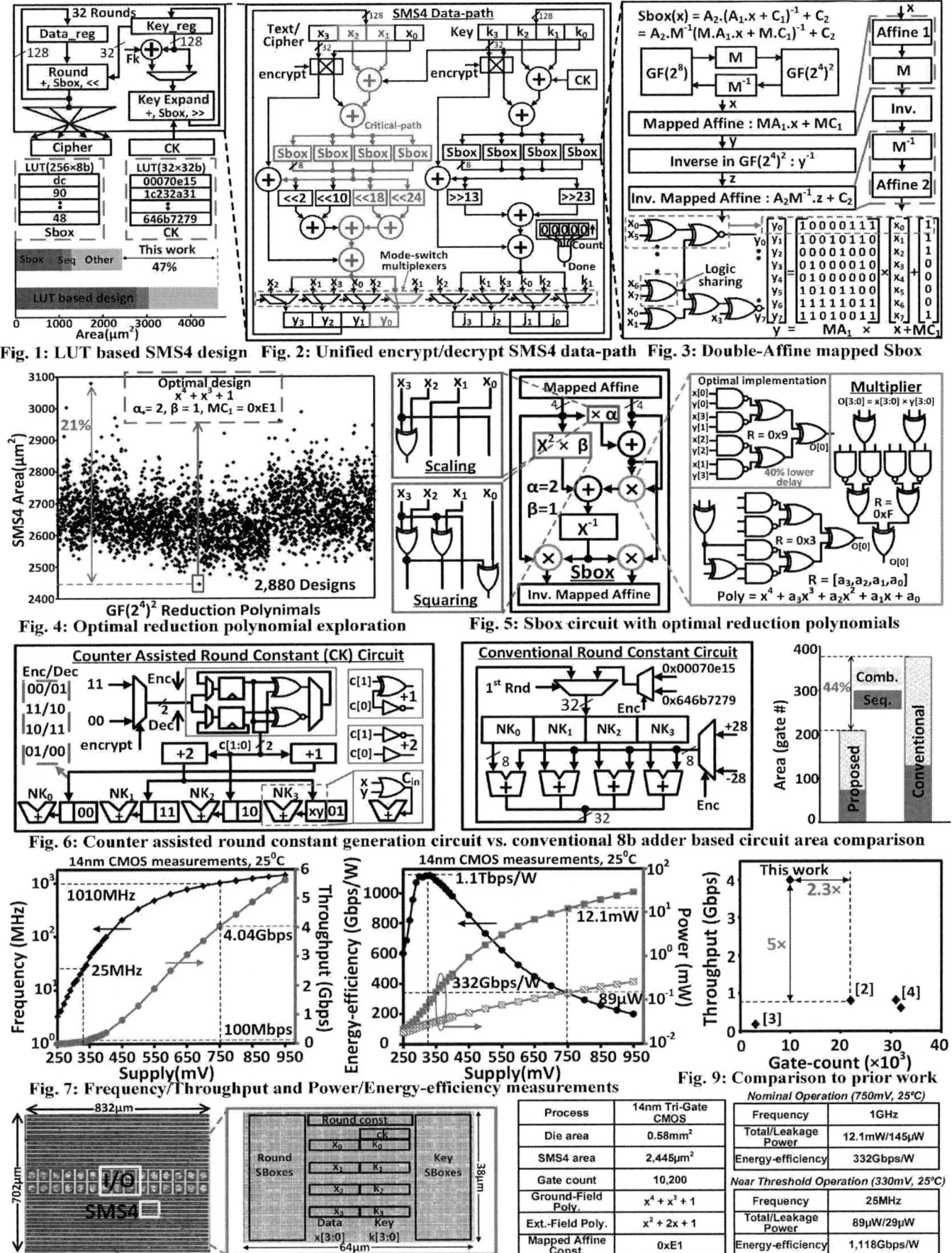

Fig. 1: LUT based SMS4 design Fig. 2: Unified encrypt/decrypt SMS4 data-path Fig. 3: Double-Affine mapped Sbox

Fig. 4: Optimal reduction polynomial exploration Fig. 5: Sbox circuit with optimal reduction polynomials

Fig. 6: Counter assisted round constant generation circuit vs. conventional 8b adder based circuit area comparison

Fig. 7: Frequency/Throughput and Power/Energy-efficiency measurements Fig. 9: Comparison to prior work

Fig. 8: Die-photo, 14nm layout and measurement summary

A Compact 446 Gbps/W AES accelerator for Mobile SoC and IoT in 40nm

Yiqun Zhang, Kaiyuan Yang, Mehdi Saligane, David Blaauw, Dennis Sylvester

University of Michigan, Ann Arbor, MI   Email: zhyiqun@umich.edu

Abstract

An AES hardware accelerator targeting energy efficient, low cost mobile and IoT applications is fabricated in 40nm CMOS. The proposed design eliminates the ShiftRow stage in conventional AES implementations and replaces flip-flops in data and key storage with latches using re-timing, saving 25% area and 69% power. Along with a 2-stage Sbox in native $GF(2^4)^2$ composite-field computation and glitch reduction techniques, this results in a compact 2228 gate design achieving 446 Gbps/W and 46.2 Mbps throughput at 0.47V.

Introduction

Security is critical for modern electronic devices with internet connectivity. Advanced Encryption Standard (AES) is a widely-used block cipher algorithm for symmetric encryption in a large range of applications. For mobile devices, silicon area (i.e., cost), throughput, and energy efficiency are all key design constraints [4]. Recently, several energy efficient implementations were presented [1,2]. However, their kbps-range throughput cannot meet the demands of mobile devices with high-speed data streaming. Highly parallelized implementations [3] provide Gbps throughput, which is critical in server applications. However, their large silicon footprint is disadvantageous in cost-sensitive mobile SoCs. This paper presents a voltage-scalable AES accelerator targeting mobile SoCs and IoT devices with ~50−500Mbps throughput, while achieving best-in-class area and energy efficiency. The proposed accelerator is fully synthesizable and implements 128-bit AES using only 2228 logic gates. By eliminating the ShiftRow and MixColumn registers and replacing data and key storage with latches, area is reduced by 41%. This, along with retiming of a 2-stage Sbox design in native $GF(2^4)^2$ composite-field computation, leads to a 3.38× energy efficiency improvement over a baseline implementation at nomial voltage with four 128-bit registers and 1-cycle $GF(2^4)^2$ Sbox methods. The proposed design achieves 1.3GHz at 0.9V, peak throughput of 494 Mbps, and peak energy efficiency of 446Gbps/W. Implemented in 40nm CMOS, the accelerator area is only 0.00429mm², marking the smallest AES accelerator considering technology scaling.

Energy Efficient AES

Fig. 1 shows the standard implementation of AES encryption using an 8-bit datapath, which was implemented in the same 40nm test chip as a baseline. Simulated power breakdown of functional modules (Fig. 1) shows that the four 128-bit registers (DataReg, MixColReg, KeyReg, and ShiftReg) constitutes ~50% of total AES power. Our approach reduces this storage to only a 128-bit latch-based DataReg, a 48-bit latch based StorageReg, and a "one-hot" indexed 128-bit latch based KeyReg; these changes reduce total sequential power by 31%.

Fig. 2 describes several ways of storing data (both input and intermediate) within AES accelerators. DataReg first stores the initial plain text and is then updated with calculated cipher text at each iteration of the algorithm. ShiftRow and MixColumn blocks compute 32-bit outputs every 4 cycles that are stored in ShiftReg and MixColReg, respectively. The authors of [4] eliminate ShiftReg by loading plaintext into DataReg in the ShiftRow byte-order.

As shown in the byte-location index in the matrix of Table 1, the data of locations L2, L5, and L8 in the 4th cycle, L1 and L4 in the 8th cycle, and L0 in the 12th cycle cannot be stored back to DataReg immediately after they are computed by the MixColumn module (highlighted in Table 1). In the proposed design, these 6 bytes are stored in a 48-bit StorageReg using the decode logic in Fig. 4. The hardwired data transfer from MixColumn output to DataReg removes the 128b ShiftReg and MixColReg (each built of 128 flip-flops) and instead uses StorageReg, consisting of only 48 registers. As a result, the total register count for the datapath is reduced to 176, compared to 384 in a conventional design and 256 in [4], marking a 30% reduction.

To further reduce sequential power and area, the design is modified to accommodate latch-based registers instead of flip-flops. This is accomplished by adding an 8-bit AdderReg (Fig. 4) and 1 additional cycle of latency (a 0.3% increase to 337 total cycles of latency), since data is bypassed to skip MixColumn at the last iteration in the AES algorithm. This change does not impact the clock frequency since DataReg is not on the critical path. Fig. 2 includes the dynamic energy and area values for each implementation. The proposed approach has a 2.66×/2.9× energy/area improvement over a conventional design and a 1.78×/1.94× energy/area improvement over [4] for these three registers. Finally, dynamic glitch power is a significant concern in AES hardware accelerators. Hence, clock gating is used in dataReg to reduce glitch power for this part by 2.74× and total power by 30%.

In addition to plain/cipher text processing, the key is also updated in each iteration of AES. The 128-bit input key is stored in the KeyReg and one byte is updated by KeyGen in each iteration. Using address generation to access the correct byte in each iteration results in a large gate count and area. Using a basic shift register reduces area but increases power. Instead, in the proposed design KeyReg is changed from a 128-bit flip-flop register to a 128-bit latch register using one-hot shift-based addressing. This design uses a cyclic address generator with a single chain of 16 single-bit registers (Fig. 4), similar to [6]. This requires 1-bit shifting rather than 128-bit shifting, reducing area by 23% and improving power by 18% (Fig. 5) compared with the conventional register implementation with decoder.

The final block to be optimized is the Sbox stage, which contributes 12% of total power (Fig. 1) and contains the accelerator's critical path. Sbox implementation choices include SRAM-based, logic based look-up table, and native composite-field $GF(2^4)^2$; these are analyzed via simulation in Table 2. A conventional single-cycle $GF(2^4)^2$ offers compact area at the expense of power and performance. This higher power is due in part to the difference in signal arrival times of fast and slow paths in the Sbox (Fig. 3), resulting in glitch power [5]. To address this, we re-time the Sbox datapath by adding 12 flip-flops before the path converges, equalizing path delays. This incurs a modest 4.3% area overhead while providing 37% power savings at the same frequency as the 1-cycle $GF(2^4)^2$ implementation. Also, splitting Sbox into two cycles shortens the critical path, decreasing clock cycle time by 28%; through voltage scaling this improves accelerator energy efficiency by 3.38× energy efficiency at iso-frequency.

Measurements & Conclusion

The proposed AES accelerator was implemented in 40nm CMOS along with a separate baseline implementation. Fig. 5 shows the simulated power breakdown of baseline and proposed designs. At 0.9V and 25°C the proposed design has a measured Fmax of 1.3GHz while consuming 4.39mW (Table. 3). The proposed design is fully synthesized, enabling operation across a wide voltage range. Fig. 6 shows the measured clock frequency, throughput, energy efficiency, and power across Vdd. At 1V, performance of 1.47GHz is obtained while peak energy efficiency of 446 Gbps/W is achieved at Vdd = 0.47V. Compared with [4], the proposed design is 41% smaller considering technology scaling and 3.1× more energy efficient at 432Mbps throughput. Overall the power consumption of the proposed design is compatible with mobile SoCs at its highest performance point (4.39mW) and offers a compelling option for IoT applications as it consumes only 100μW with 46.2Mbps throughput at sub-0.5V. Fig. 7 shows the die photo.

Acknowledgement

We thank the TSMC University Shuttle Program for chip fabrication.

References

[1] J. Myers, et al., ISSCC, 2015.
[2] W. Zhao, et al., TVLSI, 2015.
[3] S. Mathew, et al., JSSC, 2011.
[4] S. Mathew, et al., JSSC, 2015.
[5] S. Morioka, et al., CHES, 2002.
[6] D. Jeon, et al., JSSC, 2012.
[7] P. Hamalainen, et al., EUROMICRO, 2006.
[8] T. Good, et al., TVLSI, 2010

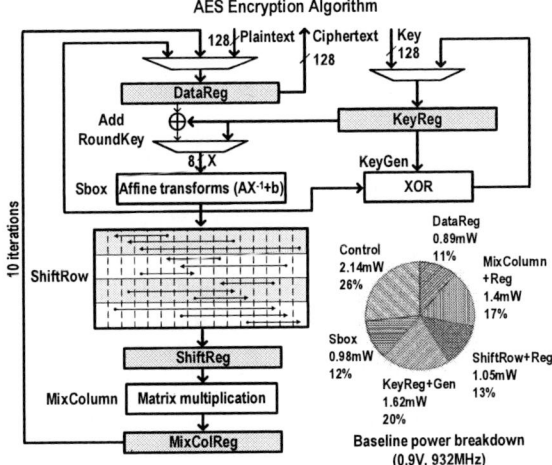

AES Encryption Algorithm

Fig.1. Standard implementation of AES encryption circuit. Total Baseline power is based upon measurement results.

Baseline power breakdown (0.9V, 932MHz)
- Control 2.14mW 26%
- DataReg 0.89mW 11%
- MixColumn +Reg 1.4mW 17%
- ShiftRow+Reg 1.05mW 13%
- KeyReg+Gen 1.62mW 20%
- Sbox 0.98mW 12%

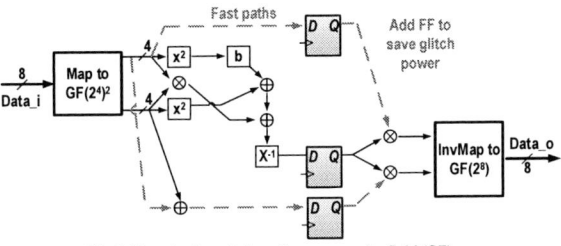

Fig.3. Sbox logic path in native composite-field (GF).

Table 2. Comparison table of Sbox implementations (based on simulation results)

S-box Architecture	Conventional			Proposed
	Look-up table (SRAM)	Look-up table (logic)	GF(2^4)^2 in 1 cycle	GF(2^4)^2 in 2 cycle
Area (µm^2)	2175	816	558	582
Power (mW)	1.7	1.15	1.42	0.9
Cycle Time (ns)	0.55	0.4	0.64	0.46

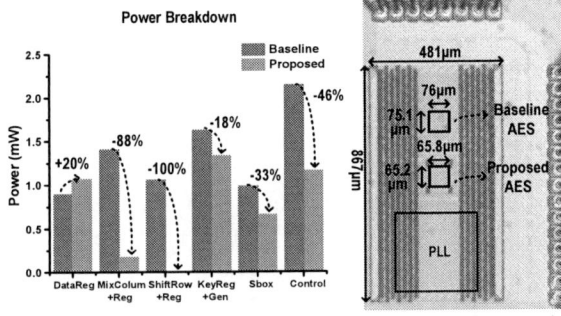

Fig.5. Simulation based power breakdown.

Fig.7. Die Photo

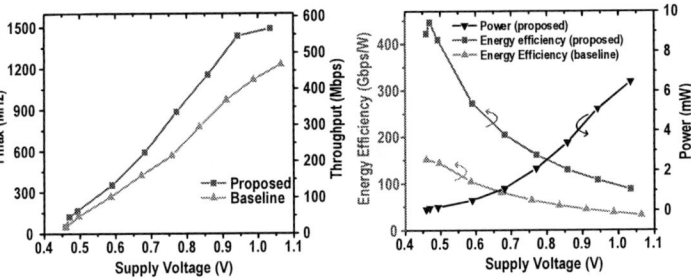

Fig.6. Frequency, throughput, power and energy-efficiency measurements.

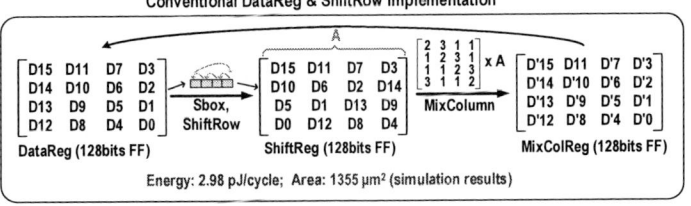

Conventional DataReg & ShiftRow implementation

Energy: 2.98 pJ/cycle; Area: 1355 µm^2 (simulation results)

[4]: Reduce ShiftReg

Energy: 1.99 pJ/cycle; Area: 903 µm^2

Location index of 16 bytes

Proposed: Further reduce MixColReg

StorageReg (48bits latch)
DataReg (128bits latch)

Energy: 1.12 pJ/cycle; Area: 466 µm^2

Fig.2. DataReg & ShiftRow implementation comparison.

Table 1. Proposed MixColumn output data & location

		D'15	D'14	D'13	D'12
4th cycle	Data	D'15	D'14	D'13	D'12
	Location	L15	L2	L5	L8
8th cycle	Data	D'11	D'10	D'9	D'8
	Location	L11	L14	L1	L4
12th cycle	Data	D'7	D'6	D'5	D'4
	Location	L7	L10	L13	L0
16th cycle	Data	D'3	D'2	D'1	D'0
	Location	L3	L6	L9	L12

Data in the byte locations of L0/1/2/4/5/8 need to be delayed and put in StorageReg

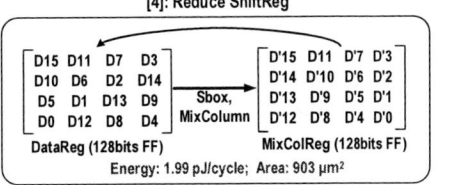

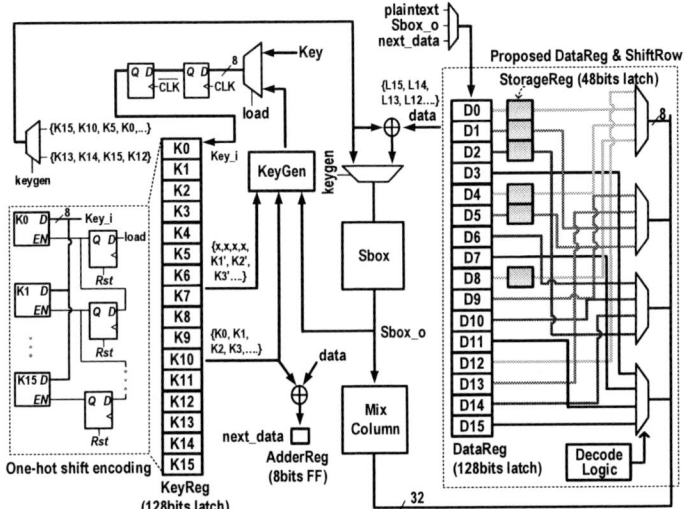

Fig.4. Proposed 8-bit datapath AES accelerator architecture

Table 3. Chip measurement summary and comparison table of AES designs

	EuroMicro'06 [7]	TVLSI'10 [8]	JSSC'11 [3]	JSSC'15 [4]	Proposed
Technology	130nm	130nm	45nm	22nm	40nm
Voltage (V)	Not Reported	0.8 / 0.75	1.1 / 0.32	0.9 / 0.43	0.9 / 0.47
Power (mW)	17.98 / 3.9	0.099 / 0.000692	125 / 0.409	13 / 0.45	4.39 / 0.10
Frequency	290 MHz / 130 MHz	12 MHz / 100 KHz	2.1 GHz / 32 MHz	1.1 GHz / 220 MHz	1.3 GHz / 122 MHz
Throughput	232 Mbps / 104 Mbps	34 Kbps / 280 bps	53 Gbps / 800 Mbps	432 Mbps / 83.6 Mbps	494 Mbps / 46.2 Mbps
Energy Efficiency (Gbps/W)	12.9 / 26.7	0.343 / 0.405	424 / 1955	33 / 186	113 / 446
Energy/bit (pJ/b)	77.5 / 37.5	2915 / 2469	2.36 / 0.512	31 / 5.38	8.85 / 2.24
Number of Gates	3200 / 3900	5500	Not Reported	1947	2228
Area (mm)^2 (Norm to 40nm)	Not Reported	< 1	0.15 (0.119)	0.0022 (0.0073)	0.00429

978-1-5090-0636-6/16 $31.00 © 2016 IEEE   197   2016 Symposium on VLSI Circuits Digest of Technical Papers

A 4fJ/bit Delay-Hardened Physically Unclonable Function Circuit with Selective Bit Destabilization in 14nm Tri-gate CMOS

Sanu Mathew, Sudhir Satpathy, Vikram Suresh, Mark Anders, Himanshu Kaul, Amit Agarwal, Steven Hsu, Greg Chen, Ram Krishnamurthy, Vivek De

Circuits Research Lab, Intel Corporation, Hillsboro, OR USA

sanu.k.mathew@intel.com

Abstract

A 1024-bit delay-hardened physically unclonable function (PUF) array is fabricated in 14nm tri-gate CMOS, targeted for on-die secure generation of a full-entropy 128bit key. Differential clock delay injection, selective destabilization of unstable bits and temporal-majority-voting (TMV) based winnowing enable 1.7x higher post-burn-in BER improvement, 50% reduction in dark-bit induced bit-errors and worst-case BER of 1.46%. Spectral analysis of unstable PUF bits show significant 1/f noise impacts below 500MHz. In-situ field aging with write feedback improves bit stability by up to 48%.

Introduction

Physically Unclonable Function (PUF) circuits represent a paradigm shift in enabling secure digital identities and die-specific cryptographic keys by providing the repeatability of non-volatile one-time-programmable fuses while overcoming their vulnerability to malicious probing attacks. While a variety of PUF circuits ranging from bias generators [1], ring-oscillators [2], current mirrors [3], pre-charged cross-coupled inverters [4, 5], oxide-breakdown [6] and SRAMs [7] have been used to harvest entropy from random process variation, there remain significant challenges to producing stable repeatable keys in the presence of process-voltage-temperature variations and aging-induced process drift.

In this paper, we present a delay-hardened PUF circuit that leverages NBTI/PBTI aging during burn-in to inject a differential delay bias into clock release times in a pre-charged cross-coupled inverter circuit. TMV-based winnowing during burn-in enables selective destabilization of highly unstable PUF bits, resulting in 50% reduction in BER without impacting tester time and cost. We also report the influence of 1/f noise on PUF bit-error rates and recommend that worst-case BERs should be characterized at frequencies below 500MHz.

PUF Array Organization

The 1024-bit PUF is organized as an 8-entry column-multiplexed array producing 128 raw PUF bits/cycle that are stabilized using TMV15 [4] and soft dark-bits [5], followed by error correction using BCH[255,129,18] code (Fig. 1). Accounting for worst-case entropy losses along the multiple post-processing datapath stages (50% loss in AES-CBC-MAC entropy extractor, 516 BCH syndrome fuse bits, non-ideal 1:0 ratio, hard-coded darkbits), a 128-bit full-entropy output key requires an input bitstream containing at least 1017 PUF bits.

Delay-hardened (DH) PUF Circuits

Raw PUF bits are generated using delay-hardened PUF circuits that employ a pair of pre-charge transistors to force bi-stable cross-coupled inverters into an unstable state (Fig. 2). At the rising clock edge, the output node resolves to one of two stable states (bit=0 or bit=1) based on mismatches in inverter and clock device characteristics due to manufacturing-time random variations. In contrast to the hybrid cell [4], the delay-hardened PUF introduces 2 additional clock delay inverters into each pre-charge path (clk0 and clk1). During accelerated aging of dies for infant mortality testing (burn-in=1), the complementary value (bit#) is differentially written back into the cell, biasing inverter devices in a direction such that NBTI/PBTI aging reinforces pre-existing biases and improves cell stability. The clock delay inverters in the DH cell are also differentially biased during burn-in such that a cell with an initial bias towards '1' self-biases and ages the appropriate clock devices to push out clk1 rise delays, while pulling in clk0 rising edges relative to a nominal 2 inverter delay (Fig. 3). In contrast, a cell with a pre-existing bias towards '0' pushes out

clk0 delays relative to clk1. Raw BER measurements of 1024-bit PUF arrays fabricated in 14nm tri-gate CMOS [8] demonstrate 1.7x higher post-burn-in BER improvement (48% vs. 28%) for the DH cell compared to the hybrid cell (Fig. 4).

Selective Bit Destabilization

While soft dark-bits offer an effective technique for sequestration of highly unstable PUF bits [5], mismatches in run-time generated masks between golden (0.65V, 70°C) and field (0.55-0.75V, 25-110°C) conditions can induce up to 92% of total bit-errors (Fig. 8a). Dark-bit induced bit-errors are caused when PUF bits identified as 'dark' during golden conditions in the tester are observed to be stable during field conditions or vice-versa (Fig. 6). TMV count measurements of the 1024-bit PUF array at field conditions show that while 948 stable bits have consistent counts <10% or >90%, dark bits are present in the entire range of 0-100%. A TMV-based winnowing technique that stabilizes PUF bits with counts <10% and >90%, while selectively destabilizing bits with counts between 10-90%, will identify 80% of dark bits from the entire operating range (Fig. 7). Destabilization of these highly unstable cells by disabling write-back paths during burn-in increases their sequestration probability at all operating conditions and reduces dark-bit induced BER by 50% (Fig. 8b), resulting in a worst-case total BER of 1.46% (Table I).

1/f Noise & Aging Measurements

Low frequency measurements of the PUF circuit show that unstable bit count increases by 4x at frequencies below 800MHz (Fig. 9a), indicating the impacts of 1/f noise as PUF devices remain in pre-charged state for longer durations. Bit stability degradation mechanisms are examined using 500-point FFT spectral analysis of 167 unstable bits from multiple tiles indicating 10x higher bit transitions, with 9.3x increase in mean-square FFT magnitude below 500MHz (Fig. 9b). To account for the impacts of 1/f noise on PUF cell behavior, all BER metrics reported in this paper are measured at 100MHz. Long-term PUF aging behavior is examined by subjecting the array to high-voltage stress for time periods ranging from 500ms to 5s (Fig. 10). Aging measurements show that in-situ field aging with activated write feedback paths reduce unstable bit count by up to 48%, indicating favorable aging that improves array stability over the lifetime of the die.

Conclusion

A delay-hardened PUF array targeted for on-die generation of stable secure keys is fabricated in 14nm tri-gate CMOS technology, with ultra-low 4fJ energy/bit (Table II) and compact cell layout of $1.84\mu m^2$ (Fig. 11). TMV-based winnowing with selective bit destabilization enables tolerance to 200mV voltage and 25-110°C temperature variation, resulting in maximum BER of 1.46%. Write feedback circuits leverage long-term process drift to improve cell stability by 48% over die lifetime.

Acknowledgments

The authors thank M. Haycock, M. Mayberry, J. Tschanz, S. Iyengar, A. Rajan, R. Parker, P. Munguia for encouragement and discussions.

References

[1] J. Li et al., *Symp. on VLSI Circuits*, pp 250-251, June 2015.
[2] K. Yang et al., *ISSCC 2015*, pp. 254-255, Feb. 2015.
[3] A. Alvarez et al., *ISSCC 2015*, pp. 256-257, Feb. 2015.
[4] S. Mathew et al., *ISSCC 2014*, pp.169-170, Feb. 2014.
[5] S. Satpathy et al., *ESSCIRC 2014*, pp. 239-242, Sept.2014.
[6] N. Liu et al., *Symp. on VLSI Circuits*, pp. 231-232, 2010.
[7] H. Fujiwara et al., *Symp. on VLSI Circuits*,pp.76-77, 2011.
[8] C-H Jan et al., *Symp. on VLSI Technology*, pp.12-13, 2015.

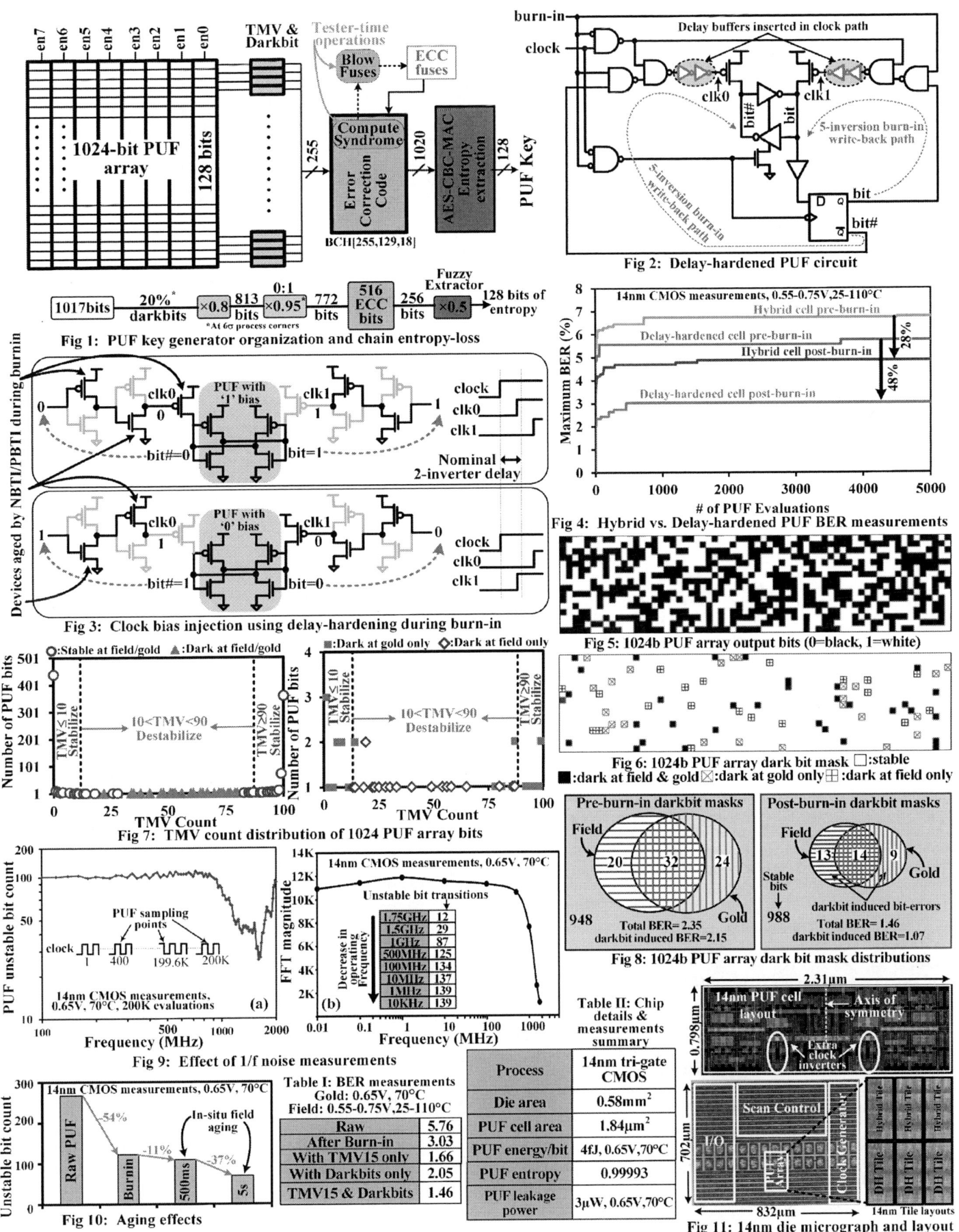

Fig 1: PUF key generator organization and chain entropy-loss

Fig 2: Delay-hardened PUF circuit

Fig 3: Clock bias injection using delay-hardening during burn-in

Fig 4: Hybrid vs. Delay-hardened PUF BER measurements

Fig 5: 1024b PUF array output bits (0=black, 1=white)

Fig 6: 1024b PUF array dark bit mask ☐:stable ■:dark at field & gold ⊠:dark at gold only ⊞:dark at field only

Fig 7: TMV count distribution of 1024 PUF array bits

Fig 8: 1024b PUF array dark bit mask distributions

Fig 9: Effect of 1/f noise measurements

Fig 10: Aging effects

Fig 11: 14nm die micrograph and layout

A 0.58mm^2 2.76Gb/s 79.8pJ/b 256-QAM Massive MIMO Message-Passing Detector

Wei Tang[1], Chia-Hsiang Chen[1,2], Zhengya Zhang[1]
[1]University of Michigan, Ann Arbor, MI, [2]Intel Labs, Santa Clara, CA

Abstract

A 0.58mm^2 40nm CMOS message-passing detector (MPD) is designed for a 256-QAM massive MIMO system supporting 32 concurrent mobile users in each time-frequency resource. Leveraging channel hardening in massive MIMO, a symbol hardening technique is proposed to reduce MPD's complexity by more than 60% with minimal SNR loss. The MPD is implemented in a 4-layer 2-way interleaved architecture to enable a 2.76Gb/s throughput (average 4.9 iterations at 27dB SNR with early termination) using 76% smaller area than a fully parallel architecture. With dynamic precision control and clock gating to exploit algorithmic properties, the energy is reduced to 79.8pJ/b (or 2.49pJ/b per TX antenna).

Introduction

Massive MIMO has been identified as a key disruptive technology for the upcoming fifth generation (5G) wireless communication systems [1]. Massive MIMO is a multi-user MIMO technique that relies on a large number, e.g., hundreds, of base station antennas to serve a multiplicity of, e.g., tens, of autonomous single-antenna users in each time-frequency resource [2]. The large number of antennas provide a high spatial multiplexing gain for an increased capacity; and the radiated energy can be focused to the intended receivers for an energy-efficient downlink transmission. However, massive MIMO requires complex and power-hungry signal processing for uplink processing in base station. Recently iterative message-passing detectors (MPD) have been proposed [3]. An MPD exploits channel hardening in a massive MIMO system, and it approximates the sum of interference using a Gaussian distribution. An MPD has a comparable complexity as an MMSE detector, and yet in a massive MIMO system, it outperforms an MMSE detector by a large margin.

Iterative Message-Passing Detection

In an $N_r \times N_t$ M-QAM massive MIMO uplink system (Fig. 1), N_t users transmit at the same time and frequency, and N_r antennas at the base station pick up not only the intended transmissions but also interference plus noise. A MIMO detector attempts to retrieve the intended transmissions by canceling the interference. An MPD uses a set of N_t interference cancellation PEs (IPE) and a set of N_t constellation matching PEs (CPE) to iteratively estimate the N_t user symbols. An IPE is connected to N_t CPEs (Fig. 3(a)), and it computes a Gaussian approximation of a user symbol after canceling the interference from the other $N_t - 1$ users. A CPE is connected to one IPE (Fig. 3(a)), and it estimates a symbol by considering its likely locations at M constellation points. In every iteration, an IPE requires $8(N_t - 1)$ real-valued multiply-accumulates (MAC) to compute the mean and variance of the Gaussian approximation; and a CPE requires $2\sqrt{M}$ Gaussian evaluations and $4\sqrt{M}$ MACs to compute the mean and variance of a symbol estimate. The complexity of an MPD grows with the number of users N_t and the order of modulation M, presenting an implementation challenge for a high-order massive MIMO system.

Complexity Reduction Leveraging Channel Hardening

In this work, we design an MPD for a 128×32 ($N_r = 128$, $N_t = 32$) 256-QAM system. With channel hardening in a massive MIMO system, the variance of symbol estimate converges at a fast pace. The fast convergence permits the use of a small fixed variance to eliminate the variance calculations, saving nearly 4K MACs in 32 IPEs and 1K MACs in 32 CPEs. The use of a small variance further reduces the CPE processing to making one hard symbol decision based on its distribution. The symbol hardening technique eliminates 1K MACs and 1K Gaussian evaluations in 32 CPEs, sacrificing 0.25dB of SNR at 10^{-4} BER, but the optimized MPD still outperforms an MMSE detector by 1dB (Fig. 2).

Layered and Interleaved Architecture

The MPD can be fully parallelized with 32 IPEs and 32 CPEs (Fig. 3(a)), requiring nearly 4K MACs and 10K interconnects. Despite the high throughput, the fully parallel architecture is dominated by global wiring, resulting in a large silicon area, low clock frequency, and high power.

Therefore we opt for a more compact design (Fig. 3(b)) that divides 32 users into 4 layers with 8 users per layer for processing, using 1/4 as many MACs in each IPE. In each layer, 32 IPEs compute the sum of interference contributed by 8 users and update the symbol estimates. The updated estimates are then forwarded to the next layer. The intra-iteration forwarding speeds up convergence by nearly 2× compared to the flooding schedule used in the fully parallel architecture. Based on trial designs, the 4-layer architecture reduces area and power by 66% and 61%, respectively, but with faster convergence, its throughput is only 28% lower (Fig. 5(a,b)).

The layered architecture imposes data dependency between layers, requiring one layer to be completed in one clock cycle. To relax the data dependency and reduce area, we halve the number of IPEs to 16 and time-multiplex their use between 2 groups in a 2-stage pipeline (Fig. 3(c)). In each cycle the symbol estimates are computed for either group 1 or group 2 users, which are interleaved to avoid pipeline stalling. Based on trial designs, the 4-layer 2-way interleaved architecture (Fig. 4) reduces area and power by 76% and 65% respectively over the baseline (Fig. 5(a,c)).

Dynamic Precision Control and Clock Gating

The datapath power is dominated by the 512 MACs. To save dynamic power, we adapt the multiplier precision dynamically to exploit the MPD's convergence behavior. In early iterations, the MPD makes coarse symbol estimates using 6b×2b low-precision multiplications; but in late iterations, the MPD fine tunes symbol estimates using 12b×4b full-precision multiplications. Each full-precision multiplier is designed to support the low-precision mode with LSBs disabled (Fig. 6(a)), saving 75% of the switching activity and the associated dynamic power.

Registers are used as data memory to support the wide data access required by the architecture. The memory access is regular (Fig. 6(b)), e.g., the 3Kb interference memory (P MEM) is updated once every 8 cycles. Therefore, we implement clock gating to turn off the clock input when the memory is not updated to save dynamic power.

Test Chip Measurement Results

A massive MIMO MPD test chip (Fig. 8) is fabricated in TSMC 40nm CMOS technology. The chip includes a 0.58mm^2 MPD core, a PLL to generate clock, a test memory to store test vectors, and scan chains for input and output. The chip is measured to run at a maximum frequency of 425MHz at the nominal supply voltage of 0.9V in room temperature, dissipating 221mW. Incorporating the proposed architecture techniques along with dynamic precision control and clock gating, the MPD's power dissipation is reduced by 70% and energy per bit reduced by 52% over the baseline (Fig. 5(a,d)). With early termination enabled on chip, detection converges in 5.7, 5.2 and 4.9 iterations on average at 23dB, 25dB and 27dB SNR, allowing a throughput up to 2.76Gb/s or 86Mb/s per mobile user (Fig. 7). A higher throughput for massive MIMO can be achieved by deploying multiple MPD modules and applying interleaving.

The results are compared with state-of-the-art MIMO detector chips in Table I. Note that all the previous designs, including SD [4]-[6] and MMSE [7], will incur much higher implementation costs in a massive MIMO system. To the best of our knowledge, this chip is the first silicon demonstration that supports large-scale multi-user MIMO detection.

Acknowledgements

The work was supported in part by NSF and Intel.

References

[1] F. Boccardi, et al., *Commun. Mag.*, 2014.
[2] E. G. Larsson, et al., *Commun. Mag.*, 2014.
[3] T. L. Narasimhan, A. Chockalingam, *J. Sel. Topics Signal Process.*, 2014.
[4] F. Borlenghi, et al., *Proc. ESSCIRC*, 2012.
[5] B. Noethen, et al., *ISSCC Dig. Tech. Papers*, 2014.
[6] M. Winter, et al., *ISSCC Dig. Tech. Papers*, 2012.
[7] C.-H. Chen, et al., *ISSCC Dig. Tech. Papers*, 2015.

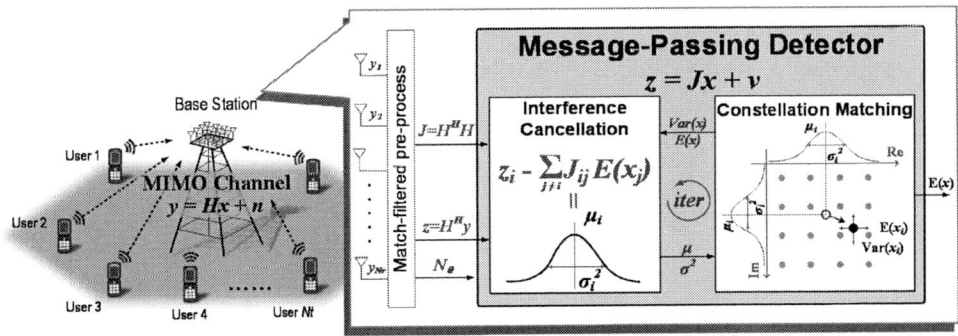

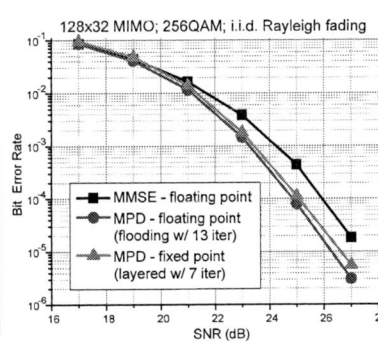

Fig. 1. Illustration of an uplink large-scale MIMO system of N_t single-antenna users and N_r antennas at base station; and a top-level block diagram of a message-passing detector (MPD).

Fig. 2. Bit error rate (BER) performance of 128x32 uplink MMSE detection and MPDs.

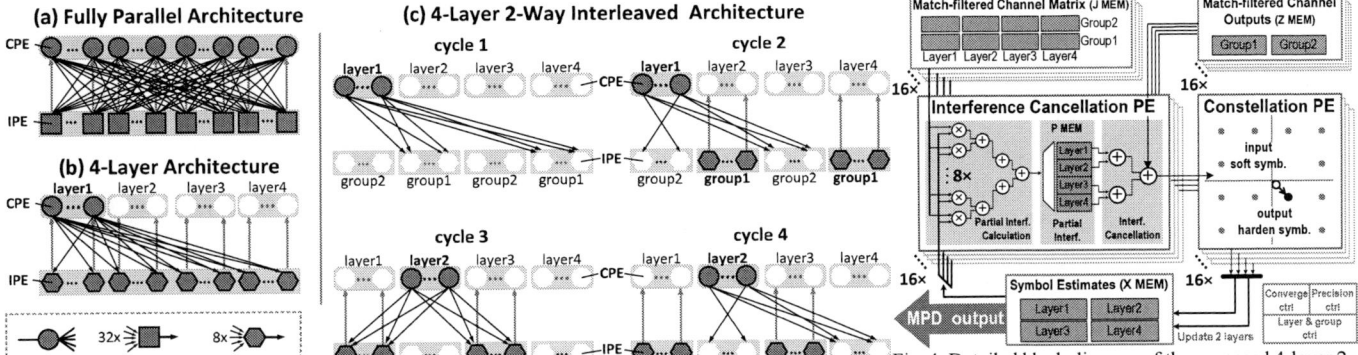

Fig. 3. Architectural optimization from (a) fully parallel architecture using a flooding schedule to (b) 4-layer architecture, and (c) 4-layer 2-way interleaved architecture (the first 4 pipeline cycles are shown).

Fig. 4. Detailed block diagram of the proposed 4-layer 2-way interleaved MPD using 16 Interference cancellation PEs (IPEs) and 16 Constellation matching PEs (CPE).

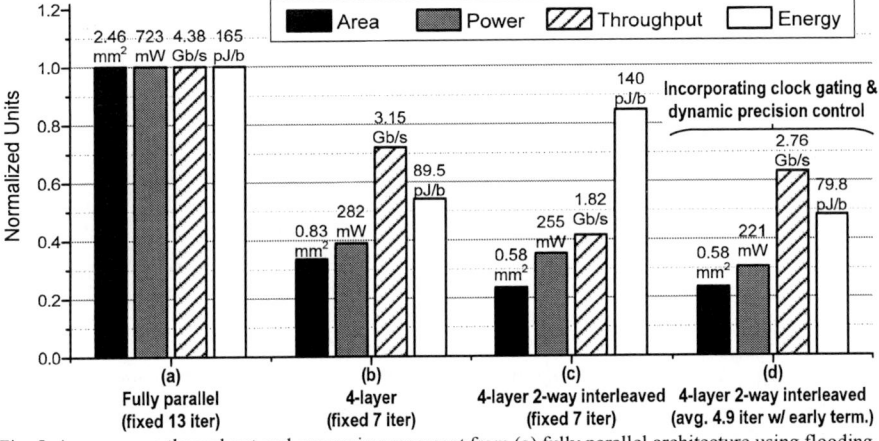

Fig. 5. Area, power, throughput and energy improvement from (a) fully parallel architecture using flooding schedule to (d) 4-layer 2-way interleaved architecture with early termination, incorporating dynamic precision control and clock gating.

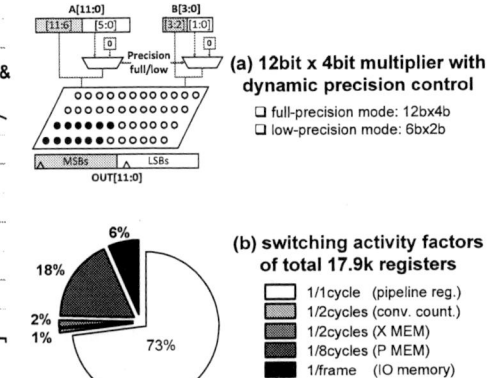

Fig. 6. Low power techniques: (a) dynamic precision multiplier, and (b) clock gating: breakdown of register usage and the corresponding update activity.

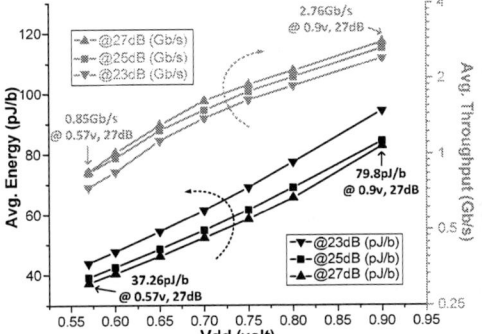

Fig. 7. Measured average throughput (red) and average energy (black) with voltage scaling at different SNRs.

Fig. 8. Chip microphotograph.

TABLE I. COMPARISON TABLE OF STATE-OF-THE-ART MIMO DETECTOR DESIGNS

Detector	Borlenghi [4]	Noethen [5]	Winter [6]	Chen [7]	This Work
Algorithm	SD [a]	SD [a]	SD [a]	MMSE	MPD [b]
MIMO size (Nr × Nt)	MIMO ≤ 4×4	MIMO ≤ 4×4	MIMO ≤ 4×4	MIMO 4×4	Massive MIMO 128×32
Modulation	≤ 64	≤ 64	≤ 64	256	256
Technology [nm]	65	65	65	65	40
Core area [mm²]	2.78	-	0.31	0.7	0.58
Frequency [MHz]	135	445	333	517	425
Power [mW]	-	87	38	26.5	220.6
Throughput [Gb/s]	0.194	0.396	0.296-0.807	1.379	2.76 [c]
Area efficiency [Gb/s/mm²]	0.07	-	0.96-2.6	1.97	4.76
Energy [pJ/b]	920	220	48	19.2	79.8
Energy efficiency [pJ/b/TX antenna]	230	55	12	4.8	2.49

(a): sphere decoding. (b): message-passing detection.
(c): early termination with average 4.92 iterations at SNR=27dB.

978-1-5090-0636-6/16 $31.00 © 2016 IEEE

A Machine-learning Classifier Implemented in a Standard 6T SRAM Array

Jintao Zhang, Zhuo Wang, and Naveen Verma

Princeton University, Princeton, NJ, USA

Abstract

This paper presents a machine-learning classifier where the computation is performed within a standard 6T SRAM array. This eliminates explicit memory operations, which otherwise pose energy/performance bottlenecks, especially for emerging algorithms (e.g., from machine learning) that result in high ratio of memory accesses. We present an algorithm and prototype IC (in 130nm CMOS), where a 128×128 SRAM array performs storage of classifier models and complete classifier computations. We demonstrate a real application, namely digit recognition from MNIST-database images. The accuracy is equal to a conventional (ideal) digital/SRAM system, yet with 113× lower energy. The approach achieves accuracy >95% with a full feature set (i.e., 28×28=784 image pixels), and 90% when reduced to 82 features (as demonstrated on the IC due to area limitations). The energy per 10-way digit classification is 633pJ at a speed of 50MHz.

System Overview

Recent trends towards in-/near-memory computing are motivated by limitations posed by memory operations [1]. For example, with 16b words and 32kB SRAMs, SRAM-access vs. multiplication consumes 17pJ vs. 1pJ in 45nm. In the demonstrated system, computation is performed in-place by the bit cells of an SRAM, avoiding energy-intensive accesses. This faces two key challenges: (1) the constrained structure of standard 6T arrays limits the computations possible; and (2) circuit non-idealities, especially high variability in bit cells, degrades the quality of outputs. To overcome (1), we exploit an idea from machine learning called boosting. In boosting, outputs from multiple *weak classifiers* are combined (e.g., via weighted voting) to form a *strong classifier* (weak/strong classifier implies inability/ability to fit arbitrary training data). We focus on the algorithm Adaptive Boosting (AdaBoost), where weak classifiers are iteratively trained to correct the fitting errors of previous iterations. Theory shows that very weak classifiers (marginally better than 50/50 guessing) can be used [2], enabling use of the limited structures possible via SRAM bit cells. To overcome (2), we employ an algorithmic extension to AdaBoost we previously developed called Error-Adaptive Classifier Boosting (EACB) [3]. EACB exploits the actual weak-classifier implementations within the iterative training to adaptively correct not only fitting errors but also errors due to any static (possibly random) non-idealities in the implementations; analysis and experimental validation show that substantial non-idealities can be overcome [3].

Fig. 1 shows the block diagram. The system operates in two modes: SRAM Mode and Classify Mode. In SRAM Mode, the system behaves as a standard SRAM, enabling read/write of the classifier model (derived from training) into bit cells. For classification, algorithms typically utilize a comparison metric between input data and the classifier model. A common metric is the inner product. A binary linear classifier is simply based on taking the inner product between an input feature vector $\vec{x}$ and a weight vector $\vec{w}$, and then performing sign thresholding: $sign(\vec{x} \cdot \vec{w})$. In Classify Mode, the system implements a very weak form of linear classifier in every column, where the elements of $\vec{w}$ are restricted to be +/-1,

according to data stored in the bit cells. First BL/BLB's are precharged. Then all the WL's are driven at once to analog voltages corresponding to the elements of $\vec{x}$ (i.e., the features); this is done via the peripheral WLDACs. Thus, each bit cell pulls a current modulated by its WL voltage from either BL or BLB, depending on its stored state. This approximates multiplication by +/-1. The bit-cell currents I_{BC} from the column then add together, as in an inner product, discharging BL or BLB. Finally, a comparator provides sign thresholding of the differential signal. Below we consider circuit non-idealities, and how these are overcome either by the architecture itself or via the training algorithm.

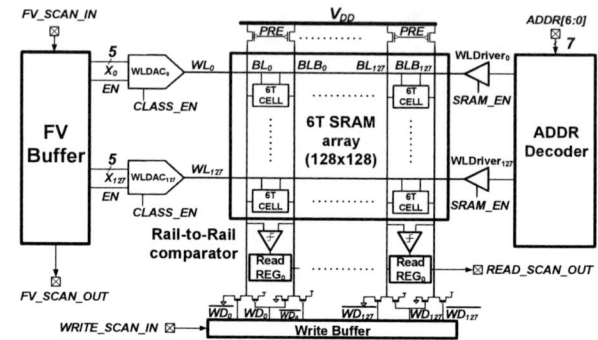

Fig. 1: System Block Diagram.

Circuit Design

For SRAM Mode, standard circuit structures are used (precharge devices, address decoders, WL drivers, write drivers). For Classify Mode, the key circuit blocks, described below, include: WLDACs, bit cells, and comparators.

Fig. 2 shows the WLDAC circuit and simulated output (WL) waveform (note, WL drivers are disabled in Classify Mode). Digital features are provided as 5 bits $X[4:0]$, to select binary-weighted PMOS current sources; an additional PMOS current source is included to enhance the linearity of the charge drawn by a bit cell ($Q_{BC} = \int I_{BC}dt$), as described below. The current from the PMOS sources is used to bias a replica of the bit cell, consisting of a V_{DD}-connected driver transistor (MD_R) and diode-connected access transistor (MA_R), which drives the WL. Thus, all bit cells in the row provide a corresponding current I_{BC} (i.e., scaled by the replica sizing ratio). With the WLDAC output impedance $\sim 1/g_{m,MAR}$, up sizing the replica easily yields the desired charging times for the WL capacitance ($\sim$120fF). Note that with all WLs driven this way, a potentially large number of bit cells drive BL/BLB at the same time (128 in the prototype). To not saturate discharge of BL/BLB capacitances, Q_{BC} of each is designed to be low <10fC, amounting to WL voltages <0.4V.

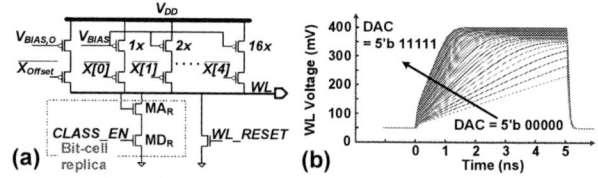

Fig. 2: (a) WLDAC circuit, and (b) WL transient voltage with different WLDAC input codes.

Bit-cell sizing is identical to standard 6T cell (for read/write margin). We note that in Classify Mode an upset condition is possible. Namely, we like to maximize the dynamic range of BL/BLB for inner-product computation. This implies that cells can be exposed to low BL/BLB voltages, as in a write condition. But, given the low WL voltages in Classify Mode, Monte Carlo simulations show that, with typical cell sizing, less stringent margin is needed than for standard read/write.

BL/BLB discharge also faces two sources of non-linearity. The first is the Q_{BC} transfer function in Fig. 3a (shown with BL/BLB statically connected to V_{DD}). This is mitigated by (1) including a constant offset current source in the WLDAC, and (2) applying the inverse of the nominal transfer function to input features. The second is non-constant BL/BLB discharge, shown in Fig. 3b. This is due to both I_{BC} reduction with reducing V_{DS} across the access transistors of bit cells pulling down, and competing charging current from bit cells pulling up. However, such monotonic effects due to the discharge level do not impact the final BL/BLB comparison required for classification (we note that with finite comparator offset there will be some impact, but this can be addressed by EACB).

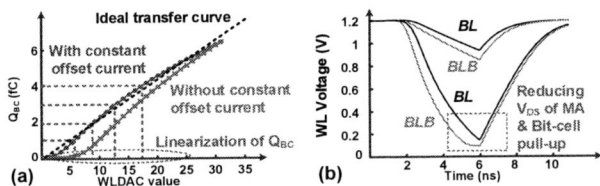

Fig. 3: (a) BL/BLB discharge (Q_{BC}) transfer function, and (b) non-constant bit-line discharge due to pull-down/-up current variation.

To support the large dynamic range on BL/BLB, the system must incorporate a comparator capable of rail-to-rail inputs. The circuit used is shown in Fig. 4a. A simple mechanism for offset compensation is described below to enhance the system.

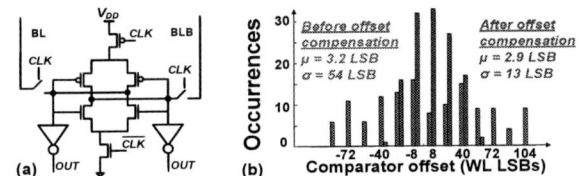

Fig. 4: (a) Rail-to-rail comparator circuit, and (b) measured offset histograms (128 comparators) before/after compensation.

Classifier Training

EACB (and AdaBoost) are meta-algorithms, which bias the way weak classifiers are trained at each iteration by increasing the emphasis on data instances that are incorrectly classified. However, an appropriate base training algorithm is needed for the weak classifiers themselves. For linear classifiers, L2 linear regression is often employed to train the weights $\vec{w}$, via the cost function: $\sum_i \| \vec{x_i} \cdot \vec{w} - y_i \|_2^2$ (where x_i are the training feature vectors and y_i are the corresponding training labels). This leads to quadratic convex optimization. However, in this system, weights can only be +/-1. Simple quantization after training would yield highly sub-optimal performance. Instead, we add an explicit optimization constraint, restricting the elements of $\vec{w}$ to be +/-1. This leads to mixed integer-programming optimization, for which fast methods exist, integrated into readily available solvers.

Though EACB has previously shown to overcome nonlinearities (as in the Q_{BC} transfer function) and offsets (as in the comparators) [3], minimizing these can reduce the number of weak-classifier iterations needed. Q_{BC} nonlinearity is compensated as described previously (Fig. 3a). Comparator offset is compensated by allocating some SRAM rows to an offset-compensation set. All rows in this set receive the same WLDAC code, resulting in the same nominal Q_{BC}. During offset compensation, other rows in the array are selected to discharge BL/BLB by a nominally equal amount. Preferably, various random selections of rows are employed and averaged, to mitigate bit-cell variability. Then, for all columns (done in parallel), data is written to bit cells in the offset-compensation set to equalize the probability of either comparator state (i.e., setting the trip point). This is done in a binary-search manner for rapid convergence. For the demonstration, 32 rows are designated to the offset-compensation set, with WLDAC code of 16. Fig. 4b shows the measured pre-/post-compensated offset over 128 comparators, in terms of the difference in WLDAC code for BL/BLB discharge at the trip point.

Measurement Results and System Demonstration

The prototype is shown in Fig. 5, with a measurement summary. Though bit-cell device sizing in the 128×128 SRAM array is that conventionally used, the cell layout is larger to meet logic design rules. MNIST digit classification is demonstrated by employing 45 binary classifiers between all pairs of digits, and performing all-vs.-all (AVA) voting. Due to prototype column limitations, 128 classifiers are tested at a time for the boosting iterations required in EACB (ideally, enough SRAM columns would be implemented for all classifiers and iterations needed). Due to prototype row limitations, the 28×28 images are down-sampled to a set of 82 pixel features (this reduces classification accuracy from 95% to 90%). Using these features, the classification accuracy vs. boosting iterations is shown for: (1) an ideal conventional SRAM/digital-MAC system based on linear classifiers, requiring 10b weights (from simulations); (2) an ideal conventional SRAM/digital-adder system based on linear classifiers restricted as proposed to +/-1 weights; and (3) the demonstrated system. The energy of each system is shown for 10-way digit classification, taking into account the differences in iterations (assuming 5.3pJ for MAC, 0.42pJ for adder, and 0.12pJ per SRAM bit access, all measured or post-layout simulated). The proposed system employs 18 iterations of EACB, achieving 113× and 12× energy savings, respectively.

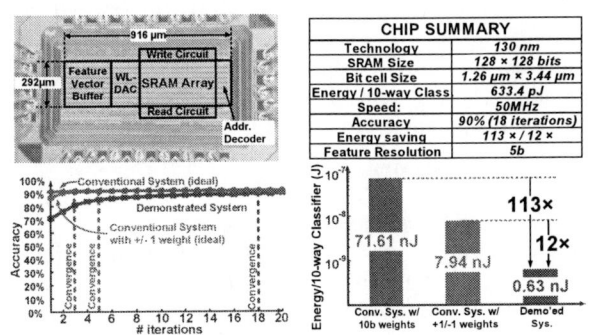

Fig. 5: Die photo and measurement summary.

[1] M. Kang, et al., *ISCAS*, pp. 2505-2508, May 2015.
[2] R. Schapire, et al., Boosting: Foundations and Algorithms, 2012.
[3] Z. Wang, et al., TCAS-I, vol.62, no.4, pp.1136-1145, April 2015.

A 16-Channel 1.1mm^2 Implantable Seizure Control SoC with Sub-µW/Channel Consumption and Closed-Loop Stimulation in 0.18µm CMOS

Mahsa Shoaran[1], Masoud Shahshahani[2], Masoud Farivar[1], Joyel Almajano[3], Amirhossein Shahshahani[2], Alexandre Schmid[2], Anatol Bragin[3], Yusuf Leblebici[2], Azita Emami[1]

[1]California Institute of Technology, Pasadena, CA, [2]EPFL, Lausanne, Switzerland
[3]University of California, Los Angeles, CA

Abstract

We present a 16-channel seizure detection system-on-chip (SoC) with 0.92µW/channel power dissipation in a total area of 1.1mm² including a closed-loop neural stimulator. A set of four features are extracted from the spatially filtered neural data to achieve a high detection accuracy at minimal hardware cost. The performance is demonstrated by early detection and termination of kainic acid-induced seizures in freely moving rats and by offline evaluation on human intracranial EEG (iEEG) data. Our design improves upon previous works by over 40× reduction in power-area product per channel. This improvement is a key step towards integration of larger arrays with higher spatiotemporal resolution to further boost the detection accuracy.

Introduction

Patients with intractable epilepsy can benefit from devices that perform automatic seizure detection and responsive stimulation [1-3]. Available EEG-based systems (e.g. [1]) are non-invasive, but suffer from limited resolution and high susceptibility to artifacts. Utilizing intracranial EEG signals [2, 3] improves the seizure detection accuracy, but inevitably poses tight power and area constraints on the implantable system. Integrating a large number of wide-BW iEEG channels discourages the application of conventional methods, such as spectral energy extraction in multiple bands or frequency spectrum computation. Specifically, extracting features from each channel separately requires either a dedicated ADC or extra TDMs and buffers. It also increases the dimensionality of feature space, potentially degrading the classification performance due to the curse of dimensionality.

In order to efficiently exploit multi-channel data in dense iEEG arrays, we combine every 16-channel x_i into a single $\tilde{x}$ using a linear spatial filter with i.i.d. random Bernoulli weights (w) [4]. The combined signal $\tilde{x}$ (Fig. 1) not only contains the essential discriminative information for epileptic activity classification, but also the crucial information for offline reconstruction of the original channel signals based on Compressive Sensing theory. Compared to [4] which relied on a single feature to detect seizures, this work integrates an optimal set of features and additionally includes a responsive closed-loop neural stimulator, with the full system verified in-vivo. This design improves the state-of-the-art by over 40× reduction in power-area product per channel and is readily scalable to larger arrays, given its small area and low power.

System Architecture

The proposed system architecture is shown in Fig. 1. Each front-end channel incorporates a dual-stage capacitive-coupled (CC) fully differential low-noise amplifier. The analog outputs of the amplifiers pass through a low-power (320nW) SC spatial filter followed by an 8b SAR ADC employing an area-efficient binary-weighted split array (total capacitance: 3.3pF). The feature extraction processor is subsequently activated. Upon seizure detection, the current-mode neural stimulator is triggered to suppress impending seizures.

Fig. 2 shows the AFE and spatial filtering stage (SFS). The first-stage CC-LNA provides a gain of 30dB at a small input capacitance (1.5pF) and bias current (700nA). A conventional CC topology with reduced-size unit capacitors is employed in the second stage (12dB). The SFS is composed of a SC summing amplifier that performs the $w^T x$ vector multiplication. The feedback attenuation factor (C_F/C_i) is adjustable by switching in the required number of unit capacitors. Through channel multiplication by 0 or 1, the dynamic range is efficiently boosted to accommodate the full-scale ADC input. Fully-differential folded-cascode OTAs are implemented in channels, while a two-stage OTA with rail-to-rail swing is utilized in SFS.

The seizure detector unit extracts four time-domain features mathematically defined in Fig. 3. In addition to widely used power and line-length features [5], the correlation of maxima and minima in successive iEEG windows is used to capture the high-frequency oscillations (HFOs). Lastly, the moving average of absolute value is computed, as a fast and sensitive but moderately specific detector. In a patient-specific training phase, the single optimal feature, or a logical combination of features is assigned to trigger the stimulator. To reduce detection delay, feature vectors are calculated for windows of length 256 (64ms at 4kS/s). The detection flag is raised upon 3 consecutive threshold crossings to limit the number of false positives. The data collector provides parallel threshold input to the arbiter, while the raw feature vector is generated at the output of a 16b PISO. Combined with stimulator logic, FE consumes 240nW at 4kHz CLK and 0.8V supply, 68% of which is leakage.

Fig. 4 shows the architecture and timing diagram of the stimulation block. Upon seizure detection from a 16-channel subset of electrodes, an adjustable charge sequence is delivered to the stimulation site. Biphasic pulses with 5b varying amplitude (<818µA), frequency (<1kHz), width, and burst duration (BD) are generated at the output of the current driver with a gain of 8, avoiding extra power consumption in the bias branch of binary DAC. In non-seizure mode, the entire current of stimulator including the static bias of DAC and output driver is turned off.

Experimental Results

The individual front-end channel draws 1µA from a 0.8V supply, has a midband gain of 42dB (Fig. 2), and exhibits an integrated input-referred noise of 5.9µV$_{rms}$ (10Hz-10kHz) and NEF of 2.94. Fig. 2 further presents the measured time-domain output of SFS for sample iEEG signals applied to the chip.

The measured output voltage of stimulator using a 1kΩ resistive load is shown in Fig. 4, with a measured current mismatch of 0.7% at 560µA. The stimulator consumes 360µW from a 3.3V supply (560µA, 200Hz) at continuous stimulation mode. Since the entire SoC operates in a closed loop, the stimulator is rarely active and has a negligible contribution to total power of the system.

Fig. 5 shows the results of an in-vivo study of the proposed system in a freely moving adult Wistar rat, implanted with 16 recording electrodes in neocortex and hippocampus and a pair of stimulating electrodes in hippocampal commissure. Epileptiform activities were evoked by injection of 0.4μg/0.2μl kainic acid into left dorsal hippocampus. Using the logical conjunction of four criteria, the seizure events are successfully detected and suppressed by responsive stimulation, as shown in Fig. 5. During a four-hour continuous iEEG monitoring, 93% of total seizures (25 of 27) were successfully detected with a maximum delay of 0.5s, and subsequently terminated by responsive stimulation (reviewed by an epileptologist).

Our seizure detection algorithm was further evaluated offline, using 420h of human iEEG data containing 23 seizures from four patients at Bern University Hospital. It achieved a sensitivity of 100% and an average false alarm rate (FAR) of 0.15/h, improving the specificity by 56% compared to [4].

Fig. 5 presents the chip micrograph and layout of a front-end channel implemented in 0.18μm CMOS. This system improves the state-of-the-art in terms of both power consumption and chip area while maintaining a reasonable accuracy, as shown in the comparison table.

References

[1] M. Altaf et al., *IEEE ISSCC*, Feb. 2015.
[2] W. Chen et al., *IEEE ISSCC*, Feb. 2013.
[3] K. Abdelhalim et al., *IEEE JSSC*, Oct. 2013.
[4] M. Shoaran et al., *IEEE TCAS-II*, Feb. 2015.
[5] L. Logesparan et al., *Med Biol Eng Comput*, 2012.

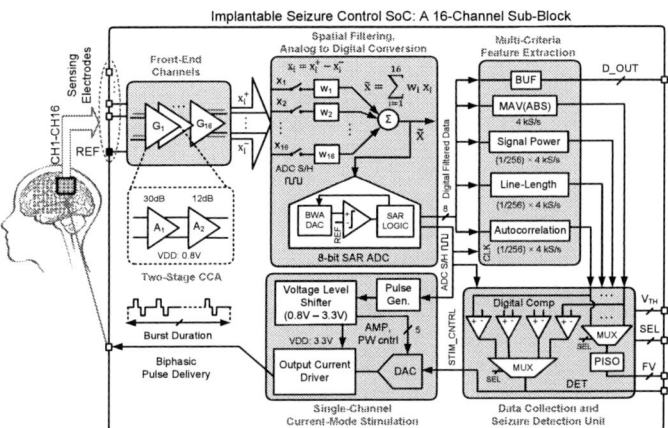

Fig. 1 Architecture of the proposed closed-loop seizure control system.

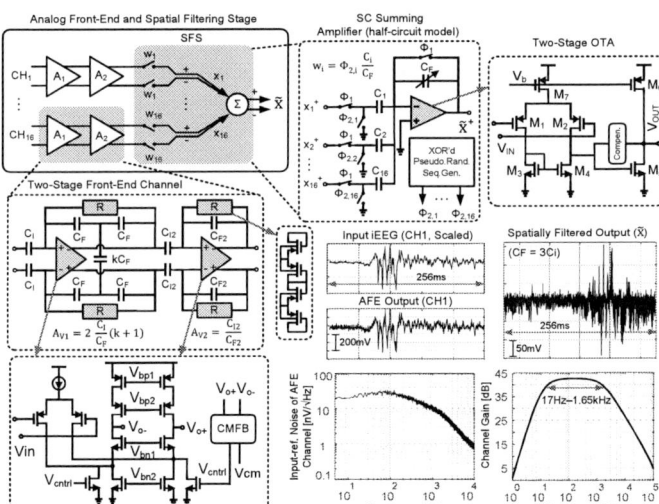

Fig. 2 AFE and SFS circuit diagrams with measured time-domain outputs, input-referred noise and gain of channels.

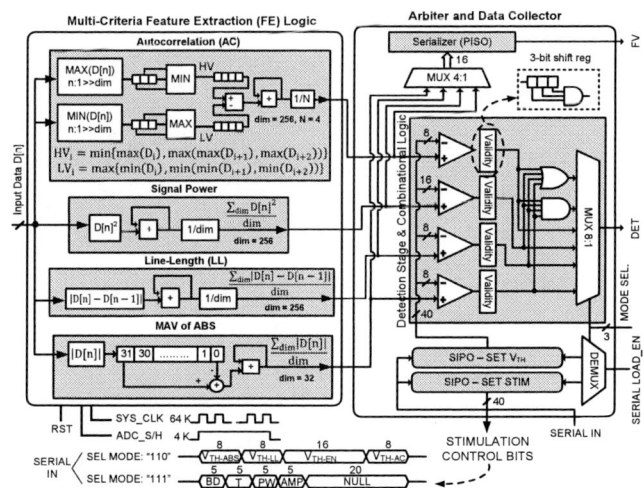

Fig. 3 Seizure detection processor including the feature extraction blocks, arbiter, and the data collector.

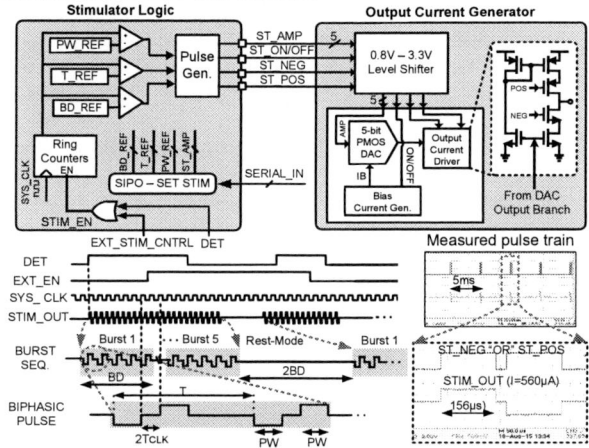

Fig. 4 Current-mode neural stimulator, the corresponding timing diagram, and measured output voltage on 1kΩ resistive load.

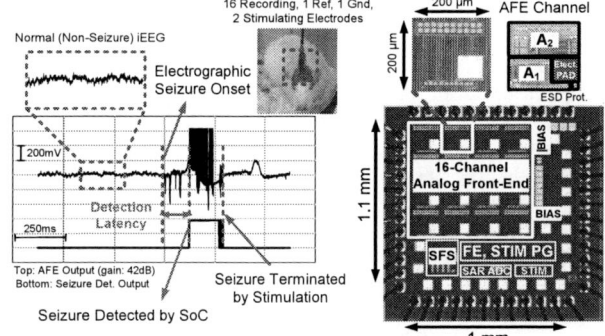

Fig. 5 (Left) In-vivo test results of the closed-loop SoC. (Right) Die micrograph and layout of front-end channels.

Table 1. Performance Summary and Comparison

Parameter	[1]	[4]	[2]	[3]	This work
Technology (CMOS)	0.18μm	0.18μm	0.18μm	0.13μm	0.18μm
Signal Modality	EEG	iEEG	iEEG	iEEG	iEEG
Chan. Count	1 to 16	16	8	64	16
Area [mm²]	25	1[d]	13.46[a]	12[b]	1.1[d]
Closed-Loop	YES	NO	YES	YES	YES
Feature Type	FTDM-BPF	LL	Entropy, Freq. Spect.	Mag, Phase, PLV	Energy, LL, AC, ABS
Multi-Feature Detection	NO	NO	YES	YES	YES
Bandwidth [Hz]	0.5–100	30–1.7k	1–7k	1–5k	17–1.65k
Sample Rate/Chan.	1kS/s	4kS/s	62.5kS/s	54kS/s	4kS/s
Power	-	13.6μW	2.798[a]mW	1.4[c]mW	14.8μW
Sensitivity (%)	95.7	100	92	100, 70	100
FAR (Specificity)	98%	0.34/h	-	1.2-2/h, 0.6/h	0.15/h

[a]Including a wireless power supply generator, TX and stimulator
[b]Including an stimulator/channel and an UWB TX
[c]Including an UWB TX   [d]Active area (excluding pads)

A Microelectrode Array with 8,640 Electrodes Enabling Simultaneous Full-frame Readout at 6.5 kfps and 112-Channel Switch-Matrix Readout at 20 kS/s

X. Yuan[1,2], S. Kim[1], J. Juyon[1], M. D'Urbino[1,3], T. Bullmann[1], Y. Chen[2],
A. Stettler[2], A. Hierlemann[2], U. Frey[1,2]

[1] RIKEN, Kobe, Japan; [2] ETH Zurich, Zurich, Switzerland; [3] TU Delft, Delft, The Netherlands
E-mail: xinyue.yuan@riken.jp

Abstract

CMOS microelectrode arrays allow for recording from neurons at thousands of sites. Here, we introduce the concept of a 'dual-mode operation' microelectrode array, leveraging the advantages of full-frame scanning and switch-matrix array architectures into a single device. The chip was fabricated in 0.18 μm CMOS technology. Measured noise levels were 11.1 μV_{rms} for full-frame scanning and 1.6 μV_{rms} for switch-matrix mode at 3.3 μW and 38.1 μW per channel power consumption. Recordings of electrical activity from cultured neurons have been successfully conducted. (Keyword: CMOS MEA, full-frame, and switch-matrix)

Introduction

Microelectrode arrays (MEAs) on planar substrates or on needle-shape probe devices are used to record from and stimulate neuronal cells *in vitro* and *in vivo*. CMOS circuitry allows for time-division multiplexing from a large number of electrodes at subcellular spatial resolution. The technology trend for *in vivo* and *in vitro* MEAs goes towards ever-larger arrays at higher electrode densities and better SNR [1]. Two principal array architectures have become established, namely active-pixel sensor (APS) arrays, featuring full-frame readout, and switch-matrix (SM) arrays, providing low-noise readout of arbitrarily selectable electrode subsets. Typically, APS devices enable simultaneous recordings from thousands of electrodes, but with sub-optimal SNR due to the limited area available for the AFE in each pixel. In contrast, SM devices enable recordings at best possible SNR, since the AFE can be placed outside the array; however, only a small subset of the electrodes can be recorded from simultaneously. As a result, applications of APS devices have been focused more on network level studies, while scientific experiments with SM devices were targeted at single-cell studies, such as investigating axonal physiology. To overcome some of these limitations, we have designed and fabricated a dual-mode MEA (DM-MEA), which leverages the advantages of both, APS and SM architectures and enables the acquisition of richer datasets. For a variety of experiments, DM is an enabling technology, as it simultaneously facilitates both, full-frame readout and high SNR.

System architecture

Fig. 1 shows the system architecture of the DM-MEA with the electrode array, SM and APS readout circuits. The array is composed of 96×96 pixels with an electrode density of 3050 mm^{-2} (pitch 18 μm). To reserve as much area as possible for the in-pixel APS AFE, a SM design with a 1b 6T-SRAM per pixel was chosen [2]. The total number of electrodes is 8,640 (hexagonal arrangement), and the array area is 1.6×1.8 mm^2.

For APS, the array is divided into two 48×96 sub-arrays, with each being split into 12 tiles of 48×8 pixels. Each pixel has one APS AFE. One half-column shares one column

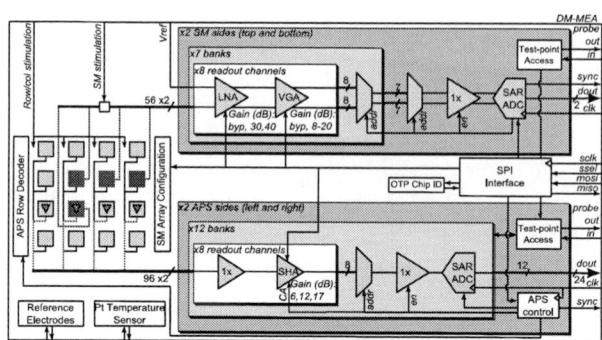

Fig. 1 Block diagram of the DM-MEA including electrode array, SM and APS readout.

circuitry, which consists of a buffer and a SHA with 6-17 dB gain (Fig. 2). After the SHA, the signal is multiplexed through an 8:1 MUX and finally digitized by an 8-bit SAR ADC with a sampling rate of 2.5 MS/s. For full-frame readout, the rows are selected by a row decoder. The most critical part of the APS circuit is the in-pixel AFE. Due to the limited pixel area, there is a direct trade-off between area and noise performance. In this design, an AFE with two single-ended amplification stages was implemented, as shown in Fig. 2. Capacitors $C_{1,2}$ provide proper biasing and couple AC signals to an input stage, which is an inverter-based amplifier, built around $M_{1,2}$ to achieve optimal SNR. M_{3-5} are used as MOS-based voltage-controlled RC LPF. The switch $\overline{\Phi_2}$ is used to avoid crosstalk between the 'select' signal (Φ_2) and the 'input' signal. For the output stage, a common-source amplifier was used, which is activated only when the specific row has been selected to reduce power consumption. A HPF is also implemented in the in-pixel AFE to suppress the typical low-frequency signal fluctuations of neuronal interfaces. It involves the periodic reset (via Φ_1 with tunable frequency $f_{\Phi1}$) of the inverter-based amplifier, during which the amplifier's input and output are shorted, and the gate of M_2 is connected to V_{reset}, thus resetting its operating points. The row decoder supports a rolling reset with adjustable delay t_{delay} and pulse width $t_{\Phi1}$.

For the SM mode, two banks of 56 channels and two 10-bit SAR ADCs have been implemented. Since the AFE is located outside the array, it is not limited in area. The SM readout circuit has a fully differential structure consisting of an LNA with a gain of 30 or 40 dB, a VGA with 0-20 dB gain, a 56:1 MUX, and a SAR ADC (Fig. 1). The LNA is a Miller-compensated closed-loop, two-stage amplifier with an AC-coupled input stage, which is designed to achieve optimal power-noise tradeoff [3].

The DM-MEA enables electrical stimulation of neurons by providing two options of routing electrodes to off-chip V/I stimulation buffers. One routing option makes use of the SM routing and can be used to flexibly address any electrode. The

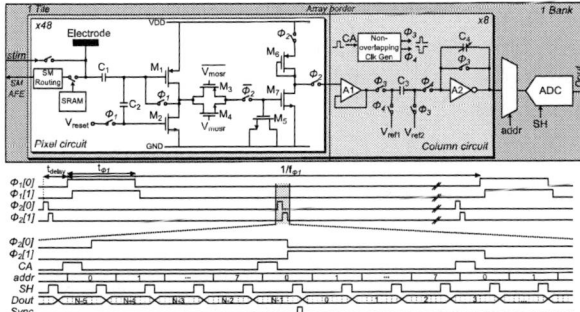

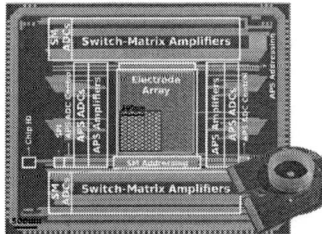

Fig. 3 Die micrograph with an inset showing an enlarged subset of the electrode array. Packaged chip with culture medium chamber at the lower right.

Fig. 2 Schematic and timing diagram of the APS readout.

second routing option is column/row-based and independent of the SM and APS wiring, and it proved to be useful not only for stimulation, but also for testability purposes.

Measurement and conclusion

The chip was fabricated in 0.18 μm 6M1P CMOS (Fig. 3) with a chip size of 6.4×5.2 mm². Pt electrodes were fabricated through post-processing, using a previously published shifted-electrode design [4]. Pt-black was deposited by electroplating to lower the electrode impedance.

Fig. 4 shows the gain and noise measurement results. Noise measurements were done for the circuits only (w/o el), and for amplifiers connected to Pt-black electrodes immersed in saline solution (w/ el), showing the cumulative noise for total band, action potential (AP) band and spike sorting (SS) band. In SM mode, the LNA and VGA were set to 40 dB and 14 dB. For the APS mode, the measurement was performed by using an off-chip ADC to test the performance of the in-pixel AFE. $f_{\Phi 1}$ was set to 8 Hz, which leads to voltage jumps in the recorded data during each reset. A simple template construction and subtraction algorithm was used to correct for these expected artifacts. The results are shown in Fig. 4 and included in the performance summary shown in Tab. 1 and compared to [4-7]. The total measured power consumption was 32.6 mW, including all circuits (IO, biasing, on-chip ADCs, etc.).

Fig. 5 shows recording results from cultured neurons (E18 Wistar rat), including a 2D activity map of spike amplitudes over the entire array, obtained with APS full-frame readout, data of a selected subset by the SM readout, the spike train measured on a selected electrode simultaneously by SM and APS mode, and the spike waveform of one neuron measured by SM readout. All results were obtained with the on-chip ADCs and a calibrated gain to get accurate results.

This first prototype demonstrates the DM-MEA concept. In APS mode, we achieve the lowest noise at the lowest power consumption per channel ever reported for devices with a similar spatial resolution. In SM mode, we achieve much lower noise for selected electrodes in comparison to the APS mode, and a performance comparable to that of other published SM devices. Both APS and SM modes are readily scalable so that we envisage their use in future MEAs with significantly more electrodes.

References

[1]. M. Obien et al., Front. Neurosci., v 8, n 423, 2015.
[2]. U. Frey et al., ISSCC Dig. Tech. Papers, pp. 158–593, 2007.
[3]. D. Han et al., ISSCC Dig. Tech. Papers, pp. 290-291, 2013.
[4]. M. Ballini et al., IEEE JSSC, v 49, pp. 2705–2719, 2014.
[5]. L. Berdondini, et al., Lab Chip, v 9, pp. 2644–51, 2009.
[6]. B. Johnson et al., BioCAS, pp. 109–112, 2013.
[7]. G. Bertotti, et al., BioCAS, pp. 304–307, 2014.

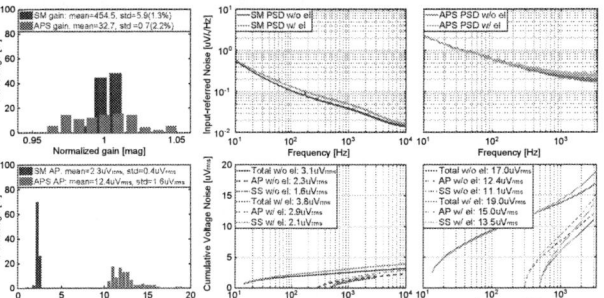

Fig. 4 Measurement and characterization results of the SM (on-chip ADC) and APS (off-chip ADC) readout channels.

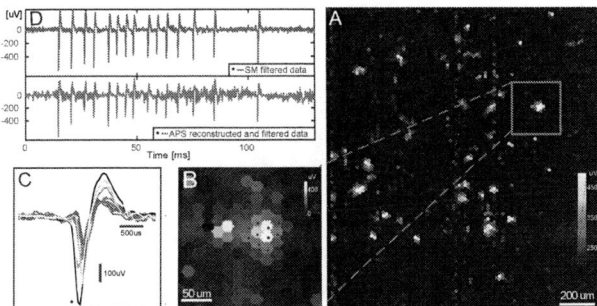

Fig. 5 Recordings from cultured neurons. A. 2D activity map (spike amplitude) by APS readout for the whole array. B. Activity map (spike amplitude) by SM readout for an array subset. C. Waveforms of a single spike recorded by different electrodes with the SM readout as marked by dots in (B). D. Spike train measured at one selected electrode (marked by a star in B) by simultaneously using SM and APS mode.

Tab. 1 Comparison to state-of-the-art CMOS-based MEAs.

		[4]	[5]	[6]	[7]	This work	
Technology [μm]		0.35	0.35	0.18	0.18	0.18	
Mode		SM	APS	APS	APS	APS	SM
No. electrodes		26400	4096	1120	4225	8640	
No. channels		1024	4096	1120	4225	9216	112
Electrode density [mm⁻²]		3265	567	400	977	3050	
Pixel pitch [μm]		17.5	42.0	50.0	16.0	18.1	
Frame rate [kfps]		20	8	20	77	6.5	20
ADC resolution		10	off-chip	10	off-chip	8	10
Input noise [μV$_{rms}$]	Total	5.9 (1Hz -10kHz)	26 (-)	4.3 (20Hz -9kHz)	52 (1Hz -10kHz)	17.0 (10Hz -3.3kHz)	3.1 (10Hz -10kHz)
	AP band	2.4 (300Hz -10kHz)	-	-	44 (300Hz -10kHz)	12.4 (300Hz -3.3kHz)	2.3 (300Hz -10kHz)
	Spike sorting band	1.8 (500Hz -3k Hz)	-	-	-	11.1 (500Hz -3kHz)	1.6 (500Hz -3kHz)
Power/ch [μW]		73	32	12.6	-	3.3	38.1

A Wearable Ear-EEG Recording System Based on Dry-Contact Active Electrodes

Xiong Zhou[1,2], Qiang Li[2], Søren Kilsgaard[3], Farshad Moradi[1], Simon L. Kappel[1] and Preben Kidmose[1]

[1]Department of Engineering, Aarhus University, Aarhus, Denmark
[2]ISL/ME/ETFID, University of Electronic Science and Technology of China, Chengdu, China
[3]Widex A/S, Copenhagen, Denmark

Abstract - This work reports an ear-EEG acquisition system with dry-contact active electrodes for future wearable applications. Employing dedicated chopper buffer in the active electrodes, a prototype fabricated in a 0.18-μm CMOS demonstrates input impedance as large as 18GΩ@DC and 6.7GΩ@50Hz, and 3.03fA/√Hz input current noise, and a total input-referred noise (IRN) of 0.67μV$_{rms}$ in 0.5-100Hz bandwidth. System's CMRR in combination with active electrodes is higher than 100dB@DC. Under large source impedance imbalance of 1MΩ, a 78-dB@50Hz CMRR is still obtained. To validate the system's ability to record EEG, an auditory stead-state response was measured, showing same SNR with wet electrodes and a commercial EEG amplifier.

Ear-EEG [1] is a method for measuring electroencephalograms (EEG) from the outer ear, and it is envisioned to enable recording of EEG in natural environments in a discreet and user-friendly way. Biopotential recording using dry-contact active electrodes (AE) is an appealing alternative to conventional wet ones, providing advantages like ease of use, less preparation and long-term usability. With AE, amplification and impedance transfer are performed at the closest point and thereby signal's acquisition robustness is enhanced [2]. However, since dry electrode on bare skin exhibits high skin-electrode impedance (SEI), practical design of dry-contact analog front end (AFE) faces challenges in several aspects. First, current noise (i_n) must be scaled down in proportion to SEI's increase. Second, the obtainable CMRR is dominantly limited by imbalances of SEI and finite input impedance (Z_{in}), rather than mismatches in AFE. Third, DC rejection range of hundreds of mV is necessary. Unfortunately, satisfying all requirements is difficult, since it's hard to realize a good trade-off between voltage-domain and current-domain accuracies. For example, [3] illustrates usage of chopper modulation to remove 1/f noise and offset will inevitably give rise to penalties like higher i_n and reduced Z_{in}. In contrast, without chopping, design [4] preserves low i_n and ultra-high Z_{in} regardless of offset; although Z_{in} is higher than TΩ, sacrifice of ESD robustness makes it relatively unsuitable for dry-contact purpose. In this work, we propose a tailored system with advanced SEI-AFE interface. Based on chopper buffer AEs, it provides state-of-art performances, making ear-EEG technique feasible for wearable applications.

System Architecture As shown in Fig.1, the system is built on a personalized ear piece, consisting of two acquisition channels. Thanks to AEs, back-end's (BE) requirements on Z_{in} and i_n are largely relaxed. So in the BE capacitively-coupled instrumentation amplifier (CCIA) with 40-dB mid-band gain is designed [5]. A programmable gain amplifier (PGA) is followed to provide 6-to-36 dB gain and also serves as 2nd-order filter to suppress high-frequency spikes. Quantization is accomplished by a 13-bit sigma-delta modulator (SDM). In addition, an optional transconductance driven-right-leg amplifier (TDRL) is implemented to further improve CMRR.

Circuits Implementation A. *Active Electrode* Although AEs with gain are more attractive from the perspective of system's energy efficiency [2], unit-gain buffer is intentionally adopted with considerations of inter-channel matching and electrodes' shielding. Despite of more power dissipation, the benefits are two-fold: external trimming free and immediate use of active shielding. Fig. 2 illustrates the interface's nonidealities and concept of shielding. For AEs with gain, Z_{in} is normally improved by impedance boosting technique; here in unit-gain buffer, the strategy is using the active shielding to neutralize crucial capacitors e.g. input pair's parasitic capacitor (C_p) between the differential nodes and external parasitic ($C_{p,ext}$) on electrodes. Therefore, Z_{in} is enhanced. Noting that i_n is highly relevant to the switch-capacitor (SC) activity, it can be simultaneously reduced. Fig. 3 shows the proposed chopper buffer. As is shown, $M_{1,2}$, $M_{1c,2c}$ and $M_{3,4}$ are devised as local source followers (blue), which keep voltage drops between C_p (red) almost constant. In other words, these parasitic capacitors on the differential pair will not load the input. What's more, for maximized Z_{in} and minimized i_n the system was designed under a low chopper frequency (f_{chop}) of 1 kHz. Since bias current (i_b) has decisive impacts on i_n, any current flow should be prevented. Therefore, bulks of input chopper switches are shorted to the input and/or buffer's output instead of VDD or VSS, eliminating leakage path via substrate. Lastly, cascode topology is adopted for high open-loop gain. Compared with [6], high gain contributes to improve the inter-channel CMRR between AEs, without concerns of gain setting components.

B. Capacitively-Coupled Instrumentation Amplifier Fig. 4 shows the CCIA. It consists of the main amplifier, ripple reduction loop (RRL), DC servo loop (DSL) as well as impedance boosting loop. In the main amplifier, the core is a two-stage folded-cascode amplifier. DSL can give DC rejection range larger than ±200mV. Since f_{chop} is as low as 1 kHz, output saturation becomes an issue under low chopper frequency especially large electrode offset is given [5]. Decreasing bias resistance can relax this issue at cost of introducing higher thermal noise. In this design, a SC bias is employed. Instead of a fixed reference voltage, one terminal of each SC resistor is connected to the output of RRL. Whenever large ripples appear at CCIA's output, the proposed bias path would release the charge at the virtual ground from RRL's output.

Measurement The prototype was fabricated in a standard 0.18μm CMOS process, and consumes a total current of 120μA from 1.8V supply. Fig. 5 shows chip micrographs of one AE and BE ASIC. Measurement results are presented in Fig. 6. Top-left plot shows IRN of BE and complete readout channel. Chopping at 1 kHz leads to a total IRN of 0.67μVrms in 0.5-100Hz. Top-right plot shows measured input impedance. Z_{in} is close to 18GΩ@DC and 6.7GΩ@50Hz, which is at least 6 times higher than cutting-edge design [7]. Bottom-left plot shows input current noise density as low as 3.03fA/√Hz, which is 6 times lower than state of the art [8]. Lastly, bottom-right plot presents measured CMRR. The CMRR with AEs are higher than 100dB@DC. Noting that under 1-MΩ impedance imbalance (ΔR), CMRR is improved from 56 dB to 78dB@50 Hz by TDRL, which is 36dB higher than [8]. The proposed design is compared with state-of-the-art biopotential recording systems in Table I. To validate the system's capability to measure EEG, an auditory steady-state response (ASSR) was measured using a 40 Hz AM noise stimulus. A view of practical experimental setup is given in Fig. 7. Fig. 8 shows power spectrum of an averaged ear-EEG signal, exhibiting a clear 40Hz component of the modulation signal. This result is on par with ASSR's SNR in previous work [1] measured with wet electrodes and a commercial amplifier.

References

[1] Preben Kidmose, et al., "A Study of evoked potentials from ear-EEG," *IEEE TBE*, 2013.
[2] Jiawei Xu, et al., "A wearable 8-channel active-electrode EEG/ETI acquisition system for body area networks," *IEEE JSSC*, 2014.
[3] Xu, J., et al., "Measurement and analysis of current noise in chopper amplifiers," *IEEE JSSC*, 2013.
[4] Yu M. Chi, et al., "Ultra-high input impedance, low noise integrated amplifier for noncontact biopotential sensing," *IEEE JESTCS*, 2011.
[5] Fan, Q., et al., "A 1.8 W 60nV/√Hz capacitively-coupled chopper instrumentation amplifier in 65 nm CMOS for Wireless Sensor Nodes," *IEEE JSSC*, 2011.
[6] Guermandi, M., et al., "Active electrode IC for EEG and electrical impedance tomography with continuous monitoring of contact impedance," *IEEE TBCAS*, 2015.
[7] Ha, U. "A wearable EEG-HEG-HRV multimodal system with real-time transcranial electrical stimulation monitoring for mental health management," *IEEE ISSCC*, 2015.
[8] Jiawei Xu, et al., "A 15-channel digital active electrode system for multi-parameter biopotential measurement, "*IEEE JSSC*, 2015.

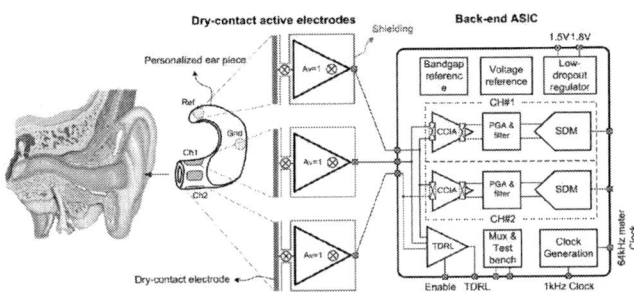

Fig. 1. System's block diagram.

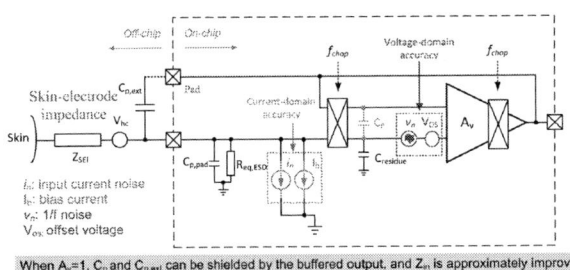

Fig. 2. Nonidealities in SEI-AFE interface and the concept of shielding.

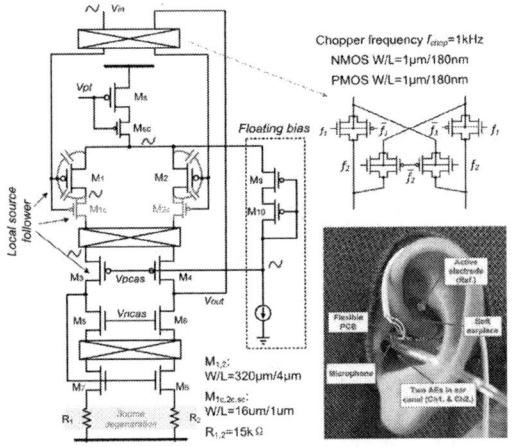

Fig. 3. Chopper buffer.

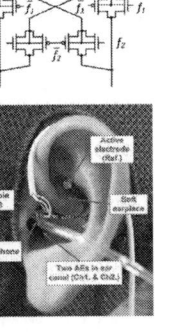

Fig. 7. Experimental setup.

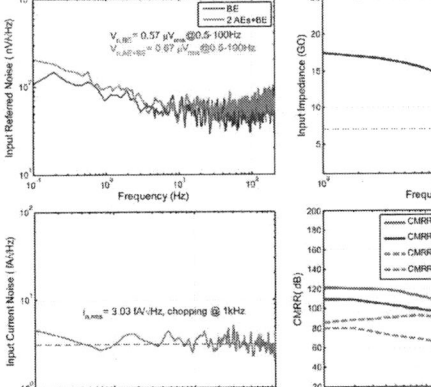

Fig. 6. Measured performance of the readout channels.

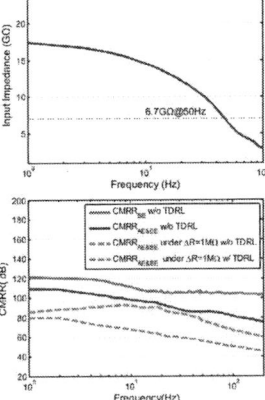

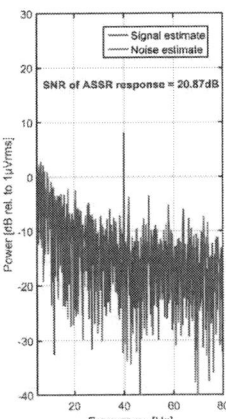

Fig.8. Measured ASSR.

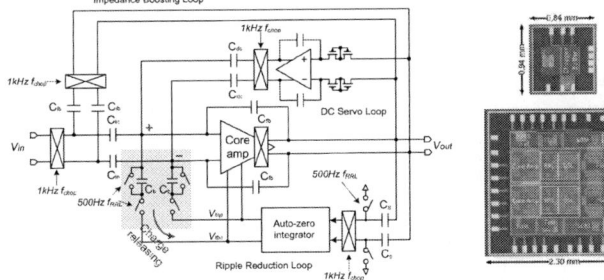

Fig. 4. CCIA's topology.

Fig.5. Die photos.

Table I. Comparison with the state of the art.

Parameters		[2]	[6]	[7]	[8]	This work
Supply (V)		1.8	3.3	1.2	1.8	1.8
Active electrode		√	√	×	√	√
IRN (µVrms)		1.75	0.28	1.2	0.65	0.67
Input impedance	@DC	-	~2GΩ	-	~1GΩ	~18GΩ
	@50Hz	~300MΩ	100MΩ	1GΩ@60Hz	~100MΩ	6.7GΩ
Current Noise (fA/√Hz)		-	-	-	20@fchop=4kHz	~3@fchop=1kHz
Offset tolerance (mV)		±250	±200	±200	±350	>±200
CMRR (dB) @50Hz	w/o CMRR enhancement	~60	64	100	<60	86
	w/ CMRR enhancement	84 (CMFF)	-	132 (CMFB)	102(CMFF)	108 (TDRL)
	w/ impedance imbalance	-	-	-	42/ΔR=800kΩ	78/ΔR=1MΩ
Current per channel (µA)		46	120	7.4	58	60
Dry-contact application		√	×	×	√	√

978-1-5090-0636-6/16 $31.00 © 2016 IEEE

A 2.048 Mb/s Full-Duplex Free-Space Optical Transceiver IC for a Real-Time In Vivo Neurofeedback Mouse Experiment Under Social Interaction

Gunpil Hwang[1], Jong-Kwan Choi[1], Jaehyeok Yang[1], Sungmin Lim[1], Jae-Myoung Kim[1], Min-Gyu Choi[1], Dae-Shik Kim[1], Kiuk Gwak[1], Jinwoo Jeon[1], Hee Sup Shin[2], Il-Hwan Choi[2], Sol Park[2], Hyeon-Min Bae[1]

[1]KAIST, Daejeon, Korea, [2]Institue for Basic Science, Daejeon, Korea E-mail: gphwang@kaist.ac.kr

Abstract

We report the first free-space optical transceiver IC for social psychological neurofeedback experiments with multiple mice. The proposed IC, which includes a neural stimulator and neural recorder, is embedded in the head-mounted module of a mouse and communicates with an optical base station by using a mandatory 615 nm visible behavior tracking LED to save power. The proposed IC fabricated in a 0.18 μm BCDMOS process consumes 90 μW for optical TX while achieving a 2.048 Mb/s transmission rate.

Introduction

Social psychological stress is considered a major factor in the development of cardiovascular disease, cancer, diabetes, depression, and immunodeficiency. In order to effectively relieve such stress, understanding the brain mechanisms during social interaction behavior is important. In vivo brain research relies solely on animal testing, and diverse brain-monitoring devices for animal brain studies have been developed [1-4]. However, prior works are inappropriate for in vivo brain study during social activity: (1) they only considered a neural recording/stimulation system for a single mouse or (2) consumed excessive power from the allocation of individual channels for data transmission (RF) and behavior tracking (visible light) [3]. In this paper, we report a low-power full-duplex free-space optical transceiver (TRX) IC that enables neurofeedback (NFB) for in vivo experiments involving multiple mice.

Architecture

A behavioral experiment using multiple mice is conducted in a 6 lux dark chamber without any interference from external light sources. This makes free-space optics an appropriate choice of communication (Fig. 1). A 940 nm infrared (IR) LED mounted on an optical base station (OBS) is used to broadcast packetized control signals to the head-mounted modules (HMMs) of two mice. The broadcasted data is also used to synchronize the system clock of each HMM, which is important for the demodulation and separation of the CDMA-modulated neural recording data transmitted from the two mice. The HMM is powered by a single Li battery (CR2016, 3 V 90 mAh). In order to reduce the power consumption of the HMM, a mandatory 615 nm visible behavior tracking (BT) LED is used to not only 1) track a mouse in a dark test environment but also 2) transmit the recorded neural signal to the OBS.

Circuit Implementation

A. Full-Duplex Free-Space Optical Transceiver

A clock-embedded packetized data optically broadcasted from the OBS is detected/amplified/sliced by the AFE in the optical RX of the HMM, as shown in Fig. 2. Then, a dual-loop reference-less CDR following the AFE extracts a system clock. In the beginning, a pilot tone is broadcasted from the OBS, and an FLL carries the frequency of the ring VCO in the vicinity of the target frequency by using a rotational frequency detector (RFD). Once the frequency acquisition is completed, the packetized data is transmitted from the OBS, and a PLL achieves phase lock by using a bang-bang phase detector (BBPD). The broadcasted data from the OBS is packetized to include an alternating sequence prior to the payload, which ensures a sufficient transition density for a stable phase lock. The bandwidth of the FLL is set significantly lower than that of the PLL to prevent lock failure. Once the 16.384 MHz system clock is phase-locked to the data packet from the OBS, a synthesized packet analyzer extracts command words including the start of recording, PGA gain, and stimulation patterns of the corresponding mouse. The 8 ch-8 bit neural recording data (Fig. 2) is serialized and then CDMA-modulated by using 2b Walsh codes. The modulated signal is level-shifted from 1.8 V to 3 V to drive a visible LED which has the threshold voltage of 2 V. The current of the LED driver is made controllable (0.75-to-2.5 mA) in the digital domain to strike a balance between power and performance under diverse experimental conditions.

B. Neural Recorder and Stimulator

A microelectrode (tetrode wires and Harlan 4 Drive, NeuraLynx, USA) is implanted into each mouse's cerebral cortex (Fig. 3) to monitor the local field potential (LFP) and extracellular action potential (EAP) caused by depolarization of the membrane. Since the impedance of microelectrodes varies because of biochemical reactions, unity gain buffers are inserted to segregate the microelectrode from the PGA. A PGA implemented with AC-coupled cascaded LNAs amplifies the LFPs and EAPs while suppressing the DC offset. The LFP and EAP have bandwidths and amplitudes on the orders of a few mHz-to-kHz and uV-to-mV, respectively. The gain of the PGA can be tuned by selecting the ratio of C1 to C2 (Fig. 3). In order to locate a lower 3 dB cutoff frequency below 1 Hz, PMOS pseudo-resistors R1 and R2 are employed [1]. The amplified LFPs and EAPs are quantized by a 16 kS/s single slope (SS) ADC. The neural spikes sampled by C3 are discharged by R3 at a rate of 610 mV/μs. A subsequent 8b counter computes the time required to reach Vcomp by using a 16.384 MHz CLKcount. The neural stimulator (Fig. 3) injects a monophasic intracranial stimulation current (MISC) into the medial forebrain bundle (MFB), which is the pleasure and reward center of the brain. The appropriate magnitude and duration of the MFB MISC are 60 μA and 100-to-200 μs, respectively. Since the impedance of the brain is on the order of 100 kΩ and the target stimulation current is 60 μA, a charge pump-based DC–DC converter [1] is integrated to boost the Li battery supply of 3 V to 7 V. The boosting clock frequency is implemented selectable from 100 k-to-1 MHz to adjust the boosting voltage depending on the impedance of the brain tissue while minimizing the power consumption. The stimulation magnitude, pulse duration, period, and number of iterations of the MISC are made controllable in the digital domain, which are 20-to-90 μA, 100-to-200 μs, 2.5 m-to-1 s, and 100-to-2700, respectively.

Measurement Results

The BER versus emitting power of LED mounted on the HMM is shown in Fig. 4. The BER is measured while changing the location (center and corner of the chamber) and the tilting angle of the HMM (0°–to–45°) which reflects the behavior and head motion of a mouse. Error free (with BER<10^{-7}) transmission is observed in overall angles in the center (0°–to–45°) and corner(<30°). In the severely tilted condition (>30°), the lens improves BER and attains 0.7–to–1.3 dBm power reduction at a BER of 10^{-6}. The 16 kHz 1024-divided system clocks from mouse 1 and mouse 2 are synchronized to the broadcasted optical data from the OBS, as shown in Fig. 5. The TX eye diagrams from the two mice show a delay of 20 ns (0.04UI) and long-term accumulated peak-to-peak jitter of 188 ns. The tunable gain of the PGA is shown in Fig. 6. The time-aligned measured EAPs can clearly be distinguished. The MISC patterns are adjustable to diverse patterns that are known to evoke the dynamic restitution of action potentials. The measured period and resolution of the MISC are 96 µs and 72 nA/LSB, respectively.

The proposed system (Fig. 6) demonstrated 1) a unique modularized RF-free NFB system that can be expanded up to 2 mice, records EAPs and LFPs (8 ch recorder), and injects the MISC (1 ch stimulator) while simultaneously chasing the behavior of the mice; and 2) a minimum TRX power consumption of 90 µW in addition to the intrinsic power consumption of a BT LED. To the best of our knowledge, the proposed system is the first wireless optical NFB system for mouse experiments with social interaction. The proposed IC is fabricated in a 0.18 µm 1P6M HV BCDMOS process.

Acknowledgements

This work was supported by Center for Cognition and Sociality, Institute for Basic Science under IBS-R001-D1.

References

[1] Azin, Meysam, et al. *JSSC*, Mar., 2011.
[2] Szuts, Tobi A., et al. *Nature neuroscience*, Apr., 2011.
[3] Triangle Biosystems International, www.trianglebiosystems.com.
[4] emka TECHNOLOGIES, telemetry.emka.fr, Mar., 2015

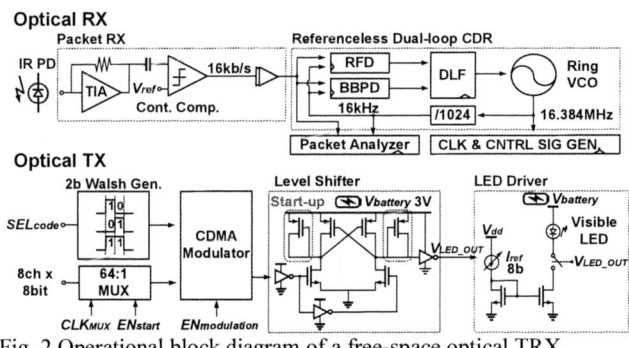

Fig. 2 Operational block diagram of a free-space optical TRX.

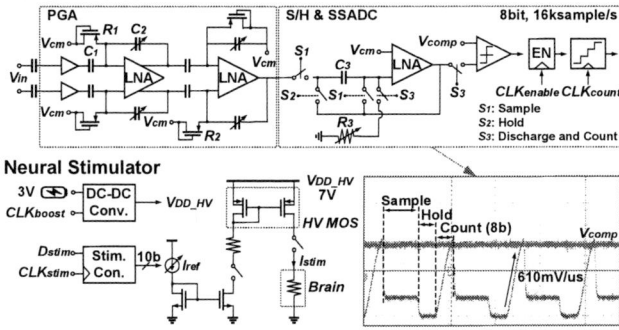

Fig. 3 Operational block diagram of a neural recorder and stimulator.

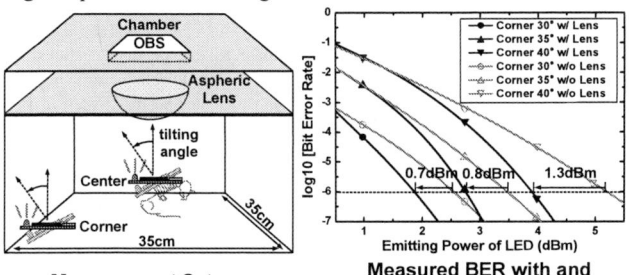

Fig. 4 Measurement results of free-space optical communication.

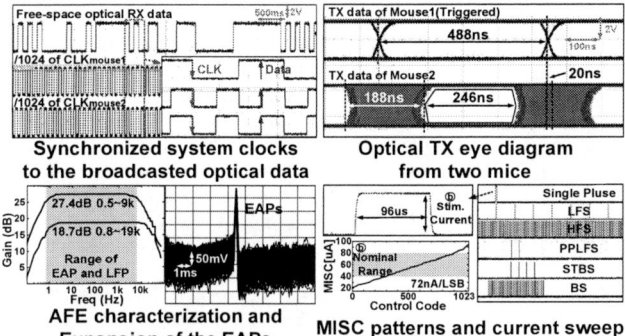

Fig. 5 Measurement results of the proposed IC.

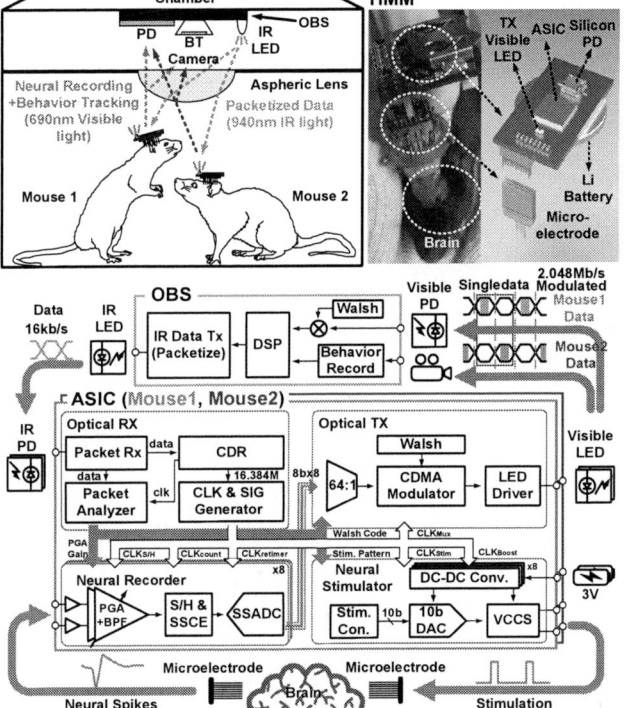

Fig. 1 Architecture of in vivo NFB system under social behavior using free-space optical communication.

Fig. 6 Chip microphotograph and performance comparison table.

978-1-5090-0636-6/16 $31.00 © 2016 IEEE   211   2016 Symposium on VLSI Circuits Digest of Technical Papers

A 450mV Timing-Margin-Free Waveform Sorter based on Body Swapping Error Correction

Seongjong Kim, Joao Pedro Cerqueira, Mingoo Seok, Columbia University, New York, NY, USA

Abstract: We propose an error detection and correction technique based on local body swapping for eliminating the worst-case margin in near/sub-V_{TH} non-instruction parallel architectures. We apply the proposed technique on an unsupervised waveform sorter for brain computer interface microsystems, improving energy-efficiency by 49.3% and throughput by 35.6% over the baseline that is margined for the worst-case variation. The area overhead is 4.1%.

Introduction

In the on-going quest to enabling energy-efficient cognitive computing, parallel, and non-instruction architectures implemented in near/sub-V_{TH} circuits emerge as a promising candidate [1,2,9]. However, the complex and parallel nature of such architectures combined with the large delay variability from near/sub-V_{TH} circuits impose prohibitive timing margin to cycle time (T_{CLK}), limiting achievable energy-efficiency and throughput.

In-situ error detection and correction (EDAC), combined with dynamic voltage and frequency scaling (DVFS), can operate the chip at the point of first failure (PoFF). This can eliminate the margins for static and slow variations (e.g. process and temperature) and fast variations (e.g. V_{DD} drop) [3-8].

However, the existing approaches [3-7], as they often target super-V_{TH} in-order microprocessors, may not be well-suited for parallel non-instruction architectures in near/sub-V_{TH} circuits. The existing approaches often use the program counter for replaying instructions to perform correction [3]. The targeted architectures, however, do not have a program counter and also have distributed memory mixed with logics. Thus, in order to use replay correction, such architectures must have additional memory to store past architectural states for rolling-back. The proposed sorter (Fig. 1), for example, needs to duplicate 80.2% of the distributed registers to store single past architectural state. We estimate this can cause >28.8% area overhead (Table. 1). Refs. [4-6] proposes EDACs for a SIMD, a NoC router and a register file. However, all of them rely on replay correction that need either program counters or additional roll-back memory. Ref. [7] proposes non-replay correction based on local clock-gating. However, the area overhead of the technique is non-negligible (up to 87% in [7]). Also, an error and correction process can spread across entire architectures, which can hurt throughput and energy efficiency particularly in parallel architectures.

Here, we propose a new EDAC design that is able to correct errors without replay and thus are more suitable for the targeted non-instruction architectures in near/sub-V_{TH} circuits. We propose three techniques: (1) body swapping correction that eliminates the need for replay correction, (2) a fully-static error-detecting (ED) latch, and (3) area-efficient 2-phase latch ED pipelines. Via these techniques, we design an unsupervised waveform sorter based on spiking neural network (SNN) for brain computer interface (BCI) microsystems (Fig. 1). At V_{DD}=0.45V, the hardware can detect and correct timing errors without stopping any of parallel pipelines, eliminating timing margins for process, voltage, and temperature (PVT) variations. This enables 49.3% higher energy efficiency and 35.6% higher throughput than the baseline margined for the worst-case variation. It requires no additional V_{DD} and causes the area overhead of only 4.1%.

Proposed Techniques and Sorter Implementation

We propose three techniques and apply them on an unsupervised waveform sorter. The sorter architecture is based on [9]. It can take spike waveform inputs, train itself based on spike-timing dependent plasticity rules, and perform clustering (Fig. 2). High-V_{TH} devices are used for low leakage.

The first proposed technique is body swapping correction, which requires no replay and also incurs very low overhead. Our recent work [8] has proposed local V_{DD} boosting for correcting errors without replay. However, as shown in Fig. 3, it needs bulky level conversion and bypass circuits (LC/bypass) and additional supply voltage (V_{DDH}). The newly proposed technique requires only a small circuit called a body controller (BC) (Figs. 4 and 6).

In this technique, if the data arrives late at the ED latches (i.e., timing error), it still enters the correction stage via cycle borrowing and the BC swaps the bodies of NMOSs and PMOSs (NB and PB) of that stage. This can induce forward body bias and accelerate the computation to prepare the error-free results before the next rising clock edge (Fig. 5). The BC (Fig. 6) is sized to make the delay from the error detection to NB/PB swapping to be sufficiently fast, which is measured to be <3% of T_{CLK} (Fig. 7). We used the self-oscillating test mode (Fig. 6) for this measurement. This body swapping can provide more speed-up than required (2.2×) for correcting the worst-case timing violation (Figs. 5 and 8). The area overhead for isolating the bodies of the correction stage is minimal (Fig. 9) since the deep-nwell boundary is within the power ring. We insert well taps every 15μm. The total area overhead of BC and the well isolation is only 1.2% in the sorter design. It requires no additional V_{DD}.

Next, we propose fully-static ED latch circuits that are more robust than the existing semi-static design with floating detection channels [8] (Fig. 10). The proposed latch also can avoid the clk-to-q delay mismatch problem [8] since it compares the data (S) stored in the shadow latch (the opposite phase with main latch) with the incoming data (D) instead of the data stored in the main latch (Q). The impact of the extra loading on D can be small as the ED latches are inserted only in the output neurons based on the sparse detection scheme [8]. The proposed ED latch passes 100k Monte-Carlo simulations with process variations and also reliably operates at as low as 0.3V.

Finally, we optimize 2-phase ED latch based pipelines for low overhead. We apply the sparse error detection scheme [8] to minimize the number of inserted ED latches [8]. The sparseness (N_{OPT}) is found to be 8 latch stages at V_{DD}=0.35V and T_{CLK} =110 FO4 delays. While the architecture has various data flow paths across training, synapse-updating, and clustering phases, we find that implementing the output neurons as our detection and correction stage allows all the data flow paths to reach ED latches while traveling <N_{OPT} latch stages. Errors are handled independently in each output neuron. In order to reduce the overhead of 2-phase latch sequencing itself, which can cause up to 13-21% area overhead over flip-flop (FF) sequencing [7,8], we remove the local clock buffers in the latches and distribute the clock with a merged (centralized) buffer via 1-level [10] clock tree. The area overhead of the 2-phase latch pipelines is 2.6% and that of the inserted 126 ED latches is 0.3%.

Measurement Results

Test chips are fabricated in 65nm (Fig. 14). The baseline has no adaptive techniques and thus needs margin for the worst-case PVT variation (defined as the slowest among 10 dies, -20°C, and -10% V_{DD} drop) even when operating at the typical condition (typical die, 25°C, and no V_{DD} drop). The proposed design, on the other hand, can operate without margins at the PoFF. The baseline minimum energy dissipation (E_{OPT}) is 132nJ/clustering at energy-optimal V_{DD} (V_{OPT}) of 0.525V and F_{CLK} of 2.36MHz (Fig. 11). The proposed design achieves E_{OPT} of 69.1nJ/clustering (49.3% smaller) and F_{CLK}=3MHz (35.6% better) at V_{OPT}=0.450V (75mV lower). At the same F_{CLK} that the baseline works at its E_{OPT}, the proposed design achieves 47.6% energy savings at 100mV lower V_{DD}. At the V_{OPT} of the baseline, the proposed design achieves 2.6× higher throughput with 42.1% less energy dissipation. We summarize the energy savings at slow, typical, and fast corners at the same F_{CLK} of the baseline at its E_{OPT} (Table. 2). Error statistics measurement (Fig. 12) show that handling errors independently in each output neuron exercises 4.6× lower error handling as compared to the conventional replay case (i.e. counter Comb.) where single error requires replaying all output neurons. As compared to [8], the proposed EDAC technique can be well-suited to parallel and non-instruction architectures with a minimal area overhead of 4.1% and without any additional V_{DD} (Table. 3).

Acknowledgements

This work was supported by Catalyst Foundation, DARPA, and NSF.

References: [1] D. Jeon, et al., JSSC, 2014. [2] V. Karkare, et al., JSSC, 2013. [3] D. Bull, et al., JSSC, 2011. [4] R. Pawlowski, et al., ISSCC, 2012. [5] S. Paul, et al., Symp. VLSI Circuits, 2013. [6] J. Kulkarni, et al., ISSCC, 2015. [7] M. Fojtik, et al., ISSCC, 2012. [8] S. Kim, et al., JSSC, 2015. [9] B. Zhang, et al., ISLPED, 2015. [10] M. Seok, et al., ISLPED, 2010.

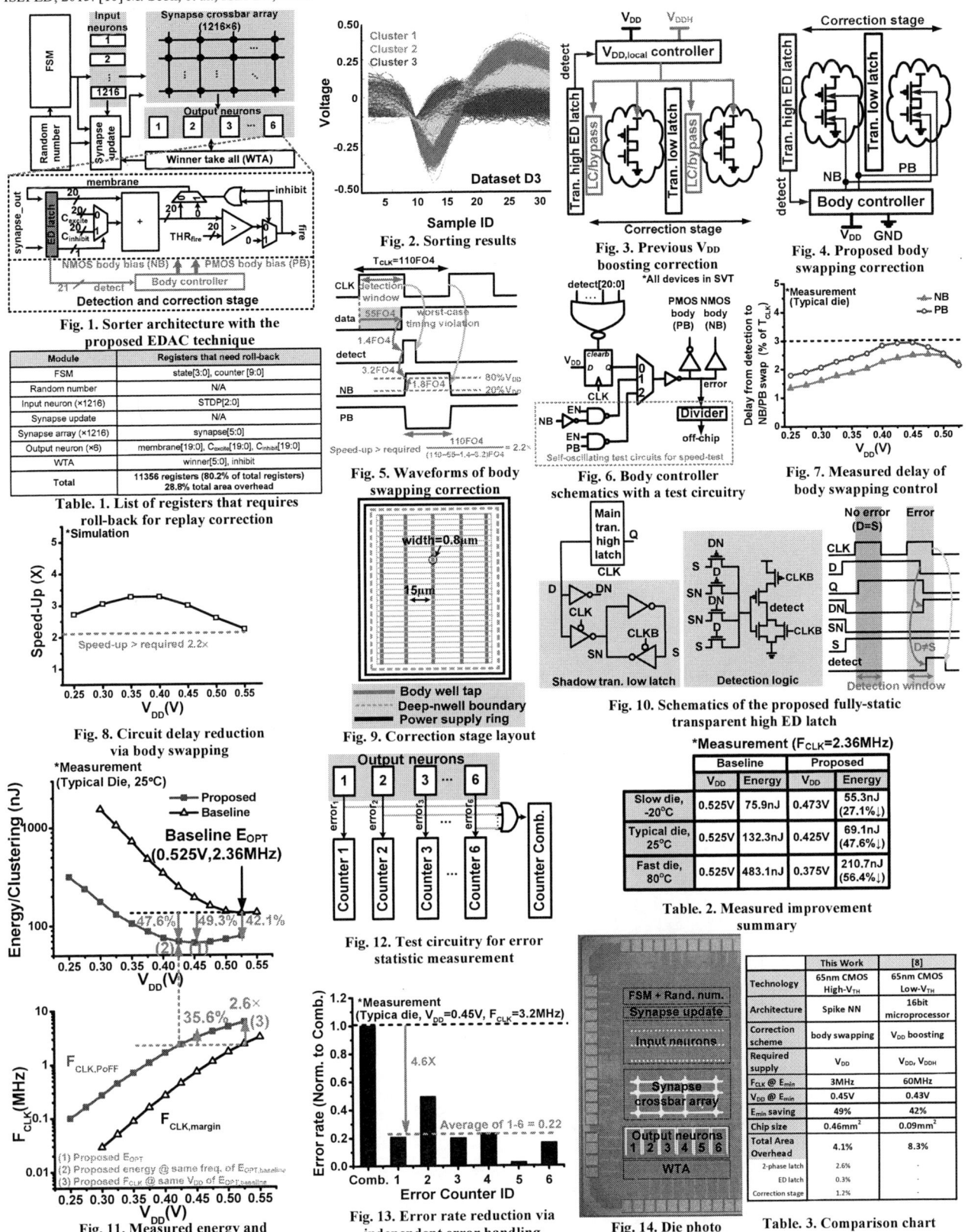

Fig. 1. Sorter architecture with the proposed EDAC technique

Table. 1. List of registers that requires roll-back for replay correction

Fig. 8. Circuit delay reduction via body swapping

Fig. 11. Measured energy and throughput improvement

Fig. 2. Sorting results

Fig. 5. Waveforms of body swapping correction

Fig. 9. Correction stage layout

Fig. 12. Test circuitry for error statistic measurement

Fig. 13. Error rate reduction via independent error handling

Fig. 3. Previous V_DD boosting correction

Fig. 6. Body controller schematics with a test circuitry

Fig. 10. Schematics of the proposed fully-static transparent high ED latch

Fig. 4. Proposed body swapping correction

Fig. 7. Measured delay of body swapping control

Table. 2. Measured improvement summary

Fig. 14. Die photo

Table. 3. Comparison chart

AUTHOR INDEX

Adeagbo, E. 112
Agarwal, A. 118, 194, 198
Ahmadi, A. 170
Aitken, R. 116
Akinin, A. 74
Akita, K. 174
Akiyama, K. 176
Alahmadi, H. 156
Alhoshany, A. 156
Ali, A. M. A. 164
Almajano, J. 204
Alon, E. 88
Anders, M. 118, 194, 198
Ando, Y. 98
Andreas, T. 112
Arimoto, H. 174
Arnaud, F. 114
Aseron, P. 60
Asuncion, S. 38
Aurangozeb 130
Ba, N. L. 182
Baas, B. 112
Baba, T. 68
Babaie, M. 50
Badaroglu, M. 120
Bae, H.-M. 210
Bae, S.-J. 138
Baek, J.-B. 44
Baek, S.-H. 110
Bai, X. 12
Balasubramanian, S. 166
Bang, J.-S. 44
Bang, S. 126, 182
Banno, N. 12
Bao, J. 120
Bardsley, S. 164
Bekele, A. 32
Bera, D. 30
Bhattacharya, U. 102
Bhavnagarwala, A. 116
Bhoraskar, P. 164
Blaauw, D. 52, 70, 78, 126, 182, 196
Blagojevic, M. 48
Blutman, K. 46
Boesch, R. 134
Bohnenstiehl, B. 112
Bol, D. 54
Boroujeni, B. K. 106
Borrelli, C. 38
Bosch, J. G. 30
Bragin, A. 204
Bright, W. 166
Bucki, B. 120
Buhler, F. N. 108
Bullmann, T. 206

Bunch, R. 164
Cacho, F. 114
Calderin, L. 88
Cao, J. 94, 132
Cao, T. 170
Carta, C. 106
Cauwenberghs, G. 74
Cerqueira, J. P. 26, 212
Cestero, A. 14
Chae, M.-K. 138
Chan, W.-M. 8
Chandra, V. 116
Chang, C.-H. 192
Chang, D. 94
Chang, Jonathan 8
Chang, James 94
Chang, K. 32, 38, 168
Chang, M.-C. F. 90, 136
Chang, Z. Y. 30
Chao, Y. 184
Chen, B. 84
Chen, C. 30
Chen, C.-H. 18, 200
Chen, G. 118, 194, 198
Chen, J.-L. 42
Chen, K.-H. 42
Chen, Mark 192
Chen, Mike S.-W. 122
Chen, R. 50
Chen, S. 32
Chen, X. 94, 184
Chen, Y. 132, 206
Chen, Y.-H. 8
Chen, Z. 30, 102
Chen, Z.-Z. 90, 136
Chi, T. 28
Chih, T.-H. 94
Chiu, Y. 160
Cho, G.-H. 44
Cho, H. C. 28
Cho, J. 32
Cho, L.-C. 50
Cho, M. 148
Cho, W.-H. 136
Cho, Y.-C. 138
Choi, I.-H. 210
Choi, J. 190
Choi, J.-H. 138
Choi, J.-K. 210
Choi, J.-M. 96
Choi, M.-G. 210
Choi, S.-W. 44
Choi, Y. 110
Chou, P.-S. 178
Ciampolini, L. 114

AUTHOR INDEX

Cochet, M. 48
Coulon, J. 166
Croain, D. 114
Danjo, T. 34
De Gyvez, J. Pineda 46
De Jong, N. 30
De Streel, G. 54
De, V. 60, 148, 198
Derounian, P. 164
Dillon, C. 164
Dinc, H. 164
Do, A. T. 24
Dommaraju, S. 94
Dong, Q. 78
Doppler, K. 98
Dorgan, V. E. 102
Dorrance, R. 144
Dosho, S. 186
Draxelmayr, D. 76
Du, J. 136
Du, Y. 136
Durant, F. 54
D'Urbino, M. 206
Dwobeng, E. 166
Eberlein, M. 154
Echeverri, J. 46
Ellinger, F. 106
Elzeftawi, M. 38
Emami, A. 172, 204
Ezaki, T. 176
Faetti, T. 106
Farivar, M. 204
Fatemi, H. 46
Feng, P. 120
Ferreira, S. B. 50
Fischer, J. 120
Flatresse, P. 48
Flynn, M. P. 108
Frans, Y. 32, 38, 168
Fredenburg, J. A. 108
Frenkel, C. 54
Frey, U. 206
Fujimoto, R. 58
Fujita, M. 98
Gaalaas, E. 84
Gao, W. 94
Gao, Y. 158
Ghesquiere, P. 106
Gill, P. 62
Giner, F. 114
Goh, W. L. 158
Golz, J. 14
Gray, B. 164
Grossnickle, V. 60
Gurné, T. 54

Gwak, K. 210
Ha, S. 74
Hada, H. 12
Haendler, S. 114
Hara, S. 186
Hassanpourghadi, M. 122
Hayashi, T. 94
He, T. 18
Hedayati, H. 38
Hierlemann, A. 206
Higashi, H. 34
Higuchi, T. 34
Hong, Y. 158
Hong, Y.-J. 56
Honkote, V. 60
Horowitz, M. 180
Hossain, A. K. M. D. 130
Hossain, M. 130
Hsieh, C.-C. 128
Hsieh, K. 192
Hsieh, S.-E. 128
Hsu, M.-C. 178
Hsu, S. 118, 194, 198
Hsueh, F.-L. 50
Hu, B. 90
Huang, C. 40, 72
Huang, G. 50
Huang, K.-K. 94
Huang, P.-T. 136
Huang, T.-C. 192
Huang, Z. 132
Huard, V. 114
Hübler, A. C. 106
Huh, Y. 44
Hung, L. D. 176
Hwang, G. 210
Hwang, H. 96
Hwang, S. 36
Ibars, S. 114
Ide, Y. 34
Ido, T. 174
Iguchi, N. 12
Ike, A. 68
Ikeda, S. 186
Im, J. 32, 38
Iqbal, I. 116
Ishida, K. 106
Ishida, M. 82
Ishihara, N. 186
Ishii, Y. 10
Ishikawa, Y. 186
Ishikuro, H. 72
Isobe, A. 98
Ito, H. 186
Iwamoto, H. 176

AUTHOR INDEX

Iyer, S. 14
Jain, A. 22
Jain, S. 60
Jang, S.-J. 138
Jang, T. 52
Jang, Y.-J. 138
Jayaraman, B. 14
Jeloka, S. 70
Jeon, D. 24
Jeon, J. 210
Jeong, J. 182
Jeong, S. 126
Jha, A. 170
Ji, S. 180
Jiang, Z. 26
Jiyang, Y. 186
Jou, C.-P. 50
Ju, Y.-M. 44
Julien, C. 114
Jung, H. G. 96
Jung, M.-Y. 44
Jung, W. 126
Juyon, J. 206
Kaibara, K. 82
Kamisuki, K. 186
Kang, H.-S. 96
Kano, H. 34
Kapoor, A. 46
Kappel, S. L. 208
Kasamatsu, A. 186
Kato, A. 176
Kato, K. 98
Kato, Y. 176
Kattamuri, R. 132
Kaul, H. 118, 194, 198
Kawai, S. 34
Kawajiri, T. 72
Kaylor, S. 166
Keller, B. 48
Keller, R. 166
Kempanna, T. 14
Khan, F. 14
Khellah, M. 148
Kidmose, P. 208
Kiefl, S. 106
Kilker, R. 14
Kilsgaard, S. 208
Kim, B. 138
Kim, C. 36, 74, 126
Kim, D.-S. 210
Kim, H.-J. 96
Kim, H.-S. 52
Kim, J. 110
Kim, J.-M. 210
Kim, K.-D. 44

Kim, M. 190
Kim, R. 60
Kim, Seongjong 26, 148, 206, 212
Kim, Soo Gil 96
Kim, Seong Joong 56
Kim, T. 24, 182
Kim, Y. 56, 136
Kinget, P. R. 92
Kinoshita, Y. 82
Kireev, V. 38
Kirihata, T. 14
Kitano, Y. 176
Knag, P. 142
Kobayashi, A. 100
Kocaman, N. 132
Koh, D. 94
Kokubo, M. 174
Kong, L. 188
Koyanagi, Y. 34
Krishnamurthy, R. 118, 194, 198
Krishnan, L. 94
Kshattry, S. 170
Kuan, Y.-C. 90
Kubendran, R. 74
Kudo, M. 34
Kulkarni, S. H. 102
Kuo, F.-W. 50
Kuo, H.-Y. 192
Kuroda, R. 178
Kwon, K.-W. 96
Kwon, S.-C. 96
Lanford, J. 164
Leblebici, Y. 204
Lecocq, C. 114
Lee, I. 52, 70
Lee, Sangheon 96
Lee, Seung Bae 66
Lee, Sheau Jiung 136
Lee, S.-G. 56
Lee, S.-J. 110
Lee, S.-M. 138
Lee, Y. 110, 174, 190
Leu, D. 14
Leuenberger, S. 104
Li, A. 184
Li, C.-C. 192
Li, Q. 208
Li, W. 66
Li, Y. 90, 136
Liao, C.-C. 192
Liao, H.-J. 8
Liao, H.-Y. 192
Liao, K. 170
Liaw, J.-J. 8
Lien, Y.-C. 124

AUTHOR INDEX

Lim, B. C. 180
Lim, S. 210
Lim, W. 52
Lim, Y. 108
Lin, C. B. 98
Lin, C.-H. 42
Lin, J.-R. 42
Lin, K.-C. 8
Lin, Y.-H. 42
Lin, Y.-Z. 162
Liu, B. 112
Liu, C. 142
Liu, X. 40
Lu, C.-H. 162
Lu, C.-T. 192
Luong, H. 184
Maclean, K. 166
Madadi, I. 50
Majumdar, A. 46
Majumder, T. 60
Makinwa, K. 46, 76, 152
Makris, Y. 170
Mallik, D. 60
Markovic, D. 144
Martinez, J. G. 46
Masu, K. 186
Mathew, S. 118, 194, 198
Matsuda, A. 34
Matsukawa, K. 16
Matsumura, H. 68
Matsuoka, Y. 174
McLeod, S. 32
McShea, M. 164
Mehta, U. 164
Meister, T. 106
Melek, D. T. 32
Mendrela, A. E. 108
Mercier, P. P. 74
Mergault, P. 114
Miki, T. 16
Miyamura, M. 12
Miyaoka, H. 34
Miyoshi, T. 68
Mohammad, M. 130
Mok, P. K. T. 40
Momtaz, A. 132
Moon, S. 36
Moon, U.-K. 104
Moons, B. 140
Moore, R. 164
Moradi, F. 208
Morgan, A. 164
Mori, T. 34
Morioka, A. 12
Morita, T. 82

Moscatelli, A. 86
Moy, D. 14
Münzenrieder, N. 106
Murmann, B. 134
Myers, J. 98
Myers, P. 126
Naka, N. 34
Nakamura, T. 176
Nakamura, W. 176
Nallapati, G. 120
Nam, J.-W. 122
Narayanan, S. 66
Narayanasetti, P. 120
Naudet, S. 114
Nebashi, R. 12
Nguyen, A. 116
Nguyen, B. 166
Nii, K. 10
Niitsuma, W. 176
Niknejad, A. 88
Nikolic, B. 48, 88
Noblet, D. 114
Noguchi, N. 186
Nomoto, T. 4
Noothout, E. 30
O, K. K. 170
Obara, T. 186
Obata, K. 16
Ogasawara, Y. 58
Ogata, Y. 34
Oh, S. 182
Ohta, J. 64
Oike, Y. 4, 176
Okamoto, K. 12
Okamoto, S. 98
Omran, H. 156
Ondricek, D. 116
Onuki, T. 98
Osawa, K. 64
Park, C. S. 56
Park, H. 132
Park, H.-J. 138
Park, Jiwoong 74
Park, Jong Seok 28
Park, S. 210
Park, S.-H. 44, 44
Parra, M. 114
Paul, S. 60
Pavan, S. 22
Periasamy, V. 94
Pertijs, M. A. P. 30
Petti, L. 106
Pham, T. 38
Pimentel, J. 112
Planes, N. 114

AUTHOR INDEX

Polley, A. 66
Ponte, J. 30
Prins, C. 30
Pu, J. 180
Puckett, S. 164
Puglielli, A. 88
Qiu, A. P. 150
Quoirin, M. 114
Raghavan, B. 132
Raghavan, R. 14
Raghunathan, S. B. 30
Raj, M. 168
Ramakrishnan, S. 88
Ramaswamy, S. 66
Ranica, R. 114
Rao, L. 132
Ravinuthula, V. 166
Razavi, B. 188
Rim, K. 120
Ruibing, D. 186
Saeedi, S. 172
Sakai, Y. 34
Sakamoto, T. 12
Sakano, Y. 176
Sakurai, H. 58
Salama, K. N. 156
Saligane, M. 196
Salvatore, G. A. 106
Sami, K. 58
Sankman, R. 60
Sano, T. 10
Sanyal, A. 20
Sato, M. 176
Satpathy, S. 118, 194, 198
Sawada, Y. 10
Schmid, A. 204
Schmidt, G. 106
Sculley, T. 66
Seok, M. 26, 212
Seong, C. 96
Sevat, L. 46
Shabanimotlagh, M. 30
Shabanpour, R. 106
Shahshahani, A. 204
Shahshahani, M. 204
Shalmany, S. H. 76, 152
Sharma, A. 66
Shen, B. 94
Shi, Y. 70
Shibasaki, T. 34
Shim, M. 126
Shimizu, T. 68
Shin, C. 44
Shin, H. S. 210
Shin, J. 38

Shin, S.-U. 44
Shirai, N. 34
Shiroshita, H. 176
Shoaran, M. 204
Shodo, K. 64
Shuai, C. C. 98
Sim, J.-Y. 138
Sohn, Y.-S. 138
Song, J. 36, 96
Song, S. C. 120
Stas, F. 54
Staszewski, R. B. 50, 192
Stettler, A. 206
Stillmaker, A. 112
Styczynski, M. 28
Su, A. 28
Su, Y. 150
Sugawa, S. 178
Sugibayashi, T. 12
Sugimura, M. 68
Sugo, H. 178
Suleiman, A. 146
Sumi, H. 178
Sun, J. 84
Sun, N. 20
Sun, R. 94
Sung, J. H. 28
Suresh, V. 194, 198
Sushihara, K. 16
Sylvester, D. 24, 52, 70, 78, 126, 182, 196
Sze, V. 146
Tada, M. 12
Takemoto, T. 174
Takeuchi, K. 100
Tamura, H. 34, 98
Tamura, S. 82
Tan, K. H. 32
Tanaka, M. 10
Tanaka, S. 10
Tang, W. 200
Taylor, G. 164
Temam, O. 1
Temes, G. C. 18
Terasawa, F. 34
Terasawa, Y. 64
Terashima, K. 34
Terrier, F. 114
Tessarolo, A. 80
Thach, D. 68
Tohidian, M. 50
Tokunaga, C. 148
Tokutomi, T. 100
Toma, G. D. 106
Tomita, Y. 68
Toyama, T. 176

AUTHOR INDEX

Tran, A. 112
Tröster, G. 106
Tsai, C.-H. 162
Tsai, T.-H. 192
Tsai, T.-Y. 42
Tschanz, J. 60, 148
Tsou, S.-C. 162
Tsuji, Y. 12
Tsukamoto, Y. 16
Tummuru, R. R. 14
Turgis, D. 114
Ueda, T. 82
Uesugi, W. 98
Ujita, S. 82
Umeda, H. 82
Unruh, G. 94
Upadhyaya, P. 32, 38, 168
Van Der Wel, A. 46
Vangal, S. 60
Varzaghani, A. 132
Venes, A. 94
Venkatachala, P. K. 104
Verhelst, M. 140
Verma, N. 202
Verweij, M. D. 30
Villaret, A. 114
Viraraghavan, J. 14
Vladimirescu, A. 48
Vogelsang, T. 62
Vos, H. J. 30
Wakabayashi, H. 4
Wakashima, S. 178
Wang, H. 28, 74
Wang, J. 120
Wang, N.-Y. 94
Wang, Q. 26
Wang, T.-J. 178
Wang, X. 150
Wang, Y. 158
Wang, Z. 202
Watanabe, Y. 68
Weaver, M. 166
Weber, O. 114
Wei, G. 94
Wey, C.-L. 42
Wong, C.-H. 90, 136
Woo, J. 96
Wu, C. 132
Wu, C.-W. 8
Wu, J. Y. 98
Wu, L. 184
Wu, S. H. 98
Wu, X. 70
Xia, G. M. 150
Xiao, J. 94

Ximenes, A. 192
Xu, B. 160
Xu, J. 120
Xu, X. 94
Xu, Yang 92, 104
Xu, Yong Ping 150
Yabuuchi, M. 10
Yahav, I. 154
Yamaguchi, H. 34
Yamasaki, H. 68
Yamashita, H. 174
Yamashita, Y. 178
Yamazaki, S. 98
Yang, D. 120
Yang, J. 210
Yang, K. 70, 78, 196
Yang, W.-H. 42
Yang, X. 132
Yao, L. 158
Yeap, G. 120, 170
Yew, T. R. 98
Yi, I.-M. 138
Yin, M. 14
Yoon, H. 190
Yousefzadeh, B. 152
Yu, Q. 164
Yuan, M.-S. 192
Yuan, X. 206
Yun, R. 84
Yun, S.-J. 56
Zeinolabedin, S. M. A. 24
Zhang, A. 122
Zhang, B. 132
Zhang, G. 38
Zhang, H. 38
Zhang, J. 202
Zhang, K. 102
Zhang, W. 32
Zhang, Y. 18, 196
Zhang, Z. 142, 146, 200
Zhao, H. 32
Zhao, J. 150
Zhao, Y. 150
Zheng, K. 134
Zhou, L. 38
Zhou, X. 208
Zhou, Y. 160
Zhu, C. 28
Zhu, J. 120
Zhu, N. 164